Sustainable Procurement
in Supply Chain Operations

Mathematical Engineering, Manufacturing, and Management Sciences

Series Editor

Mangey Ram, Professor, Assistant Dean (International Affairs)
Department of Mathematics,
Graphic Era University, Dehradun, India

The aim of this new book series is to publish the research studies and articles that discuss the latest developments and research applied to mathematics and its applications in the manufacturing and management sciences areas. Mathematical tools and techniques are the strength of engineering sciences. They form the common foundation of all novel disciplines as engineering evolves and develops. The series will include a comprehensive range of applied mathematics and its application in engineering areas such as optimization techniques, mathematical modeling and simulation, stochastic processes and systems engineering, safety-critical system performance, system safety, system security, high assurance software architecture and design, mathematical modeling in environmental safety sciences, finite element methods, differential equations, reliability engineering, etc.

Sustainable Procurement in Supply Chain Operations
Edited by Sachin K. Mangla, Sunil Luthra, Suresh Kumar Jakhar, Anil Kumar, and Nripendra P. Rana

Sustainable Procurement
in Supply Chain Operations

Edited by
Sachin K. Mangla
Sunil Luthra
Suresh Kumar Jakhar
Anil Kumar
Nripendra P. Rana

CRC Press
Taylor & Francis Group
Boca Raton London New York

CRC Press is an imprint of the
Taylor & Francis Group, an **informa** business

CRC Press
Taylor & Francis Group
6000 Broken Sound Parkway NW, Suite 300
Boca Raton, FL 33487-2742

First issued in paperback 2020

© 2019 by Taylor & Francis Group, LLC
CRC Press is an imprint of Taylor & Francis Group, an Informa business

No claim to original U.S. Government works

ISBN-13: 978-1-138-60815-3 (hbk)
ISBN-13: 978-0-367-77967-2 (pbk)

Visit the Taylor & Francis Web site at
http://www.taylorandfrancis.com

and the CRC Press Web site at
http://www.crcpress.com

Contents

SECTION I   Sustainable Procurement/Sourcing

SECTION II   Sustainable Supplier Selection

SECTION III   Enablers/Barriers for Sustainable Procurement Operations

SECTION IV *Research Methods in Sustainable Procurement Operations*

Preface

The complete work of this book is divided into four sections. The first section titled *Sustainable Procurement/Sourcing* includes all the chapters related to sustainable procurement/sourcing applications. The second section titled *Sustainable Supplier Selection* contains all the chapters related to sustainable supplier selection. The third section titled *Enablers/Barriers for Sustainable Procurement Operations* contains all the chapters related to enablers/barriers for sustainable procurement operations. The last section titled *Research Methods in Sustainable Procurement Operations* includes chapters by authors who applied different methods in sustainable procurement. A brief description of each section follows.

The first section titled *Sustainable Procurement/Sourcing* contains fours chapters. In the first, Chapter 1, Sengupta and Shukla attempt to analyze the current state of academic research in the domain of sustainable procurement by conducting a systematic literature review of relevant articles with a focus on performance indicators, measures, and practices. Thirty-four sustainability reports were shortlisted from the top ten global manufacturing (steel) companies from the Forbes 2000 list. Using qualitative content analysis, indicators and practices were identified and segregated into social, economic, and environmental indicators. The indicators retrieved from the academic and practitioner literature were then mapped to the 17 Sustainable Development Goals in order to understand the current gap in the practices of manufacturing industries in comparison to the Sustainable Development Goals. In addition, we analyzed the theory-practice divide between academic and practitioner literature from the perspective of different goals. A conceptual framework is proposed based on the research and practice gaps inferred from the content analysis. Future research directions and limitations are discussed based on the conceptual framework.

To contribute the same section objective, in Chapter 2, Yerpude and Singhal discuss that the Internet of things (IoT) is one of the prominent digital technology from the Industry 4.0 revolution. Cyber-physical systems are the main contributors to Industry 4.0. IoT has entered almost all business process. In this chapter, we evaluate the implementation in one of the supply chain management constructs, i.e., Procurement. Fundamentally, the process that links all the members in the supply chain assuring and managing the quality of supplies as per the business process and customer expectations is termed Procurement. A procurement process deployed successfully can become a game-changer for an organization bringing in the competitive advantage for the organization amidst volatile customer demands. The entire information required for procurement to take action needs to be made available on a common platform where each of the stakeholders is able to comprehend their part and take appropriate action. There is an indisputable need of real-time data to flow right from the customer to the supplier. With the advent of IoT, the physical world can now be connected to all the systems with Internet. IoT implemented in the procurement function proves as an enabler, transforming it into a digital workspace creating intelligent spaces. The chapter constitutes an introduction to the subject with a literature review that helps hypothesis formation. Research methodology conducted includes measurement model analysis with a structured equation model. Discussion section reveals the importance of the implementation of IoT and real-time data. Limitations, future study, and conclusion help understand the subject and the next steps of the case.

In Chapter 3, Mor, Bhardwaj, Singh and Kharub talked about framework for measuring the procurement performance in the dairy supply chain. They explain that the dairy industry has a significant role in the Indian economy as well as for the rural development. For business operations in the dairy industry, the quick and safe procurement of milk from the farmer is of crucial importance. The current chapter is aimed to develop a framework for evaluating the performance of procurement practices in the dairy industry. An optimum competitive procurement model has been developed through factor analysis and exploratory structural equation modelling techniques. The model is developed based on responses collected from dairy industries from northern India. The results of hypotheses testing suggest that all the four factors (derived through factor analysis) influence the procurement performance positively. The outcome of this chapter reveals that the milk processing industry needs substantial development in procurement practices to meet the best product quality, supply chain sustainability, and food safety norms of the global market. This study is useful for the dairy industry to handle the traceability issues, execution of effective information systems, the overall quality of procured milk, build supplier trust, and hence, to achieve higher procurement performance.

In Chapter 4, Garg discuses modeling the sustainable procurement initiatives using fuzzy-based Multi-Criteria Decision-Making (MCDM) framework. In the current environment, many firms recognize that implementation of sustainable practices not only provides a competitive edge but also mandatory for a firm's survival. Moreover, the growth of human population all around the world has massively increased resource consumption and production activities. It is therefore imperative to encourage sustainability philosophy in the Supply Chain (SC) activities of a company. Sustainable procurement is the first step toward implementation of a sustainable supply chain (SSC). This has been done by selecting suitable sustainable outsourcing suppliers/partners. To do this, a focal firm should concentrate on sustainable procurement initiatives. This work proposes a fuzzy-based MCDM model for evaluation and ranking of the sustainable procurement initiatives in order to achieve SSC. A case of the Indian electronic industry demonstrates the application of the proposed framework. This study may benefit managers and business professionals not only in sustainable procurement (SP) initiatives evaluation process, but also in the selection of the efficient sustainable suppliers.

Section II contains all the chapters related to the sustainable supplier selection. To follow this, in Chapter 5, Bag and Sahu discuss green procurement research: a review and modeling Supplier Selection Problem (SSP). They explain that green procurement research has gained momentum in recent years due to increased sustainability-related concerns. The purpose of the study is to conduct a methodical literature review and understand past, present, and future trends in green procurement research. The list of papers was downloaded using Scopus (https://www.scopus.com), which is the largest abstract and citation database for scientific research journals, books, and conference proceedings. Through strict screening, only journal papers were selected for conducting the review of literature. The extant literature is categorized according to multiple classification schemes. A bibliometric analysis of the reviewed literature is also presented. The review findings show that a total 87 papers have been published over a span of 20 years. The number of publishing started going up from the year 2008 and significantly from the year 2015. The paper further presents an illustration of the green SSP through application of GREY-based methodology. The study finally concludes with a key takeaway for green procurement professionals.

To follow the same objective, in Chapter 6, Mariouli and Abouabdellah explain that the evaluation and selection of suppliers has become a key player in the sustainable development policy of all companies aiming to build a sustainable supply chain. The level of sustainability of a company is linked with its suppliers. Upstream operations and relationships with suppliers are critical to ensure downstream quality with customers (distribution); hence, the importance of evaluating and measuring the performance of suppliers. In our research, we studied the issue of sustainable supplier choice and presented a new model for managers to measure supplier performance for sustainable purchasing. Our model is based on the Fuzzy DEMATEL method, which is used to calculate the weights of sustainability criteria, then these weights are used in our mathematical model to calculate the supplier's sustainability index.

Contributing to the same section objective, in Chapter 7, Jahan and Panahande talk about supplier selection for protective relays of power transmission network with the fuzzy approach. Various products should be provided for customer demands in the global field of competition. An increase in customer demands provides an opportunity for the supplier to fulfill the biggest part of the job. In this case, the company will be more dependent on its suppliers. This dependence requires more coordination between the company and the suppliers, which highlights the importance of supply chain management (SCM). SCM is used with the aim of creating more efficient processes and, at the same time, decreasing the risk of chain disconnection

In Chapter 8, Umarusman discuses that supplier selection is one of the crucial components of Supply Chain Management where mistakes should be avoided, as it may be impossible to compensate for them. The decision of supplier selection is a part of managing a supply chain, and it is one of those decisions which plays an important role in the achievement of business. Supplier Selection Criteria (SSC) differ based on the characteristic structures of companies. In addition to their needs, each and every company may have different principles and policies. SSC vary depending on the characteristics of companies. The criteria that companies use in their supplier selection directly affect their preference. Therefore, it is important that SSC should be defined as realistically as possible. The Supplier Selection Problem (SSP), which targets to detect the most convenient suppliers in terms of the demands of a business, is a Multi-Criteria Decision-Making (MCDM) problem with many conflicting quantitative and qualitative criteria.

Section III entitled *Enablers/Barriers for Sustainable Procurement Operations* contains all the chapters related to find the enablers and barriers for sustainable procurement operations. To support this in Chapter 9, Prakash, Satydev, Mathiyazhagan, Dwivedy, and Narula discuss Sustainable Competitive Advantages (SCA) is essential for the firms to thrive in global competition. This study starts with the identification of the drivers of SCA particularly related to the Indian manufacturing organization. These drivers are the motivating factors that have a direct or indirect influence on the entire organizational systems and direct the firms forward toward positive growth. In this study, a total 13 factors are identified from the extant literature covering the aspects of SCA in organizations from assets to features for capabilities that are unique with promising benefits. Industry experts vet this vast pool of select drivers and prioritization is performed. The interpretive structural modeling (ISM) is used for modeling and analysis of these identified SCA drivers. The MICMAC analysis is also performed for driving and dependence power of the identified drivers. The results indicate that tracking the development of directives, green design and purchasing, information system, and collaborative R&D with suppliers are potentially associated with SCA for a manufacturing organization and main areas for strategic focus. The other vital drivers include improvement in innovative product, managerial capacity, environmental education, and training. The outcome of the study is well-grounded theoretical justification for the critical drivers for achieving SCA over competitors in Indian manufacturing sector related organizations.

Contributing to the same section objective, in Chapter 10, Vimal, Sasikumar, Mathiyazhagan, Nishal and Sivakumar explain that the sustainable development in manufacturing can be achieved by imbibing sustainability starting with procurement activities. However, the drivers of sustainable procurement and its interrelationships have not been thoroughly investigated in the existing literature. Thus, this chapter attempts to examine the common drivers identified through literature reviews and opinions from the subject expert. The initial relationship matrix was developed by collecting the inputs from 15 experts through a face-to-face interview. The Decision making trial and evaluation laboratory (DEMATEL) approach is applied for the average of the initial relationship matrix collected from the experts. By applying the DEMATEL approach, skillful policymaking is identified as the significant drivers and stakeholder encouragement as the least significant drivers. This study will help the practicing managers to identify the important driver for better adoption of sustainability concepts in procurement.

Chapter 11 is all about barriers on sustainable supply chain management implementation for Turkish SME suppliers, and where Türk and Çelik discuss that supply chain management is crucial for businesses to acquire competitive advantage and supply chain management process implementation. It is particularly important to examine environmental efforts on business performance and to encourage the corresponding efforts for Small- and Medium-Sized Enterprises (SMEs) and large enterprises. However, businesses may encounter barriers throughout supply chain management implementation that are reasonably challenging to overcome simultaneously. Therefore, it is relatively important for businesses to prioritize their strategies and to acquire a competitive advantage during the transition process of supply chain management implementation. The primary purpose of this chapter is to understand how to prioritize the significance of barriers that SME suppliers might encounter, with respect to their supply chain management implementation using a fuzzy analytic hierarchy process (AHP) approach. Following the existing literature, the corresponding barriers are classified into five main categories including finance, involvement and support, technology, knowledge, and outsourcing. This study was carried out by 10 SME suppliers' expert opinions currently working at the Small and Medium Enterprises Development Organization (KOSGEB) in Turkey. The empirical results reveal that the most significant barrier of Turkish SME suppliers on supply chain management implementation was the lack of knowledge. The other barriers were arranged as involvement and support, finance, and outsourcing, respectively.

In Chapter 12, Zarei, Carrasco-Gallego, and Ronchi discuss that humanitarian supply chains are characterized by uncertainty and unpredictability of demand and volatility of the context. Despite such specificities, integrating sustainability into humanitarian procurement seems imperative due to the dire need stated by research and past antecedents of unsustainable procurement. This study identifies the barriers and enablers on the way of integrating sustainability into humanitarian procurement. The ideas of three humanitarian experts were elicited for identification of the barriers and enablers. Next, a group analytic hierarchy process (AHP) was applied to aggregate the ideas of experts and prioritize the barriers based on their potential for improvement. The results show that "local procurement" is the most important category of barriers, followed by "funding environment," and "inter- and intra-organizational barriers." Finally, the barriers and enablers within each category are discussed in detail and possible ways to address them are suggested.

Section IV contains all chapters related to *Research Methods in Sustainable Procurement Operations*. The first chapter included in the section is "Text mining applied to literature on sustainable supply chain (1996–2018): an analysis based on Scopus." In this chapter, the authors discuss text mining methods applied to the literature on sustainable supply chain, this is Chapter 13. Chapter 14 is all about multi-criteria decision-making techniques to address sustainable procurement in supply chain operations. The investigation of a sustainable procurement system to satisfy Turkey's energy demand by means of multi-period interval parameter integer linear programming (MP-IPILP) is illustrated in Chapter 15. To contribute the same section objective, in Chapter 16, Khazaelpour and Chini discuss a mixed-integer sustainable public transportation model of procurement purposes surrounding a construction site. Chapter 17 is all about a sustainable inventory optimization problem under process flexibility. The final chapter is all about sustainable procurement through advanced machine learning algorithms in manufacturing.

Thanks
On behalf of the Editors

Acknowledgements

The editors would like to acknowledge the help of all the people involved in this project and, more specifically, to the authors and reviewers that took part in the review process. Without their support, this book would not have become a reality.

We would like to thank each one of the authors for their contributions. The editors wish to acknowledge the valuable contributions of the reviewers regarding the improvement of quality, coherence, and content presentation of the chapters. Most of the authors also served as referees; we highly appreciate their double task.

We are grateful to all members of CRC Press, Taylor & Francis Group publishing house for their assistance and timely motivation in producing this volume.

We hope the readers will share our excitement with this important scientific contribution to the body of knowledge about various applications of Sustainable Procurement in Supply Chain Operations.

Dr. Sachin K. Mangla
Plymouth Business School
Plymouth University
Plymouth, United Kingdom

Dr. Sunil Luthra
Department of Mechanical Engineering
State Institute of Engineering and Technology
Nilokheri, India

Dr. Suresh Kumar Jakhar
Indian Institute of Management
Lucknow, India

Dr. Anil Kumar
Centre for Supply Chain Improvement
The University of Derby
Derby, United Kingdom

Dr. Nripendra P. Rana
Department of Business
School of Management
Swansea University
Swansea, United Kingdom

Editors

Dr. Sachin K. Mangla is a lecturer of Knowledge Management and Business Decision-making, in Plymouth Business School, University of Plymouth, Plymouth, United Kingdom. He is working in the field of Green Supply Chain; Circular Economy and Sustainability; Cross Disciplinary Research in Supply and Operations Management; Knowledge Management based Decision-Making; Smart Manufacturing/Industry 4.0; Machine Learning; Risk Management; Simulation; Optimization; Reverse Logistics; Renewable Energy; Empirical research. He did his doctorate (specialization Operations and Supply Chain Management) from Indian Institute of Technology (IIT), Roorkee. He loves to write research papers and projects. He has published/presented several papers in repute (ABS and ABDC indexed) international/national journals. He has an h-index 21, i10-index 24 and Google Scholar Citations of more than 1300. He has received 2017 Most Cited Paper Award for his paper entitled "Risk analysis in green chain using fuzzy AHP approach: a case study." Currently, he is editing a book "Sustainable Procurement in Supply Chain Operations" published under CRC Press (Taylor & Francis Group). He is also currently editing a Special issue as a Lead Guest Editor in Production Planning & Control: The Management of Operations and Resources, Recycling and Conservation on "Industry 4 and Circular Economy" and "Operational Excellence and SSC Performance Improvement."

Dr. Sunil Luthra is working as an assistant professor, State Institute of Engineering and Technology (formerly known as Government Engineering College), Nilokheri, Haryana, India. He has been associated with teaching for the last 15 years. He has contributed over hundred research papers in international referred and national journals, and conferences at international and national level. His scholarly work has also been acknowledged in several International journals of repute such as the JCP, IJPE, PPC, IJPR, RSER, RCR, EGY, JRPO and many more, and conference of repute like SOM-14, NITIE—POMS, AGBA, GLOGIFT 14 and GLOGIFT 15 etc. His research is in spotlight. His works got more than 2000 Citations (h-index = 24). His RG score is higher than 85% of Research Gate members. His specific areas of interest are operation management; green supply chain management; sustainable supply chain management; sustainable consumption and production; reverse logistics; renewable/ sustainable energy technologies and business sustainability etc.

Dr. Suresh Kumar Jakhar is a faculty in the Area of Operations at IIM Lucknow. Jakhar holds his doctorate from IIT Roorkee, Master of Technology from IIT Delhi. Before joining to IIM Lucknow he worked at IIM Rohtak, Symbiosis Center for Management and Human Resources Development (SCMHRD) as an assistant professor in the area of Operations. Jakhar secured All Indian Rank 07 in Graduate Aptitude Test in Engineering 2008, jointly conducted by all IITs and IISc. Jakhar has also consulted TATA Chemicals LTD for Warehouse Improvement and Performance Evaluation. He has published research papers in International

Journals and conferences. His research work appeared in International Journals like, Journal of Cleaner Production, Production Planning & Control: The Management of Operations, Sustainable Development, International Journal of Productivity and Performance Management, Global Journal of Flexible Systems Management, International Journal of Agile Systems and Management and several papers are under review.

Dr. Anil Kumar is a Post-Doctoral Research Fellow in area of Decision Sciences at Centre for Supply Chain Improvement, University of Derby, United Kingdom (UK). For the last 8 years, he has been associated with teaching and research, which he loves to do. He has contributed over 40 + research papers in international referred and national journals, and conferences at international and national level. He has sound analytical capabilities to handle commercial consultancy projects and to deliver business improvement projects. He has skills and expertise of Advance Statistics Models, Multivariate Analysis, Multi-Criteria Decision Making, Fuzzy Theory, Fuzzy Optimization, Fuzzy Multi-Criteria Decision Making, GREY Theory and Analysis, Application of Soft-Computing, Econometrics Models. His area of research are: sustainability, green/sustainable supply chain management, sustainable and green manufacturing, customer retention, green purchasing behavior, sustainable procurement, sustainable development, circular economy, industry 4.0, performance measurement, human capital in supply chain and operations; decision modelling for sustainable business, and integration of operation area with other areas.

Dr. Nripendra P. Rana is Professor in Information Systems in the School of Management at Swansea University, UK. With an academic and professional background in Mathematics and Computer Science and with PhD in Information Systems, his current research interests focus primarily upon adoption of emerging and cutting-edge technology, e-government, m-government, e-commerce, and m-commerce systems. His work has been published in leading academic journals including European Journal of Marketing, Information Systems Frontiers, Government Information Quarterly, Production Planning and Control, Journal of Business Research, Public Management Review, and Computers in Human Behavior. He has also presented his research in some of the prominent international conferences of information systems across the world.

Contributors

Abdellah Abouabdellah
Systems Engineering Laboratory,
   MOSIL, ENSAK
University Ibn Tofail
Kenitra, Morocco

Doris Aguilera
Escuela de Ciencias Administrativas
Corporación Universitaria del Meta
Villavicencio, Colombia

Surajit Bag
Post Graduate School of Engineering
   Management
Faculty of Engineering and the Built
   Environment
University of Johannesburg
Johannesburg, South Africa

Manuel Ignacio Balaguera
Centro de Investigaciones de la Escuela
   de Negocios
Fundación Universitaria Konrad Lorenz
Bogotá, Colombia

Sedat Belbağ
Department of Business Administration
Gazi University
Ankara, Turkey

Cristian Beltrán
Centro de Investigaciones de la Escuela
   de Negocios
Fundación Universitaria Konrad Lorenz
Bogotá, Colombia

Arvind Bhardwaj
Department of Industrial & Production
   Engineering
National Institute of Technology
Jalandhar, India

Carlos Bouza
Departamento de Matemáticas
Universidad de La Habana
La Habana, Cuba

Ruth Carrasco-Gallego
Department of Organization
   Engineering, Business
   Administration and Statistics
Escuela Técnica Superior de Ingenieros
   Industriales
Universidad Politécnica de Madrid
Madrid, Spain

Ali Kemal Çelik
Department of Quantitative Methods
Ardahan University
Ardahan, Turkey

Mahdi Chini
Road Housing and Urban Development
   Research Center (BHRC)
Tehran, Iran

Mustafa Çimen
Management Science
Hacettepe University
Ankara, Turkey

Oussama El Mariouli
Systems Engineering Laboratory,
   MOSIL, ENSAK
University Ibn Tofail
Kenitra, Morocco

Miraç Eren
Faculty of Economics and
   Administrative Sciences
Ondokuz Mayıs University
Samsun, Turkey

Maheshwar Dwivedy
BML Munjal University
Gurgaon, India

Mercedes Gaitán-Angulo
Centro de Investigaciones de la Escuela
    de Negocios
Fundación Universitaria Konrad Lorenz
Bogotá, Colombia

Chandra Prakash Garg
Department of Transportation
    Management
School of Business
University of Petroleum & Energy
    Studies
Dehradun, India

Ali Jahan
Department of Industrial Engineering
Semnan Branch
Islamic Azad University
Semnan, Iran

Isabel M. João
ISEL, Instituto Superior de Engenharia
    de Lisboa
Instituto Politécnico de Lisboa
and
CEG – IST, Instituto Superior Técnico
Universidade de Lisboa
Lisbon, Portugal

K. E. K. Vimal
Department of Mechanical Engineering
National Institute of Technology
Patna, India

Manjeet Kharub
Department of Mechanical Engineering
CVR College of Engineering
Hyderabad, India

Payam Khazaelpour
Department of Mathematic and
    Statistics
Curtin University
Perth, Western Australia, Australia

Jenny-Paola Lis-Gutiérrez
Centro de Investigaciones de la Escuela
    de Negocios
Fundación Universitaria Konrad Lorenz
Bogotá, Colombia

Melissa Lis-Gutiérrez
Facultad de Ciencias Ambientales e
    Ingenierías
Universidad de Ciencias Aplicadas y
    Ambientales
Bogotá, Colombia

Kaliyan Mathiyazhagan
Department of Mechanical Engineering
Amity University
Noida, India

Rahul S. Mor
Department of Food Engineering
National Institute of Food Technology
    Entrepreneurship and Management
Sonepat, India

Somvir Singh Nain
Department of Mechanical Engineering
Centre for Materials and Manufacturing
CMR College of Engineering &
    Technology
Hyderabad, India

Sanjiv Narula
BML Munjal University
Gurgaon, India

M. Nishal
Department of Mechanical Engineering
Sri Venkateswara College of
    Engineering
Chennai, India

Erkan Oktay
Faculty of Economics and
    Administrative Sciences
Atatürk University
Erzurum, Turkey

Alireza Panahande
Department of Industrial Engineering
Semnan Branch
Islamic Azad University
Semnan, Iran

Surya Prakash
BML Munjal University
Gurgaon, India

Stefano Ronchi
School of Management
Politecnico di Milano
Milan, Italy

Anoop Kumar Sahu
Post Graduate School of Engineering
    Management
Faculty of Engineering and the Built
    Environment
University of Johannesburg
Johannesburg, South Africa

P. Sasikumar
Department of Industrial Engineering
    Technology
Abu Dhabi Women's College
Higher Colleges of Technology
Abu Dhabi, United Arab Emirates

Satydev
Motilal Nehru National Institute of
    Technology
Allahabad, India

Ümran Şengül
Faculty of Political Sciences
Çanakkale Onsekiz Mart University
Çanakkale, Turkey

Tuhin Sengupta
Indian Institute of Management
Indore, India

Suwarna Shukla
Indian Institute of Management
Indore, India

Sarbjit Singh
Department of Industrial & Production
    Engineering
National Institute of Technology
Jalandhar, India

Tarun Kumar Singhal
Symbiosis Centre for Management
    Studies, NOIDA
Constituent of Symbiosis International
    (Deemed University)
Noida, India

K. Sivakumar
Department of Production
    Engineering
National Institute of Technology
Tiruchirappalli, India

Mehmet Soysal
Operations Management
Hacettepe University
Ankara, Turkey

Bahar Türk
Ondokuz Mayıs University
Samsun, Turkey

Nurullah Umarusman
Department of Business
    Administration
Faculty of Economics & Administrative
    Sciences
Aksaray University
Aksaray, Türkiye

Samir Yerpude
Faculty of Management
Symbiosis Centre of Research and
    Innovation
Symbiosis International (Deemed
    University)
Pune, India

Mohammad Hossein Zarei
Department of Organization
    Engineering, Business
    Administration and Statistics
Escuela Técnica Superior de Ingenieros
    Industriales
Universidad Politécnica de Madrid
Madrid, Spain

and

School of Management
Politecnico di Milano
Milan, Italy

Section I

Sustainable Procurement/Sourcing

1 Conceptual Framework in Sustainable Procurement

Sustainable Development Goals (SDGs) Focused Content Analysis Approach

Tuhin Sengupta and Suwarna Shukla

CONTENTS

1.1 INTRODUCTION

Owing to the growth of the population, in the late twentieth century, we observed a substantial increase in the demand of various products owing to augmentation of consumption for making the survival easy. This enforced the organizations to increase their production for two reasons: (1) to meet the increased demand and (2) to earn profits. The objective behind fulfilling the above conditions made the firms to opt for unsustainable yet profitable production processes (Rajeev et al., 2017). These modes of production had an instant payoff as desired but had trade-offs in the long run. The long-term effects had an impact on the society as well as on the environment, thus leading to some of the fatal industrial accidents such as the Bhopal gas tragedy in India in 1984, Exxon Valdez oil spill in the US in 1989, and many more. These fatal accidents, casualties, and pollutants are alarming; hence, there is need for reconsideration of the industrial practices and consumption patterns for sustainable development.

The Brundtland report primarily focused on achieving sustainable development and defined it as "the development that meets the needs of present without compromising the ability of future generations to meet their own needs" (Brundtland, 1985). The studies relating to sustainability has gained momentum in business-related streams as well. This definition had obvious ambiguities and vagueness because a clearer interpretation of the definition was required while attempting to apply the sustainability principles in real-life (Ahi and Searcy, 2013). Earlier studies in sustainability were on environmental issues, but with time, these studies focused on the *Triple Bottom Line* (TBL) goal. TBL can be understood as the sustainable consumption and production practices (SCP from now onwards) that take care of the environmental and natural resources and cater to the needs of poor people and society; these form the pillar for TBL approach (Freeman, 2010; Crespin-Mazet and Dontenwill, 2012; Joyce and Paquin, 2016; Rajeev et al., 2017). The research further suggests that the adoption of a comprehensive TBL roadmap by a firm often leads to a competitive advantage of the firm (Walker et al., 2008).

To achieve a comprehensive TBL performance approach, firms need to foster new types of social, economic, and environmental partnerships in the long run. The partnerships will realize that the traditional tasks can now be performed more efficiently and concede the fact that the far-reaching goals can now be effectively achieved in the near future (Elkington, 1998). From the TBL perspective, sustainability is quite often misunderstood as three disparate goals that need to be achieved or benchmarked in organizations. However, social, economical, and environmental sustainability must be well understood by replacing them with 3P's, i.e., People, Profit, and Planet. The objective is not to score more points in each category; rather, it is to balance the three objectives so that an individual action should not destroy the other (Elkington, 2013).

It is of utmost importance that firms must incorporate the TBL thinking by encompassing planet, people, and profit into the culture and strategic operations of an organization (Kleindorfer et al., 2005). However, the sole incorporation of the TBL perspective in a firm's strategic functions and culture would not suffice in achieving the objective of sustainability. It is, therefore, important to operationalize sustainability in the day-to-day operations of organizations. Sustainability must be understood through the lens of operations management (OM) because of specific reasons (Gimenez et al., 2012). First, firms are accountable for consuming energy and resources to fulfill their essential activities, which include sourcing, manufacturing, logistics, and reuse of products in their offerings. Hence, it is very evident that OM can contribute toward reducing such resource footprints. Second, organizations need to function in a sustainable manner to ensure the health and safety of its employees as well as welfare of the external community. OM is one of those domains, where significant numbers of personnel are employed; hence, this domain contributes to the highest footprint of the society. It is, therefore, prudent to study sustainability from the perspective of OM (Golini et al., 2010; Gimenez et al., 2012).

The incorporation of sustainability in OM is basically the integration of skills and leverages needed to align the business processes of a firm for better performance (Kleindorfer et al., 2005). However, a close analysis reveals that OM as a field has evolved by exchanging goods and services beyond the reach and control of the organization. Therefore, it is important to study the extended supply chain as recommended by Corbett and Kleindorfer (2001) and supported by Srivastava (2007), which leads to the evolution of the term *Sustainable Supply Chain Management* (SSCM). Sustainable supply chains imply that firms need to improve both environmental and social performance without neglecting profits (de Ron, 1998). Therefore, firms need to come up with mechanisms that will not only align the internal processes to meet the sustainability objectives, but will also improve sustainability of the suppliers and customers processes (Seuring and Müller, 2008; Pagell and Gobeli, 2009).

Our study attempts to understand the current state of affairs of sustainability practices in a firm with special focus on its supply chain. Procurement is a crucial part of sustainability because activities need to look beyond the organization's control and boundaries, thereby incorporating the entire supply chain (Walker and Phillips, 2008; Meehan and Bryde, 2011). Sustainable Procurement (SP) is defined as the effort of sustainable development practices and objectives that encompass the sourcing and supply processes (Walker et al., 2012). SP "is consistent with the principles of sustainable development, such as ensuring a strong, healthy and just society, living within environmental limits, and promoting good governance" (Walker and Brammer, 2009). It is a growing field, and many special issues have been published in the past couple of years (Walker et al., 2012) in different reputed journals such as IJOPM (International Journal of Operations and Production Management), POM (Production and Operations Management), JOM (Journal of Operations Management), JSCM (Journal of Supply Chain Management), and JCP (Journal of Cleaner Production) among others (Walker et al., 2012). This indicates that SP is a current and relevant topic, which reflects the need of practitioner's concerns. We, therefore, address the following research questions (RQs) given below:

RQ 1: What is the current knowledge gap between the academic literature on sustainable procurement and the sustainability reports presented by the practitioners?

RQ 2: What is the current knowledge gap between the academic literature on sustainable procurement and the Sustainable Development Goals (SDGs) proposed by the United Nations?

RQ 3: How can we align our future research directions through the academic literature for providing a platform to the practitioners so that sustainable procurement practices as a guideline keeping in mind the SDG?

The remainder of the chapter is as follows: Section 1.2 presents a broad overview of SDG and its implication on our study. Section 1.3 presents the literature review of sustainable procurement by highlighting the performance measures and metrics discussed in the literature. Section 1.4 presents the content analysis of sustainability reports by highlighting the presence of sustainable procurement of firms. Section 1.5 presents the mapping of academic literature on SDG, practitioner literature (content analysis) on SDG, and academic literature with practitioner literature, thereby presenting the broad research gaps and themes for future study. Section 1.6 discusses the ways in which the chapter addresses the research objectives as stated in the introduction and presents the different limitations of the paper.

1.2   SUSTAINABLE DEVELOPMENT GOALS (SDGs)

The SDGs, sometimes termed as *Global Goals* as mentioned in the UN website, act as a motivation to galvanize the nations and people for contributing towards poverty eradication, protecting the planet, and ensuring the peace and harmony among the people.[*]

The SDGs came into existence at "United Nations Conference on Sustainable development" in 2012, and finally, it came into effect in January 2016. The main objective was to tackle the environmental, social, and economical challenges that needed immediate intervention owing to some common universal targets and goals. Prior to the adoption of the SDGs, there were *Millennium Development Goals* (MDGs), which started in 2000 with the global effort to tackle the rising concerns of poverty. For the last 15 years, with the start of MDGs, there was a significant progress related to the requirements of its agenda such as the fight against many deadly diseases like HIV/AIDS, improvement in maternal health, provision of water and sanitation, etc. Therefore, the success graph of MDGs gave a kick-start to tread on newer goals, which led to the formation of SDGs. SDGs are bold commitments with some more pressing goals and challenges,[†] and they are an expansion of MDGs.

Approximately 17 SDGs came into existence after the success of MDGs, with few inclusions in the existing ones. These 17 SDGs are interconnected; hence, dealing with one of the proposed goals leads to the disturbance in managing the other goals because of their inter-relatedness.[‡] The proposed 17 SDGs are the powerful tools that cater to the TBL approach. They mainly focus on specific broad spectrums, which are eradication of poverty and hunger; promotion of sustainable agriculture; assurance of healthy lives; equitable quality of education for all; water management and sanitation; accessible, sustainable, and affordable modern energy; gender equality and empowering women; sustainable production and consumption using today's resources without depriving the future generations; sustained economic growth and productive employment; conservation of lives on land, water, and air; promotion of sustainable societies; revitalization of global partnerships; and resistance to climate changes and its impacts.[¶] We present the contribution of different SDG to the TBL objective in Table 1.1.

---

[*]   http://www.undp.org/content/undp/en/home/sustainable-development-goals.html
[†]   http://www.undp.org/content/undp/en/home/sustainable-development-goals/background.html
[‡]   http://www.undp.org/content/undp/en/home/sustainable-development-goals.html
[¶]   https://www.theguardian.com/global-development/2015/jan/19/sustainable-development-goals-united-nations

TABLE 1.1
Sustainable Development Goals

Goals	Description
No Poverty	End extreme poverty in all forms by 2030.
Zero Hunger	End hunger, achieve food security, and improved nutrition, and promote sustainable agriculture, all to be accomplished by 2030.
Good Health and Well-Being	A good health is more than just a wish. Ensuring good health and healthy lives is the essence. Also promoting well-being for all age groups is the motto.
Quality Education	Ensuring equitable and quality education for all and promoting lifelong learning experience to everyone.
Gender Equality	Achieving gender equality by empowering women and girls and leveling the inequalities in terms of work, wages, and education.
Clean Water and Sanitation	By 2030, the goal is to make clean, safe, and affordable drinking water to all by sustainable management of water and sanitation.
Affordable and Clean Energy	By using solar and wind energy, meeting the electricity requirement in a sustainable and more energy efficient way is the major goal to ensure the access to affordable, sustainable, and reliable energy for all.
Descent Work and Economic Growth	Since job growth is not keeping pace with the growing labor force, it gives an alarm to create more job opportunities. The aim is to promote sustainable economic growth and decent employment prospects for all by 2030.
Industry, Innovation, and Infrastructure	By investing in scientific research and innovation and promoting sustainable and inclusive infrastructure will lead to the sustainable development, this technological progress would help in tackling the big challenges like lack of jobs and inefficient manner of energy consumption.
Reduced Inequalities	Income inequality has been a global problem that calls for a global solution. It is very much needed to reduce the inequalities and create possibilities for everyone within and among the countries.
Sustainable Cities and Communities	Cities are the centers for culture, business and life where a lot of people settle down, they also accommodate extreme poverty. Therefore, it is very much important to make cities sustainable for all by creating goals and affordable human settlements and public housing.
Responsible Consumption and Production	In order to make the resources and available for present generation without depriving the future generations from them, we need to use the resources in a very sustainable manner so that everyone can meet their basic needs. This goal can be achieved by managing natural resources more efficiently; disposing toxic waste properly, businesses and consumers can reduce and recycle waste.
Climate Action	There is an urgent need to take actions by the countries to work together in order to combat the worst effects of climate change and its impact like earthquake, flood, tsunami, tropical cyclones, etc.
Life Below Water	It is a big target to manage and protect the life below water by conserving and sustainably using the oceans and its resources for the sustainable development.
Life on Land	There is an immense need to protect, conserve, and restore the terrestrial ecosystem such as forests, wetlands, dry lands and mountains in order to combat desertification, biodiversity loss and land degradation, by 2030.

(Continued)

TABLE 1.1 (*Continued*)
Sustainable Development Goals

Goals	Description
Peace, Justice, and Strong Institutions	The goal is to promote peaceful and inclusive societies for sustainable development; there should be one goal for peace and justice for all countries. The idea is to make all countries strive towards the aim of reducing the violence of all forms, strengthening the rules of law, reducing the flow of illicit arms, and bringing the developing countries into the center of institutions of global governance.
Partnerships for the Goals	Strengthening the means of implementation and revitalizing the goals for sustainable development by bringing all the nations together so they collectively strive towards the common goal. The partnerships among the countries would lay the path of sustainable development and would help in achieving all other goals.

1.3   LITERATURE REVIEW

For this study, the articles were selected based on the citation count of each article in Google Scholar with keywords *Sustainable Procurement* in the title of the article. We first shortlisted 387 articles as per the keyword search. Subsequently, we followed a set of delimiting criteria—Stage 1 and Stage 2—to funnel our final selection. In Stage 1, we shortlisted relevant papers from the 387 articles published in top-tier journals (ABS 3, ABS 4, ABDC A, and ABDC A*). To ensure the inclusion of important and relevant papers, we ascertained that all articles with positive citation count are shortlisted. In Stage 1, 113 articles were shortlisted, but in Stage 2, we carefully carried out a manual content analysis of abstract, introduction, and conclusion to ensure that the articles discussed SP indicators. Finally, we chose 50 articles for our review and further summarization of SP indicators as provided in Appendix 1.1 in decreasing order of citation count. While selecting the final set of papers for review, we considered two things that are worth mentioning. First, very few papers discussed the SP indicators in detail. Most of the papers provided general discussion about SP without specifically positioning around indicators or metrics or practices at a micro level. Second, even after conducting a manual content analysis on abstract, introduction, and conclusion, many papers among the list of 50 had little or no mention of SP indicators.

Grob and McGregor (2005) wrote one of the first few articles that highlighted the need for a greater role of organizations in favor of sustainable development through SP. The paper analyzed organizations from a phased framework of sustainability and validated that *rejection, non-responsiveness, compliance, efficiency, strategic pro-activity*, and *sustainable organization* are part of the phases involved in developing a sustainable organization. Walker and Brammer (2009), Preuss (2009), and Rimmington et al. (2006) examined the SP practices in the UK government sector and proposed the principles of sustainable

APPENDIX 1.1
Sustainability Indicators from Academic Literature

Author (Year)	Social	Economic	Environmental
Walker and Brammer (2009)	Purchases from local suppliers, suppliers comply with child labor laws, suppliers' locations are operated in a safe manner, suppliers to pay a *living wage* greater than a country's or region's minimum wage	Reduced packaging material, Training for procurement, Procurement staff incentives	Suppliers commit to waste reduction goals, design of products for recycling or reuse, uses a life-cycle analysis to evaluate the environmental friendliness of products and packaging, ISO14000/1 / EMS
Brammer and Walker (2011)	Has a formal minority and women-owned business enterprise (MWBE) supplier purchase program, ensures the safe, incoming movement of product to our facilities, ensures that suppliers' locations are operated in a safe manner, visits suppliers' plants to ensure that they are not using sweatshop labor, ensures that suppliers comply with child labor laws	NA	Life-cycle analysis to evaluate the environmental friendliness of products and packaging, suppliers to commit to waste reduction goals, reduces packaging material, participates in the design of products for disassembly, participates in the design of products for recycling or reuse
Preuss (2009)	Certification in Fair Trade Agreement, contracting with local suppliers	Local procurement	Green design, Environmental Risk Assessment
Meehan and Bryde (2011)	Sustainability policy, trains staff in sustainability, sustainability action plan, selection of sustainable suppliers, safe incoming movement of products, purchase from small suppliers, purchase from social and charitable enterprises	Sustainability policy, sustainability action plan, waste reduction plan	Sustainability policy, sustainability action plan, environmental management system accreditation, waste reduction plan, life-cycle impact of its products/service
Thomson and Jackson (2007)	NA	NA	Presence of an environmental management system (EMS), compliance with health, safety and environmental legislation, recycled material, recyclable material, green electricity in production, waste disposal, collection, recycling and reuse of waste, packaging and recovery of packaging

(Continued)

APPENDIX 1.1 (*Continued*)
Sustainability Indicators from Academic Literature

Author (Year)	Social	Economic	Environmental
Walker and Brammer (2012)	Purchases from local suppliers, ensures that suppliers comply with child labor laws, donates to philanthropic organizations	Reduces packaging material	Reduces packaging material, ensures that suppliers' locations are operated in a safe manner, asks suppliers to commit to waste reduction goals
Walker and Phillips (2008)	Senior management commitment, educate suppliers, SMEs, include sustainable criteria in contracts, assess suppliers, identify sustainable supply risks, investigate alternate sourcing	Senior management commitment	Waste, energy efficient buildings, transport and emissions, senior management commitment
Rimmington et al. (2006)	Avoid purchase from the sources that are known to damage the human health excessively, purchase from local suppliers, ways to meet the needs of local or regional suppliers	Waste reduction	Avoid purchase from the sources that are known to damage the environment excessively, ensuring product processing in with reduced water, energy consumption and minimized waste
Walker et al. (2012)	Local charities, purchase from small suppliers, no use of sweatshop labor, wage greater than the minimum wage of the country or region, donations, purchase from local suppliers, comply with child labor laws, suppliers' locations should be operated in safe manner	Recycling and reuse, waste reduction, reduces packaging material	Environmental friendliness of products and packaging, recycling and reuse, waste reduction
Haake and Seuring (2009)	NA	NA	NA
Crespin-Mazet and Dontenwill (2012)	Involving suppliers with natural products, cooperation with non-business actors, involving the weaker actors in supply chain	NA	NA
Kaye Nijaki and Worrel (2012)	NA	NA	Reduction in chemical fertilizers and deforestation

(*Continued*)

APPENDIX 1.1 (*Continued*)
Sustainability Indicators from Academic Literature

Author (Year)	Social	Economic	Environmental
Oruezabala and Rico (2012)	NA	Reduce number of orders, reduced supplier base, decreasing the role of small suppliers	Minimize environmental impact, improve energetic balance
Brammer and Walker (2007)	Has a formal minority and women-owned business enterprise (MWBE) supplier purchase program, ensures the safe, incoming movement of product to our facilities, ensures that suppliers' locations are operated in a safe manner, visits suppliers' plants to ensure that they are not using sweatshop labor, ensures that suppliers comply with child labor laws	NA	Life-cycle analysis to evaluate the environmental friendliness of products and packaging, suppliers to commit to waste reduction goals, reduces packaging material, participates in the design of products for disassembly, participates in the design of products for recycling or reuse
Sourani (2011)	NA	NA	NA
Walker et al. (2009)	Involvement of local suppliers	Waste management, reduced coat via e-procurement	Reuse, recycle, waste management, carbon reduction
Dawson and Probert (2007)	Procurement programs, training programs, development programs	Welsh procurement initiative to improve value for money, recycling	Green waste compost, recycling
Erridge and Hennigan (2012)	Contracts to smaller lots, increasing business with social economy enterprises, reducing inequality, e.g., long-term disabled, unemployed	Analysis of barriers to local suppliers, providing guidance to supply chain members	Low emission, reducing energy usage, carbon reduction, renewable and recycled products, energy efficiency
Ruparathna and Hewage (2015)	Safety procedures for workers, create employment opportunities, fair wages, transparency, worker training	Recycling, waste management	LEED certification, use of less toxic materials, recycling, energy use standards

(Continued)

APPENDIX 1.1 (*Continued*)

Sustainability Indicators from Academic Literature

Author (Year)	Social	Economic	Environmental
Conner et al. (2016)	Procuring locally grown fresh produce and benefitting the local farmers, fair price to farmers, involving small scale farmers	Knowledge of distributors' capacity	NA
Smith et al. (2016)	Public education, training the staff	NA	Reduced CO_2 emissions, protecting groundwater from pesticide residues, climate protection program
Grob and Benn (2014)	Sustainable Contracting process, supplier assessment programs	Sustainable Contracting process	Sustainable Contracting process, environmental regulatory compliance, supplier assessment programs, environmental Management Systems
Melissen and Reinders (2012)	NA	NA	NA
McMurray et al. (2014)	Procurement from local and small firms, philanthropy, human rights, safety, diversity	NA	Environmental disclosure, environmental impact assessment
Grandia et al. (2014)	Top management support, commitment, expertise	NA	NA
Australian Procurement and Construction Council (2007)	NA	NA	NA
Grandia (2015)	NA	NA	NA
Yeow et al. (2011)	NA	NA	NA
Ghadimi et al. (2016)	Health and safety, employment practices, local communities influence, contractual stakeholders influence	NA	Supplier performance towards implementing environmental policies, adapting environmental certificates and regular environmental quality audits are measured, waste water, solid wastes, resource consumption, use of harmful materials

(*Continued*)

APPENDIX 1.1 (*Continued*)
Sustainability Indicators from Academic Literature

Author (Year)	Social	Economic	Environmental
Meehan and Bryde (2015)	Has a sustainability policy, trains staff in sustainability, has a sustainability action plan, purchases from small- to medium-sized suppliers, purchases from local suppliers	Has a sustainability policy, trains staff in sustainability, has a sustainability action plan	Has a sustainability policy, trains staff in sustainability, has a sustainability action plan, assesses the life-cycle impact of its products/service provision
Jelodar et al. (2013)	NA	NA	NA
Jefferies et al. (2006)	NA	NA	NA
Witjes and Lozano (2016)	Engaging stakeholders for making societies more sustainable	Reductions in raw material utilization	Reductions in waste generation
Mansi (2015)	Human rights, labor conditions and decent work, diversity, philanthropy, community development, health and safety	Product responsibility, economic viability	Environmental disclosure
Kaur and Singh (2016)	NA	NA	NA
Grob and McGregor (2005)	Public Reporting of Purchasing, Stakeholder Engagement to develop new products and services, Sustainable Procurement Policy, Occupational Health and Safety in Purchasing Decisions	Sustainable Procurement Policy	Life-Cycle Assessment for all products, Sustainable Procurement Policy, Waste Reduction, Environmental Management System
Islam and Siwar (2013)	Purchases from local suppliers, Ensures that suppliers' location are operated in a safe manner, Asks suppliers to pay a *living wage* greater than a country's or region's minimum wage, Ensures the safe incoming movement of product to our facilities	Reduces packaging material, Asks suppliers to commit to waste reduction goals	Uses a life-cycle analysis to evaluate the environmental friendliness of products and packaging

(*Continued*)

APPENDIX 1.1 (*Continued*)
Sustainability Indicators from Academic Literature

Author (Year)	Social	Economic	Environmental
Roman (2017)	Transformational leadership, Procurement organizational strategic role, Stakeholders' expectations, Interdepartmental collaboration	Innovativeness, Centralization, Procurement organizational strategic role, Interdepartmental collaboration	Procurement organizational strategic role, Interdepartmental collaboration
Testa et al. (2016)	NA	NA	Training on Green Procurement practices, Knowledge of GPP toolkits and guidelines, Awareness on GPP procedures, Certified EMS adoption, Structure of purchasing process
Amman et al. (2014)	Promoting employment opportunities, Promoting decent work, Supporting social inclusion and promoting social economy organization	Reducing waste generation	Ensuring biodiversity, Reducing emission to air/water, Reducing energy and water consumption, Reducing chemical consumption
Hasselbalch et al. (2014)	Sustainable Procurement Manual, verify suppliers' reports on sustainability criteria	Sustainable Procurement Manual, Life-Cycle Costing, verify suppliers' reports on sustainability criteria	Sustainable Procurement Manual, Life-Cycle Costing, verify suppliers' reports on sustainability criteria
Wilkinson and Kirkup (2009)	Physical health, emotional and mental health, self care and living skills	Gross Income, Direct Spending, Total Spending	Life-cycle assessment, organizational footprint, activity emission tool
Large et al. (2013)	Permanent improvement of working conditions, Enhancement of qualified employment	Reduction of land use	Reduction of transport intensity and emission, Choice of carrier under considerations of sustainable aspects

(*Continued*)

APPENDIX 1.1 (*Continued*)
Sustainability Indicators from Academic Literature

Author (Year)	Social	Economic	Environmental
Lund-Thomsen and Costa (2011)	Local purchasing policies	NA	Commitment to environmental protection
Steurer et al. (2007)	Employment, improved working conditions, equal opportunity and accessibility, fair wages, employment opportunities to disabled workers, Fair Trade, safeguarding human and labor rights	Innovation or the diversity of supplier markets	Green product design, green transport
Aragão and Jabbour (2017)	NA	NA	NA
Goldschmidt et al. (2013)	NA	NA	NA
Molenaar et al. (2010)	NA	Quality of the product, Cost of the Product, Time of Delivery, Adherence to specific design requirements, qualification of purchase department employees	NA
Gromly (2014)	NA	Better Value for Money of the Purchased Product	Green Tenders, Recycling Product
Hussein and Shale (2014)	Supplier Involvement, Supplier Ethical Practices	Product Re-Usability	Supplier Involvement, Organization Commitment to green purchasing

food procurement (Smith et al., 2016) as well as identified the key performance metrics for *UK Public Sector Food Procurement Initiative*, respectively. Similar contextual study was also conducted under various themes in sustainable procurement practices in UK (Molenaar et al., 2010; Brammer and Walker, 2011; Kaye Nijaki and Worrel, 2012; Walker and Brammer, 2012). Dawson and Probert (2007) emphasized the need for committing to sustainable procurement to reduce waste in Wales, UK. Jefferies et al. (2006) highlighted the current approaches involved in the identification of risks associated with the bidding process of the PPP model, which is an alternative of the traditional procurement practices. Thomson and Jackson (2007) explained the green procurement practice in accordance with the UK local government by highlighting the role of legislation, the commitment of different stakeholders, and the barriers involved in implementing the initiatives. Brammer and Walker (2007) documented their study findings through a survey of SP practices from 280 public procurement practitioners of 20 nations. The Australian Procurement and Construction Council (2007) implemented the Australian and New Zealand sustainable procurement framework and listed the benefits derived from the implementation of the framework across the government sector. Steurer et al. (2007) presented the government-driven SP initiatives in the EU member states (Large et al., 2013; Amann et al., 2014) with the objective of submitting the report to EU high-level group on corporate social responsibility. Similar study was further conducted on behalf of the United Nations (Lund-Thomsen and Costa, 2011; Hasselbalch et al., 2014). Through a focus group discussion approach, Walker and Phillips (2008) identified four major sustainable procurement themes. The first theme was to shift the current focus toward social and economic sustainability in comparison to environmental sustainability. The second theme is about innovation in the interface of procurement and sustainability. The third and fourth themes revolve around ethical supply and other issues related to the measures of sustainable procurement (Wilkinson and Kirkup, 2009).

Haake and Seuring (2009) and Walker et al. (2012) identified the key shortcomings related to research in sustainability, particularly SP. Walker et al. (2009) first introduced an online SP course for managers from different industrial sectors with a focus on the increasing environmental concerns. Similar studies have been done with the intention for teaching classroom students in a university setting (Goldschmidt et al., 2013). Meehan and Bryde (2011) investigated the SP practices in the UK housing association and found that the failure to overcome inertia was one of the key barriers in the implementation of SP practices. Later, both authors further conducted a similar study incorporating the role of regulation and procurement consortia in the adoption of SP practices in the UK housing sector (Meehan and Bryde, 2015). Conner et al. (2016) documented the efforts of mid-scale farms in adopting SP operations, as it can provide healthy food to district schools. Crespin-Mazet and Dontenwill (2012) showed the ways in which firms develop legitimacy in SP through evolution in their supply networks. Oruezabala and Rico (2012) studied the impact of procurement practices in the French public hospitals on supplier management Sourani (2012) and identified the barriers faced by public clients in addressing SP issues in the construction sector

(Jelodar et al., 2013; Ruparathna and Hewage, 2015). Erridge and Hennigan (2012) found a strong correlation between SP practices and efficiency in the health and social care sector of Northern Ireland. A similar study has been conducted for the Irish commercial, semi-state bodies (Gormly, 2014). Melissen and Reinders (2012) reviewed the Dutch SP program and evaluated its overall contribution to sustainable development. Yeow et al. (2011) emphasized the need for innovation as a means to achieve SP. Grob and Benn (2014) conceptualized the adoption of SP through the lens of institutional theory. From the perspective of religious biasness, McMurray et al. (2014) analyzed the sustainable procurement practices among the managers and directors in both the public and private sectors of Malaysia. Grandia et al. (2014) investigated the sustainable public procurement perspective from an organization's point of view for understanding the functioning of the organizations and their prioritization of practices (Islam and Siwar, 2013) in the entire value chain. Hussein and Shale (2014) studied the impact of SP practices on organizational performance in the manufacturing sector. Grandia (2015) emphasized the role of change agents in project teams for achieving successful sustainable public procurement projects. Mansi (2015) analyzed the ways through which central public sector enterprises in India operationalized and disclosed SP practices in the reports.

Ghadimi et al. (2016) reviewed the literature on the buyer-supplier dyadic relationship from the perspective of the SP context. Kaur and Singh (2016) minimized carbon emission in the procurement and logistics stage for enhancing the resilience of the supply chain. Witjes and Lozano (2016) emphasized the need for resource-based efficiency in relation to products and, hence, proposed a framework to include product design specifications. Roman (2017) explored the conditions on the basis of which a firm adopts sustainable procurement practices under the influence of a specific style of top management leadership. Testa et al. (2016) identified the opportunities and drawbacks of green public procurement practices, the absence of which can be an effective tool for achieving sustainable development. Aragão and Jabbour (2017) emphasized the need of appropriate human resource practices and training for incorporating green procurement practices in the Brazilian public sector. We summarized the entire literature review by identifying relevant indicators and segregated them into social, economical, and environmental categories as presented in Appendix 1.1.

1.4 QUALITATIVE CONTENT ANALYSIS

We chose manufacturing firms for the purpose of our study primarily because of two reasons. First, sustainable procurement can be better studied from the perspective of manufacturing firms, as these firms typically have bulk procurement contract from its suppliers. Second, it is easier to observe the upstream value chain as well as downstream value chain from the perspective of manufacturing firms. SP is an emerging concept in supply chain and operations management literature especially in accordance with the adoption of the new manufacturing policy in the present scenario. Over the past few years, the manufacturing industry has progressed from a cost-centric approach to a customer-centric approach. SP is slowly positioning itself as an important and integrated part of the entire supply chain, because players are acknowledging the needs and regulations of the customers.

We chose the manufacturing steel plants for our study and segregated our rationale from the perspectives of environmental, economical, and social sustainability.[*]

https://www.worldsteel.org/ In the twenty-first century, climate change is a major issue for the steel industry. The reduction of carbon emissions is a top priority and efforts are made to develop technologies and innovation for causing substantial reduction in CO_2 emissions. Further, according to a report provided by the World Steel Association,[†] https://www.worldsteel.org/en/dam/jcr:0191b72f-987c-4057-a104-6c06af8fbc2b/fact_technology%2520transfer_2018.pdf efforts are made in four directions, i.e., carbon, hydrogen, biomass, and carbon capture and storage technology (CCS) for reducing the overall environmental burden. For instance, when carbon is being used as a reducing agent, it is crucial to develop technologies that can capture CO_2 and store it in an efficient manner. On similar lines, hydrogen is seen as a cleaner reducing agent in comparison to carbon, because the reaction only produces water vapor. Hydrogen, after reformed through methane, and natural gas can be used in conventional reactors. Similar logic and efforts stand for biomass, which can be further used as a reducing agent. From the perspective of economic sustainability, steel is the most common material in the world. Our dependency on steel is immense, even for other sectors such as housing, automobiles, construction, transport, food, utility, and health care. In the developing countries, steel is necessary to build roads, bridges, railways, etc. Social sustainability in steel is important mainly because of its role in employment in the steel industry as well as because of health and safety issues related to accidents and injuries. Further, steel has allied investments in other sectors such as power/ utility and different ancillary units, which are required for the survival of steel manufacturing plants. Overall, a steel plant directly and indirectly affects the welfare of a large proportion of the local community.

By shifting our focus towards the importance of SP in steel industries, we believed that a responsible procurement policy in the sector is important to incorporate overall sustainability dimension in the entire supply chain. Typically, there are many small and fringe players in the steel industry, and these players are not bound to follow sustainable practices unless the initiative is implemented from the manufacturing player. For instance, Tata Steel developed a unified procurement policy in 2011, thus ensuring that their suppliers are evaluated as per the five dimensions, which are health and safety of workforce, fair business practices, environmental protection, human rights and local community development. Although the procurement processes in the steel industry are structured, the thorough evaluation of the suppliers ensures the sustainability of the supply chain. Another prominent example would be that most manufacturing steel plants procure iron-ore from mines that are owned by different parties. It is important to ensure effective management of resources, preservation of biodiversity, and implementation of a structured health and safety management system by suppliers and mine owners.

---

[*] https://www.worldsteel.org/

[†] https://www.worldsteel.org/en/dam/jcr:0191b72f-987c-4057-a104-6c06af8fbc2b/fact_technology%2520transfer_2018.pdf

TABLE 1.2
Descriptive Statistics of Firms

Company	Forbes 2000 Rank	Country	Sales	Profit	Market Value
Vale	156	Brazil	$27.1B	$3.8B	$45.4B
Arcelor Mittal	216	Luxembourg	$56.8B	$1.8B	$25.1B
Posco	299	South Korea	$45.8B	$1.1B	$19.2B
Nippon Steel & Sumitomo Metal	440	Japan	$41.6B	$477M	$20.2B
Fosun International	448	China	$11.1B	$1.5B	$13.1B
Nucor	657	United States	$16.2B	$731M	$19.4B
JFE Holdings	819	Japan	$29.4B	$273M	$9.4B
Novolipetsk Steel	1009	Russia	$7.6B	$907M	$10.2B
Severstal	1022	Russia	$5.9B	$1.6B	$12.3B
Tata Steel	1076	India	$16.1B	$444M	$7.5B

Refer to Table 1.2 for the list of companies chosen from Forbes 2000 Global Companies list based on their ranking. We purposefully chose 10 companies to make our study robust. The 10 companies that we chose reside in different parts of the globe, i.e., North America, South America, Europe, Central Asia, Indian Sub-Continent, and South East Asia. Table 1.3 provides the details of the sustainability reports selected for our study, which includes the name of the reports downloaded and its corresponding range of timeline. Our final analysis sample consisted of 34 sustainability reports from 10 firms.

We adopted a thorough qualitative content analysis for our study (Schreier, 2012). Content analysis (Graneheim and Lundman, 2004) is the most commonly used technique in qualitative research studies. There are broadly three approaches to qualitative content analysis, namely, *conventional*, *directed*, and *summative* (Hsieh and Shannon, 2005). All three approaches adhere to the naturalistic paradigm and are different in terms of coding schemes, origin of codes, and threats to trustworthiness. In the domain of supply chain management, the qualitative content analysis has been widely used in the literature (Seuring and Müller, 2008; Stock and Boyer, 2009; Gold et al., 2010; Seuring and Gold, 2012; Brandenburg et al., 2014). Following Stock and Boyer (2009), we adopted the content analysis technique for answering our research questions. The unit of analysis for conducting the content analysis was sustainability reports taken from the steel sector. Since our objective is to understand the ways in which firms report their sustainability practices in line with SDGs, we decided to demarcate sustainability practices into social, environmental, and economical sustainability.

To ensure that our analysis and inference would be *trustworthy*, the trustworthiness component of the research results was based on three factors: *transferability*, *credibility*, and *dependability* (Graneheim and Lundman, 2004). In terms of credibility,

TABLE 1.3
Sustainability Report of Selected Firms

Company	Name of Reports Downloaded	Year
Vale	Sustainability-report	2014–2016
Arcelor Mittal	ArcelorMittal-CR-Report-2011single	2011–2014
	ArcelorMittal-CR-Report-2011single	
	crreport-16april	
	global-corporate-responsibility-report-2014	
Posco	2013_SR_eng	2013–2015
	poscoreport_2014_eng_web	
	2015_POSCO_Report_EN	
Nippon Steel &Sumitomo Metal	report2015_all	2015–2017
	report2016_all	
	report2017_all	
Fosun International	report-2013z-en	2013
	report-2015z-en	2015
	Corporate_Social_Responsibility_Report_2016	2016
Nucor	Nucor_SustainabilityReport09	2009
	Nucor_SustainabilityReport11	2011
	Nucor_SustainabilityReport13	2013
JFE Holdings	csr2015e	2015–2017
	csr2016e	
	csr2017e	
Novolipetsk Steel	Corporate Social Responsibility Report	2010–2012
Severstal	Severstal_CSR_Report_2014	2014–2016
	Severstal_CSR_Report_2015_eng	
	Severstal_2016-eng-web-0610	
Tata Steel	csr-2014-15	2014–2015
	csr-2015-16	2015–2016
	csr-2016-17	2016–2017

all the sustainability reports were downloaded from the company's website. We believed that our inferences are credible as per the assumption that the company's policies are ethical in reporting their practices. Dependability is based on the degree of variability of data with time and the interpretation of the data in consideration. The inclusion of the multiple sustainability reports from different years for the same company in our analysis helped in arguing the dependability of our results, as we reported the data based on the triangulation from multiple sustainability reports. The transferability refers to the extent to which the findings can be transferred to other contexts. Since a common method was employed for evaluating the data of the manufacturing sector comprising of multiple sustainability reports over the years, it can be reasonably assumed that the adoption of the same approach in different settings (at least the results would be reliable in other industry where the primary function of firms lies in manufacturing) would produce reliable results. We summarize our entire qualitative content analysis in Table 1.4.

TABLE 1.4

Qualitative Content Analysis of Manufacturing Firms from the Perspective of Sustainable Procurement

Company	Social	Economic	Environmental
Vale	Hires local workforce and suppliers, selection of suppliers is based on human right practices, safety and corruption aspects, collaboration towards socioeconomic development, focus on individual actions with communities on basic health care, basic education, job and income generation, social protection, suppliers development program and their training, suppliers compliance with the code of ethics, measures taken to effectively abolish the child labor, forced labor, start of family agriculture program and purchase the produce.	Developing projects/programs related to social needs with the long-term economic development vision, focus on science and technology, contributing to urban territory management and development of new economic vocations that promote sustainability of regions in the long run, due to adoption of agro forestry techniques there was significant increase in the production.	Collaboration towards environmental development, focus on environmental conservation, promotes research and development of new technologies to mitigate and evaluate climate change, waste disposal quantity increased, access to environmental information by investors and buyers through global disclosure system, they cover emissions related to the purchase of electric energy and process steam, they have a voluntary clause related to the purchase of most emitting materials, purchase of Amazonia S.A. for preservation of green areas and recovery of deteriorated areas.
Arcelor Mittal	Assessment of 243 suppliers, compliance conformity with suppliers as per code of business conduct, sustainability reporting compliance with suppliers, guidance document to ensure responsible sourcing with suppliers, in January 2013 they added a new clause in their contract so that all the suppliers may comply with relevant health and safety regulations, ensuring the online training of all their buyers in their code, REACH and GHS regulations are designed to protect human health.	Assessment of 243 suppliers, training of buyers for integrating with company's product codes, sustainability reporting compliance with suppliers, guidance document to ensure responsible sourcing with suppliers, technology innovation for economic efficiency, S-in-motion range of steel solutions meets the targets for fuel economy by reducing the weight of the car by 19% without compromising strength and safety, integration with vehicle manufacturers to integrate S-in-motion steel solution in their products.	Partnership with their suppliers to meet their environmental goals, assessment of 243 suppliers, sustainability reporting compliance with suppliers, guidance document to ensure responsible sourcing with suppliers, technology innovation for reduced waste, partnership with suppliers for energy efficiency and recycling, in January 2013 they added a new clause in their contract so that all the suppliers may comply with relevant environmental regulations, S-in-motion range of steel solutions contributes to the carbon dioxide reduction by 13.5% and other equivalent emissions, REACH and GHS regulations are designed to protect environment.

(Continued)

TABLE 1.4 (Continued)
Qualitative Content Analysis of Manufacturing Firms from the Perspective of Sustainable Procurement

Company	Social	Economic	Environmental
Posco	Attaining shared growth by fostering competitiveness and growth potential of suppliers via win-win cooperation, promotion of local economies by steady investments like wages for employees and placing orders with suppliers, enhancing suppliers' CSR competitiveness, operating an ethical clause in the company policy, developed a behaviour codes for its suppliers, educations programs for employees, partners and suppliers, benefit sharing between Posco and suppliers, respect for the employees, supplier shall not hire anyone below 15 years.	Awards for quality management, profits from technological development and improvement activities, providing the best quality and service realizing the World Best Supply Chain, supplier should participate in win-win growth situation for healthy corporate ecosystem and contribute to the economic development of the region.	Environmental management training is provided, emission of green house gases is checked, electric power network for energy efficiency, safe handling of harmful substance that can pollute environment, recycling waste, reusing, they conduct eco friendly supply chain, management of carbon emissions, target to reduce carbon dioxide emission, use of eco friendly slag cement.
Nippon Steel & Sumitomo Metal	Close communication with customers and suppliers, ensuring social concerns at all levels of supply chain from procurement to production sales, person to person interaction with young people regarding manufacturing and environment initiatives, personnel policies for fair treatment of personnel, implementation of various health and safety measures, use of domestic resources gives employment opportunities in local community.	Energy efficient technologies with supplier information, awards to suppliers for cost improvement and procurement.	Person to person interaction with young people regarding environment initiatives, eco friendly purchasing for prevention of resources, carbon dioxide reduction, control on environmental burden substances.

(Continued)

TABLE 1.4 (*Continued*)

Qualitative Content Analysis of Manufacturing Firms from the Perspective of Sustainable Procurement

Company	Social	Economic	Environmental
Fosun International	Openness, fairness, equity, exchange of discussions between various companies for flow of ideas, safe working environment for employees, protect employees, provide customers with safe products and services, strive together to achieve excellence.	Cost reduction, efficiency enhancement, suppliers should meet technical requirements, suppliers should be able to win the bid at the lowest price, integrated with other companies to improve business, constantly improve management and use advanced technologies.	Comply with environment protection laws and regulations, following government's emission standards and requirements, reduce waste gas, reduce waste water, management of solid waste and green house gases, use of environment friendly raw materials.
Nucor	Code of conduct to be followed by suppliers to comply with high standards of social responsibility, prohibits forced and child labor, fairness, lawful wages, treat employees with respect and dignity, awareness and education is provided.	Largest producer of steel in US.	All the contractors, vendors, suppliers need to comply with environment laws, recycling and reusing the materials, conservation of natural resources.
JFE Holdings	Nurturing the trust and understanding of suppliers, purchasing and procurement policies to be followed by suppliers, observe human rights, work environment, occupational health and safety, meeting customer needs.	Largest supplier of high strength round steel pipes for Tokyo Sky tree construction, contributed to the reduction of procurement costs, one of the world's greatest integrated steel producer, highly competitive production system, use of sophisticated technologies.	DBJ environmentally related loan, awarded for best company for excellent and advanced environmental initiatives, use of materials that do not impact environment, preventing global warming and carbon dioxide emission, pollution prevention, resource recycling, increased environmental R&D investment.
Novolipetsk Steel	Compliance with the laws related to social development, funds for maintenance and renovation of educational facilities, computer and sporting goods, medical center for employees and local residents, residential building for employees, cooperating with business partners (customers and suppliers).	Compliance with the laws related to economic development, Russia's largest steel producer.	Compliance with the laws related to environmental protection.

(*Continued*)

TABLE 1.4 (Continued)
Qualitative Content Analysis of Manufacturing Firms from the Perspective of Sustainable Procurement

Company	Social	Economic	Environmental
Severstal	Purchase of goods and services from local suppliers, make efforts for social stability in the regions, ensure safety of all participants, meeting customer expectations, employee training program, involvement of local farmers and purchase the produce, giving high quality food to their employees, cooperation with suppliers transportation, compliance with labor legislation, small and medium business development, housing program for employees, purchase equipments for educational institutions, employee health and medical facilities, purchase of electronic learning aids for abandoned and orphan children, donation to children's hospital.	Makes efforts for economical stability in the regions, large supplier of resources to the external clients, Industry leading supplier and developer of innovative technologies, attention on quality management, largest Russian supplier of largest diameter pipes, new automated supplier management system, purchase from local farmers drives regional economy, centralized diagnostic service that monitors the quality, defect elimination, streamlined manufacturing, range expansion, high electrical power generation, reduced electricity purchase from third parties, largest Russian producer of iron ore and coking coal, leading producer of automotive products.	Reduction in green house gases emission, protection of water resources, waste management, suppliers to comply with corporate anti-corruption policy for environment protection, suppliers should comply all environmental laws in the contract, environmental safety by employee training program.
Tata Steel	Adhering to corporate ethics and building long term relationship with partners, community development activities like education, health, safety and infrastructure, strong relationship with suppliers, encourages entrepreneurs from socially disadvantaged communities, work opportunity for marginalized local communities and women.	Supply high quality steel to the customers, adoption of best available technologies to improve their processes, improved operational efficiencies and business excellence, reduce procured raw materials and service cost, lowest cost steel producers in the world.	Focus on water consumption, carbon dioxide emission, waste management, biodiversity action plan to ensure no harm to the environment, plantation, increase recycling.

1.5 CROSS MAPPING OF ACADEMIC LITERATURE, PRACTITIONER LITERATURE, AND SDGs

We obtained different SP indicators from academic literature under each of the TBL category, i.e., social, economical, and environmental. However, from practitioner reports, it is crucial to see the relation of these indicators to the different SDG. The purpose of this analysis is to understand whether the firm's actions and initiatives toward different sustainability factors are in line with the United Nations global benchmark. This will give us clarity about two things: first, it helps us to know the current state of affairs for sustainability reporting standards from the perspective of procurement, and second, the ways of improving the reporting standards for covering most of the sustainable development goals. The mapping of practitioner literature (sustainability reports) and SDGs is provided in Table 1.5. There are three key highlights of the mapping. First, the high-prioritized indicators typically cater to SDG 3 and SDG 17, which essentially means that firms are dominantly emphasizing the need for healthy lives and well-being of individuals and community, while dealing with suppliers and ensuring the implementation of initiatives through relevant partnerships. Under SDG 3, we observed that supply firms primarily concentrate on either employees or the community at large. For instance, the relevant indicators include high-quality food for employees, employee health and medical facilities, housing facilities for employees, donations for children's hospitals, electronic learning aids for orphaned children, and so on. Correspondingly, understanding the focus of SDG 17, we observed that firms typically report on how partnerships with different stakeholders ensure compliance and implementation of specific initiatives. For instance, indicators include partnerships with suppliers for energy efficiency and recycling, compliance with labor regulations, etc. Second, the focus was also on SDGs 1, 8, 9, 10, 11, 12, and 13, which primarily constitute poverty, inclusive growth and economic development, innovation and industrialization, inequality, safety of population settlements, and climate change. It is logical because poverty, inclusive growth, inequality, and human settlements are commonly covered under social sustainability initiatives for most firms. The environmental aspects are covered under the climate change initiatives and the economic sustainability initiatives that include innovation and industrialization and sustainable production and consumption. Almost all the firms, irrespective of the sector or industry, cover these initiatives. Finally, it is important to highlight the development goals, i.e., SDG 2, 4, 5, 6, 7, 14, 15, and 16, that are underrepresented in the sustainability practices of firms. These include hunger, food security, equitable quality education, women empowerment, access to affordable and sustainable energy for all, conservation of marine resources, conservation of ecosystem, and inclusive institutions thereby providing justice to all. Apart from conservation of marine resources, it can be argued that it is possible to represent most of the goals from the perspective of suppliers.

TABLE 1.5

Mapping of Indicator Themes from Academic Literature and Practitioner Reports to SDG Goals

SDG Goal	Practitioner Reports	Academic Indicators	SDG Goals Description
1	Purchase parts from local producers, involvement of local farmers, purchase equipments for educational institutions, purchase of electronic learning aids for abandoned and orphaned children, donation to children's hospital, funds for maintenance and renovation for educational facilities, investment in CSR for poverty elimination	Purchase from local suppliers, purchase from small suppliers, purchase from social and charitable enterprises, donation to charitable organizations, diversity of supplier markets, employment, improved working conditions, equal opportunity, employment opportunities for disabled worker	End poverty in all its forms everywhere
2	Use of agro-forestry technique for increasing production, start of family agriculture program	Reducing chemical consumption	End hunger, achieve food security and improve the nutrition and promote sustainable agriculture
3	Giving high quality food to employees, housing programs for employees, employee health and medical facilities, purchase equipments for educational institutions, purchase of electronic learning aids for abandoned and orphaned children, donation to children's hospital, funds for maintenance and renovation for educational facilities, safe working conditions for employees, provide customers with safe products	Suppliers high living wage rate, donation to charitable organizations, improve working conditions, physical, mental and emotional health, self-care and living skill, reducing chemical consumption, occupational health and safety in purchasing decisions, reduction of solid waste and harmful consumption materials	Ensure healthy lives and promote well-being for all at all ages
4	Employee training program, training of buyer for integrating company product codes	Training for procurement, qualification of purchase department employees, enhancement of qualified employers	Ensure inclusive and equitable quality education and promote lifelong learning opportunities for all

(Continued)

TABLE 1.5 (*Continued*)

Mapping of Indicator Themes from Academic Literature and Practitioner Reports to SDG Goals

SDG Goal	Practitioner Reports	Academic Indicators	SDG Goals Description
5	Work opportunity for marginalized local communities and women	Equal opportunity	Achieve gender equality and empower all women and girls
6	Waste management, biodiversity action plan	NA	Ensure availability and sustainable management of water and sanitation for all
7	NA	NA	Ensure access to affordable, reliable, sustainable and modern energy for all
8	Purchase parts from local producers, involvement of local farmers, funds for maintenance and renovation for educational facilities, donation to children's hospital, funds for maintenance and renovation for educational facilities, prohibition of forced child labor, lawful wages, treat employees with respect and dignity	Purchase from local suppliers, suppliers comply with child labor laws, purchase from small suppliers, purchase from social and charitable enterprises, equal opportunity, employment, employment opportunities for disabled worker, promoting social inclusion	Promote sustained, inclusive and sustainable economic growth, full and productive employment and decent work for all
9	Encourage entrepreneurs from socially disadvantaged communities, development of innovative technologies, attention on quality management, defect elimination, increased environmental R&D investment, use of sophisticated technologies, energy efficient technology with supplier information, technology innovation for reducing waste	NA	Build resilient infrastructure, promote inclusive and sustainable industrialization and foster innovation

(*Continued*)

TABLE 1.5 (Continued)
Mapping of Indicator Themes from Academic Literature and Practitioner Reports to SDG Goals

SDG Goal	Practitioner Reports	Academic Indicators	SDG Goals Description
10	Purchase parts from local producers, involvement of local farmers, funds for maintenance and renovation for educational facilities, donation to children's hospital, funds for maintenance and renovation for educational facilities, selection of suppliers based on human right practices	Purchase from local suppliers, suppliers high living wage rate, purchase from small suppliers, purchase from social and charitable enterprises, equal opportunity, employment opportunities for disabled worker, safeguarding human and labor rights	Reduce inequality within and among countries
11	Waste management, biodiversity action plan, increase recycling, resource recycling, reduced waste gas, reduce waste water, management of solid waste, environmentally friendly raw material	Suppliers locations are operated in a safe manner, suppliers commit to waste reduction goal, ensure the safety of incoming product to facilities, waste disposal, collection recycling and reuse of waste	Make cities and human settlements inclusive, safe, resilient, and sustainable
12	High product quality to customers, use of technology for business process improvement, procure lowest cost product from suppliers, focus on water consumption, protection of water resources, cost reduction, efficiency improvement, product quality	Reduced packaged material, design of products for recycling or reuse, life-cycle analysis of product, environmental impact assessment, participate in design of products for disassembly, green design, environmental risk assessment, presence of environmental assessment system, green electricity in production, quality of product, cost of product, time of delivery and adherence to specific design requirements, reduction of land use, gross income, direct spending, total spending, organizational footprint, activity emission tool, life-cycle costing, reducing emission to air and water, reducing energy and water consumption, product responsibility, and economic viability	Ensure sustainable consumption and production patterns

(Continued)

TABLE 1.5 (*Continued*)

Mapping of Indicator Themes from Academic Literature and Practitioner Reports to SDG Goals

SDG Goal	Practitioner Reports	Academic Indicators	SDG Goals Description
13	Reduce carbon dioxide emission, reduce green house gases, use of non-hazardous materials, pollution prevention, environmental management training, safe handling of harmful substances	NWBE supplier purchase program, supplier and organizational commitment to green purchasing, organizational commitment to environmental protection, green product design, green transport, reduction of transport intensity and emission, choice of carrier for environmental sustainability, organizational footprint, activity emission tool, reducing emission to air and water, reduce carbon dioxide emission, protecting groundwater from pesticides residues, climate protection program, energy use standards, reduction in chemical fertilizers and deforestation	Take urgent action to combat climate change and its impacts
14	Protection of water resources	Reducing emission to air and water	Conserve and sustainably use the oceans, seas and marine resources for sustainable development
15	Waste management, biodiversity action plan, plantation, conservation of natural resources, eco-friendly purchasing for prevention of resources	Ensuring biodiversity, reducing emission to air and water, protecting groundwater from pesticides residues, reduction in chemical fertilizers and deforestation	Protect, restore and promote sustainable use of terrestrial ecosystems, sustainably manage forests, combat desertification, and halt and reverse land degradation and halt biodiversity loss
16	Reduce supplier corruption and bribery, ethical code of conduct to be followed by suppliers	Supplier involvement in supplier ethical practices	Promote peaceful and inclusive societies for sustainable development, provide access to justice for all and build effective, accountable and inclusive institutions at all levels

(*Continued*)

TABLE 1.5 (*Continued*)
Mapping of Indicator Themes from Academic Literature and Practitioner Reports to SDG Goals

SDG Goal	Practitioner Reports	Academic Indicators	SDG Goals Description
17	Adhering to corporate ethics, building long term relationships with partners, community development activities, strong relationships with suppliers, compliance with environmental laws, cooperation with suppliers transportation, compliance with labor legislation, small and medium business development, compliance with laws related to economic development, close communication with customers, supplier, person to person interaction regarding environmental initiatives, partnership with supplier for energy efficiency and recycling, access of environmental information to investors and buyers through global disclosure system	Certification in fair trade agreement, contracting with local suppliers, sustainability policy, sustainability action plan, environmental assessment goal accreditation, compliance with health safety and environmental legislation, green tender, sustainable procurement manual, verify suppliers report on sustainability criteria, knowledge of GPP toolkit and guidelines, certifies environmental management system (EMS) adoption, structure of purchasing process, transformational leadership, organizational strategic role, inter-departmental collaboration, public reporting of purchasing, stake holder engagement to develop new products and services, environmental disclosure, contractual stakeholder influence, local communities influence, suppliers performance towards implementing environmental policy, environmental quality audits, supplier assessment program	Strengthen the means of implementation and revitalize the Global Partnership for Sustainable Development

The next step of our analysis is to understand the degree of relevance of indicators from academic literature in terms of its emphasis on SDGs. Table 1.5 links the different indicators retrieved from the academic literature to SDGs. The table helps us to understand the value of SDGs as given by the researchers across the globe, and whether the research is focused at a particular direction or not. In comparison to the overrepresentation of practitioner reports on SDGs, we observed that SDG 17 is given the highest priority, because the initial assessment shows that forging partnerships for sustainability and ensuring compliance with suppliers is crucial for both the academic as well as the practitioner literature. The appropriate representation of SDGs is similar in both the academic and practitioner literature, except for SDG 9, which has no representation in the academic literature. This depicts the dearth of research related to the impact of firm practices on innovation and industrialization. This research gap is contrary to the actual belief that the academic literature might have focused on the firms' utilization of innovative technologies for imparting sustainability into their business processes. The underrepresented goals are similar in both the academic and practitioner literature, except SDG 9 that has to be added to the underrepresented category in the academic literature.

The third part of our analysis revolves around whether there is synchronization between academic literature and practitioner literature, while capturing the indicators. It has been already observed that while mapping the academic and practitioner literature, all the goals related to sustainable development are appropriately represented in varying degrees as discussed in the earlier paragraphs. The mismatch between academic and practitioner literature shows a different aspect of research gaps, where the academic literature can provide better understanding of firms that are currently lagging behind in their sustainability initiatives. A careful scrutiny provides us the useful inferences. For instance, under the SDG 2, the academic literature emphasizes on the need to reduce chemical consumption of farmland, and the practitioner literature stresses the promotion of family agricultural program and agroforestry techniques to improve production. However, the steps to improved agricultural production without the use of chemical consumption are unclear in both the literature. Under SDG, the academic literature is silent about the ways in which sustainable procurement initiatives can improve water management and sanitation. On the contrary, the practitioner literature emphasizes waste water management and biodiversity action plan. The broad research gaps and themes are presented in Figure 1.1, and a conceptual map is presented in Figure 1.2, which depicts a pictorial diagram of future research directions with focus on SDGs.

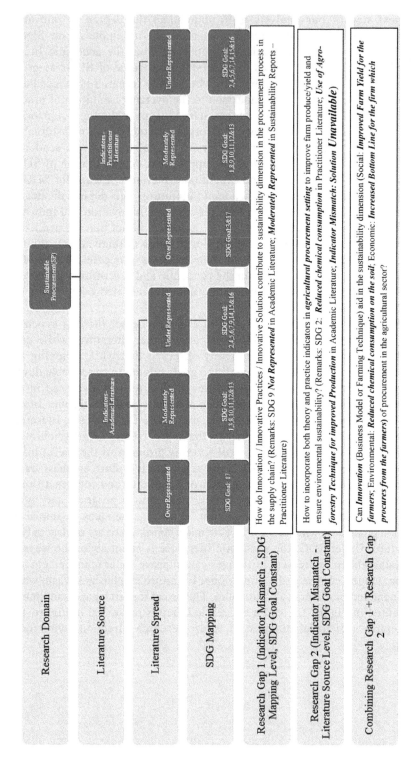

FIGURE 1.1   Research themes and broad research gaps identified from academic and practitioner literature.

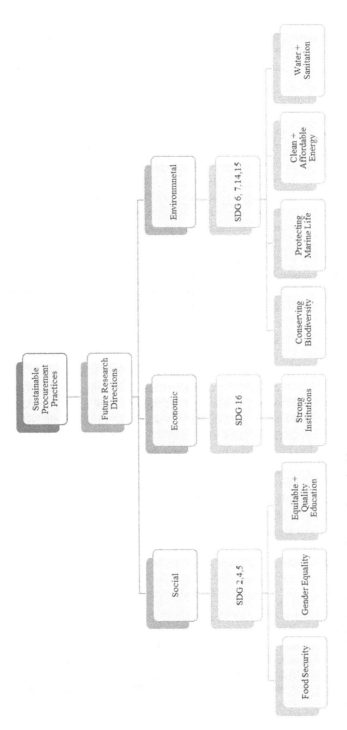

FIGURE 1.2 SDG focused conceptual map of sustainable procurement.

1.6 DISCUSSION AND LIMITATIONS

The chapter discusses the relevance of SP in the interface of supply chain and sustainability literature through adoption of a three-pronged approach toward SP. The first approach was to understand the current state of affairs on SP from the academic literature by documenting the key indicators and segregating them under social, economical, and environmental dimensions of the TBL approach in sustainability. The indicators were then analyzed and compared with the United Nations Sustainable Development Goals, and it posited our understanding about the current research trends in SP and its relevance and focus toward SDGs, thus our first research objective was answered (RQ 1). The second approach was to document indicators from practitioner reports and to analyze the same with the SDGs. This ensured our understanding about the current SP practices of manufacturing firms and the alignment of these practices with the benchmarked dimension of SDGs. Therefore, we could compare the difference between academic literature and practitioner literature (sustainability reports) pertaining to the SP indicators, thus providing answer to our second research objective (RQ 2) on the current knowledge gap between theory and practice at the reporting standards of research and practice. Based on the understanding and inference drawn from RQ 1 and RQ 2, we generated broad research themes and gaps as a conceptual framework, which can be pursued in future research studies keeping the organization as the source of data for validating the results from such study. Our third research objective of the chapter (RQ 3) is, thus, answered and highlights future research directions and provides an opportunity for contributing to the body of knowledge of SP.

Like most research studies, our study also has certain limitations. First, the literature review is based on 50 research articles selected from Google Scholar as per their citation count. However, a thorough literature review on SP would provide a rich knowledge of indicators paving the way for more in-depth analysis. Second, the practitioner literature contains sustainability reports from only one sector, i.e., steel industries. The study can be made more robust by selecting different industries under the manufacturing sector. For instance, we could have included automobiles, food, textiles, etc., under the overall umbrella of manufacturing industries. Third, the entire conceptual model could be validated in a manufacturing unit to understand the operationalization of the indicators at the field level and to subsequently posit relevant directives to align their practices in line with the SDGs.

REFERENCES

Ahi, P., & Searcy, C. (2013). A comparative literature analysis of definitions for green and sustainable supply chain management. *Journal of Cleaner Production, 52*, 329–341.
Amann, M., Roehrich, J. K., Eßig, M., & Harland, C. (2014). Driving sustainable supply chain management in the public sector: The importance of public procurement in the European Union. *Supply Chain Management: An International Journal, 19*(3), 351–366.
Aragão, C. G., & Jabbour, C. J. C. (2017). Green training for sustainable procurement? Insights from the Brazilian public sector. *Industrial and Commercial Training, 49*(1), 48–54.
Australian Procurement and Construction Council. (2007). *Australian and New Zealand Government Framework for Sustainable Procurement*. Australian Procurement and Construction Council, Deakin West, Australia.

Brammer, S., & Walker, H. L. (2007). Sustainable procurement practice in the public sector: An international comparative study. University of Bath, School of Management. Working Paper.

Brammer, S., & Walker, H. (2011). Sustainable procurement in the public sector: An international comparative study. *International Journal of Operations & Production Management, 31*(4), 452–476.

Brandenburg, M., Govindan, K., Sarkis, J., & Seuring, S. (2014). Quantitative models for sustainable supply chain management: Developments and directions. *European Journal of Operational Research, 233*(2), 299–312.

Brundtland, G. H. (1985). World commission on environment and development. *Environmental Policy and Law, 14*(1), 26–30.

Conner, D. S., Nowak, A., Berkenkamp, J., Feenstra, G. W., Kim, J. V. S., Liquori, T., & Hamm, M. W. (2016). Value chains for sustainable procurement in large school districts: Fostering partnerships. *Journal of Agriculture, Food Systems, and Community Development, 1*(4), 55–68.

Corbett, C. J., & Kleindorfer, P. R. (2001). Environmental management and operations management: Introduction to part 1 (manufacturing and ecologistics). *Production and Operations Management, 10*(2), 107–111.

Crespin-Mazet, F., & Dontenwill, E. (2012). Sustainable procurement: Building legitimacy in the supply network. *Journal of Purchasing and Supply Management, 18*(4), 207–217.

Dawson, G. F., & Probert, E. J. (2007). A sustainable product needing a sustainable procurement commitment: The case of green waste in Wales. *Sustainable Development, 15*(2), 69–82.

de Ron, A. J. (1998). Sustainable production: The ultimate result of a continuous improvement. *International Journal of Production Economics, 56*, 99–110.

Elkington, J. (1998). Partnerships from cannibals with forks: The triple bottom line of 21st-century business. *Environmental Quality Management, 8*(1), 37–51.

Elkington, J. (2013). Enter the triple bottom line. In *The Triple Bottom Line* (pp. 23–38). Routledge, London, UK.

Erridge, A., & Hennigan, S. (2012). Sustainable procurement in health and social care in Northern Ireland. *Public Money & Management, 32*(5), 363–370.

Freeman, R. E. (2010). *Strategic Management: A Stakeholder Approach.* Cambridge University Press, Cambridge, UK.

Ghadimi, P., Azadnia, A. H., Heavey, C., Dolgui, A., & Can, B. (2016). A review on the buyer–supplier dyad relationships in sustainable procurement context: Past, present and future. *International Journal of Production Research, 54*(5), 1443–1462.

Gimenez, C., Sierra, V., & Rodon, J. (2012). Sustainable operations: Their impact on the triple bottom line. *International Journal of Production Economics, 140*(1), 149–159.

Gold, S., Seuring, S., & Beske, P. (2010). Sustainable supply chain management and interorganizational resources: A literature review. *Corporate Social Responsibility and Environmental Management, 17*(4), 230–245.

Goldschmidt, K., Harrison, T., Holtry, M., & Reeh, J. (2013). Sustainable procurement: Integrating classroom learning with university sustainability programs. *Decision Sciences Journal of Innovative Education, 11*(3), 279–294.

Golini, R., Cagliano, R., & Longoni, A. (2010). The role of NFWO in sustainability strategies: An OM perspective. In *International Annual EurOMA Conference.* Porto, Portugal.

Gormly, J. (2014). What are the challenges to sustainable procurement in commercial semi-state bodies in Ireland? *Journal of Public Procurement, 14*(3), 395–445.

Grandia, J. (2015). The role of change agents in sustainable public procurement projects. *Public Money & Management, 35*(2), 119–126.

Grandia, J., Groeneveld, S., Kuipers, B., & Steijn, B. (2014). Sustainable procurement in practice: Explaining the degree of sustainable procurement from an organisational perspective. In *Public Procurement's Place in the World* (pp. 37–62). Palgrave Macmillan, London, UK.

Graneheim, U. H., & Lundman, B. (2004). Qualitative content analysis in nursing research: Concepts, procedures and measures to achieve trustworthiness. *Nurse Education Today, 24*(2), 105–112.

Grob, S., & Benn, S. (2014). Conceptualising the adoption of sustainable procurement: An institutional theory perspective. *Australasian Journal of Environmental Management, 21*(1), 11–21.

Grob, S., & McGregor, I. (2005). Sustainable organisational procurement: A progressive approach towards sustainable development. *International Journal of Environment, Workplace and Employment, 1*(3–4), 280–295.

Haake, H., & Seuring, S. (2009). Sustainable procurement of minor items–exploring limits to sustainability. *Sustainable Development, 17*(5), 284–294.

Hasselbalch, J., Costa, N., & Blecken, A. (2014). Examining the relationship between the barriers and current practices of sustainable procurement: A survey of UN organizations. *Journal of Public Procurement, 14*(3), 361–394.

Hsieh, H. F., & Shannon, S. E. (2005). Three approaches to qualitative content analysis. *Qualitative Health Research, 15*(9), 1277–1288.

Hussein, I. M. R., & Shale, I. N. (2014). Effects of sustainable procurement practices on organizational performance in manufacturing sector in Kenya: A case of Unilever Kenya Limited. *European Journal of Business Management, 1*(11), 417–438.

Islam, M. M., & Siwar, C. (2013). A comparative study of public sector sustainable procurement practices, opportunities and barriers. *International Review of Business Research Papers, 9*(3), 62–84.

Jefferies, M. C., McGeorge, D., Chen, S. E., & Cadman, K. (2006). *Sustainable Procurement: A Contemporary View on Australian Public Private Partnerships (PPPs)*. British University of Dubai, Dubai, United Arab Emirates.

Jelodar, M. B., Yiu, T. W., & Wilkinson, S. (2013). Stirring sustainable procurement by conceptualizing relationship quality in construction. In *World Building Congress 2013*. Brisbane Convention & Exhibition Centre, Queensland, Australia.

Joyce, A., & Paquin, R. L. (2016). The triple layered business model canvas: A tool to design more sustainable business models. *Journal of Cleaner Production, 135*, 1474–1486.

Kaur, H., & Singh, S. P. (2016). Sustainable procurement and logistics for disaster resilient supply chain. *Annals of Operations Research*, 1–46.

Kaye Nijaki, L., & Worrel, G. (2012). Procurement for sustainable local economic development. *International Journal of Public Sector Management, 25*(2), 133–153.

Keeble, B. R. (1988). The Brundtland report: 'Our common future'. *Medicine and War, 4*(1), 17–25.

Kleindorfer, P. R., Singhal, K., & Van Wassenhove, L. N. (2005). Sustainable operations management. *Production and Operations Management, 14*(4), 482–492.

Large, R. O., Kramer, N., & Hartmann, R. K. (2013). Procurement of logistics services and sustainable development in Europe: Fields of activity and empirical results. *Journal of Purchasing and Supply Management, 19*(3), 122–133.

Lund-Thomsen, P., & Costa, N. (2011). Sustainable procurement in the United Nations. *The Journal of Corporate Citizenship, 42*, 55.

Mansi, M. (2015). Sustainable procurement disclosure practices in central public sector enterprises: Evidence from India. *Journal of Purchasing and Supply Management, 21*(2), 125–137.

McMurray, A. J., Islam, M. M., Siwar, C., & Fien, J. (2014). Sustainable procurement in Malaysian organizations: Practices, barriers and opportunities. *Journal of Purchasing and Supply Management, 20*(3), 195–207.

Meehan, J., & Bryde, D. (2011). Sustainable procurement practice. *Business Strategy and the Environment, 20*(2), 94–106.

Meehan, J., & Bryde, D. J. (2015). A field-level examination of the adoption of sustainable procurement in the social housing sector. *International Journal of Operations & Production Management, 35*(7), 982–1004.

Melissen, F., & Reinders, H. (2012). A reflection on the Dutch sustainable public procurement programme. *Journal of Integrative Environmental Sciences, 9*(1), 27–36.

Molenaar, K. R., Sobin, N., & Antillón, E. I. (2010). A synthesis of best-value procurement practices for sustainable design-build projects in the public sector. *Journal of Green Building, 5*(4), 148–157.

Oruezabala, G., & Rico, J. C. (2012). The impact of sustainable public procurement on supplier management: The case of French public hospitals. *Industrial Marketing Management, 41*(4), 573–580.

Pagell, M., & Gobeli, D. (2009). How plant managers' experiences and attitudes toward sustainability relate to operational performance. *Production and Operations Management, 18*(3), 278–299.

Preuss, L. (2009). Addressing sustainable development through public procurement: The case of local government. *Supply Chain Management: An International Journal, 14*(3), 213–223.

Rajeev, A., Pati, R. K., Padhi, S. S., & Govindan, K. (2017). Evolution of sustainability in supply chain management: A literature review. *Journal of Cleaner Production, 162*, 299–314.

Rimmington, M., Carlton Smith, J., & Hawkins, R. (2006). Corporate social responsibility and sustainable food procurement. *British Food Journal, 108*(10), 824–837.

Roman, A. V. (2017). Institutionalizing sustainability: A structural equation model of sustainable procurement in US public agencies. *Journal of Cleaner Production, 143*, 1048–1059.

Ruparathna, R., & Hewage, K. (2015). Sustainable procurement in the Canadian construction industry: Current practices, drivers and opportunities. *Journal of Cleaner Production, 109*, 305–314.

Schreier, M. (2012). *Qualitative Content Analysis in Practice*. Sage Publications, London, UK.

Seuring, S., & Gold, S. (2012). Conducting content-analysis based literature reviews in supply chain management. *Supply Chain Management: An International Journal, 17*(5), 544–555.

Seuring, S., & Müller, M. (2008). From a literature review to a conceptual framework for sustainable supply chain management. *Journal of Cleaner Production, 16*(15), 1699–1710.

Smith, J., Andersson, G., Gourlay, R., Karner, S., Mikkelsen, B. E., Sonnino, R., & Barling, D. (2016). Balancing competing policy demands: The case of sustainable public sector food procurement. *Journal of Cleaner Production, 112*, 249–256.

Sourani, A. (2011). Barriers to addressing sustainable construction in public procurement strategies. Loughborough University Institutional Repository.

Srivastava, S. K. (2007). Green supply-chain management: A state-of-the-art literature review. *International Journal of Management Reviews, 9*(1), 53–80.

Steurer, R., Berger, G., Konrad, A., & Martinuzzi, A. (2007). Sustainable public procurement in EU member states: Overview of government initiatives and selected cases. Final Report to the EU High-Level Group on CSR. European Commission, Brussels, Belgium.

Stock, J. R., & Boyer, S. L. (2009). Developing a consensus definition of supply chain management: A qualitative study. *International Journal of Physical Distribution & Logistics Management, 39*(8), 690–711.

Testa, F., Annunziata, E., Iraldo, F., & Frey, M. (2016). Drawbacks and opportunities of green public procurement: An effective tool for sustainable production. *Journal of Cleaner Production, 112*, 1893–1900.

Thomson, J., & Jackson, T. (2007). Sustainable procurement in practice: Lessons from local government. *Journal of Environmental Planning and Management, 50*(3), 421–444.

Walker, H., & Brammer, S. (2009). Sustainable procurement in the United Kingdom public sector. *Supply Chain Management: An International Journal, 14*(2), 128–137.

Walker, H., & Brammer, S. (2012). The relationship between sustainable procurement and e-procurement in the public sector. *International Journal of Production Economics, 140*(1), 256–268.

Walker, H., Di Sisto, L., & McBain, D. (2008). Drivers and barriers to environmental supply chain management practices: Lessons from the public and private sectors. *Journal of Purchasing and Supply Management, 14*(1), 69–85.

Walker, H., & Phillips, W. (2008). Sustainable procurement: Emerging issues. *International Journal of Procurement Management, 2*(1), 41–61.

Walker, H., Miemczyk, J., Johnsen, T., & Spencer, R. (2012). Sustainable procurement: Past, present and future. *Journal of Purchasing and Supply Management, 18*, 201–206.

Walker, H. L., Gough, S., Bakker, E. F., Knight, L. A., & McBain, D. (2009). Greening operations management: An online sustainable procurement course for practitioners. *Journal of Management Education, 33*(3), 348–371.

Wilkinson, A., & Kirkup, B. (2009). *Measurement of Sustainable Procurement.* East Midlands Development Agency, Nottingham, UK.

Witjes, S., & Lozano, R. (2016). Towards a more Circular Economy: Proposing a framework linking sustainable public procurement and sustainable business models. *Resources, Conservation and Recycling, 112*, 37–44.

Yeow, J., Uyarra, E., & Gee, S. (2011). Sustainable innovation through public procurement: The case of 'closed loop' recyled paper (No. 615). Manchester Business School Working Paper.

2 The Internet of Things-Enabled SMART Procurement

The Next Horizon

Samir Yerpude and Tarun Kumar Singhal

CONTENTS

2.1 INTRODUCTION

Cyber physical systems represent the Industrial revolution 4.0 (Jazdi, 2014). The prediction for future includes modular with least batch sizes of production yet maintaining the economies of scale (Milgrom & Robberts, 1990). High level of collaboration is expected from the system, as isolation will be disastrous. The pressing need is to connect the physical and the virtual world. Revolution in the network domain has given rise to innovations, such as the Internet of things (IoT), due to the pervasive presence of the Internet. IoT in the current era is dominating almost all the business processes, while most important one being supply chain management (Gubbi et al., 2013). Researchers, *vide* this study, propose the implementation of IoT for the procurement function of supply chain management. Profoundly, any person or organization that provisions for goods or services for a customer (may not be the end customer) is called a Supplier (Canada Business Network, 2009), while a Vendor is any person or organization that sells goods or service to the end customer (Haleem, 2015). The development of plans for the suppliers that are strategic in nature and align to the organizational objectives is termed the Supplier Relationship Management. The organization involved in the development of supplier plans could be manufacturing or retailing (Lambert & Cooper, 2000). The most crucial link of the Supplier relationship management is procurement, which is a business function comprising of two significant parts; sourcing and purchasing. While sourcing identifies the source of supply, purchasing is responsible for ordering, receiving, and supplier payment processing. Procurement links business expectations with the supplier (Novack & Simco, 1991). It is termed as one of the most complicated tasks in the supply chain primarily accountable for the cost cutting and aligning the objectives of the organization such as profit margins and bottom line contribution (Samper, 2017). For an organization, procurement becomes a game-changer when deployed successfully (Lambert & Cooper, 2000). Supplier relationship management teaches new ways of collaborating with the suppliers with higher levels of transparency and trust (Carter, 2003a). In the past few decades, Information Technology (IT) has made great in-ways. The Internet has become the fastest medium for the transfer of data (Yerpude & Singhal, 2017d). There are many disruptions recorded in the field of the Internet that have helped IT professionals innovate and convert the Internet from a micro to macro global network connecting billions of things (Kopetz, 2011). In 1999, Kevin Ashton coined the term, Internet of Things (IoT), which is fundamentally a collection of sensors connected over the Internet capable of transmitting data. The sensors, which were meant to deliver just the information of state change, now are capable of transferring the data over the Internet making them smarter (Yerpude & Singhal, 2018a). This complete data is stored in a single repository and used for real-time decision-making (Davenport & Harris, 2007). Meaningful dashboards are generated aiding in fact-based decisions (Yerpude & Singhal, 2018d). Organizations in the current era are facing tough competition from global players and are under continuous pressure to remain competitive. Suppliers of these organizations are fine-tuned for combating the cost pressures and bringing about cost reductions by creating alternatives and value-adds (Carter, 2003b). The critical link in the complete supply chain is data that proves to be a key in the various organizational initiatives such as cost reduction. (Ranganathan et al., 2011).

2.2 REVIEW OF LITERATURE

2.2.1 PROCUREMENT

Organizations should treat procurement as an important function among the strategic functions since it links all the core functions. Business strategy and procurement are linked directly, and therefore, procurement planning is extremely crucial. Half of the business revenues are utilized in the procurement and activities related to procurement (Fournier, 2015). Buyers in a procurement organization are making extremely crucial decisions; hence, data is the key for making fact-based decisions. Innovations in IT have made the Internet the quickest medium for data transfer. Due to its ubiquitous presence, the Internet is making huge in-roads in businesses and its processes (Yerpude & Singhal, 2017c). Further disruptions in the Internet space have given rise to IoT, wherein the physical world and the virtual world are brought together with the sensing layer forming the vital layer of the complete IoT landscape (Dohr et al., 2010). Every sensor in the IoT landscape can be identified uniquely such that the data source can be traced back. Real-time analytics is enabled with the huge data that is collected in the single repository via the business models (Sagiroglu & Sinanc, 2013). Concept mapping between the procurement and the IoT constructs has unearthed the fact that IoT is not yet implemented in the procurement domain. Although there are some evidences sighted, there are not enough cases to prove the adequacy of the research done in this domain. Therefore, we pursue this case study forward.

2.2.2 AMALGAMATION OF PROCUREMENT AND THE INTERNET OF THINGS

Procurement is a business function, wherein a series of actions are performed by the individuals of an organization to obtain products or services from the best sources at the best possible prices (Fournier, 2015). While the procurement costs are close to 50% of the revenues, even a small difference in cost creates a larger impact and influences the organizational profit (Bendorf & Kolodisner, 2017). Procurement aligns with the business strategy and, therefore, it cannot work in isolation. The procurement policy for an organization must be made in conjunction with the corporate strategy (Watts et al., 1995). In larger organizations, the procurement function is branched into two halves; direct and indirect procurement. Products that are consumed directly in the manufacturing processing form the part of direct procurement, while products supporting the manufacturing operations form a part of the indirect procurement. Products such as consumables that are not processed in the manufacturing operations directly are bought as a part of indirect procurement (Gebauer & Segev, 2001). The procurement function is also classified on the basis of its reach. Long-term procurement and short-term procurement types are defined based on the reach. Generally, when the strategic part of buying is discussed, it qualifies to be long-term including sourcing, selection of supplier, negotiation, and contracting with the overall governance definition. It also encompasses designing of the entire engagement model for better supplier relationship management. Short-term includes requisition driven transactions where activities such as one-time negotiation of rates, raising purchase order and receiving delivery of goods, completing the payment process, etc. are undertaken. It also comprises of purchase arrangements, such as Just-in-Time (JIT), wherein goods are received as and when required. It aids monetary benefits such as improved cash flows due to negligible inventory holding costs (Gebauer & Segev, 2001). Significant benefits in the area of procurement are promised with the implementation of new technologies (Neef, 2001). Significant new wealth creation opportunities are created with a complete transformation of business (Amit & Zott, 2001). New technologies require an extensive exchange of data, transfer of information, and communication between all the relevant stakeholders (Gebauer & Segev, 2001). The Internet, being the fastest mode of data transmission, assists in the exchange of data and ensures near real-time communication (Yerpude & Singhal, 2018d). There is a phenomenal growth of the Internet observed over the past few decades resulting in a series of innovations. One such innovation being IoT, which connects billion of physical things, i.e., sensors and transmits the data over the Internet. The things behave in an interconnected way with the objects possessing pervasive intelligence (Xia et al., 2012). Decision-making is aided by the business analytics derived from the IoT origin real-time data that is collated and stored in a central repository (Yerpude & Singhal, 2017d). Real-time data provokes the requisite agility into business process that helps address customer demands. Data is exchanged seamlessly over the Internet linking the customers to the suppliers and bringing uniformity amongst the understanding of all the stakeholders (Kopetz, 2011).

Technology embraced by the procurement organization helps seamless execution of the business processes in a cost-effective way (Panayiotou et al., 2004). As mentioned earlier, the procurement function has two integral parts; sourcing and purchase. Purchase is related to buying of goods at a negotiated price and as per the budget allocations (Argentus, 2017). It's all about provisioning the right part at the right time within the agreed budgetary constraints (Fournier, 2015). There is a large differential recorded in the process efficiency because of the transparency added by the IT application interventions (Gebauer & Segev, 2001). An early and complete adoption of the IT system provides businesses the upper edge by reducing the transactional costs and offering more value due the fact-based decision-making (Panayiotou et al., 2004).

2.2.3 Can Internet of Things Become a Differentiator

The organization considered for this discussion includes a medium- to large-sized organization. In such a scale of operations, the procurement function is split into long-term, i.e., strategic, and short-term, i.e., transactional function. Procurement of direct material because of its criticality to the organization is linked to the strategic procurement where businesses maintain long-term relationships with the suppliers (Lutz & Bayer, 2001). Indirect procurement, in comparison, can be attributed to *ad hoc* and one-time procurement where the entire focus is on reducing the cost of the transaction (Gebauer & Segev, 2001). Short-term procurement changes for different organizations. Purchasing strategies such as JIT, vendor-managed inventories for automatic replenishments, etc. are the focus areas (Gebauer & Segev, 2001). Impetus in both types of provision can be achieved with the implementation of IoT, as data is the most important and inherent need of the function that aids the requisite agility to service customer demands. It also assists the organization with cost savings and on-time deliveries simultaneously aiding the control of the complete process.

Researchers had conducted focus group discussions in order to validate the influence of IoT on the complete procurement process. A total of 28 executives involved in procurement activity within their respective organizations were subjected to a questionnaire. The responses taken due the discussions were recorded and appropriate inferences were drawn around the application of IoT in different domains. Case studies were also used to support the inferences and draw conclusions about the implementation of IoT in the procurement domain that would in effect augment the effectiveness of supplier relationship management. The need for data is indisputable in both the types of procurement, i.e., strategic or transactional. **S**trategic, **M**easurable, **A**ligned, **R**esults-based, and **T**ime-bound (SMART) measures store shelves, in-transit supplies, etc. and transmit the much needed real-time data to the organization. The SMART shelves comprise of sensors that track the binning and retrieving of products. These sensors are capable of sending this real-time information over the Internet to a central database. (Bi et al., 2014). In similar fashion, the in-transit inventory information data feeds as a stream for the purchasing activity. Lucrative realizations are observed with the amalgamation of IoT and procurement making the implementation and eminent choice for the organizations. In the next section, we deal with some of the prominent transactional and strategic procurement activities to visualize the benefits of implementation of IoT in the procurement domain.

2.2.4 TRANSACTIONAL PROCUREMENT

2.2.4.1 Customer Demand Forecasting

Anticipating the occurrence of an activity based on a scientific calculation before it actually happens is termed as forecasting (Yerpude & Singhal, 2017b). It profoundly is a proactive step to envisage the demand before customers actually ask for a product or service. The scientific calculations mentioned are the quantitative forecasting methods deployed by business. There are two types of models for the same. The process of consuming the past data for predicting the future is termed as Time Series Forecasting (Chatfield, 2000). While discovering causal relationships between the independent and the dependent variables using statistical analysis is termed as Trend Method (Feige & Pearce, 1979). The time series method of forecasting is deployed to anticipate the future by analyzing series of data gathered in a particular frequency. The trend method is fundamentally the dissimilarity in the series that predicts the future (Chatfield, 2000). In the case of a causal method, a model is positioned; that is, best determining the relationship between the independent variables and the dependent variables (Sani & Kingsman, 1997). While organizations expect to forecast with minimum errors, data plays a vital role in the overall process. Agility to respond to market variability is imparted with the introduction or real-time data (Talbot, 2015). It further contributes to growth in revenue since the service levels improve (Crates, 2014). Improved accuracy in forecasting also aids in cost savings (Stark & Croushore, 2002). SMART shelves, i.e., IoT-enabled shelves, transmit the consumption data in real time, which is then fed into the forecasting model. It further supports real-time analytics systems that send alerts to the managers for rapid action based on the exceptions defined in the system (Yerpude & Singhal, 2017c).

2.2.4.2 Vendor Managed Inventory

The difference between a vendor and a supplier is mentioned in the prevision sections. Vendor Managed Inventory (VMI) is fundamentally an arrangement between the vendor and the procurement organization where the vendor uses an allotted space at the buyers end as his extended warehouse. The entire tracking and monitoring of inventory, including replenishment at the appropriate levels, is managed by the vendor as per the agreed norms (Özgen, 2010). The enablement of a payment transaction against the supplies made is done only when the inventory is consumed from the extended warehouse. The most important task of optimizing the inventory is also with the vendor in case of VMI (Yerpude & Singhal, 2017d). Transparency and right-time communication is the key to a successful VMI, which is imperative. To enable quick, real-time communication researchers recommend the implementation of IoT for enabling real-time communication between the relevant stakeholders. The discussion with the digital experts further echoed similar sentiments about the usage of IoT enabling VMI. Some of the experts further extrapolated the flow of communication with the logistics provider too for aiding better planning of vehicles for the supply of products. The stock information, consumption, etc. is transmitted transparently over the Internet for two purposes; one is payment processing and second for the material requirement planning at the vendor end (Yerpude & Singhal, 2017a). Organizations with the implementation of IoT are better equipped to combat the customer demand volatility and market turbulence due to the agility and flexibility imparted (Baarlid & Claesson, 2015). With techniques such as VMI, the inventory carrying cost of the buyers' organizations reduces drastically improving the performance of the supply chain significantly. Information of stock, sales, forecasts, orders, etc. are the few important data points shared with the vendor on the common platform (Liu & Sun, 2011). Following paybacks can be documented with the help of IoT implementation in the VMI system compared to the pain points of traditional supply chain (Yerpude & Singhal, 2017a) (Table 2.1).

TABLE 2.1
Comparison of the Pain Points of Supply Chain vs IoT-Based VMI

Pain Points of Supply Chain	Benefits of using IoT-augmented VMI System
• Reactive management	• Proactive approach
• Volatile demand and speculative ordering	• High forecasting accuracy
• Costly year-end write-offs	• Lesser write-offs and better cash flows
• Time spent in planning and ordering	• Planning and ordering time is saved
• Delayed information resulting in loss of sale due to stock outages and discounted sales	• Huge trust because of transparency in the system
• Lower inventory turns	• Higher inventory turns and better inventory management
• Rigid to evolving customer demand	• Higher customer responsiveness, higher customer satisfaction

Source: Yerpude, S. and Singhal, T. K., *Int. J. Appl. Bus. Econ. Res.*, 15, 469–482, 2017.

2.2.5   PROCUREMENT OPERATIONS

JIT, order-based replenishments, and safety stock level buying are some of the major processes of the transactional procurement. Considerable benefits are achieved with the implementation of IoT since the monitoring of stocks, consumption, etc. is automatically ensured. (Kinder, 2014). The point of consumption directly connects to the point of manufacture seamlessly with the data transmitted without any human intervention (Yerpude & Singhal, 2018d). Automation of mundane tasks and aiding predictive decision-making are the two most important outcomes of the implementation of IoT-enabled technologies apart from the enormous opportunities unearthed in the ordering and buying processes. There is a paradigm shift observed from the transactional to strategic management that adds significant value to the organization (Kinder, 2014). Proactive management of customer orders is enabled with automated systems managing the ordering of parts from the vendor directly transmitting the inventory position in real time (Bhandari, 2014). Transparency in the entire system imbibes trust in the buyer-supplier relationship strengthening it further (Wigan, 2004). Implementation of IoT automates procurement tasks such as communication of pre-purchase information, aids approval workflows, auto replenishments, and monitoring of contracts and service level agreements (SLA). It emphatically replaces tasks such as purchase order raising and monitoring, while reducing the errors due to manual operations. Smart devices, i.e., sensors deployed within the IoT architecture, monitor the direct and indirect parameters related to procurement. Within the warehouse, weight, volume, quantity and quality of a product, part movement data, etc. are captured by these sensors that facilitate the creation of automated supplier orders (York, 2015). Therefore, proactive approach is impressed by the IoT implementation on the transactional procurement driving agility in the end-to-end procurement process.

2.2.6   STRATEGIC PROCUREMENT

Supplier relationship management solely depends on management and monitoring of suppliers in a strategic way. There are various tools available within the organizations that serve this purpose. Based on the data collected and the parameters agreed on with the vendors, a scorecard is created. Researchers have evaluated the Balanced ScoreCard (BSC) as a tool for vendor evaluation and, recommend the usage of IoT origin real-time data for reporting the BSC. BSC is widely used by decent-sized organizations. BSC as a tool providing a comprehensive framework for transforming the organizational strategy into a rational set of performance measures (Schneier et al., 1995). BSC enables the managers of an organization to look forward and beyond. The vision of the organization is converted into specific, measurable parameters to gauge the performance of the vendor. There are four parameters termed as balanced parameters based on which the vendor performance is judged. The four parameters are financial, customers, internal business processes, and learning and growth. Financial controls are inherited with the implementation of BSC along with other measures such as building capabilities and acquiring the intangible assets required for future growth (Kaplan & Norton, 2001). An indisputable impression is created by IoT implementation in the procurement domain with a real-time supplier balanced scorecard. IoT origin real-time data brings in the requisite agility with enablement of correction in direction if required during the process (Yerpude &

Singhal, 2018). Previous studies have successfully articulated a *supply chain management scorecard*, based on the corporate balanced scorecard (Gunasekaran et al., 2001). Researchers have identified the performance parameters of the procurement function in each of the balanced perspective from the BSC. IoT plays a vital role in collating the underlying data requirement. Sensors in the IoT landscape play a dual role: first, identifying the change of state and, second, transmitting this information over the Internet. All the heterogeneous applications working cohesively in the landscape transmit data seamlessly to be used for the vendor scorecard generation (Sundmaeker & Guillemin, 2010). The enormous data collected from all the sensors is stored in a central repository and consumed to calculate the scorecard parameter values, which form an integral part of the balanced scorecard. Certain parameters hugely depend upon the direct consumption of data, while for others it is a calculated value. For example, delivery performance, and delivery reliability or rejection rate that form the part of the financial balanced perspective are directly achieved from the product movement data, while productivity ratio, cost per operation hour, and information carrying costs are calculated by the business analytics models. Similar examples, to quote from the customer-balanced perspective, are that of range, order lead time, delivery lead time, and delivery reliability that are direct deliverables; however, the flexibility of service system is derived by expending the material movement data in the analytics mode. A similar comparison can be drawn further for the internal processes and learning and growth (Table 2.2).

TABLE 2.2
Balanced Scorecard

Financial	Customer	Internal Processes	Learning and Growth
Net profit versus productivity ratio	Range of products and services	Total supply chain cycle time	Capacity utilization
Delivery performance	Order lead time	Flexibility of service systems to meet particular customer needs	Accuracy of forecasting techniques
Delivery reliability	Flexibility of service systems to meet particular customer needs	Supplier lead time against industry norms	Flexibility of service systems to meet particular customer need
Cost per operation hour	Delivery lead time	Level of supplier's defect-free deliveries	Range of products and services
Information carrying cost	Delivery reliability	Accuracy of forecasting	
Supplier rejection rate	Responsiveness to urgent deliveries	Purchase order cycle time	
	Quality of delivered goods defect-free deliveries	Total inventory cost	
		Supplier rejection rate	
		Frequency of delivery	

Source: Gunasekaran, A. et al., *Int. J. Oper. Prod. Manag.*, 21, 71–87, 2001.

Researchers further directed a statistical analysis for endorsing the important parameters of procurement and their relevant cross-functional linkages.

2.3   CONCEPTUAL RESEARCH MODEL FOR SMART PROCUREMENT

2.3.1   HYPOTHESIS FORMATION

Forecasting: Profoundly forecasting is all about determining the future requirement of product or service with the help of analytical models. Forecasting further aids warehouse and logistics planning for the enablement of dispatches against customer orders so that the order servicing is within the committed time (Chen et al., 2000). Huge inventory levels due to incorrect forecasting increase the inventory holding costs and block the warehouse space (Yerpude & Singhal, 2018d). On the other hand, shortage of products needed for servicing customer orders results in customer dissatisfaction and lower customer retention (Kotler & Gary, 1991; Anderson & Sullivan, 1993). Accurate forecast help the organizations plan better to fulfill the market demands comprising of volatile customer orders and achieve customer excellence (Yerpude & Singhal, 2018c). Therefore, the researchers propose:

H_1: Forecasting has a positive impact on procurement

Purchasing contracts: Purchasing contracts are agreements, which intimate the suppliers to build capacity to supply the product developed. The contract generally marks the beginning of a long-term engagement of the buyer and the supplier organization (Taylor & Plambeck, 2007). Contract defines the nature of engagement and the terms of engagement for business. The contract also has a binding on the buyer organization to purchase an agreed quantity within a stipulated period (Wang et al., 2012). Contracts help in streamlining the procurement function and iron our issues, which can surface out of the long-term engagement and changing business scenarios (Handfield & McCormark, 2007). Hence, researchers posited:

H_2: Purchasing contracts have a positive impact on procurement

Supplier rating system: In the current era of intense global competition, it is very important to respond to the industry with better quality products, consistently establishing mechanisms, and directly linking the customer to the supplier (Chen et al., 2000). Supplier rating system is one such tool that establishes the relationship between the organizational objectives basis the customer service and the expectations from the supplier (Choy et al., 2004). The entire supply chain gets converted in a value chain generating value for the respective stakeholders (Kramer, 2011).

Customer-supplier relationship management has become crucial for the longevity of the organization (Lambert & Cooper, 2000). Therefore, the researchers propose:

H₃: Supplier rating system has a positive impact on procurement

Real-time information: In the past decade, there has been a paradigm shift observed in the way businesses are managed across the globe (Yerpude & Singhal, 2018b). The organizations have realized that to remain in business goals need to be defined along with critical success factor matrix (Wurster, 1999). Fast pace changes are experienced by organizations in terms of environments resulting in the impressing need of decision-support systems. Business analytics driven intelligence aids businesses in this direction so that appropriate controls can be exercised on the different processes in the organization may it be customer side or supplier side (Sahay & Ranjan, 2008). Operational and transactional efficiency in the procurement domain require strong simulation categories such as *What if*, so that the key functions could be reconfigured quickly and effectively (Reddy, 2006). Real-time information-supported business intelligence gives the requisite agility to the business to combat the fierce competition in the market (Sahay & Ranjan, 2008). Therefore, researchers posited:

H₄: Real-time information has a positive impact on procurement

Effective procurement function balances the demand versus supply in the most efficient manner retaining its customers in the competitive market (Novack & Simco, 1991). Customer satisfaction is the key to customer retention (Kotler & Gary, 1991). The essentials to increase customer satisfaction include reliability of supplies, better quality products, and the highest level of customer service. These are possible with a proficient supplier base and a good supplier relationship management. Procurement function links the demand and supply, aiding the organization in achieving the business objectives (Chen et al., 2000). Hence, the researchers proposed the following:

H₅: Effective procurement positively influences customer satisfaction
H₆: Effective procurement positively influence and improves the supplier relationship

Conceptual Research Model is formulated basis the above literature review to corroborate the understanding of cross-functional integrations and determine the importance of each in the procurement process (Figure 2.1).

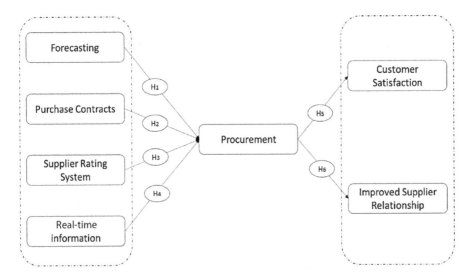

FIGURE 2.1   Conceptual research model.

2.4   RESEARCH METHODOLOGY: SAMPLE CHARACTERISTICS AND INSTRUMENT DEVELOPMENT

2.4.1   SURVEY QUESTIONNAIRE METHOD

Researchers collected 270 responses to test the different propositions developed basis the conceptual model, through a structured questionnaire. Likert scale (Albaum, 1997) was deployed for gathering the responses from the respondents.

2.4.2   SAMPLING TECHNIQUE

Deliberate sampling technique, also known as purposive sampling, was used for the study. The purpose of the sample determined the choice of respondents. Since this case pertained to the procurement and supplier relationship aspects, the sample chosen consisted of managers from and related to the procurement department in their respective organizations. The focus was on profile of the respondents pertaining to the supply chain management and procurement function so as to get valid inputs. The respondents within procurement were a mix of procurement function catering to manufacturing as well as to the retail segment covering both the aspect in procurement.

2.4.3   BIAS HANDLING

Typically, errors such as Sampling error, Non-response error, and Coverage error get introduced during the data collection and must be cured prior to the statistical tests. Researchers have ensured unbiased responses by not hiring an interviewer for the data collection so that the respondents can record their responses with any pressure. The respondents were reached at convenient times, and sufficient time was awarded for the responses, which safeguarded the researchers against a hassled retort.

2.4.4 RELIABILITY TESTING OF THE INSTRUMENT

Reliability testing is all about testing the reproducibility of the instrument. The measures of the same are Cronbach's Alpha and Split-Half correlation. In post calculations, the value of Cronbach's Alpha is recorded as 0.91, thereby indicating a higher level of reliability. We also tested for Split-Half correlation and Split-half with Spearman-Brown adjustment. The values of 0.87 and 0.93 were observed, respectively. The calculated values confirm the reliability of the instrument.

2.5 DATA ANALYSIS

Statistical analysis was conducted on SPSS 22 and AMOS 21 for the data collated with the help of self-administered survey. The statements are cited from studies in the strategic and transactional procurement studies. For extracting the factors from the data without imposing any constraint/limitation on the outcome, the researchers conducted Exploratory Factor Analysis (EFA) (Child, 1990). The extraction method used for the factors was Principal Component Method with Varimax rotation. Two-stage validation of the measurement model is carried out. The first stage-objective is to substantiate the measurement model with the help of confirmatory factor analysis (CFA). The second stage consists of Structural equation modeling (SEM). The analysis of the relationships among various variables was considered in the study by scrutinizing the regression weights of each on the dependent variable.

2.5.1 EXPLORATORY FACTOR ANALYSIS

As per the standard practice testing of sampling adequacy is quite important prior to the EFA. The results recorded a Kaiser–Meyer–Olkin's (KMO) value of 0.744. Another value, i.e., the Barlett's test is observed significant at 0.000 levels ($p<0.000$). The calculated values of EFA are recorded higher than the recommended tolerance value. Hair et al. (2013) recommended 0.5, and 0.6 was suggested by Fidell et al. (2013). The factors extraction was basis eigenvalue more than 0.6.

2.5.2 MEASUREMENT MODEL ANALYSIS

As mentioned earlier, confirmatory factor analysis is conducted as the first step towards the validation of the measurement model. The results were not encouraging, inferring a poor fit of the initial model as mentioned in Table 2.3. The measure, i.e., ratio of chi-square and degree of freedom, is observed greater than the acceptable value of 3, i.e., ($\chi2/df \sim 5$). The other indices such as goodness-of-fit [GFI = 0.7] and adjusted goodness-of-fit [AGFI = 0.59] are recorded lower than the acceptable value of 0.90. Another measure of model fitness, i.e., root mean square error of approximation (RMSEA), is calculated at 0.117, which is higher than the tolerance of 0.05. Hence, we reject the first model as the estimations have gone haywire. To improve the model further, the researchers validated all the statements for the factor loadings as per the procedure. From the study as per the process, removal of statements

TABLE 2.3
CFA Estimates

Goodness-of-Fit Indices	Recommended Value	Observed Values	Initial CFA Model	Revised CFA Model
Chi Sq/Df	1–3	2.566	5	~3
GFI	> = 0.9	0.86 ~ 0.9	0.7	0.87 ~ 0.9
AGFI	> = 0.9	0.81	0.6	0.8
RMSEA	< = 0.05	0.044	0.117	~0.05

Source: SPSS.

below the tolerance level of 0.7 was conducted. The model significantly improved with better readings of fitness indices. The chi-square/degree of freedom calculated, i.e., (χ2/df ~ 3), is as per the prescribed value, and all the model fit indices, such as [GFI = 0.87], [AGFI = 0.8], and [RMSEA ~ 0.05] are close to the threshold level. Hence, the researchers stated the successful closure of the statistical analysis.

Nunnally and Bernstein (1994) and Hair et al. (2013) have clearly stated in their research that the standardized cut-off used for testing internal consistency of each statement is 0.70. Next, the convergent validity of statements was tested. Method used for testing convergent validity of statements is average variance explained (AVE) (Table 2.4). We can indicate the presence of construct validity in the survey instrument

TABLE 2.4
Construct Reliability and Convergent Validity

Latent Variable	Observed Variable	Standard Loadings (>0.7)	CR (>0.7)	AVE (>0.5)
Forecasting	FR1	0.764	0.85	0.6
	FR3	0.757		
	FR4	0.84		
	FR5	0.7		
Purchasing contracts	PC1	0.81	0.82	0.6
	PC2	0.8		
	PC3	0.7		
Supplier rating system	SR1	0.824	0.85	0.65
	SR2	0.84		
	SR4	0.74		
Real-time information	RT1	0.8	0.8	0.66
	RT3	0.814		

Source: SPSS.

TABLE 2.5
Discriminant Validity

	Forecasting	Purchasing Contracts	Supplier Rating System	Real-Time Information
Forecasting	0.60			
Purchasing contracts	0.29	0.60		
Supplier rating system	0.41	0.17	0.65	
Real-time information	0.29	0.22	0.39	0.66

Source: SPSS.

as the values of AVE are recorded greater than 0.5. Fornell and Larcker (1981) standards recommend check of the AVE values. As per them, it should be greater than the squared correlation between the constructs to test Discriminant validity (Table 2.5).

Since all the values of squared correlations are less than the average variance, it indicates a better model fit.

2.5.3 ANALYSIS OF STRUCTURAL MODEL

Researchers led the second stage of analysis with the causal path analysis, *vide* the structural equation model. It was recorded that all the parameters positively influenced procurement while the regression weightages were found varying. Hair et al. (2013) suggested that model fit indices values from 0.5 to 0.8 are largely good for social science studies. The values reported in the current analysis were recorded well within the tolerance limits indicating a very good fit of the model. The model explained 75% of variance as the value of R^2 is observed 0.75.

2.6 DISCUSSION

Procurement function is one of the strategic functions of an industry. Literature review in this domain unearthed four important parameters that affect the procurement function. Influence of each of them was validated as a part of the statistical tests with the resultant influence on the customer satisfaction and supplier relationship.

A. *Forecasting:* The activity of forecasting is profoundly determining the demand in advance before the occurrence. A lot of emphasis is put on demand forecasting in the recent times due to the volatility in the customer demand. The complete supply chain gets unsettled due to the demand volatility, especially in case of manufacturing. Incorrect forecasting gives rise to huge inventories of unwanted products blocking the cash flows. Demand forecasting is tested as a part of this study for its impact on the procurement function. Even though relatively lesser weightage was calculated, it positively affects procurement. Effective forecasting tends to reduce the stress in the supply chain along with contribution towards customer satisfaction as well as better supplier relations (Taylor & Plambeck, 2007).

B. *Purchase contracts:* Effective purchase contracts are the ones that clearly define the terms of requirement, quality, timelines expected, etc. Contracts are fine-tuned to create a win-win agreement between the supplier and the buyer organization. The impact of a contract is seen in long-term over the tenure of the contract, because later the contracting terms become standard operating principles (Taylor & Plambeck, 2007). While our study reflected a positive impact of purchase contracts on procurement, the lowest regression weight was calculated for the link between contracts and procurement.

C. *Supplier rating system:* To have a positive control on the procurement function it is vital that organizations are well equipped with data. Organizations generally convert all the facts related to the agreement into measurable constructs (Presutti, 2003). Researchers have mentioned about one such tool earlier, i.e., BSC. The overall supplier ratings are emerging from this data and study too reveals that it is one of the most influential parameter for efficiency in the procurement function. Basis the data organizations divide the supplier base into segments. Basis the segments, different strategies are applied to different suppliers. Therefore, impact of these ratings is very high with a positive impact on procurement function.

D. *Real-time information:* A paradigm shift is observed in the customer behavior during the recent decade, reflecting in volatile customer demands. To cater varying customer demands, it is becoming difficult for the organizations especially along with the cost pressures. To build agile response systems for the market and equally swift workforce is the only way organizations have. Real-time data becomes inevitable, without which the organization cannot survive (Gebauer et al., 1998). The statistical analysis conducted by the researchers for this study too echoes similar sentiment indicating a positive impact on procurement function and accords highest regression weight to the real-time information parameter. The consolidated details of the findings are reported in Table 2.6.

TABLE 2.6
Path Analysis

Path	Standard Regression Weight	Significance Level	Status
Forecasting procurement	0.42	***	Supported
Purchasing contracts procurement	0.31	0.023	Supported
Supplier rating system procurement	0.52	***	Supported
Real-time information procurement	0.67	0.042	Supported

Source: AMOS.

2.7   MANAGERIAL AND PRACTICAL IMPLICATIONS

The propositions in the study have direct managerial and practical implications that impact the longevity of an organization. The initial study in our research recommends the usage of IoT in the procurement space. The benefits of this amalgamation are mentioned in primarily two areas; transactional procurement and strategic procurement. Transactional procurement explores the usage of real-time data in processes such as demand forecasting and techniques like vendor management inventory and procurement activities. Strategic procurement encompasses the implementation of overall control and governance of the supply relationship management, *vide* the balanced scorecard tool. The statistical analysis conducted revealed that all the parameters considered in the study such as forecasting, purchase contracts, supplier rating system, and real-time information have a positive influence on the procurement function. Maximum regression weight is assigned to the real-time information. Organizations need to take this study forward and implement IoT in the procurement domain to enable real-time decision-making. Since procurement as a business function links the customer demand to the supplies and ensures a positive customer experience, it is imperative for the organization to invest in digital technologies such as IoT. Better customer experience results in customer retention; hence, the implications of not adopting technologies that serve the customer can be grave. On the other hand, customer orders serviced in time with real-time tracking visibility, due to the real-time data availability, imbibe trust in the minds of the customer. This step strengthens the customer relationship and assists in transforming the customer into a brand ambassador. Suppliers, as well, due to the availability of real-time information feel involved and aid the organizations to be flexible to the volatile customer demands. Supplier governance is strengthened ensuring better supplier relationship. Thus, researchers, basis the managerial and the practical implications strongly recommend the implementation of IoT in the procurement function.

2.8   LIMITATIONS/WAY FORWARD

The researchers conducted focused group discussion as well as primary data analysis in this study to establish the implications of implementation of IoT in the procurement function. Rigorous methodology was followed to arrive at the findings, yet there are some limitations. The present study is conducted with four parameters as an input to the procurement functions, i.e., forecasting, purchasing contracts, supplier rating system, and real-time information. The impact of other positive parameters can be undertaken in the future, and they could include psychological drivers, payment terms, and logistics arrangements such as milk runs, etc. The objective of the current study was not to confine the findings but to focus on the influence of IoT in the procurement function covering transactional and strategic parameters. Other researchers considering different parameters influencing the procurement function can carry out similar studies. Digital technologies are undergoing a revolution, and there is an implicit need for further research in this area to upkeep the present implementation. IoT of things can be leveraged further to build an artificial intelligence (AI) ecosystem automating quite a few functions. AI automatically ensures execution of processes along with exception handling. The amalgamation has the potential to create a market turnaround for the early movers. Further studies can be directed towards finalizing the architecture of IoT and AI applicable in the procurement function and implementation of the same. An event driven service oriented architecture can be evaluated as a way ahead in this study.

2.9 CONCLUSION

The impressive need of implementing IoT in the procurement domain is brought to light by the researchers. It augments the effectiveness of constraints driven supplier relationship management. Researchers critically examined the applicability of IoT in the procurement function. Serious deliberations in both transactional and strategic procurement have been carried out in this research work. The parameters influencing the procurement function such as forecasting, purchase contracts, supplier rating system, and real-time information were considered for the statistical testing. Evidently, all had a positive impact on procurement function with real-time data significantly leading with the highest regression weight. The strategic relationship with the supplier/vendor is explained with a BSC tool. The purpose for emphasizing the BSC was to set the right foundation for enabling a long-term relationship with the vendor/supplier.

Further, this study aids other researchers with some specific performance measures that are a part of each of the balanced perspective. The selection of the measures is the basis of their applicability to the procurement function. The applicability of IoT is verified for the parameters with the digital experts during the focus group discussions and direct face-to-face interviews.

When the BSC is applied in the procurement domain, it is interesting to observe the contradiction between the measures of the same or different balanced perspective, such as supplier rejection rate and buyer-supplier partnership level. These contradict each other when viewed from the balanced perspective of finance. Similarly, the range of products and flexibility to meet customer demands probably look like two different directions. It is already established that the parameter range affects the supply chain efficiency adversely. Managers and strategic thinkers of an organization have to handle these contradictions effectively to take appropriate decisions (Bhagwat & Sharma, 2007).

SMART performance measures for the supplier determine effective business relationships. A strategic governance agenda is the first step in defining SMART measures in the procurement function. For governing the relationship, clear segregation of roles and responsibilities is required. Between the buyer-supplier organization, measurable key performance indicators are agreed for monitoring the relationship. Periodic reporting of the same is safeguarded and reviewed to analyze the health of the relationship. Alignment with the corporate goals is ensured by designing the goals for the relationship cautiously. The accountability resides directly with the individuals, even though there is a presence of a complete team of professionals who are working on each side. Focusing on the outcome, the assignment of accountability is result-based. In addition, finally time-bound plans are chalked out along with target setting. Hence, the above method is termed as SMART performance measures, i.e., Strategic, Measurable, Aligned, Results-based, and Time-bound measures (Crates, 2014). Organized collaboration and disciplined business approaches are the specific outcomes of this methodology. Supplier participation in innovation and value-adds shows an increasing trend further strengthening the relationship. The internal resources in the organization have to collaborate seamlessly for an effective engagement with the supplier and to build a strong relationship with the supplier. Generally, the tool used by the buyers to assess the strengths possessed by the supplier along with the other measures, such as weakness, opportunity, and threat, is called SWOT (Strength, Weakness, Opportunity, and Threat). It is imperative in a procurement process that it is vital to understand the strengths possessed by the supplier so that they can be used towards the benefit of the buying organization for remaining competitive in the market gaining the upper edge (Panayiotou et al., 2004). While there are numerous benefits that researchers have mentioned in the previous section, the most critical ones spelled out during the focused group discussion are:

- Streamlining of fulfillment process
- Better monitored and controlled vendor relationships
- Purchase order efficiency is improved multifold
- Reduction in inventory cost
- Reduction in order cycle time

The primary data analysis and secondary data, including the focus group discussions, have confirmed a good fit of IoT for the procurement domain integrating all the aspects of supply chain, collaborating with all constructs and enabling the whole system with a proactive response. (Panayiotou et al., 2004).

REFERENCES

Albaum, G. (1997). "The likert scale revisited." *Market Research Society Journal* 39(2): 1–21. doi:10.1177/147078539703900202.
Amit, R. and Zott, C. (2001). "Value creation in e-business." *Strategic Management Journal* 22(6–7): 493–520.
Anderson, E. W. and Sullivan, M. W. (1993). "The antecedents and consequences of customer satisfaction for firms." *Marketing Science* 12(2): 125–143.
Argentus. 2017. "What's the difference between procurement and purchasing? (Infographic)— Supply Chain Game Changer." *Supply Chain Game Changer*. https://supplychain gamechanger.com/whats-the-difference-between-procurement-and-purchasing-infographic/.
Baarlid, A. and Claesson, A. (2015). "Aligning strategy, structure and performance in supply chain management: A case study in the power systems industry." Masters, Copenhagen Business School.
Bendorf, R. and Kolodisner, M. (2017). "5 Procurement benchmarking references." https://www.purchasing-procurement-center.com/procurement-benchmarking.html.
Bhagwat, R. and Sharma, M. K. (2007). "Performance measurement of supply chain management: A balanced scorecard approach." *Computers & Industrial Engineering* 53(1): 43–62.

Bhandari, R. (2014). "Impact of technology on logistics and supply chain management." *IOSR Journal of Business and Management* 2: 17. Retrieved from http://www.iosrjournals.org/iosr-jbm/papers/7th-ibrc.

Bi, Z., Da Xu, L., and Wang, C. (2014). "Internet of things for enterprise systems of modern manufacturing." *IEEE Transactions on Industrial Informatics* 10(2): 1537–1546.

Canada Business Network. 2009. "Manage your suppliers." *Infoentrepreneurs.org.* http://www.infoentrepreneurs.org/en/guides/manage-your-suppliers.

Carter, T. C. (2003a). *Supplier Relationship Management: Models, Considerations and Implications for DOD.* Washington, DC, Defense Technical Information Center. http://www.dtic.mil/docs/citations/ADA422054.

Carter, T. C. (2003b). *Supplier Relationship Management: Models, considerations and Implications for DOD.* Washington, DC, Industrial Coll of the Armed Forces.

Chatfield, C. (2000). *Time-series Forecasting.* CRC Press.

Chen, F., Drezner, Z., Ryan, J. K., and Simchi-Levi, D. (2000). "Quantifying the bullwhip effect in a simple supply chain: The impact of forecasting, lead times, and information." *Management Science* 46(3): 436–443.

Child, D. (1990). *The Essentials of Factor Analysis.* London, UK: Cassell Educational Ltd.

Choy, K., Bun Lee, W., and Lo, V. (2004). "Development of a case based intelligent supplier relationship management system: Linking supplier rating system and product coding system." *Supply Chain Management: An International Journal* 9(1): 86–101. doi:10.1108/13598540410517601.

Crates, L. (2014). "Key benefits of real-time data: Voluntas|One place for customer intelligence." *Voluntas|One Place for Customer Intelligence.* http://www.voluntas.co.uk/key-benefits-real-time-data/.

Davenport, T. H. and Harris, J. G. (2007). *Competing on Analytics.* 1st arg. Boston, MA, United States of America: Harvard Business School Press.

Dohr, A., Modre-Opsrian, R., Drobics, M., Hayn, D., and Schreier, G. (2010). "The Internet of Things for ambient assisted living." In *Information Technology: New Generations (ITNG), 2010 Seventh International Conference on*, pp. 804–809. IEEE.

Feige, E. L. and Pearce, D. K. (1979). "The casual causal relationship between money and income: Some caveats for time series analysis." *The Review of Economics and Statistics*: 521–533.

Fidell, S., Tabachnick, B., Mestre, V., and Fidell, L. (2013). "Aircraft noise-induced awakenings are more reasonably predicted from relative than from absolute sound exposure levels." *The Journal of the Acoustical Society of America* 134(5): 3645–3653.

Fornell, C. and Larcker, D. F. (1981). "Structural equation models with unobservable variables and measurement error: Algebra and statistics." *Journal of Marketing Research*: 382–388.

Fournier, J. (2015). "What is procurement?" *Hcmworks.com.* https://www.hcmworks.com/blog/what-is-procurement.

Gebauer, J., Beam, C., and Segev, A. (1998). "Impact of the internet on procurement." *Acquisition Review Quarterly* 5(2): 167–184.

Gebauer, J. and Segev, A. (2001). "Changing shapes of supply chains-how the internet could lead to a more integrated procurement function." *In Supply Chain Forum: An International Journal* 2(1): 2–9.

Gubbi, J., Buyya, R., Marusic, S., and Palaniswami, M. (2013). "Internet of things (IoT): A vision, architectural elements, and future directions." *Future Generation Computer Systems* 29(7): 1645–1660. doi:10.1016/j.future.2013.01.010.

Gunasekaran, A., Patel, C., and Tirtiroglu, E. (2001). "Performance measures and metrics in a supply chain environment." *International Journal of Operations & Production Management* 21(1/2): 71–87.

Hair, J. F., Ringle, C. M., and Sarstedt, M. (2013). "Partial least squares structural equation modeling: Rigorous applications, better results and higher acceptance." *Long Range Planning* 46(1–2): 1–12.

Handfield, R. and McCormack, K. P. (2007). *Supply Chain Risk Management: Minimizing Disruptions in Global Sourcing.* Auerbach Publications.

Haleem, H. H. (2015). "What is the difference between the vendor and the supplier? Bayt.com specialties." *Bayt.com.* https://www.bayt.com/en/specialties/q/179619/what-is-the-difference-between-the-vendor-and-the-supplier/.

Jazdi, N. (2014). Cyber physical systems in the context of Industry 4.0. In *Automation, Quality and Testing, Robotics, 2014 IEEE International Conference on* pp. 1–4, IEEE.

Kaplan, R. S. and Norton, D. P. (2001). "Transforming the balanced scorecard from performance measurement to strategic management: Part I." *Accounting Horizons* 15(1): 87–104.

Kinder, P. (2014). "The internet of things: Positive for procurement? Supply Management. https://www.cips.org/supply-management/opinion/2014/july/the-Internet-of-things-positive-for-procurement/.

Kopetz, H. (2011). *Real-time Systems: Design Principles for Distributed Embedded Applications.* Springer Science & Business Media.

Kotler, P. and Gary, A. (1991). *Principles of Marketing*, 5th ed. Englewood Cliffs, NJ, Prentice Hall.

Kramer, M. (2011). *The Big Idea: Creating Shared Value.* pp. 4–17. Harvard Business Review. Retrieved from http://www.nuovavista.com/SharedValuePorterHarvardBusinessReview.PDF.

Lambert, D. M. and Cooper, M. C. (2000). "Issues in supply chain management." *Industrial Marketing Management* 29(1): 65–83. doi:10.1016/s0019-8501(99)00113-3.

Liu, X. and Sun, Y. (2011). "Information flow management of vendor-managed inventory system in automobile parts inbound logistics based on internet of things." *JSW* 6(7): 1374–1380.

Lutz, J. and Bayer, A. G. (2001). Procurement strategies for eBusiness. *Presentation at the Haas School of Business.*

Milgrom, P. and Roberts, J. (1990). "The economics of modern manufacturing: Technology, strategy, and organization." *American Economic Review* 80(3): 511–528.

Neef, D. (2001). *E-Procurement: From Strategy to Implementation.* FT Press.

Novack, R. A. and Simco, S. W. (1991). "The industrial procurement process: A supply chain perspective." *Journal of Business Logistics* 12(1): 145.

Nunnally, J. C. and Bernstein, I. H. (1994). "The theory of measurement error." *Psychometric Theory* 209–247.

Özgen, C. (2010). "Inventory management through vendor managed inventory in a supply chain with stochastic demand." PhD dissertation, Middle East Technical University.

Panayiotou, N. A., Gayialis, S. P., and Tatsiopoulos, I. P. (2004). "An e-procurement system for governmental purchasing." *International Journal of Production Economics* 90(1): 79–102. doi:10.1016/s0925-5273(03)00103-8.

Presutti Jr., W. D. (2003). "Supply management and e-procurement: Creating value added in the supply chain." *Industrial Marketing Management* 32(3): 219–226.

Ranganathan, C., Teo, T. S. H., and Dhaliwal, J. (2011). "Web-enabled supply chain management: Key antecedents and performance impacts." *International Journal of Information Management* 31(6): 533–545. doi:10.1016/j.ijinfomgt.2011.02.004.

Reddy, S. B. (2006). "Strategic flexibility and information technology properties: Competitive advantage and asset specificity." *Journal of Competitiveness Studies* 14(1): 16.

Sagiroglu, S. and Sinanc, D. (2013). "Big data: A review." In *Collaboration Technologies and Systems (CTS), 2013 International Conference on*, pp. 42–47. IEEE.

Sahay, B. S. and Ranjan, J. (2008). "Real time business intelligence in supply chain analytics." *Information Management & Computer Security* 16(1): 28–48. doi:10.1108/09685220810862733.

Samper, H. M. (2017). "How to improve supplier relationship management in the life sciences." *Pharmaceuticalonline.com.* https://www.pharmaceuticalonline.com/doc/how-to-improve-supplier-relationship-management-in-the-life-sciences-0001.

Sani, B. and Kingsman, B. G. (1997). "Selecting the best periodic inventory control and demand forecasting methods for low demand items." *Journal of the Operational Research Society* 48(7): 700–713.

Schneier, C. E., Shaw, D. G., Beatty, R. W., and Baird, L. S., eds. (1995). *Performance Measurement, Management, and Appraisal Sourcebook.* Human Resource Development.

Stark, T. and Croushore, D. (2002). "Forecasting with a real-time data set for macroeconomists." *Journal of Macroeconomics* 24(4): 507–531.

Sundmaeker, H., Guillemin, P., Friess, P., and Woelfflé, S. (2010). "Vision and challenges for realising the Internet of Things." *Cluster of European Research Projects on the Internet of Things, European Commision* 3(3): 34–36.

Talbot, M. (2015). "What a business that leverages real-time data looks like: GoSpotCheck". *GoSpotCheck.* https://www.gospotcheck.com/2015/07/13/what-a-business-that-leverages-real-time-data-looks-like/.

Taylor, T. A. and Plambeck, E. L. (2007). "Supply chain relationships and contracts: The impact of repeated interaction on capacity investment and procurement." *Management Science* 53(10): 1577–1593. Institute for Operations Research and the Management Sciences (INFORMS). doi:10.1287/mnsc.1070.0708.

Wang, Q., Li, J., Ross, W., and Craighead, C. (2012). The interplay of drivers and deterrents of opportunism in buyer–supplier relationships. *Journal of the Academy of Marketing Science* 41(1): 111–131. doi:10.1007/s11747-012-0310-9.

Watts, C. A., Kim, K. Y., and Hahn, C. K. (1995). "Linking purchasing to corporate competitive strategy." *International Journal of Purchasing and Materials Management* 31(1): 2–8.

Wigan, R. T. (2004). "Business-to-business electronic commerce: The convergence of relationships, networked supply chains and value webs." In *E-Life after the Dot Com Bust,* pp. 137–155. Physica, Heidelberg, Germany.

Wurster, T. S. *Blown to Bits: How the New Economics of Information Transforms Strategy.* Harvard Business School Press, 1999.

Xia, F., Yang, L. T., Wang, L., and Vinel, A. (2012). "Internet of things." *International Journal of Communication Systems* 25(9): 1101–1102.

Yerpude, S. and Singhal, T. K. (2017a). "Augmentation of effectiveness of Vendor Managed Inventory (VMI) operations with IoT data: A research perspective." *International Journal of Applied Business and Economic Research* 15(16): 469–482.

Yerpude, S. and Singhal, T. K. (2017b). *Customer Service Excellence through Internet of Things.* 1st arg. Germany, Lambert Academic Publishing.

Yerpude, S. and Singhal, T. K. (2017c). "Impact of Internet of Things (IoT) data on demand forecasting." *Indian Journal of Science and Technology* 10(15).

Yerpude, S. and Singhal, T. K. (2017d). "Internet of things and its impact on business analytics." *Indian Journal of Science and Technology* 10(5).

Yerpude, S. and Singhal, T. K.. (2018a). "Customer service enhancement through 'On-Road Vehicle Assistance' enabled with Internet of Things (IoT) solutions and frameworks: A futuristic perspective." *International Journal of Applied Business and Economic Research* 15(16–Part II).

Yerpude, S. and Singhal, T. K. (2018b). "Enhancing new product development effectiveness with internet of things origin real time data." *Journal of Cases on Information Technology (JCIT)* 20(3): 21–35.

Yerpude, S. and Singhal, T. K. (2018c) "Internet of things based customer relationship management: A research perspective." *International Journal of Engineering & Technology* 7(2.7): 444–450.

Yerpude, S. and Singhal, T. K. (2018d). "SMART warehouse with internet of things supported inventory management system." *International Journal of Pure and Applied Mathematics* 118(24C): 1–15.

York, M. (2015). "What will the internet of things mean for procurement? Part I « CPO Rising – The site for Chief Procurement Officers & Leaders in Supply Management". Cporising.Com. http://cporising.com/2015/03/19/what-will-the-internet -of-things-mean-for-procurement-part-i/.

3 Framework for Measuring the Procurement Performance in the Dairy Supply Chain

Rahul S. Mor, Arvind Bhardwaj, Sarbjit Singh, and Manjeet Kharub

CONTENTS

3.1   INTRODUCTION

India is the leading milk producer globally and has perceived about a 4% improvement in the production of milk per annum, which surpasses the typical global improvement of approximately 1% in the past three decades. The primary goal of the dairy industry in India is to boost the milk production as well as to upgrade the milk processing sector. The milk is being treated and sold by 170 milk producer cooperative (co-op) unions consisting of 22 State Co-op Federations in India.

Milk unions protect the milk producers from unfair trade practices of *intermediaries*, *dudhiyas* (milkmen in the unorganized sector), and milk contractors thereby improving their economic condition tremendously. However, the organized sector handles only 20% of the milk and rest is controlled by the unorganized sector of the industry. Milk unions provide a ready-market to the milk producers for sale of milk in villages through cooperatives and works to provide wholesome hygienic good quality processed milk to the urban consumers at a remunerative price. The dairy supply chain starts by procuring the milk from the farmers, transportation to the plant, milk processing, packaging, and distribution to the retailers and finally to the consumer. Technological innovations, supply chain integration and collaboration, and eradication of uncertainties help to achieve sustainability in the food processing sector. Further, it needs to set-up priority for R&D, investment, governance, and trade policies (Mor et al., 2015, 2017, 2018e). The decentralized supply chain is the indication of distortion in food quality (Chen et al., 2014; Bhardwaj et al., 2016; Mor et al., 2018a, 2018b, 2018c). Some major players serve the dairy market in India, namely; Nestle, Amul, Mother Dairy, Verka, Britannia, Vita, Lakshaya, Nandini, SuperHF, etc.

Structural equation modeling (SEM) is a useful multivariate method, and its usage has been growing since its development in the 1980s. SEM is used to calculate the reliability and validity of the constructs of agile manufacturing predominant in the automotive industries (Roberts et al., 2010). The academicians favor the SEM as it estimates the interrelated dependence in a precise analysis. It is a group of statistical methods for the analysis of associations between multiple predictor and response variables (Bagozzi and Yi, 1988). A linear structural method can be used to confirm the enablers of sustainable practices developed through interpretive structural modeling (Thirupathi and Vinodh, 2016). The SEM methodology based on Malcolm Baldrige principles can be used to estimate the environmental performance of SMEs (Hussey and Eagan, 2007). The SEM methodology consists of two sorts of variables, *viz.*, endogenous and exogenous (Vinodh and Joy, 2012a). The latent variables having abstract nature are difficult to recognize directly, whereas the observed variables are more accessible to explore (Tenenhaus et al., 2005; Xiong et al., 2015).

Vinodh et al. (2012b) focused on turbulent market changes through an agile model using the SEM. Sitek and Wikarek (2015) demonstrated a hybrid structure through the mathematical programming to improve the sustainable supply chain resolutions. The statistical modeling indicators are indeed the beginning of all-inclusive methods for the techniques like SEM and multilevel (regression) modeling (Goldstein, 1986).

The rest of this chapter is designed in the following way. Section 3.2 is the comprehensive literature review, whereas, Section 3.3 consists of the problem formulation detailing the emergence of theme and objectives. Section 3.4 covers the methodology, and Section 3.5 comprises of the analysis, results, and discussion part. The final production model is developed in Section 3.6 along with the hypotheses testing and significance of the results obtained. Finally, Section 3.7 consist of the conclusion and the scope for future work in continuation of this research work.

3.2 LITERATURE REVIEW

A comprehensive literature review of the dairy supply chain and structural equation modeling is presented in this section, as follows.

3.2.1 DAIRY SUPPLY CHAIN

Mangla et al. (2016) suggested to focus on the critical success factors to improve the organizational performance and for achieving sustainability in the food supply chain. Okano et al. (2014) focused on consolidating the domains of the dairy supply chain through performance indicators to benchmark the supply chain practices and become sustainable. Lemma et al. (2015) offered the modeling approach for short life cycle food supply chain by focusing on the product's perishability and the waste and loss assessment. Bamgboje-Ayodele et al. (2014) recognized the research issues in the approaches for unpacking and knowledge optimization in the food supply chain. García et al. (2014) emphasized the strategic inferences in the context of packaging design for achieving a competitive advantage in sustainable supply chain operations. Ghosh et al. (2014) analyzed the enablers of risk management for the dairy industry using the interpretive structural modeling (ISM). Kumar (2014) aimed to evaluate the effectiveness of a theoretical model of the dairy supply chain and proposed an integrated model for performance measurement of the dairy industry. Prakash and Pant (2013) presented the case of Indian milk processing sector and demonstrated how balanced scorecard methodology could help to assess the supply chain performance. Singh and Javadekar (2011) found that a loss of 72% of perishable food products due to no use of information technology by the retailers of perishable food products in the unorganized sector. Kumar et al. (2011) found that the traceability and food safety helps in strengthening the modern milk supply chain. Punjabi (2009) considered different factors influencing the effectiveness of the dairy sector through performance analysis. Gupta et al. (2012) untaken an assessment of the profits gained from vertical coordination through regression analysis and field survey. Glover et al. (2014) concluded that a broader systemic way is desired to encourage the sustainable practices in the dairy supply chain for the energy reduction.

3.2.2 Structural Equation Modeling

Thirupathi and Vinodh (2016) explored the usage of ISM methodology with the SEM to establish a structural relationship among the enablers of sustainable supply chain practices. Hussey and Eagan (2007) established a model to evaluate the performance of green aspects in small and medium enterprises through SEM methodology and found the model as valid. Vinodh and Joy (2012c) developed a theoretical model comprising of the enablers as criteria and attributes for the manufacturing industry. Hou et al. (2014) discovered the relationship between sustainable operations and the issues responsible for behavior changes in the manufacturing industry. Vinodh and Joy (2012a) studied the usage of SEM procedures to evaluate the lean supply chain practices through empirically collected data from manufacturing industries in Tamil Nadu. The conventional SEM models have been generalized to accommodate different responses. Eid (2009) discovered the interactions among variables and explained the requisite to check the factors affecting world-class manufacturing empirically. Mor et al. (2016, 2018a, 2018b, 2018c, 2018d) worked on the supply chain management practices of food processing industry using ISM and SEM methods and found that the usage of these tools helps to establish a contextual relationship among the factors affecting the performance of supply chain practices.

3.2.3 Research Gaps

After a comprehensive literature review, it is revealed that no research yet relates to the assessment of the performance of procurement practices in Indian dairy industry. Further, the SEM techniques have broad applications, but its use to study the procurement practices in the context of the dairy industry has not been tried. Thus, this is a unique study that explores the procurement performance in the context of the dairy industry, and all these causes encouraged the authors to discover the procurement concerns in this chapter.

3.3 PROBLEM FORMULATION

The increasing competition and global quality standards have stirred the industries towards managing their supply chain activities effectively. On the other hand, the pricing, order-fill-rate, and delivery requirements of food sector are the critical concerns pooled by every stakeholder in the supply chain. Thus, it has been realized from the literature review that most of the studies researching the dairy sector focus on dairy science, raw milk quality, and technology development, and does not consider the procurement practices, especially in the Indian context. Moreover, the existing studies put forward the high-tech support to the farmers for networking them up with markets. Further, the literature review reveals that only a few dairy industries or regions have been covered in India and many states, like northern India, being the focal point of the White Revolution in the 1970s, are still not explored. Hence, this research attempts to address various procurement issues in dairy supply chains. An optimum competitive procurement model has been proposed by applying the structural equation modeling methodology to the responses obtained from a questionnaire survey of respondents in dairy industries from the northern region of India.

3.3.1 OBJECTIVE OF THE STUDY

- To bring out various issues related to the procurement practices of the dairy supply chain.
- To develop an optimum competitive supply chain model for the procurement practices in Indian dairy industry using SEM techniques.

3.4 METHODS

Both the primary and secondary type of data is used in this chapter. The primary data composed of a well-structured questionnaire including crucial supply chain issues and other organizational information. The secondary information has been collected from various published as well as unpublished research articles, websites, press notes, media clippings and official reports of National Dairy Development Board (NDDB) as well as the internationally acclaimed organizations. SEM has been used to study the structural relationship and interaction between measured variables and latent constructs of supply chain practices of the dairy industry. A scale was developed to quantify the supply chain performance. A number of measures were used in developing the scale. The survey questionnaire was prepared as per the comprehensive literature review of the supply chain operations and dairy industry. The questionnaire was improved in consultation with managers and academicians. Morgan (1993) mentions such modification when the target population is new, as in this case. The first section of the questionnaire involves the statements related to procurement, processing and distribution practices, and outcome variables measuring overall performance. The second section is dedicated to accumulating the demographic data of the industry. In advance of distribution, the questionnaire was authenticated via pilot study as proposed by Robson (2002). The pilot study enabled to diagnosing the inconsistency or lack of understanding. The suggestions of the experts were incorporated, and the questionnaire was revised accordingly. Figure 3.1 represents the research methodology flowchart for the chapter.

In order to evaluate the procurement performance of the dairy industry, both cooperative and private industries of northern India were nominated. The questionnaire was distributed to 504 persons with an appeal to submit their visions about the issues highlighted in the questionnaire. The questionnaire was also prepared online through Google Forms. The answers were collected over both methods, i.e., by personal visits to the dairy industries as well as via online (Google Forms). The snowball sampling method was employed to enable the direct contact with respondents (Nargundkar, 2004). The respondents were requested to enter their perceptions of procurement issues in the dairy supply chain on a 5-point Likert scale. Initially, the introductory e-mails and letters to grant permission for the exploratory study were sent to the general manager or plant head of each selected industry. The unit-heads referred the researcher to key respondents for filling out the questionnaire. The responses were collected through both personal visits to industries and online via Google Forms after verifying their identity. First, the respondents were briefed about the issues and the Likert scale to obtain more reliable responses. The respondent's privacy was retained surreptitious in order to achieve the unbiased answers (Robson, 2002; Saunders et al., 2009). Most of the persons filled and returned the questionnaire

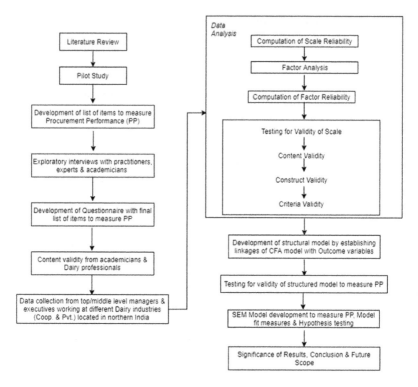

FIGURE 3.1   Research methodology flowchart.

on-spot, while others returned it in continual calls for getting the high response rate as well as the quality of the collected figures. This method of data collection was validated by other researchers like Flynn et al. (1990) and Forza (2002). The replies were received from almost all levels, i.e., managing directors, senior managers, plant heads, general managers, shop-floor executives, head of departments, etc. About 45% of responses were received from cooperatives and shop-floor executives, 25% from middle management, and 30% from the top management level. Additionally, 21% replies were received from logistics and transportation department, 28% from incoming-quality control department, 19% from administration and management information system (MIS) department, and 30% from the team dealing with procurement activities at the various co-op societies. Out of the expected responses, only 265 valid retorts were chosen for the next level of exploration.

3.5   ANALYSIS, RESULTS, AND DISCUSSION

3.5.1   PILOT STUDY

A pilot study was conducted to assess the validity and practicality of the questionnaire as well as to enrich the quality of questionnaire (Forza, 2002). Accordingly, the pilot study was performed by distributing the questionnaire to ten industry experts and two from an academic organization to get a view on questionnaire design. The views of experts were incorporated in order to make minor changes. The questionnaire was thus made ready-for-distribution after modifications.

3.5.2 RELIABILITY ANALYSIS

The reliability of the issues related to procurement practices was assessed using the *Cronbach alpha coefficient* after screening the missing values. Cronbach alpha describes the reliability through internal consistency method, which redirects the *homogeneity and inter-correlation* of procurement issues. The Cronbach alpha coefficient was carried out by IBM SPSS v22 software and it comes out to be 0.837, which indicates the highly reliable data with reference to Lee et al. (2000) and Cronin and Taylor (1992). The reliability parameters of all constructs have been verified (Cronbach, 1951).

3.5.3 FACTOR ANALYSIS

The exploratory factor analysis (EFA) has been performed in various steps. All the 20 procurement issues, taken as variables, were taken for EFA. Initially, *Bartlett test of Sphericity* was performed to check the importance of factors and it was judged by the correlation matrix of the data (Hair et al., 2005). At the same time, the suitability of sampling ($N = 265$ in current study) was referred by KMO (Kaiser-Meyer-Olkin) value. The KMO varies from zero to one, and it comes out to be 0.831 in this case, which reflects the significance of factor analysis. The respective scores of *Bartlett test of Sphericity* and *KMO* are as follows: Chi-Square: 5324.516; df: 378; Sig.: 0.000. The results of factor analysis are indicating the suitability of factor analysis (Hair et al., 2005). EFA was directed using principal component analysis (PCA) method with Varimax rotation and Kaiser Normalization (Eigenvalue > 1) in the SPSS software. The maximum likelihood (ML) estimation method was used in this analysis, which confirmed that there exist no skewness or kurtosis in the joint distribution of variables and all the variables are continuous (Hoyle, 2011; Kline, 2010). The Harman's one-factor-test was adopted to test the common method bias in the current analysis. At first, an analysis was performed with a four-factor model, and another analysis was carried out with a one-factor model consisting of all measured variable loadings. The value without rotation came out to be 0.71 for the proposed four-factor model and 0.29 for the one-factor model. Thus, the common method bias problem was also confirmed (Kharub et al., 2018; Kharub & Sharma, 2018). Initially, 38 questions were selected in the questionnaire to conduct EFA, which was later reduced to 20 due to various reasons like a low factor score, commonality below 0.5, cross-loading, etc. Four factors were obtained explaining approximately 69.79% of the total variance. The individual factor explained 20.969%, 17.977%, 17.810%, and 13.037% of variance correspondingly. All the factor loading values were reliable with the proposed factor structure (Table 3.1).

Now, the factors extracted through EFA were entitled as "Inventory management, Quality Management, Supplier Management, and Technological Innovations" based on the subjective opinion of the investigator and experts. The communalities state the amount of variance derived through the four factors. All the statements have

TABLE 3.1

Exploratory Factor Analysis

Factor No.	Statements (Name & Label)	Commonality	Factors				Measurement on 5-Point Likert Scale		Overall Score of Factor	
			F1	F2	F3	F4	Mean	Standard Deviation	Mean	Standard Deviation
F1	**Inventory Management (F1) (Rate followings as how do you manage the surplus or shortage of milk):**								3.81	1.37
	PRC34 Convert the value-added products into milk	0.73	0.80				3.89	1.31		
	PRC28 Convert milk into other value-added products	0.79	0.85				3.80	1.36		
	PRC36 Buy milk from other units	0.87	0.86				3.94	1.35		
	PRC33 Suppliers make-up for the shortage by increasing supply of milk	0.62	0.78				3.78	1.39		
	PRC35 Convert the stored powder into milk	0.55	0.65				4.06	1.30		
	PRC32 Stores the milk	0.50	0.69				3.38	1.53		
F2	**Quality Management (F2) (Rate followings as how do you perceive the wastages of milk due to):**								3.51	1.34
	PRC20new Improper milk handling at source/site	0.79		0.87			3.40	1.25		
	PRC21new Unhygienic practices at source/site	0.82		0.87			3.71	1.39		
	PRC23new Improper quality checks	0.70		0.82			3.95	1.34		
	PRC19new Deterioration in milk quality due to temperature variation	0.68		0.70			3.18	1.44		
	PRC18new Delayed shipping due to poor road conditions/poor infrastructure	0.50		0.69			3.32	1.30		

(Continued)

TABLE 3.1 (Continued)
Exploratory Factor Analysis

Factor No.	Statements (Name & Label)	Commonality	Factors				Measurement on 5-Point Likert Scale		Overall Score of Factor	
			F1	F2	F3	F4	Mean	Standard Deviation	Mean	Standard Deviation
F3	**Supplier Management (F3)**									
	PRC9 You are satisfied with the exchange of information and level of cooperation with suppliers	0.56			0.69		4.27	0.86	4.19	0.95
	PRC1 You have installed the automatic milk quality testing machines at each society	0.80			0.88		4.16	0.99		
	PRC2 You have implemented the traceability system for identification of poor-quality milk suppliers	0.75			0.86		4.23	0.91		
	PRC4 You are satisfied with the consistency of milk quality supplied by suppliers	0.67			0.79		4.12	0.98		
	PRC7 You include the key suppliers in your planning and goal-setting activities	0.66			0.77		4.16	0.99		

(Continued)

TABLE 3.1 (Continued)
Exploratory Factor Analysis

Factor No.	Statements (Name & Label)	Commonality	Factors				Measurement on 5-Point Likert Scale		Overall Score of Factor	
			F1	F2	F3	F4	Mean	Standard Deviation	Mean	Standard Deviation
F4	**Technological Innovations (F4)**									
	PRC12 You have installed the vehicle tracking system like GPS/GIS/others for procurement	0.82				0.89	3.04	1.50	2.90	1.52
	PRC13 You have installed the Lid Opening Sensors in procurement vehicles	0.75				0.84	2.62	1.53		
	PRC14 You have installed the independent BMCs (Bulk Milk Chillers) at each society	0.79				0.89	2.91	1.48		
	PRC17 You have installed the ERP system for dealing with suppliers	0.61				0.64	3.02	1.55		
Reliability (Cronbach Alpha[c] value) of identified factors			0.886	0.881	0.877	0.861				

Note: Factor Extraction Mode: Principal Component Analysis.
Rotation Mode: Varimax with Kaiser Normalization.
[a] Rotation congregated in 5 iterations.
[b] Cut-off point for loadings is 99% substantial and is measured using $2.58/\sqrt{n}$ (Pitt et al., 1995), where $(n = 20)$ is the no. of statements, F1 to F4 signifies the factors.
[c] α values ≥ 0.70 are adequate (Nunnally, 1978).

substantial communalities, i.e., ≥ 0.50 (Hair et al., 2005) (Table 3.1). Individual factor values signify the associations between the item and the respective factor. Each item represents the factor loading of >0.5 and is acceptable (Pitt et al., 1995). The internal reliability of factors is perceived through the Cronbach constants, which come out to be 88.6%, 88.1%, 87.7%, and 86.1% (Table 3.1) and hence, is acceptable (Bagozzi and Yi, 1988; Nunnally, 1978).

3.5.4 CONFIRMATORY FACTOR ANALYSIS

Confirmatory factor analysis (CFA) method is used to authenticate the developed measure for evaluating the performance of procurement practices. As CFA analyzes the model fit, the CFA model is developed using SPSS AMOS v21 software for all the four factors with their respective items. The model fit has been studied for each factor with reference to the procedure recommended by Sureshchandar et al. (2002) and Bienstock et al. (1997), Table 3.2.

As shown above, all the goodness-of-fit index (GFI) values are above the threshold value, i.e., 0.9, which indicates the confirmation of the individual factor (Hair et al., 1998). The final CFA model comprising of the four factors and 20 sub-factors, or items, is represented in Figure 3.2.

3.5.4.1 Model Fit Measures

The GFI is measured by running the developed CFA model through AMOS v21 software. The Normed Chi-square value comes out to be 2.481, which implies a decent model fit. The standard value of Normed Chi-square value, as suggested by Bollen and Ting (1993); Tanaka (1987), ranges from 3 or even 5. Further, the GFI, CFI (comparative fit index), and NFI (normed fit index) value come out to be 0.879, 0.927, and 0.884, respectively. The Root Mean Square Estimation of Approximation (RMSEA) value of the model, i.e., 0.07, displays a logical model fit. Based on the CFI, NFI, GFI score, it is concluded that the developed model signifies an adequate fit for the problem undertaken.

3.5.4.2 Explanation of Factor Structure

The total variance offered by four factors was 69.793%. These outcomes indicate that these factors can illuminate the procurement issues considerably. The average score and standard deviation were calculated using IBM SPSS v22 software (Table 3.1). The extracted factors are labeled as follows.

TABLE 3.2
Key Fit Indices for CFA model of Procurement Practices

Factors	(χ^2)/df = Cmin/df	RMR	GFI	NFI	CFI	RMSEA
F1: Inventory management	1.921	0.046	0.980	0.985	0.993	0.059
F2: Quality management	1.196	0.029	0.993	0.994	0.999	0.027
F3: Supplier management	2.742	0.035	0.979	0.936	0.957	0.081
F4: Technological innovations	1.403	0.041	0.995	0.990	0.999	0.039

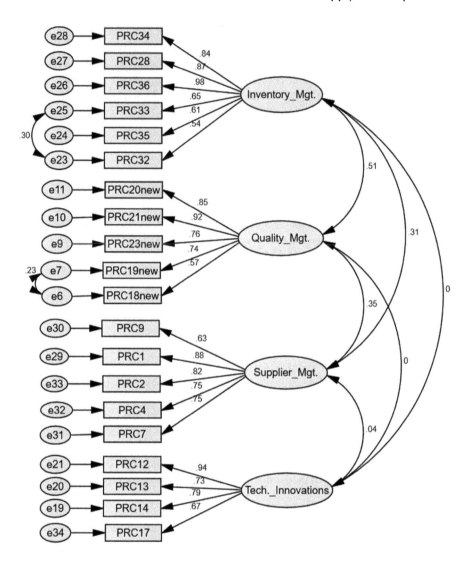

FIGURE 3.2   CFA model development for procurement practices.

The first factor labeled as *Inventory management* explains 20.969% variance. Six statements supporting this factor have the loadings in the range of 0.54–0.98 (Table 3.1). Based on the first factor, the following hypothesis has been established to measure its impact on the outcome of the SEM model:

H1: "Inventory management" has a positive impact on procurement performance of the dairy supply chain.

The second factor labeled as *Quality management* explains 17.810% variance. Five statements supporting this factor have the loadings in the range of 0.57–0.92

(Table 3.1). Based on the second factor, the following hypothesis has been established to measure its impact on the outcome of the SEM model:

H2: "Quality management" has a positive impact on procurement performance of the dairy supply chain.

The third factor labeled as *Supplier management* explains 17.977% variance. Five statements supporting this factor have the loadings in the range of 0.63–0.88 (Table 3.1). Based on the third factor, the following hypothesis has been established to measure its impact on the outcome of the SEM model:

H3: "Supplier management" has a positive impact on procurement performance of the dairy supply chain.

The fourth factor labeled as *Technological Innovations* explains 13.037% variance. Four statements supporting this factor have the loadings in the range of 0.67–0.94 (Table 3.1). Based on the fourth factor drawn, the following hypothesis has been established to measure its impact on the outcome of the SEM model:

H4: "Technological innovations" have a positive impact on procurement performance of the dairy supply chain.

3.5.5 VALIDITY OF CONSTRUCT

3.5.5.1 Face Validity

The face validity of the model is a subjective technique, and it is measured by considering the model *on-its-face*. The model illustrates a good image of the procurement issues (Trochim, 2007).

3.5.5.2 Content Validity

The content validity is reasonably measured through many discussions with the researchers and academicians in this area, the perceptions resulting from previous studies, and the scholar's individual understanding (Trochim, 2007). Consequent modification in the scale has been confirmed via group conversation with the dairy professionals. The instrument thus has strong content validity for procurement practices.

3.5.5.3 Construct Validity

Construct validity is calculated in three phases, as follows.

1. *Uni-dimensionality:* The value of CFI relates the developed structure with a null model supposing that there exist no associations among the measures. The developed CFA model represents the CFI value of 0.927, which entails a strong uni-dimensionality (Bollen and Ting, 1993; Byrne, 1994).
2. *Convergent Validity:* The convergent validity determines the degree to which dissimilar methodologies of developing a theory offers the similar output (Ahire et al., 1996). As recommended by Chin et al. (2003), a value of ≤ 0.5

exhibits a robust convergent validity. In this case, all the factor loading ranges from 0.54 to 0.94, and thus, the items have a strong convergent validity.

3. *Discriminant Validity:* The discriminant validity demonstrates the amount to which a model and its statements vary from a different (Bagozzi and Yi, 1988). Fornell and Larcker (1981) proposed that the square root of the mean ought to be superior to the total value of homogenous correlation with any other construct in a study. Table 3.3 denotes the square root of average variance explained (AVE) for respective factor as diagonal cell and correlation number as non-diagonal. The discriminant validity has been calculated by the *StatToolPackage* proposed by Prof. James Gaskin. In this chapter, the square root of AVE is more than the correlation number of the specific factor compared to other factors. Thus, the AVE values reinforced the discriminant validity in the current study.

4. *Predictive Validity:* The predictive validity is recognized if a measure exterior to the scale is associated with factor configuration (Nunnally, 1978). It was evaluated by recognizing the interrelationship of each factor with the average score of items using the Pearson correlation value. All the correlation values were found substantial (Table 3.4) in the current study.

TABLE 3.3

Results of Discriminant Validity for CFA Model of Procurement Practices

Inventory management	0.829			
Quality management	−0.287	**0.765**		
Supplier management	0.046	0.312	**0.771**	
Technological innovations	−0.256	0.505	0.348	**0.776**

Note: √AVE is represented as diagonal cells and correlation in other cells.

TABLE 3.4

Results of Correlation between Dimensions and Procurement Practices

Dimension	Correlation with Procurement Practices
Inventory management	0.776[a]
Quality management	0.744[a]
Supplier management	0.912[a]
Technological innovations	0.832[a]

[a] 0.05 level (2-tailed).

3.6   PROCUREMENT MODEL

After confirming the reliability and validity measures, a final model has been developed in order to evaluate the procurement practices of the dairy industry by linking the CFA model with the three-outcome variable. The outcome variables were considered with the consultation of experts from academics and dairy industry sector. Relevant information was collected from 265 participants, as a part of the questionnaire itself, for the outcome variables. The selected outcome variables for measuring the procurement performance are as follows.

- PRC_out1: Traceability in milk quality issues for procured milk
- PRC_out2: Overall quality level for procured milk (Reduction in total milk adulteration, wastages during transportation and rejections due to various procurement issues)
- PRC_out3: Supplier trust in the organization

The final procurement model consisting of the four factors (with 20 statements) and the three outcome variables defines the procurement performance in the dairy supply chain (Figure 3.3).

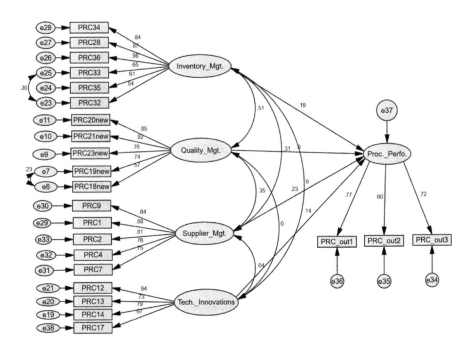

FIGURE 3.3   SEM model for measuring procurement performance.

3.6.1 Validity of Model

The validity of the model is assessed as follows.

Discriminant Validity of Model: The discriminant validity has been calculated by the *StatToolPackage* and the *AMOS Plugin* proposed by Prof. James Gaskin. All the respective values of Maximum Shared Variance (MSV) are lower than AVE, all the AVE values are higher than 0.5, and the critical ratio (CR) is higher than AVE. The square root of AVE for all factors comes out to be higher than the correlation constant of an individual factor with others. Thus, these values verified the discriminant validity of the model (Table 3.5). References to the thresholds proposed by Hu and Bentler (1999), all the values are within limits. Hence, there are no validity concerns here in the proposed procurement model.

3.6.2 Model Fit Measures

The model fit measure has been carried out as follows.

The model fit measures were calculated by the *StatToolPackage* proposed by Prof. James Gaskin. All the respective values come out to be as: Cmin/DF: 2.234, CFI: 0.923, SRMR: 0.073, RMSEA: 0.068, and PClose: 0.06. In reference to the thresholds given by Hu and Bentler (1999), all the respective values are within the specified limits (Table 3.6).

Hence, the developed SEM model is acceptable for measuring the performance of procurement practices in the dairy industry.

3.6.3 Hypotheses Testing

The prime aim of SEM is to check the validity of the hypothetical model by identifying, estimating, and assessing the linear interactions among observed and unobserved variables (Panuwatwanich et al., 2008). The hypotheses tested in SEM

TABLE 3.5

Results of Discriminant Validity for Measuring PP

	CR	AVE	MSV	MaxR(H)	Tech._Innovations	Inventory_Mgt.	Supplier_Mgt.	Quality_Mgt.
Tech._Innovations	0.812	0.503	0.081	0.917	**0.709**			
Inventory_Mgt.	0.864	0.507	0.255	0.978	−0.285	**0.712**		
Supplier_Mgt.	0.847	0.504	0.122	0.982	0.043	0.313	**0.710**	
Quality_Mgt.	0.829	0.503	0.255	0.985	−0.256	0.505	0.349	**0.709**

Source: Hu, L.T., and Bentler, P.M., *Struct. Equ. Model.: Multidiscip. J.*, 6, 1–55, 1999; Gaskin, J., and Lim, J., *Master Validity Tool*. AMOS Plugin, Gaskination's StatWiki, 2016. Available: http://statwiki.kolobkreations.com/.

TABLE 3.6
**Results of Model Fit Measure for Measuring
Procurement Performance (PP)**

Measure	Estimate	Threshold	Interpretation
Cmin	486.915	—	—
DF	218	—	—
Cmin/DF	2.234	Between 1 and 3	Excellent
CFI	0.923	>0.95	Acceptable
SRMR	0.073	<0.08	Excellent
RMSEA	0.068	<0.06	Acceptable
PClose	0.06	>0.05	Acceptable

Source: Hu, L.T., and Bentler, P.M., *Struct. Equ. Model.: Multidiscip. J.*,
6, 1–55, 1999; Gaskin, J., and Lim, J., *Master Validity Tool.*
AMOS Plugin, Gaskination's StatWiki, 2016. Available: http://
statwiki.kolobkreations.com/.

are often more positive in nature and considered as much more definitive than other correlational analyses methods (Collis and Rosenblood, 1985; Crosbie, 1986; Cudeck and O'Dell, 1994; Larzelere and Mulaik, 1977). Here, the predictors are the four factors drawn during EFA (Table 3.1) and the outcome indicates the *outcome variable* for the SEM model, and S.E. is the standard error. From the values of CR and p values given in Table 3.7, all the CRs are above 1.96 (95% confidence) and p values are <0.05 indicating a significance. Thus, all the four factors have positive impact on procurement performance of dairy supply chain and the hypotheses are accepted.

Further, to bring out the order of importance of four factors comprising various statements for measuring the procurement performance, the regression analysis was performed by considering the ratings of procurement practices as the dependent variable and the mean scores on the four factors as independent variables. The standardized coefficient beta (β) of the individual dimension represented their importance (Parasuraman et al., 1985, 1988) as presented in Table 3.7. The results clearly show the significance of the overall regression model ($p<0.05$) with 69.79% of the variance in procurement practices explained by the independent variables. The significant factors that remained in the equation in the procurement practices and are shown in order of their importance based on a standard estimate or β coefficient. Higher the standardized β coefficient, the more the factor contributes to explaining the dependent variable (Lee et al., 2000). The factor *quality management* emerges to be the most important dimension followed by the others as inventory management, supplier management, and the technological innovations.

TABLE 3.7
SEM Model for Measuring Procurement Performance

Sr. No.	Outcome	Predictor	Estimate	S.E.	C.R.	p-Value	Status	Order of Importance
1	Procurement performance	Inventory management	0.138	0.065	2.133	a	Significant	2
2		Quality management	0.123	0.072	3.321	a	Significant	1
3		Supplier management	0.188	0.068	2.760	a	Significant	3
4		Technological innovations	0.172	0.040	3.074	a	Significant	4

Note: Significance.
[a] $p < 0.05$.

3.6.4 SIGNIFICANCE OF RESULTS

The chapter has shown how the model for procurement practices was built and expressed its usefulness for managers in the dairy industry. The developed SEM model could be used by the researchers and dairy professionals/managers in several ways as follows.

1. The research can assist the scientists and dairy professionals in getting the issues related to procurement practices of the dairy industry.
2. The model offers four dimensions to measure the procurement performance in dairy industry, *viz.* inventory, quality, supplier, and the technology.
3. The scores on individual statements illustrate the proposals for improving the procurement practices.
4. The results of hypotheses testing suggest that all the four factors influence the procurement performance positively. The developed model may be used to benchmark the supply chain practices through the milk unions.

3.7 CONCLUSIONS

The current research is intended to fill the research gap for measuring the performance of procurement practices in Indian dairy industry. Additionally, the research proposes a framework for measuring the procurement performance of the dairy industry through SEM techniques. So, the research examines the procurement practices not only from an academic viewpoint but also from practical validation by developing a structural model in a real-time industrial scenario. The information, in the form of a pre-tested questionnaire, was collected from various dairy industries located in northern India. The 20 issues, called as statements, related with procurement practices were selected for the factor analysis. Factor analysis condensed the 20 issues into four factors *viz.* inventory management; quality management; supplier management; and technological innovations. The relative importance of the factors has also been carried out. Out of these four factors, *quality management* seems to be the most significant, followed by *inventory management, supplier management, and technological innovations.* The CFA model has been developed by interlinking the statements. A final procurement model has been developed after verifying the reliability and validity of the CFA model. The results of hypotheses testing suggest that all the four factors influence the procurement performance positively for the dairy industry. Finally, the SEM model has been developed by linking the CFA model with three outcome variables. The SEM model was checked for discriminant validity and model fit indices, and the authors found it acceptable for measuring the procurement performance of Indian dairy industry.

The results obtained in this chapter are in-line with previous studies in literature, such as Subburaj et al. (2015), Vinodh and Jay (2012c), Prakash and Pant (2013), Mor et al. (2018a, 2018b), Kumar et al. (2011), Singh and Javadekar (2011), and Kharub and Sharma (2018). For instance, the research conducted by Singh and Javadekar (2011) reported a loss of 72% of perishable food products due to non-usage of information technology by the retailers of perishable food products. Mor et al. (2015, 2018a, 2018b) found that the technological innovations and information technology lead to the enhanced supply chain performance of food processing industry. Further, Subburaj et al. (2015) focused on implementing the dynamic milk procurement methods and strengthening cooperative societies. Prakash and Pant (2013) recommended that the transparency and traceability would transform competence of the Indian dairy sector. Kumar et al. (2011) also focused on the traceability, food safety, and transportation facilities as critical issues for the dairy industry. Thus, the current study confirms the findings of previous research in this context.

The outcome of the current chapter depicts that the procurement performance of the dairy supply chain can be evaluated using the proposed factors and sub-factors. The procurement practices in the dairy supply chain are highly affected by the abnormal wastage and poor handling at the source and site. More wastages occur because of multiple points of handling and unhygienic practices. These challenges further compel dairy industries for significant development in their procurement system. The complexity of Indian co-op, the dairy sector with the milk unions at farmer's level also requires advanced methods for effective procurement management. Moreover, the technological innovations, automatic milk quality testing technology, effective BMC at milk collection centres, traceability in poor quality issues, and the execution of effective information systems can help the dairy sector to accomplish its long-term corporate goals. Thus, the proposed framework will support the dairy professionals to plan their procurement practices competently to fulfill the product quality, supply chain sustainability, and food safety and security standards of the global marketplace.

3.7.1 Limitations and Future Scope

The interpretations of current research rely on the cooperation of academicians and managers from the dairy industry. The proposed model can be applied to other milk procuring and processing firms across different states and countries. Some cause and effect relationships can also be conducted. Further, this procedure may be executed to other perishable food processing industries like meat, bakery, poultry, fishery, and so forth.

REFERENCES

Ahire, S. L., Golhar, D. Y., & Waller, M. A. (1996). Development and validation of TQM implementation constructs. *Decision Sciences, 27*(1), 23–56.

Bagozzi, R. P., & Yi, Y. (1988). On the evaluation of structural equation models. *Journal of the Academy of Marketing Science, 16*(1), 74–94.

Bamgboje-Ayodele, A., Ellis, L., & Turner, P. (2014, October). Identifying key research challenges in investigating knowledge optimization strategies in perishable food chains. *Proceedings of the 11th International Conference on Intellectual Capital, Knowledge Management and Organisational Learning*, Sydney, Australia, pp. 48–56.

Bhardwaj, A., Mor, R. S., Singh, S., & Dev, M. (2016). An investigation into the dynamics of supply chain practices in Dairy industry: A pilot study. *Proceedings of the 2016 International Conference on Industrial Engineering and Operations Management*, Detroit, MI, pp. 1360–1365.

Bienstock, C. C., Mentzer, J. T., & Bird, M. M. (1997). Measuring physical distribution service quality. *Journal of the Academy of Marketing Science, 25*(1), 31–44.

Bollen, K. A., & Ting, K. F. (1993). Confirmatory tetrad analysis. *Sociological Methodology, 23*, 147–175.

Byrne, B. M. (1994). *Structural Equation Modeling with EQS and EQS/Windows: Basic Concepts, Applications, and Programming*. Sage, Thousand Oaks, CA.

Chen, C., Zhang, J., & Delaurentis, T. (2014). Quality control in food supply chain management: An analytical model and case study of the adulterated milk incident in China. *International Journal of Production Economics, 152*, 188–199.

Chin, W. W., Marcolin, B. L., & Newsted, P. R. (2003). A partial least squares latent variable modeling approach for measuring interaction effects: Results from a Monte Carlo simulation study and an electronic-mail emotion/adoption study. *Information Systems Research, 14*(2), 189–217.

Collis, B. A., & Rosenblood, L. K. (1985). The problem of inflated significance when testing individual correlations from a correlation matrix. *Journal for Research in Mathematics Education, 16*(1), 52–55.

Cronbach, L. J. (1951). Coefficient alpha and the internal structure of tests. *Psychometrika, 16*(3), 297–334.

Cronin, J. J. Jr., & Taylor, S. A. (1992). Measuring service quality: A reexamination and extension. *The Journal of Marketing, 56*, 55–68.

Crosbie, J. (1986). A Pascal program to perform the Bonferroni multistage multiple-correlation procedure. *Behavior Research Methods, 18*(3), 327–329.

Cudeck, R., & O'dell, L. L. (1994). Applications of standard error estimates in unrestricted factor analysis: Significance tests for factor loadings and correlations. *Psychological Bulletin, 115*(3), 475–487.

Eid, R. (2009). Factors affecting the success of world class manufacturing implementation in less developed countries: The case of Egypt. Journal of Manufacturing Technology Management, 20(7), 989–1008.

Flynn, B. B., Sakakibara, S., Schroeder, R. G., Bates, K. A., & Flynn, J. B. (1990). Empirical research methods in operations management. Journal of Operations Management, 9(2), 250–284.

Fornell, C., & Larcker, D. F. (1981). Evaluating structural equation models with unobservable variables and measurement error. *Journal of Marketing Research*, *18*(1), 39–50.

Forza, C. (2002). Survey research in operations management: A process-based perspective. International Journal of Operations & Production Management, 22(2), 152–194.

García-Arca, J., González-Portela, A. T., & Prado-Prado, J. C. (2014). Packaging as source of efficient and sustainable advantages in supply chain management: An analysis of briks. International Journal of Production Management and Engineering, 2(1), 15–22.

Gaskin, J., & Lim, J. (2016). *Master Validity Tool, AMOS Plugin*, Gaskination's StatWiki. Available: http://statwiki.kolobkreations.com/.

Ghosh, A., Sindhu, S., Panghal, A., & Bhayana, S. (2014). Modelling the enablers for risk management in milk processing industry. *International Journal of Management and International Business Studies*, 4(1), 9–16.

Glover, J. L., Champion, D., Daniels, K. J., & Dainty, A. J. D. (2014). An Institutional Theory perspective on sustainable practices across the dairy supply chain. *International Journal of Production Economics*, *152*, 102–111.

Goldstein, H. (1986). Multilevel mixed linear model analysis using iterative generalized least squares. Biometrika, 73(1), 43–56.

Gupta, K., & Roy, D. (2012). Gains from coordination in milkfed dairy in Punjab. Journal of Agribusiness in Developing and Emerging Economies, 2(2), 92–114.

Hair, J. F., Jr., Anderson, R. E., Tatham, R. L., & Black, W. C. (1998). *Multivariate Data Analysis*. 5th ed. Prentice-Hall, Upper Saddle River, NJ.

Hair, J. F. Jr., Anderson, R. E., Tatham, R. L., & Black, W. C. (2005). *Multivariate Data Analysis*. 5th ed. Pearson Education, New Delhi, India.

Hou, D., Al-Tabbaa, A., Chen, H., & Mamic, I. (2014). Factor analysis and structural equation modelling of sustainable behaviour in contaminated land remediation. *Journal of Cleaner Production*, *84*, 439–449.

Hoyle, R. H. (2011). *Structural Equation Modeling for Social and Personality Psychology*. Sage Publications, Thousand Oaks, CA.

Hu, L. T., & Bentler, P. M. (1999). Cutoff criteria for fit indexes in covariance structure analysis: Conventional criteria versus new alternatives. *Structural Equation Modeling: A Multidisciplinary Journal*, 6(1), 1–55.

Hussey, D. M., & Eagan, P. D. (2007). Using structural equation modeling to test environmental performance in small and medium-sized manufacturers: Can SEM help SMEs? *Journal of Cleaner Production*, *15*(4), 303–312.

Kharub, M., Mor, R. S., & Sharma, R. K. (2018). The relationship between cost leadership strategy and firm performance: A mediating role of quality management. *Journal of Manufacturing Technology Management*, in press.

Kharub, M., & Sharma, R. (2018). An integrated structural model of QMPs, QMS and firm's performance for competitive positioning in MSMEs. *Total Quality Management & Business Excellence*, 1–30. doi:10.1080/14783363.2018.1427500.

Kline, R. B. (2010). *Principles and Practice of Structural Equation Modeling*. Guilford Press, New York.

Kumar, A., Staal, S. J., & Singh, D. K. (2011). Smallholder dairy farmers' access to modern milk marketing chains in India. *Agricultural Economics Research Review*, 24, 243–253.

Kumar, R. (2014). Performance measurement in dairy Supply chain management. *Indian Journal of Research*, 3(3), 100–101.

Larzelere, R. E., & Mulaik, S. A. (1977). Single-sample tests for many correlations. *Psychological Bulletin*, *84*(3), 557–569.

Lee, H., Lee, Y., & Yoo, D. (2000). The determinants of perceived service quality and its relationship with satisfaction. *Journal of Services Marketing, 14*(3), 217–231.

Lemma, H. R., Singh, R., & Kaur, N. (2015). Determinants of supply chain coordination of milk and dairy industries in Ethiopia: A case of Addis Ababa and its surroundings. *SpringerPlus, 4*(1), 1–12.

Mangla, S. K., Sharma, Y. K., & Patil, P. P. (2016). Using AHP to rank the critical success factors in food supply chain management. *International Conference on Smart Strategies for Digital World-Industrial Engineering Perspective*, Vol. 58, Nagpur, India.

Mor, R. S., Bhardwaj, A., & Singh, S. (2018a). Benchmarking the interactions among Performance Indicators in dairy supply chain: An ISM approach. *Benchmarking: An International Journal*, 25(9), 3858–3881.

Mor, R. S., Bhardwaj, A., & Singh, S. (2018b). A structured-literature-review of the supply chain practices in dairy industry. *Journal of Operations and Supply Chain Management, 11*(1), 14–25.

Mor, R. S., Bhardwaj, A., & Singh, S. (2018c). Benchmarking the interactions among barriers in Dairy supply chain: An ISM approach. *International Journal for Quality Research, 12*(2), 385–404.

Mor, R. S., Bhardwaj, A., & Singh, S. (2018d). A structured literature review of the supply chain practices in food processing industry. *Proceedings of the 2018 International Conference on Industrial Engineering & Operations Management*. Bandung, Indonesia, March 6–8, pp. 588–599.

Mor, R. S., Jaiswal, S. K., Singh, S., & Bhardwaj, A. (2018e). Demand forecasting of short-lifecycle Dairy products. *Understanding the Role of Business Analytics*. doi:10.1007/978-981-13-1334-9_6.

Mor, R. S., Singh, S., & Bhardwaj, A. (2016). Learning on lean production: A review of opinion and research within environmental constraints. *Operations and Supply Chain Management: An International Journal, 9*(1), 61–72.

Mor, R. S., Singh, S., Bhardwaj, A., & Bharti, S. (2017). Exploring the causes of low-productivity in Dairy industry using AHP. *Jurnal Teknik Industri, 19*(2), 83–92.

Mor, R. S., Singh, S., Bhardwaj, A., & Singh, L. P. (2015). Technological implications of supply chain practices in agri-food sector: A review. *International Journal of Supply and Operations Management, 2*(2), 720–747.

Morgan, D. L. (1993). Qualitative content analysis: A guide to paths not taken. *Qualitative Health Research, 3*(1), 112–121.

Nargundkar, R. (2004). *Marketing Research: Test and Cases*, 2nd ed., Tata McGraw-Hill, New Delhi, India.

Nunnally, J. C. (1978). *Psychometric Theory*. 2nd ed. McGraw-Hill, New York.

Okano, M. T., Vendrametto, O., & Santos, O. S. (2014). How to improve dairy production in Brazil through indicators for economic development of milk chain. *Modern Economy, 5*(6), 663–669.

Panuwatwanich, K., Stewart, R. A., & Mohamed, S. (2008). The role of climate for innovation in enhancing business performance: The case of design firms. *Engineering, Construction and Architectural Management, 15*(5), 407–422.

Parasuraman, A., Zeithaml, V. A., & Berry, L. L. (1985). A conceptual model of service quality and its implications for future research. *Journal of Marketing, 49*(4), 41–50.

Parasuraman, A., Zeithaml, V. A., & Berry, L. L. (1988). SERVQUAL: A multiple-item scale for measuring consumer perception. *Journal of Retailing, 64*(1), 12–37.

Pitt, L. F., Watson, R. T., & Kavan, C. B. (1995). Service quality: A measure of information systems effectiveness. *MIS Quarterly, 19*, 173–187.

Prakash, G., & Pant, R. R. (2013). Performance measurement of a dairy supply chain: A balance scorecard perspective. *Proceedings of the 2013 IEEE-IEEM*, pp. 196–200.

Punjabi, M. (2009). India: Increasing demand challenges the dairy sector, Smallholder dairy development: Lessons learned in Asia. *RAP Publication 2009/02*, assessed July 15, 2017, http://www.fao.org/docrep/011/i0588e/I0588E05.htm.

Roberts, N., Thatcher, J. B., & Grover, V. (2010). Advancing operations management theory using exploratory structural equation modelling techniques. *International Journal of Production Research*, *48*(15), 4329–4353.

Robson, C. (2002). *Real World Research*. 2nd ed. Blackwell, Oxford, UK.

Saunders, M., Lewis, P., & Thornhill, A. (2009). *Research Methods for Business Students*. 5th ed. Person Education, Essex, UK.

Singh, N., & Javadekar, P. (2011). Supply chain management of perishable food products: A strategy to achieve competitive advantage through knowledge management. *Indian Journal of Marketing*, *41*(10), 32–44.

Sitek, P., & Wikarek, J. (2015). A hybrid framework for the modelling and optimization of decision problems in sustainable supply chain management. *International Journal of Production Research*, *53*(21), 6611–6628.

Subburaj, M., Babu, T. R., & Subramonian, B. S. (2015). A study on strengthening the operational efficiency of dairy supply Chain in Tamilnadu, India. *Procedia-Social and Behavioral Sciences*, *189*, 285–291.

Sureshchandar, G. S., Rajendran, C., & Anantharaman, R. N. (2002). Determinants of customer-perceived service quality: A confirmatory factor analysis approach. *Journal of Services Marketing*, *16*(1), 9–34.

Tanaka, J. S. (1987). "How big is big enough?" Sample size and goodness of fit in structural equation models with latent variables. *Child Development*, *58*(1), 134–146.

Tenenhaus, M., Vinzi, V. E., Chatelin, Y. M., & Lauro, C. (2005). PLS path modeling. *Computational Statistics & Data Analysis*, *48*(1), 159–205.

Thirupathi, R. M., & Vinodh, S. (2016). Application of interpretive structural modelling and structural equation modelling for analysis of sustainable manufacturing factors in Indian automotive component sector. *International Journal of Production Research*, *54*(22), 6661–6682.

Trochim, W. M. (2007). *Research Methods*. Biztantra, New Delhi, India.

Vinodh, S., & Joy, D. (2012a). Structural equation modelling of lean manufacturing practices. *International Journal of Production Research*, *50*(6), 1598–1607.

Vinodh, S., & Joy, D. (2012c). Structural equation modeling of sustainable manufacturing practices. *Clean Technologies and Environmental Policy*, *14*(1), 79–84.

Vinodh, S., Aravindraj, S., Pushkar, B., & Kishore, S. (2012b). Estimation of reliability and validity of agility constructs using structural equation modelling. *International Journal of Production Research*, *50*(23), 6737–6745.

Xiong, B., Skitmore, M., & Xia, B. (2015). A critical review of structural equation modeling applications in construction research. *Automation in Construction*, *49*, 59–70.

4 Modeling the Sustainable Procurement Initiatives Using Fuzzy-Based MCDM Framework

Chandra Prakash Garg

CONTENTS

4.1 INTRODUCTION

In today's business environment, government legislation, customer pressure, carbon emission, and green issues force the focal firms to incorporate sustainable business practices into their supply chain operations irrespective of manufacturing and service industries (Garg et al., 2017).

Earlier supply chain operations have focused on optimization of resources and achieving effciency and effectivenss in Supply Chain (SC) operations for focal firms (Gunasekaran et al., 2001). That has enlarged the problem of available scarce resources, carbon emission, waste management, and ecological balance (Kumar & Garg, 2017). The evidence of this can be understood by the fact that more than 20% of global carbon emissions are generated through supply chain operations of top the 2500 global firms (Dubey et al., 2017). Consumers' and external stakeholders' considerations towards green and social concern, competitive business dynamics, and enforced legislation, all play a central role in the way businesses are created and managed in the global market (Govindan & Popiuc, 2014, Luthra et al., 2017). It is, therefore, imperative to encourage sustainability philosophy in SC activities for an organization. To start with SC process, sustainable procurement is one of the grave decisions encountered by supply chain managers and experts in order to incorporate sustainable philosophy into it as well as assist organizations to improve their competitive positions (Govindan et al., 2013, Luthra et al., 2017). The sustainable procurement directly affects sustainability implementation in downstream, upstream, and reverse supply chain operations. Due to firms' dependency on resources (physical as well as intangible), organizations used to outsource the procurement activities to suppliers. Therefore, these suppliers play a crucial role in supply chain, and their selection directly affects implementation of sustainable initiatives. Such outsourcing suppliers/partners could be beneficial to focal firms to achieve the goal of sustainable initiatives throughout the supply chain activities.

Sustainable procurement (SP) is the initial step towards implementing sustainability into supply chain and would depend on many attributes apart from capability, competency, and ability of suppliers (Luthra et al., 2018). Focal firms need to asses these attributes in order to achieve sustainable procurement into supply chain activities.

The Indian electronics sector is the fastest growing market. This sector is contributing highly in overall production and consumption. High GDP growth, the presence of large middle-class population, a rising younger generation, and increasing disposable income has derived the high growth of consumer electronics items in India. Moreover, government initiatives such as MAKE IN INDIA, an e-governance scheme, promote an increase in the production of electronics items/products and gadgets in India and sell them locally and globally to increase global market share. Sustainability in supply chain initiates with sustainable procurement decision that affect later stages of the supply chain process; in this line sustainable procurement initiatives (SPIs) are very important for Indian electronics firm so negative impact of such products on environment and society can be minimized apart from economic benefits. Thus, it is evident that there is a need to evaluate to these initiatives.

The remaining part of this study is structured as follows. The background of the study is discussed in Section 4.2. Section 4.3 represents the problem definition. The Fuzzy Analytic Hierarchy Process (FAHP) model is presented in Section 4.4. The results and discussion sections are given in Section 4.5. The conclusion is reported in Section 4.6.

4.2   LITERATURE REVIEW

Recently, researchers, and practitioners have understood, identified, and addressed the sustainability requirements in supply chain operations (Carter & Rogers, 2008, Gobbo et al., 2014, Govindan et al., 2016, Luthra & Haleem, 2015, Prakash & Barua, 2015, 2016a). Firms have dependency for resources (physical as well as intangible) to the external environment; therefore, they outsource their major supply chain activities to outsourcing partners/suppliers (Luthra et al., 2017). Procurement is the first activity that affects the later stages of the supply chain process (Prakash & Barua, 2016c). Sustainability in supply chain operations can be achieved if all the members or partners of the supply chain work together and share their responsibilities, rewards, and risks towards implementation sustainable initiatives in all the supply chain operations (Govindan et al., 2013). Many researchers have suggested that sustainable procurement is the first step towards making supply chain sustainable (Büyüközkan & Çifçi, 2011,

Sengar et al., 2017, Trapp & Sarkis, 2016). Companies need to implement sustainable procurement initiatives through their suppliers (Luthra et al., 2017). In addition to this, partners need to be educated, trained, and guided to implement sustainable procurement initiatives, environmental management systems, and green and technological innovations in their operations (Luthra et al., 2018). Govindan et al. (2013) observed that a sustainable outsourcing partner (SOP) can implement green innovations in their operations that would support focal firms to attain sustainable supply chain. Bai & Sarkis (2010) analyzed that a supplier must have environmental certification and the ability to invest in renewable energy initiatives and research for better and more sustainable products. Büyüközkan and Çifçi (2011) studied that the partners must provide green items, which can be reused, recycled, and disposed in an environment-friendly manner and utilize non-hazardous materials. Prakash and Barua (2017a, 2017b) analyzed that partners must able to provide efficient service and delivery and maintain flexibility in requirements of products, components, and parts. Luthra et al. (2017) suggested that partners must provide economic, social, and environmental sustainability in the supply chain operations. Gold et al. (2010) suggested that there is a lack of empirical-based studies, especially in electronics industry, which have suggested SPIs implementation. Hence, there is a need of such studies which can support policy makers and experts to assist in further decision-making.

4.2.1 RESEARCH GAPS

Sustainable procurement initiatives implementation decisions are crucial because they affect the effectiveness and responsiveness of supply chain activities. SPI implementation is not only addressing economic goals of the firm, but also incorporates social and ecological concerns. Therefore, focal firms have to choose and evaluate their partners' not only considering traditional selection criteria, but also need to add more SPIs implementation-based criteria to achieve success (Azadi et al., 2015). Based on past literature, this study tries to explore these research gaps. There is a dearth of qualitative papers from growing nations, particularly from India, related to SPIs evaluation and assessment. The SPIs may vary from country to country, region to region, and firm to firm because of different market dynamics, varied nature of resources, competencies, and capabilities of the firms (Luthra et al., 2017, Prakash & Barua, 2016a). The Indian electronics sector is the fastest growing market. This sector is contributing highly in overall production and consumption. In this line, SPIs are very important for Indian electronics firms so negative impact of such products on environment and society can be minimized apart from economic benefits. Thus, it is evident that there is a lack of studies in the spectrum of SPIs for Indian electronics industry.

4.3 PROBLEM DEFINITION

The serious problem that the Indian electronic industry is facing is implementation of sustainable initiatives into supply chain operations so ecological impact can be minimized. Rapidly changing technologies and advancements, which have increased obsolescence rate and decreased the life cycle of the many electronic products,

TABLE 4.1
Recognized SPIs for Electronic Firms

S. No.	Sustainable Procurement Initiatives (SPIs)	Code
1	Partners must provide green items which can be reuse, recycle and disposed in environment friendly manner and utilized non-hazardous materials	SPI1
2	Capability of suppliers to provide cost efficient, quality and green materials, components and parts	SPI2
3	Partners must able to provide efficient service and delivery and maintain flexibility in requirements of products, components and parts	SPI3
4	Partners must able to implement environment management initiatives, technological innovations, carbon emission reduction strategies and green operations	SPI4
5	Partners must have environmental certification and ability to invest in renewable energy initiatives and research for better and more sustainable products	SPI5
6	Partners must able to maintain good collaboration and relationship with social and environmental groups along with inter and intra organizational collaborations.	SPI6
7	Partners must able to manage social welfare and values and provide supports for community development	SPI7
8	Collaboration and selection of supplier/outsourcing partner considering economic, social and environmental sustainability dimensions	SPI8

Note: Combined results of the studies of Luthra et al. (2017), Luthra et al. (2015), Büyüközkan & Çifçi (2011), Gold et al. (2010), Luthra et al. (2018), Bai & Sarkis (2010), Prakash & Barua (2016), Govindan et al. (2013), & Decision-making team.

has created e-waste problems (Prakash & Barua, 2016b). The increasing substantial amount of e-waste is a major problem in India. E-waste contains a variety of materials including hazardous substances such as PBR, lead, mercury, chromium, cadmium, etc., which are extremely dangerous to human health and environment. This issue can be minimized at an early stage of the supply chain process, if organizations can implement sustainable initiatives in procurement process. To do this, various SPIs are needed to be identified and assessed. Identified SPIs are highlighted in Table 4.1. In this work, the evaluation of various SPIs is done, which is considered as a multiple attribute for decision-making problems.

4.4 SOLUTION METHODOLOGY

This study recognized eight SPIs for electronic firms (See Table 4.1). The diagram of the proposed research is highlighted through Figure 4.1, and the process of FAHP has been illustrated in the subsequent part.

4.4.1 FUZZY AHP

The analytic hierarchy process (AHP) model is pioneered by Saaty (1980). It is a quantitative approach of the multi-criteria decision-making (MCDM) model.

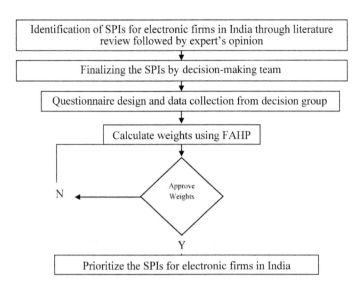

FIGURE 4.1   Proposed FAHP methodology for SPIs assessment.

AHP application has few drawbacks due to its uses in imprecise environment, varied assessment rating, and its subjectivity (Mangla et al., 2014, 2015, Prakash et al., 2015, Prakash & Barua, 2015, 2016b, 2016c). This necessitates the introduction of fuzzy theory to overcome these issues (Garg, 2016, Garg et al., 2017, Gupta et al., 2017, Vishwakarma et al., 2016, 2017). This fuzzy-based AHP approach reduces by giving freedom in the assessment scale to decision makers (Amrita et al., 2018, Mahtani & Garg, 2018, Sengar et al., 2017). This would be done by using triangular fuzzy numbers (TFNs) as highlighted in Table 4.2.

Chang's extent analysis (1992) is the FAHP process, according to this approach; the values of the extent method for each criterion g_i are obtained by using the following notation.

TABLE 4.2
TFN Matrix

Assessment Rating	Corresponding TFN
Likewise	(1, 1, 1)
Very poor	(1, 2, 3)
Poor	(2, 3, 4)
Average	(3, 4, 5)
High	(4, 5, 6)
Very high	(5, 6, 7)
Excellent	(7, 8, 9)

Step 1: The fuzzy synthetic extent value (S_i) with respect to the ith criterion is defined as,

$$S_i = \sum_{j=1}^{m} M_{gi}^{j} \times \left[\sum_{i=1}^{n} \sum_{j=1}^{m} M_{gi}^{j} \right]^{-1} \tag{4.1}$$

where l denotes lower limit, m is the average, and u is the higher limit.

Step 2: The value of possible degree is determined by

$$P_2 = (l_2, m_2, u_2) \geq P_1 = (l_1, m_1, u_1) \text{ is calculated as}$$

$$V\ (P_2 \geq P_1) = {}_{y \geq x}^{sup}[\min(\mu_{P_1}(x), \mu_{P_2}(y)]$$

and x and y are the membership values of each attributes.

Step 3: The value of possible degree for P with respect to higher k TFN P_i $(i = 1,2,.....,k)$ is calculated by

$$V\ (P \geq P_1, P_2, ..., P_k)$$

$$= V\ [(P \geq P_1) \text{ and } (P \geq P_2) \text{ and}.....\text{and } (P \geq P_k)]$$

$$= \min V(P \geq P_i),\ i = 1,2,.....,k \tag{4.2}$$

$$\text{Suppose that } d'(A_i) = \min\ V(P_i \geq P_k)$$

For $k = 1, 2,.....,n$, $k \neq i$, and weight values are calculated by Eq. 4.3 as,

$$W' = (d'(A_1), d'(A_2),.....d'(A_m))^T \tag{4.3}$$

Step 4: Normalized weight values are obtained by using Eq. 4.4 as,

$$W = (d\ (A_1), d\ (A_2),.....d\ (A_m))^T \tag{4.4}$$

4.4.2 CALCULATION OF WEIGHTED VALUES

The expert team consists of four domain specialists, as discussed above, to assess SPIs for the electronics industry. These selected domain specialists have more than 10 years of working experience. The group of experts consists of four people: one industrial specialist dedicated to working on effective adoption and implementation of sustainable development initiatives in electronics supply chain of the firm, one SC consultants, one representative from the Indian Ministry of Environment Forest and Climate Change (MoEF & CC) involved in e-waste regulation issues, and one academician actively indulged in research in the area of supply chain, e-waste management, and operations management issues. The panel of experts has finalized eight SPIs in electronics supply chain in the India. Overall, all the specialists are agreed with the given list and confirmed to not include any other additional initiatives.

The assessment ratings for SPIs are given as per rating given in Table 4.3. Hereafter, weight values of each SPI are calculated using above discussed steps.

TABLE 4.3
Assessment Rating of SPIs

	SPI1	SPI2	SPI3	SPI4	SPI5	SPI6	SPI7	SPI8
SPI1	(1, 1, 1)	(0.25, 0.33, 0.5)	(3, 4, 5)	(0.33, 0.5, 1)	(3, 4, 5)	(0.25, 0.33, 0.5)	(1, 2, 3)	(0.2, 0.25, 0.33)
SPI2	(2, 3, 4)	(1, 1, 1)	(0.33, 0.5, 1)	(0.2, 0.25, 0.33)	(0.33, 0.5, 1)	(3, 4, 5)	(0.25, 0.33, 0.5)	(0.25, 0.33, 0.5)
SPI3	(0.2, 0.25, 0.33)	(1, 2, 3)	(1, 1, 1)	(3, 4, 5)	(0.25, 0.33, 0.5)	(0.25, 0.33, 0.5)	(3, 4, 5)	(2, 3, 4)
SPI4	(1, 2, 3)	(3, 4, 5)	(0.2, 0.25, 0.33)	(1, 1, 1)	(1, 2, 3)	(0.2, 0.25, 0.33)	(0.25, 0.33, 0.5)	(2, 3, 4)
SPI5	(0.2, 0.25, 0.33)	(1, 2, 3)	(2, 3, 4)	(0.33, 0.5, 1)	(1, 1, 1)	(3, 4, 5)	(0.25, 0.33, 0.5)	(0.2, 0.25, 0.33)
SPI6	(2, 3, 4)	(0.2, 0.25, 0.33)	(2, 3, 4)	(3, 4, 5)	(0.2, 0.25, 0.33)	(1, 1, 1)	(0.33, 0.5, 1)	(0.25, 0.33, 0.5)
SPI7	(0.33, 0.5, 1)	(2, 3, 4)	(0.2, 0.25, 0.33)	(2, 3, 4)	(2, 3, 4)	(1, 2, 3)	(1, 1, 1)	(0.25, 0.33, 0.5)
SPI8	(3, 4, 5)	(2, 3, 4)	(0.25, 0.33, 0.5)	(0.25, 0.33, 0.5)	(3, 4, 5)	(2, 3, 4)	(2, 3, 4)	(1, 1, 1)

The fuzzy synthetic extent value of first SPI is calculated by using Eq. (4.1).

$$S\ (\text{SPI1}) = (9.03, 12.42, 16.33) \otimes [52.72, 73.75, 97.5]^{-1}$$

$$= (0.093, 0.168, 0.309)$$

Similar steps have been undertaken to determine fuzzy synthetic extent values of other SPIs. Minimum possible degree and V values are obtained through Eqs. 4.2 and 4.3, respectively.

$$m(\text{SPI1}) = \min V(P_1 \geq P_k) = 0.6691$$

and other values are $m(\text{SPI2}) = 0.6453$, $m(\text{SPI3}) = 0.8178$, $m(\text{SPI4}) = 0.7029$, $m(\text{SPI5}) = 0.6001$, $m(\text{SPI6}) = 0.6620$, $m(\text{SPI7}) = 0.7252$ and $m(\text{SPI8}) = 1$
Obtained weight values are

$$W_V = (0.669, 0.645, 0.817, 0.703, 0.600, 0.662, 0.725, 1)^T$$

Normalized weights are obtained

$$W = (0.114, 0.110, 0.140, 0.120, 0.103, 0.113, 0.124, 0.171)$$

Final weights of SPIs and their ranking are reported in Table 4.4.

4.5 ANALYSIS OF RESULTS AND FINDINGS

The priority rating of the SPIs has been calculated on the basis of higher weightage value, which implies that SPI8 > SPI3 > SPI7 > SPI4 > SPI1 > SPI6 > SPI2 > SPI5 are in decreasing orders (see Table 4.4). It indicates the following: SPI8, i.e., Collaboration and selection of supplier/outsourcing partner considering economic, social and environmental sustainability dimensions obtained first position as SPI with maximum weights (0.1717); SPI3, i.e., Partners must able to provide efficient service and delivery and maintain flexibility in requirements of products, components and parts received second priority rank with weights (0.1405); and SPI7, i.e., Partners must able to manage social welfare and values and provide supports for community development gained third priority rank with weights (0.1246). In the priority list, the last rank is received by SPI5, i.e., partners must have environmental certification and ability to invest in renewable energy initiatives and research for better and more sustainable products with weights (0.1031). The remaining SPIs priority ranks are given in Table 4.4. This model enables decision makers to evaluate which SPIs are influential in the implementation sustainable procurement in electronic firms in India.

The findings of the analysis indicate that collaboration and selection of supplier/outsourcing partner consider economic, social, and environmental sustainability dimensions are the most influential SPI among all the SPIs. It suggests that the focal firm must choose their supplier based on economic, social, and environmental initiatives competencies. It provides economic and competitive price, maintains social concern, and adheres to environmental obligations (Luthra et al., 2017). Suppliers

TABLE 4.4
Ranking of SPIs

S. No.	SPIs	Code	Final Weights	Rank
1	Partners must provide green items which can be reuse, recycle and disposed in environment friendly manner and utilized non-hazardous materials	SPI1	0.1149	5
2	Capability of suppliers to provide cost efficient, quality and green materials, components and parts	SPI2	0.1108	7
3	Partners must able to provide efficient service and delivery and maintain flexibility in requirements of products, components and parts	SPI3	0.1405	2
4	Partners must able to implement environment management initiatives, technological innovations, carbon emission reduction strategies and green operations	SPI4	0.1207	4
5	Partners must have environmental certification and ability to invest in renewable energy initiatives and research for better and more sustainable products	SPI5	0.1031	8
6	Partners must able to maintain good collaboration and relationship with social and environmental groups along with inter and intra organizational collaborations.	SPI6	0.1137	6
7	Partners must able to manage social welfare and values and provide supports for community development	SPI7	0.1246	3
8	Collaboration and selection of supplier/outsourcing partner considering economic, social and environmental sustainability dimensions	SPI8	0.1717	1

must be able to provide fast and flexible service as and when focal firms need. Since electronic products are highly technical and advances in nature, suppliers must able to maintain efficient service delivery to get competitive advantage (Prakash & Barua, 2016). Due to communal and NGOs pressures, focal firms are joining their hands with such suppliers who are proactively working for social sustainability (Kumar & Garg, 2017). Therefore, apart from efficient service, suppliers of focal firms must also concentrate on social development and focus on social welfare. Owing to government regulations, e-waste management, and handling rules, manufacturers of electronics are under tremendous pressure from MoEF & CC to implement environmentally friendly or green products, which have minimum environmental impact on society and human health (Prakash & Barua, 2016). To achieve this, currently, electronic firms are selecting suppliers who maintain sustainability at procurement phase or early stages of supply chain process. Firms prefer those suppliers who are offering sustainable procurement initiatives (SPIs), so later on firms can achieve the goal of sustainable business practices into supply chain process. There are many sustainable procurement initiatives (SPIs); focal firms can implement them on priority basis as obtained in the analysis.

4.6 CONCLUSION

Owing to environmental degradation, government and communal pressures, and e-waste issues, the electronic firms of India are aiming to implement sustainable practices into their supply chain operations. These practices can be implemented at the procurement phase, as it is the early stage of the supply chain. These sustainable procurement initiatives (SPIs) are adopted by firms through suppliers/partners. In this work, many SPIs that can be implemented by suppliers are recognized and prioritized in order to understand their importance. The FAHP model is used to assess these SPIs. The results of this analysis indicates that collaboration and selection of supplier/outsourcing partner considering economic, social and environmental sustainability dimensions obtained first position in the priority list. The priority lists of other SPIs are demonstrated in Table 4.4.

Firms must work on implementing these sustainable initiatives in terms of choosing partners so their suppliers can redesign the products with minimum carbon footprint impact and cost to customers. Businesses must participate proactively in addressing their e-waste and returns effectively and create and raise awareness to reduce waste that will lead to more economic value. This work is extremely important for electronics firms who are concerned with the adoption of SPIs by analyzing the criticality of the main factors through the suggested model.

The proposed approach considered impreciseness of the decision-making process. It indicates that the used method is significant decision-making tool. The extension of this work can be done by incorporating more numbers of SPIs. Many other MCDM approaches can be used to assess these SPIs.

REFERENCES

Amrita, K., Garg, C. P., & Singh, S. (2018). Modelling the critical success factors of women entrepreneurship using Fuzzy AHP framework. *Journal of Entrepreneurship in Emerging Economies, 10*(1), 81–116.
Azadi, M., Jafarian, M., Saen, R. F., & Mirhedayatian, S. M. (2015). A new fuzzy DEA model for evaluation of efficiency and effectiveness of suppliers in sustainable supply chain management context. *Computers & Operations Research, 54*, 274–285.
Azadi, M., Jafarian, M., Saen, R. F., & Mirhedayatian, S. M. (2015). A new fuzzy DEA model for evaluation of efficiency and effectiveness of suppliers in sustainable supply chain management context. *Computers & Operations Research, 54*, 274–285.
Bai, C. A., & Sarkis, J. (2010). Integrating sustainability into supplier selection with grey system and rough set methodologies. *International Journal of Production Economics, 124*(1), 252–264.
Büyüközkan, G., & Çifçi, G. (2011). A novel fuzzy multi-criteria decision framework for sustainable supplier selection with incomplete information. *Computers in Industry, 62*(2), 164–174.
Carter, C. R., & Rogers, D. S. (2008). A framework of sustainable supply chain management: Moving toward new theory. *International Journal of Physical Distribution & Logistics Management, 38*(5), 360–387.
Dubey, R., Gunasekaran, A., Childe, S. J., Papadopoulos, T., & Fosso Wamba, S. (2017). World class sustainable supply chain management: Critical review and further research directions. *The International Journal of Logistics Management, 28*(2), 332–362.

Garg, C. P. (2016). A robust hybrid decision model for evaluation and selection of the strategic alliance partner in the airline industry. *Journal of Air Transport Management, 52*(2016), 55–66.

Garg, C. P., Sharma, A., & Goyal, G. (2017). A hybrid decision model to evaluate critical factors for successful adoption of GSCM practices under fuzzy environment. *Uncertain Supply Chain Management, 5*(1), 59–70.

Gobbo, S. C. D. O., Fusco, J. P. A., & Gobbo Junior, J. A. (2014). An analysis of embeddedness in the value creation in interorganisational networks: An illustrative example in Brazil. *International Journal of Advanced Operations Management, 6*(2), 178–198.

Gold, S., Seuring, S., & Beske, P. (2010). Sustainable supply chain management and inter-organizational resources: A literature review. *Corporate Social Responsibility and Environmental Management, 17*(4), 230–245.

Govindan, K., Garg, K., Gupta, S., & Jha, P. C. (2016). Effect of product recovery and sustainability enhancing indicators on the location selection of manufacturing facility. *Ecological Indicators, 67*, 517–532.

Govindan, K., Khodaverdi, R., & Jafarian, A. (2013). A fuzzy multi criteria approach for measuring sustainability performance of a supplier based on triple bottom line approach. *Journal of Cleaner Production, 47*, 345–354.

Govindan, K., & Popiuc, M. N. (2014). Reverse supply chain coordination by revenue sharing contract: A case for the personal computers industry. *European Journal of Operational Research, 233*(2), 326–336.

Gunasekaran, A., Patel, C., & Tirtiroglu, E. (2001). Performance measures and metrics in a supply chain environment. *International Journal of Operations and Production Management, 21*(1–2), 71–87.

Gupta, H., Prakash, C., Vishwakarma, V., & Barua, M. K. (2017). Evaluating TQM adoption success factors to improve Indian MSMEs performance using fuzzy DEMATEL approach. *International Journal of Productivity and Quality Management, 21*(2), 187–202.

Kumar, D., & Garg, C. P. (2017). Evaluating sustainable supply chain indicators using Fuzzy AHP: Case of Indian automotive industry. *Benchmarking: An International Journal, 24*(6), 1742–1766.

Luthra, S., Garg, D., & Haleem, A. (2015). Critical success factors of green supply chain management for achieving sustainability in Indian automobile industry. *Production Planning & Control, 26*(5), 339–362.

Luthra, S., Govindan, K., Kannan, D., Mangla, S. K., & Garg, C. P. (2017). An integrated framework for sustainable supplier selection and evaluation in supply chains. *Journal of Cleaner Production, 140*, 1686–1698.

Luthra, S., & Haleem, A. (2015). Hurdles in implementing sustainable supply chain management: An analysis of Indian automobile sector. *Procedia-Social and Behavioral Sciences, 189*, 175–183.

Luthra, S., Mangla, S. K., Shanker, R., Garg, C. P., & Jakhar, S. K. (2018). Modelling critical success factors for sustainability initiatives in supply chains in Indian context using Grey-DEMATEL. *Production Planning & Control*, 1–24.

Mahtani, U. S., & Garg, C. P. (2018). An analysis of key factors of financial distress in airline companies in India using fuzzy AHP framework. *Transportation Research Part A: Policy and Practice, 117*(2018), 87–102.

Mangla, S. K., Kumar, P., & Barua, M. K. (2014). Flexible decision approach for analyzing performance of sustainable supply chains under risks/uncertainty. *Global Journal of Flexible Systems Management, 15*(2), 113–130.

Mangla, S. K., Kumar, P., & Barua, M. K. (2015). Risk analysis in green supply chain using fuzzy AHP approach: A case study. *Resources Conservation Recycling, 104*(B), 375–390.

Prakash, C., & Barua, M. K. (2015). Integration of AHP-TOPSIS method for prioritizing the solutions of reverse logistics adoption to overcome its barriers under fuzzy environment. *Journal of Manufacturing System, 37*(2015), 599–615.

Prakash, C., & Barua, M. K. (2016a). An analysis of integrated robust hybrid model for third-party reverse logistics partner selection under fuzzy environment. *Resources, Conservation & Recycling, 108*(2016), 63–81.

Prakash, C., & Barua, M. K. (2016b). A combined MCDM approach for evaluation and selection of third-party reverse logistics partner for Indian electronics industry. *Sustainable Production and Consumption, 7*(2016), 66–78.

Prakash, C., & Barua, M. K. (2016c). A robust multi-criteria decision making framework for evaluation of the airport service quality enablers for ranking the airports. *Journal of Quality Assurance in Hospitality & Tourism, 17*(3), 351–370.

Prakash, C., & Barua, M. K. (2017a). Flexible modelling approach for evaluating reverse logistics adoption barriers using fuzzy AHP and IRP framework. *International Journal of Operational Research, 30*(2), 151–171.

Prakash, C., & Barua, M. K. (2017b). A multi criteria decision making approach for prioritizing reverse logistics adoption barriers under fuzzy environment: Case of Indian electronics industry. *Global Business Review, 17*(5), 1–18.

Prakash, C., Barua, M. K., & Balon, V. (2015). Prioritizing TQM enablers to improve Indian airlines performance under fuzzy environment. *Industrial Engineering Journal, 8*(8), 28–34.

Saaty, T. L. (1980). *The Analytic Hierarchy Process: Planning, Priority Setting, Re-sources Allocation*. McGraw-Hill, New York.

Sengar, V. S., Garg, C. P., & Raju, T. B. (2017). Assessment of sustainable initiatives in Indian ports using AHP Framework. *International Journal of Business Excellence, 16*(1), 110–126.

Trapp, A. C., & Sarkis, J. (2016). Identifying Robust portfolios of suppliers: A sustainability selection and development perspective. *Journal of Cleaner Production, 112*, 2088–2100.

Vishwakarma, V., Prakash, C., & Barua, M. K. (2016). A fuzzy-based multi criteria decision making approach for supply chain risk assessment in Indian pharmaceutical industry. *International Journal of Logistics Systems and Management, 25*(2), 245–265.

Vishwakarma, V., Prakash, C., & Barua, M. K. (2019). Modeling the barriers of Indian pharmaceutical supply chain using fuzzy AHP. *International Journal of Operational Research 34*(2), 240–268.

Section II

Sustainable Supplier Selection

5 Green Procurement Research

A Review and Modeling Supplier Selection Problem Using GREY Method

Surajit Bag and Anoop Kumar Sahu

CONTENTS

5.1 INTRODUCTION

Business actions are vastly changing the natural environment and can endanger Mother Earth. Traditional business models did not take environmental thresholds into consideration and increasingly consumed natural resources without applying the principles of recycling, reusing, and remanufacturing (Dubey et al. 2016). In recent years, increased awareness has increased the amount of pressure on local and national government bodies to create environmental protection policies and enforce them strictly to prevent further environmental degradation (Mani et al. 2016). This has led to the evolution of green practices, and firms have started integrating green procurement initiatives with other business management functions (Bag and Gupta 2017; Tseng et al. 2017).

5.1.1 BACKGROUND

Procurement decisions have gained significance for multiple reasons; one of them is the considerable effect of vendors on the business performance of organizations (Dobos and Vörösmarty 2018). This has increased attention of buyers towards supplier evaluation and selection methods aimed at minimizing operational costs and enhancing competitiveness in global marketplace to create business opportunities (Bakeshlou et al. 2017; Yazdani et al. 2017). Selecting suppliers involves strategic decision-making in establishing successful adoption of green supply chain management practices (Gupta and Barua 2017; Yu et al. 2018). Thus, supplier selection plays a vital role in sustainable development of organization (Lo et al. 2018).

In today's volatile business environment with increasing customer demands, organizations generally consider outsourcing decisions rather than manufacturing all components in-house to meet the dispatch requirements of customers. Therefore, selecting right suppliers is key to business success (Wetzstein et al. 2016).

The literature review reveals that the green supplier selection problem is a popular strategy for making outsourcing decisions based on environmental safeguards in times of limited resource availability and increased pollution levels from various industrial sources (Bakeshlou et al. 2017).

Supplier selections entail trial and error method practiced by organizations, intending to improve greenness and manufacture eco-friendly goods. Literature review reports various criteria mainly categorized under economic characteristics (pricing, quality, throughput times) and environmental characteristics (greenness, resource consumption, pollution emission) to conduct supplier selections (Yu et al. 2018). However, supplier selection criteria varies in different countries, industry to industry, and is mainly influenced by culture, government regulations, and political and environmental scenarios (Mathiyazhagan et al. 2018).

For example, in South Africa, both public and private organizations purchasing actions are regulated by Preferential Procurement Policy Framework Act (PPPFA) under Broad-Based Black Economic Empowerment (B-BBBE) codes of good practice. It involves a point system, such as 80/20 preference point system from South African Rand 30,000 and up to Rand 50 million and 90/10 preference point system above Rand 50 million. This initiative is a part of sustainable development strategy and forces every organization to source products and services from organizations with high B-BBBE position. B-BBBE basically has a filter-down effect, which creates pressures on all vendors to conform the set standards. This cascading implementation on procurement develops sustainability for the entire nation (www.dti.gov.za).

Wetzstein et al. (2016) also pointed out that sustainable procurement research is at a nascent stage of the research phase. Progress in this field can happen with increased focus towards contextual outlook.

Few key points revealed from review of extant literature are: Sustainable procurement practices require changing gears to cope up with the dynamic transformation in business environment and nation's political environment (Mathiyazhagan et al. 2018), which creates a necessity to conduct an updated review of sustainable procurement research. Therefore, current research aims to provide an update on green procurement research conducted in the past twenty years.

Second, multi-criteria decision-making (MCDM) tools including multi-attribute decision-making (MADM) (AHP, ANP, BWM, TOPSIS, DEMATEL, VIKOR, PROMETHEE, ELECTRE) or multi-objective decision-making (MODM) (linear programming, mixed integer linear programming, goal programming, and fuzzy goal programming) are commonly used in solving sustainable supplier selection problems (Lo et al. 2018). However, it would be interesting to use GREY-based theory and select green suppliers in context to South African economy.

The research objectives are (i) to understand the progress in the field of green procurement research; (ii) to identify the key factors influencing green procurement programs in South African context; and (iii) to provide an illustration on green supplier selection using GREY-based methodology.

The remaining paper is organized as follows. Section 5.2 provides review of literature and key findings; Section 5.3 presents and illustration on green supplier selection using GREY-based methods. Section 5.4 presents the discussion based on findings of GREY-based application; and the final section presents the conclusion, limitations of the study and future research directions.

5.2 LITERATURE REVIEW

The systematic literature review methodology was adopted based on the guidelines advised by Tranfield et al. (2003). The key steps in the Systematic Literature Review (SLR) are (i) planning the strategy for literature review, (ii) conducting the literature review, and (iii) reporting the findings from literature review.

The literature review sub-sections are as under:

5.2.1 PLANNING THE REVIEW OF LITERATURE

Keywords used to search the academic literature for these two concepts are provided in Table 5.1.

The literature search process on Scopus database was conducted as an independent process by means of using the "or" operator for individual keyword. The search syntax used in this study is tabulated in Table 5.2.

TABLE 5.1
Keywords Used for Searching Relevant Literature

Keywords
"Green procurement"
"Sustainable procurement"

TABLE 5.2
Detailed Search Syntax

Source of Data Collection	Search Syntax
Search Performed on Scopus on 22nd Jan 2018 Website: https://www.scopus.com	(TITLE-ABS-KEY ("green procurement") OR TITLE-ABS-KEY ("sustainable procurement")) AND (EXCLUDE (PUBYEAR, 2018)) AND (LIMIT-TO (DOCTYPE, "ar") OR LIMIT-TO (DOCTYPE, "re") OR LIMIT-TO (DOCTYPE, "ip")) AND (LIMIT-TO (SUBJAREA, "BUSI")) AND (LIMIT-TO (LANGUAGE, "English"))

Source: Author's own compilation.

The search process can be re-run on Scopus website by simply copying and pasting the referred syntax under the advanced search option. However, the output may vary due to the constant updating happening on the digital database. The data was obtained on Monday, January 22, 2018, at 5 PM, South African time.

The flowchart showing steps of selecting the green procurement or sustainable procurement research papers for literature review is presented in Figure 5.1.

5.2.2 CONDUCTING THE LITERATURE REVIEW

This section details the classification of research papers. The pie chart shows the range of research papers, which were obtained after stage two in the paper search

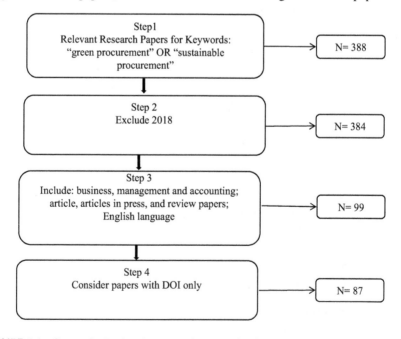

FIGURE 5.1 Steps of selecting the research papers for literature review. (*Source*: Author's own compilation.)

process. Research team found that maximum number of papers on green procurement or sustainable procurement is published in the domain of business management followed by engineering and other disciplines (refer to Figure 5.2). This means that researchers are attempting to handle green procurement more from management perspectives.

Figure 5.3 provides the details on the publishing trend. The number of publishing went up from 2008 onward and significantly from year 2015, which means increasing

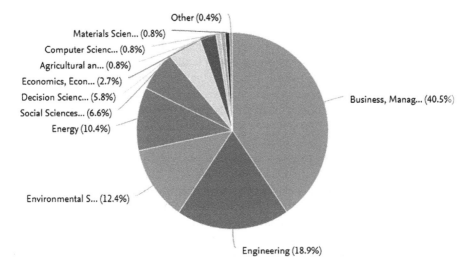

FIGURE 5.2 Range of research papers after stage four. (*Source*: Author's own compilation.)

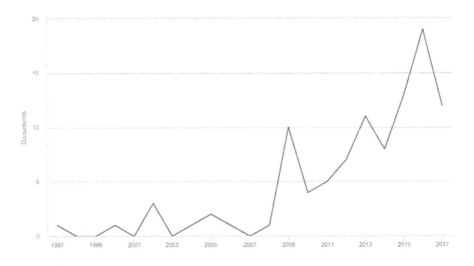

FIGURE 5.3 Publishing trends.

attention of researchers and academicians towards green procurement due to its ability in influencing business performance.

The authors' further conducted an analysis to find out the number of journals that published green procurement articles (refer Table 5.3). The result of this analysis is presented in Table 5.3. It can be seen that a total of 41 journals have published topics on green procurement or sustainable procurement. A total of 87 papers have been published over a span of 20 years. However, it is interesting that the *Journal of Cleaner Production* alone published 26% of the total publications.

The research team further performed the keyword analysis (refer Table 5.4) for identifying the most popular keywords used in green procurement or sustainable procurement research. It was found that a total of 434 keywords were used in 87 research papers. Most of the keywords appeared multiple times in different publications over the last 10 years. After sorting, it was found that there are 290 keywords. The research team tabulated (refer to Table 5.4) the top 10 popular keywords having the highest frequency of appearance in various research publications.

The research team performed citation analysis (refer to Table 5.5) and found that out of total 87 papers only 77 papers have attracted citation, falling between the ranges of 1 and 99 citations; 7 papers between the ranges of 100 and 199; 2 papers between the ranges of 200 and 299 and 1 paper above 300 citations. Finally, the research team have presented 10 papers that attracted the highest number of citations and can be considered as seminal papers in the field of green procurement or sustainable procurement.

Based on the analysis of author affiliation and country (refer Table 5.6), it was found that six papers were communicated by authors from United Kingdom based institutions; two papers from Germany; one paper from Sweden and one paper from Belgium. This means that the majority of contributions are happening from the Western part of the globe.

5.2.3 Reporting the Findings from Literature Review

Based on the review of prior literature, an attempt has been taken to present initial and ongoing cost taxonomy for sustainable procurement programs.

Table 5.7 showcases that six costs are initial costs and nine costs are ongoing costs for the organization undertaking sustainable procurement practices.

The initial costs and ongoing costs in green procurement are largely influenced through selection of green suppliers (Dubey et al. 2013). This motivated the authors to identify the key factors influencing supplier selection in green procurement. The literature reported multiple criteria that are considered by past researchers in the supplier selection problem.

Bakeshlou et al. (2017) considered cost, quality, service, technological capability, and environmental aspects. Whereas, Gupta and Barua (2017) considered collaboration, environmental investments, resource accessibility, environmental management, research and design, green procurement capabilities, regulatory pressures, and customer demands. Recently, Islam et al. (2018) considered green supplier selection using environmental criteria, supplier using non-hazardous chemicals in manufacturing goods, energy saving and waste reduction initiatives by suppliers, usage of eco-friendly transport by suppliers, suppliers operating with environmental objectives, suppliers operating with a target of

TABLE 5.3

Volume of Publications in Last Twenty Years

Journal	1997	2002	2005	2006	2009	2010	2011	2012	2013	2014	2015	2016	2017	Total
Action Learning: Research and Practice												1		1
Business Strategy and the Environment							1							1
Construction Economics and Building											1			1
Construction Management and Economics		1												1
Corporate Environmental Strategy		1												1
Corporate Social Responsibility and Environmental Management			1	1										2
Decision Sciences Journal of Innovative Education									1					1
Developments in Corporate Governance and Responsibility												1		1
Engineering, Construction and Architectural Management	1				1					1				3
Industrial and Commercial Training													1	1
Industrial Marketing Management								1		1				2
Information Technology and Management											1			1
Innovation											1			1
International Journal of Enterprise Information Systems											1			1
International Journal of Innovation and Sustainable Development						1								1

(Continued)

TABLE 5.3 (Continued)
Volume of Publications in Last Twenty Years

Journal	1997	2002	2005	2006	2009	2010	2011	2012	2013	2014	2015	2016	2017	Total
International Journal of Logistics Management													1	1
International Journal of Operations and Production Management							1				1			2
International Journal of Procurement Management					1			2				2	1	8
International Journal of Production Economics								1			1			2
International Journal of Production Research								1	1	1		1		4
International Journal of Project Management									1					1
Journal of Business Ethics												1		1
Journal of Change Management									1					1
Journal of Cleaner Production		1			1				3	1	4	6	7	23
Journal of Consumer Policy												1		1
Journal of Corporate Real Estate													1	1
Journal of Human Resources in Hospitality and Tourism									1				1	1
Journal of Industrial Engineering and Management										1				1
Journal of Management Education					1									1

(Continued)

TABLE 5.3 (Continued)
Volume of Publications in Last Twenty Years

Journal	1997	2002	2005	2006	2009	2010	2011	2012	2013	2014	2015	2016	2017	Total
Journal of Purchasing and Supply Management								1		2	1	1		5
Management Research Review						1								1
Periodica Polytechnica Social and Management Sciences										1				1
Proceedings of Institution of Civil Engineers: Management, Procurement and Law					1				1			1		3
Production and Operations Management												1		1
Public Management Review												1		1
Public Money and Management								1			1			2
Society and Economy							1							1
Supply Chain Management					2									2
Technology in Society						1								1
Vision													1	1
Worldwide Hospitality and Tourism Themes							1							1
Total	1	3	1	1	7	3	4	7	11	8	12	17	12	87

TABLE 5.4
Keywords

Keywords	Count
Green procurement	24
Sustainable procurement	22
Sustainability	16
Procurement	10
Public procurement	6
Environmental management	5
Structural equation modeling	5
Sustainable public procurement	5
Barriers	4
Drivers	4

TABLE 5.5
Ten Papers with the Highest Number of Citations

Authors	Cited by
Walker and Brammer (2009)	349
Brammer and Walker (2011)	290
Preuss (2009)	218
Hollos et al. (2012)	165
Meehan and Bryde (2011)	147
Varnäs et al. (2009)	133
Günther and Scheibe (2006)	112
Walker and Brammer (2012)	110
Blome et al. (2014)	102
Walker and Phillips (2009)	101

carbon reduction, collaboration with suppliers, transfer knowledge and technical know-how with suppliers, and supplier environmental conformance. Another study conducted by Mathiyazhagan et al. (2018) considered management, technology, manufacturing, and cost as key criteria in supplier selection problem.

The research team referred extant literature to identify the criteria for performing this research study and further consulted with industry experts having a professional association with the Mining Industry Association of Southern Africa to finalize the parameters that are relevant from a South African context. Finally, nine criteria are found to be playing an important role in selection of green suppliers in South African economy and further discussed below.

Supplier understanding of green policy: It is essential for suppliers to fully be aware of the green policy and procedures of customer firms. The green policy is mainly aligned with the business objectives and operational objectives

TABLE 5.6
Author Affiliation and Country for Top Ten Papers

Authors	Affiliations
Walker and Brammer (2009)	Operations Management Group, Warwick Business School, University of Warwick, Coventry, United Kingdom
Brammer and Walker (2011)	Strategic Management Group, Warwick Business School, University of Warwick, Coventry, United Kingdom
Preuss (2009)	School of Management, Royal Holloway, University of London, Egham, United Kingdom
Hollos et al. (2012)	Supply Chain Management Institute, EBS Business School, Wiesbaden, Germany
Meehan and Bryde (2011)	Liverpool Business School, Liverpool, United Kingdom
Varnäs et al. (2009)	Royal Institute of Technology, Land and Water Resources Engineering, SE-100 44 Stockholm, Sweden
Günther and Scheibe (2006)	Technische Universität Dresden (TUD), Department of Business Management and Economics, Professorship of Business Administration Esp. Environmental Management, Germany
Walker and Brammer (2012)	Logistics and Operations Management Section, Cardiff Business School, Cardiff University, Colum Drive, Cardiff CF10 3EU, United Kingdom
Blome et al. (2014)	Louvain School of Management and CORE, Université Catholique de Louvain, Louvain-la-Neuve, Belgium
Walker and Phillips (2009)	Centre for Research in Strategic Purchasing and Supply, School of Management, University of Bath, Claverton Down, Bath, BA2 7AY, United Kingdom

of the firm keeping in view of political, economic, social, and technological environment. Green policy is used by green procurement professionals to determine the green procurement process and procedures suitable for the firm in meeting the sustainability goals (Ahsan and Rahman 2017).

Supplier knowledge of environmental impact of products: Supplier knowledge of the life cycle of the product and its impact on the surrounding environment is helpful for developing green products with a minimum environmental impact. The customer must provide the right specifications and technical know-how for enhancing suppliers' knowledge and conform quality parameters (Ahsan and Rahman 2017; Bakeshlou et al. 2017).

Supplier firm senior management support for green initiatives: Senior management involvement is essential for successful adoption of green procurement programs. Quick approval of budget and funds without much delay are essential for upgrading infrastructure that perfectly fits the green procurement requirements (Blome et al. 2014; Mosgaard 2015).

Supplier's clear strategic goals on green program: Supplier firm's green procurement policy and performance measures must be aligned with strategic vision and mission of the company. Otherwise the mismatch will cost the firm more money and attract losses. The purchasing structure must also fit the strategic goals of the firm to reap benefits from such green

TABLE 5.7
Initial and Ongoing Cost Taxonomy

Initial Costs	Description
Initial Audits	Initially start with risk profiling and gap analysis followed by framing green procurement policy and procedures
Preliminary capability assessment	Assess the capability of vendors for fitment under green procurement program
Vendor visits	Visit to vendor manufacturing units for quality audit
Trial order placement	Placing trial order on the supplier
Investment in pollution prevention technologies	Replace obsolete machineries with modern technologies
Appointment of environmental manager	Appointing an environmental manager for monitoring progress and ensuring firm complies with green programs

Ongoing Costs	Description
Ongoing training	Continuous training and education of suppliers
Support	Cost of deploying engineer at supplier premise for green component development
Environmental awareness	Environmental awareness programmes involving suppliers
Buy green Inputs	Procurement of eco-friendly raw material at higher costs for production
Green packaging	Use of eco-friendly packing material for monthly customer dispatches
Solid waste management and waste water treatment	Incurring recurring costs for scientific way of treating industrial wastes
ISO 14001 certification	Implementation of ISO 14001 and annual audits
Environmental management system	Implement environmental management system and annual audits
Green supplier audits	Hire third-party audit firms for annual audits of green suppliers

procurement programs. Supplier firms using third-party audits has largely benefitted directly and indirectly. Third-party audits specifically focused on green procurement goals, policies, and performance standards are essential for the progress of green programs. Third-party audits help firms to identify the gaps and hurdles faced by firms, threats and vulnerabilities, audit how the mitigation strategies will be evaluated by firms, audit the crisis management strategies, and finally assess the firm's insurance coverage. This will ensure safety, security, and sustainability (Ahsan and Rahman 2017).

Supplier's internal environmental coordinator: The responsiveness of the supplier's internal environmental coordinator plays a key role in satisfying customers' environmental requirements. The internal environmental coordinator must ensure that the firm complies with state and national level government regulations to make green programs successful (Akenroye et al. 2013; Mosgaard 2015).

Supplier's competent green operations professionals: It is essential that supplier have sufficient manpower with green competency for supporting green procurement programs, which are complex in nature and require high-level technical skills apart from coordination, communication, and collaboration (Ahsan and Rahman 2017).

Supplier flexibility: The ability of suppliers to change the production line rapidly with low volume and high variety products is an important requirement to survive under this current volatile business environment. To achieve success in green procurement programs, it is necessary for the supplier to have flexibility in terms of technological change and attitudinal change (Bag 2017b).

Supplier's collaboration with tier 2 and tier 3 suppliers: Technological collaboration with tier two and tier three specialized suppliers results in improved innovation and costs savings in the green supply network (Ahsan and Rahman 2017; Gupta and Barua 2017; Islam et al. 2018).

Supplier's environmental management systems: Implementation of environmental management systems is the basic level of requirement in any green program. Supplier firms must ensure that such systems are in place and policies communicated to all employees in the organization for creating green awareness. Moreover, supplier firms must comply with customer green requirements and set performance standards to measure the variance against targets (Fet et al. 2011; Islam et al. 2018).

The research framework is presented in Figure 5.4, which provides the green procurement enablers leading to sustainability. Here, the literature is classified into nine building blocks of the sustainability framework.

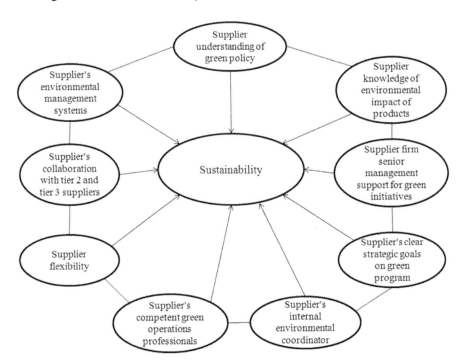

FIGURE 5.4   Research framework.

5.3   GREEN SUPPLIER SELECTION USING GREY-BASED METHODS: AN ILLUSTRATION

The research team have used an innovative MCDM technique utilizing GREY methodology for supplier selection decision-making. This technique is helpful in selecting the right combination of parameters under uncertain environment with incomplete data.

5.3.1   THEORY OF GREY NUMBERS: MATHEMATICAL BASIS

GREY theory was developed by Prof. Deng in 1982. It is an effective technique towards solving the problems that include uncertainty or dealing with discrete data/incomplete information. GREY theory has been implemented in various areas such as decision-making, system control, and computer graphics. There are a few basic principles regarding the mathematical background of the GREY system associated with GREY theory, which can be seen as follows.

Definition 1:

A GREY system is defined as a system involving the uncertain data presented by GREY set and GREY variables. The concept of the GREY system is shown in Figure 5.5.

Definition 2:

Let be the universal set. Then a GREY set G of X is defined by its two mappings

$$\begin{cases} \bar{\mu}_G(x): x \to [0,1] \\ \underline{\mu}_G(x): x \to [0,1] \end{cases} \qquad (5.1)$$

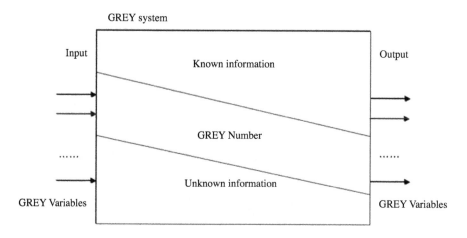

FIGURE 5.5   The concept of a GREY system.

$\bar{\mu}_G(x) \geq \mu_{\underline{G}}(x), x \in X$, $X = R$, $\bar{\mu}_G(x)$ and $\mu_{\underline{G}}(x)$ are the upper and lower membership functions in G, respectively. When $\bar{\mu}_G(x) = \mu_{\underline{G}}(x)$, the GREY set G is like a fuzzy set. It depicts that GREY theory considers the condition of fuzziness and possesses the flexibly when dealing with the fuzziness.

Definition 3:

A GREY set is the not exact value, where the upper and/or the lower limit is set up generally as GREY number is written as $\left(\otimes G = G \big|_\mu^\mu \right)$.

Definition 4:

If solely the lower limit of G is possibly calculated and G is defined as lower limit GREY set.

$$\otimes G = \left[\underline{G}, \infty \right] \tag{5.2}$$

Definition 5:

If solely the upper limit of G is possibly calculated and G is defined as lower limit GREY number.

$$\otimes G = \left[-\infty, \bar{G} \right] \tag{5.3}$$

Definition 6:

If the lower and upper limits of G is estimated and G is set up as interval GREY number.

$$\otimes G = \left[\underline{G}, \bar{G} \right] \tag{5.4}$$

Definition 7:

The fundamental operations of GREY sets $\otimes x_1 = \left[\underline{x}_1, \bar{x}_1 \right]$ and $\otimes x_2 = \left[\underline{x}_2, \bar{x}_2 \right]$ is expressed as follows:

$$\left. \begin{aligned}
\otimes x_1 + \otimes x_2 &= \left[\underline{x}_1 + \underline{x}_2, \bar{x}_1 + \bar{x}_2 \right] \\
\otimes x_1 - \otimes x_2 &= \left[\underline{x}_1 - \underline{x}_2, \bar{x}_1 - \bar{x}_2 \right] \\
\otimes x_1 \times \otimes x_2 &= \left[\underline{x}_1 \underline{x}_2, \bar{x}_1 \bar{x}_2 \right] \\
\otimes x_1 \div \otimes x_2 &= \left[\underline{x}_1, \bar{x}_1 \right] \times \left[\frac{1}{\underline{x}_2}, \frac{1}{\bar{x}_2} \right]
\end{aligned} \right\} \tag{5.5}$$

Signed distance: Let $\otimes x_1 = \left[\underline{x}_1, \overline{x}_1\right]$ and $\otimes x_2 = \left[\underline{x}_2, \overline{x}_2\right]$ be two positive interval GREY numbers. In this case, the distance between $\otimes x_1$ and $\otimes x_2$ is calculated as the signed difference between its centers is shown below:

$$d(\otimes x_1, \otimes x_2) = \frac{\underline{x}_1 - \overline{x}_1}{2} - \frac{\underline{x}_2 - \overline{x}_2}{2} = \frac{1}{2}[(\underline{x}_1 - \underline{x}_2) + (\overline{x}_1 - \overline{x}_2)] \qquad (5.6)$$

5.3.2 THE MULTI-OBJECTIVE OPTIMIZATION BY RATIO ANALYSIS (MOORA) METHOD

Multi-objective Optimization by Ratio Analysis (MOORA) technique was introduced by Brauers and Zavadskas (2006). The technique begins with a matrix of responses of different choices on different objectives:

$$X = [x_{ij}]_{m \times n} \qquad (5.7)$$

Here, x_{ij} as the response versus choices j on objective i; $i = 1, 2, ..., n$; as the objectives; and $j = 1, 2, ..., m$ as the alternatives/choices.

The MOORA technique consists of two parts: the ratio system and the reference point approach (Brauers and Zavadskas 2009).

When the determination of the overall preference index is based on the ratio system approach of the MOORA for optimization, the research team starts from the formula:

$$y_j^* = y_j^+ - y_j^-,$$

$$y_j^+ = \sum_{i \in \Omega_C^+} s_i x_{ij}^* + \sum_{i \in \Omega_G^+} \otimes s_i x_{ij}^*,$$

where,

$$y_j^- = \sum_{i \in \Omega_{\overline{C}}} s_i x_{ij}^* + \sum_{i \in \Omega_{\overline{G}}} \otimes s_i x_{ij}^*, \qquad (5.8)$$

Here, y_j^* as the overall ranking index of choices j; y_j^+ and y_j^- as total sums of maximizing and minimizing responses of choices j to objectives respectively; s_i as significance coefficient of objective i; x_{ij}^* and $\otimes x_{ij}^*...$ as the normalized responses of choices j on different objectives, which are represented in the form of crisp or interval GREY numbers; Ω_C^+ and Ω_G^+ assets of objectives to be maximized, shown on crisp or interval GREY set. The ranking is done based on maximum value is better.

5.3.3 GREY-FULL MULTIFICATION FORM

The overall ranking index is based on the full multification form of the MOORA technique and, for optimization, it is based on the multification. We start from the formula, where

$$y_j^+ = \frac{\displaystyle\prod_{i \in \Omega_C^+} \otimes s_i \, x_{ij}^*}{\displaystyle\prod_{i \in \Omega_{\overline{C}}} \otimes s_i \, x_{ij}^*} \qquad (5.9)$$

Here, y_j^* as the overall ranking index of choices j; y_j^+ and y_j^- as total sums of maximizing and minimizing responses of choices j to objectives respectively; s_i as significance coefficient of objective i; x_{ij}^* and $\otimes x_{ij}^*$... as the normalized responses of alternative j on different objectives. The ranking is done based on minimum value is better.

Stanujkic et al. (2012) proposed the three conditions of objectives depending on subjective perception of experts.

i. When objectives possess the same significance ($\lambda = 0$):

$$y_j^* = (1-\lambda)\left(\sum_{i \in \Omega_G^+} x_{ij}^* - \sum_{i \in \Omega_G^-} x_{ij}^*\right) + \lambda\left(\sum_{i \in \Omega_G^+} \overline{x}_{ij}^{-*} - \sum_{i \in \Omega_G^-} \overline{x}_{ij}^{-*}\right), \quad (5.10)$$

ii. When the decision maker possesses no preferences ($\lambda = 0.5$)

$$y_j^* = \frac{1}{2}\left(\sum_{i \in \Omega_G^+} s_i \, \underline{x}_{ij}^* - \sum_{i \in \Omega_G^-} s_i \, \underline{x}_{ij}^*\right), \quad (5.11)$$

iii. When the decision maker has no preference, and objectives have the same significance ($\lambda = 1$):

$$y_j^* = \lambda\left(\sum_{i \in \Omega_G^+} \underline{x}_{ij}^* - \sum_{i \in \Omega_G^-} \underline{x}_{ij}^*\right) + (1-\lambda)\left(\sum_{i \in \Omega_G^+} \overline{x}_{ij}^{-*} - \sum_{i \in \Omega_G^-} \overline{x}_{ij}^{-*}\right), \quad (5.12)$$

During the problem solution, i.e., ranking of alternatives, the attitude of the professionals can lie between pessimistic and optimistic, and the whitening coefficient, λ, allows expression of the professionals' degree of optimism or pessimism.

5.3.3.1 Analysis

The presented work deals with a real case study considering Tega Industries Africa Pty Limited (TIAL), located in South Africa. This company produces different varieties of rubber products that are an integral part of a mine's grinding mill. TIAL desires to choose a green supplier based on nine evaluation parameters. To make a favorable decision, the company develops a model by using nine significant parameters.

Next, the company uses the GREY set to decide by considering partial information gathered from their own company employees. The model with the proposed GREY set helps the company to evaluate the most valuable raw material supplier considering the environmental parameters for enhancing business performance. The decision-making evaluation steps are discussed below:

Step 1: In the preliminary stage, a group of five experts are formed considering members from production, purchase, logistics, health safety environment, and marketing department.

Step 2: A green supplier evaluation framework is set up based upon a conducted literature review and after analyzing the standard green supplier evaluation frameworks, as shown in Figure 5.4.

Step 3: The research team adapted an appropriate GREY linguistic scale to assign solely rating vs evaluation of green supplier parameters to further include in model, as represented in Table 5.9.

Step 4: The GREY operators are used, i.e., Eq. 5.5, and explored to summarize the GREY ratings vs evaluation of green suppliers' parameters corresponding to suppliers' alternatives are shown in Tables 5.10 through 5.12.

Step 5: Then, the GREY rating of 1st level from 2nd level is evaluated by using GREY global equation proposed by Sahu et al. (2015). The matrix is presented in Table 5.13. Then, normalized and a weighted normalized matrix is constructed by referring paper (Stanujkic et al. 2012), depicted in Tables 5.14 and 5.15.

Step 6: The Eqs. 5.8 and 5.10 through 5.12 are applied in order to evaluate the ranking orders of raw material suppliers considering environmental parameters and presented in Table 5.16.

Step 7: Then Eqs. 5.9 and 5.10 through 5.12 are applied to evaluate the ranking orders of raw material suppliers considering environmental parameters and presented in Table 5.17.

Step 8: Eventually, a comparative analysis is applied to evaluate the results.

The findings and discussion are provided in the next section.

TABLE 5.8
Green Supplier Evaluation Platform

Goal	Green Supplier Parameters (C_j)
Green Supplier Selection	Supplier understanding of green policy
	Supplier knowledge of environmental impact of products
	Supplier firm senior management support for green initiatives
	Supplier's clear strategic goals on green program
	Supplier's internal environmental coordinator
	Supplier's competent green operations professionals
	Supplier flexibility
	Supplier's collaboration with tier two and tier three suppliers
	Supplier's environmental management systems

TABLE 5.9
The Scale of Attribute Ratings

Scale	$\otimes r$
Very Poor (VP)	[0, 1]
Poor (P)	[1, 3]
Medium Poor (MP)	[3, 4]
Fair (F)	[4, 5]
Medium Good (MG)	[5, 6]
Good (G)	[6, 9]
Very Good (VG)	[9, 10]

TABLE 5.10
Appropriateness GREY Rating against Green Supplier Parameters for A_1

(C_j)	Parameters (C_{jk})		D1	D2	D3	D4	D5
(C_1)	Supplier understanding of green policy ($C_{1,1}$)	+	G	VG	G	VG	G
	Supplier knowledge of environmental impact of products ($C_{1,2}$)	+	VG	MP	VG	VG	VG
	Supplier firm senior management support for green initiatives ($C_{1,3}$)	+	MP	MP	F	MG	F
(C_2)	Supplier's clear strategic goals on green program ($C_{2,1}$)	+	MP	VG	F	G	F
	Supplier's internal environmental coordinator ($C_{2,2}$)	+	MP	F	F	VG	F
	Supplier's competent green operations professionals ($C_{2,3}$)	+	MP	F	P	MG	F
(C_3)	Supplier flexibility ($C_{3,1}$)	+	VG	F	VG	MG	MP
	Supplier's collaboration with tier 2 and tier 3 suppliers ($C_{3,2}$)	+	MG	G	F	MG	MP
	Supplier's environmental management systems ($C_{3,3}$)	+	MG	G	F	F	MP

TABLE 5.11
Appropriateness GREY Rating against Green Supplier Parameters for A_2

(C_j)	Parameters (C_{jk})		D1	D2	D3	D4	D5
(C_1)	Supplier understanding of green policy ($C_{1,1}$)	+	MG	G	MG	F	G
	Supplier knowledge of environmental impact of products ($C_{1,2}$)	+	G	G	MG	G	MG
	Supplier firm senior management support for green initiatives ($C_{1,3}$)	+	G	G	MG	MG	MP
(C_2)	Supplier's clear strategic goals on green program ($C_{2,1}$)	+	G	G	F	MG	MP
	Supplier's internal environmental coordinator ($C_{2,2}$)	+	G	MG	F	MG	F
	Supplier's competent green operations professionals ($C_{2,3}$)	+	MG	MG	VG	VG	F
(C_3)	Supplier flexibility ($C_{3,1}$)	+	MG	MG	F	F	VG
	Supplier's collaboration with tier 2 and tier 3 suppliers ($C_{3,2}$)	+	G	F	F	MG	MP
	Supplier's environmental management systems ($C_{3,3}$)	+	G	MG	VG	MG	MP

TABLE 5.12

Appropriateness GREY Rating against Green Supplier Parameters for A_3

(C_j)	Parameters (C_{jk})		D1	D2	D3	D4	D5
(C_1)	Supplier understanding of green policy $(C_{1,1})$	+	G	MG	F	MG	MP
	Supplier knowledge of environmental impact of products $(C_{1,2})$	+	G	MG	F	VG	MP
	Supplier firm senior management support for green initiatives $(C_{1,3})$	+	G	MG	F	F	VG
(C_2)	Supplier's clear strategic goals on green program $(C_{2,1})$	+	VG	MG	G	F	F
	Supplier's internal environmental coordinator $(C_{2,2})$	+	VG	VG	P	MG	F
	Supplier's competent green operations professionals $(C_{2,3})$	+	MG	G	VP	F	F
(C_3)	Supplier flexibility $(C_{3,1})$	+	MG	G	VP	F	VG
	Supplier's collaboration with tier 2 and tier 3 suppliers $(C_{3,2})$	+	MG	G	G	G	F
	Supplier's environmental management systems $(C_{3,3})$	+	MG	VG	G	MG	VG

TABLE 5.13

GREY Global Rating Matrix for Supplier Performance Measurement

Goal (C)	Suppliers	(C_1)	(C_2)	(C_3)
Green Supplier Selection	A_1	[6.2670,7.6670]	[4.667,5.800]	[5.0000,6.2670]
	A_2	[5.2607,7.2000]	[5.333,6.733]	[5.1330,6.4000]
	A_3	[5.2000,6.6000]	[5.0000,6.3330]	[5.6670,7.3330]

TABLE 5.14

GREY Global Rating Normalized Matrix

Goal (C)	Suppliers	(C_1)	(C_2)	(C_3)
Green Supplier Selection	A_1	[0.3620,0.4420]	[0.260,0.323]	[0.3204,0.4060]
	A_2	[0.3040,0.4150]	[0.297,0.375]	[0.3302,0.4104]
	A_3	[0.3000,0.3810]	[0.2078,0.353]	[0.3067,0.4705]

TABLE 5.15
Weighted Normalized Matrix

Goal (C)	Suppliers	(C_1)	(C_2)	(C_3)
Weights = 1		0.081	0.228	0.691
		max	max	max
Green Supplier	A_1	[0.029029,0.03580]	[0.0748,0.0930]	[0.22360,0.28030]
Selection	A_2	[0.020462,0.03370]	[0.0855,0.1080]	[0.22950,0.28620]
	A_3	[0.02043,0.03080]	[0.08000,0.10160]	[0.25304,0.32580]

TABLE 5.16
Ranking Results Obtained Using MOORA Technique for $\lambda = 0, 0.5, 1$

λ	$\lambda = 0$		$\lambda = 0.5$		$\lambda = 1$	
Alternatives	y_j^*	Ranking	y_j^*	Ranking	y_j^*	Ranking
A_1	−0.303	1.000	−0.269	1.000	−0.337	1.000
A_2	−0.326	2.000	−0.290	2.000	−0.361	2.000
A_3	−0.354	3.000	−0.309	3.000	−0.399	3.000

TABLE 5.17
Ranking Results Obtained Using Full Multification Form Technique for $\lambda = 0, 0.5, 1$

λ	$\lambda = 0$		$\lambda = 0.5$		$\lambda = 1$	
Alternatives	y_j^*	Ranking	y_j^*	Ranking	y_j^*	Ranking
A_1	1.562	1.000	1.750	1.000	1.375	1.000
A_2	1.171	2.000	1.254	2.000	1.089	2.000
A_3	1.061	3.000	1.196	3.000	0.926	3.000

5.4 RESULT AND DISCUSSION

After evaluating the ranking orders of raw material suppliers under environmental concern, the performance scoring is evaluated by applying GREY MOORA and Full Multification Form and compared with each other considering the pessimistic, moderate, and optimistic scenarios. TIAL determines the A_1 raw material supplier as the best supplier from a green perspective for future use. The A_1 raw material supplier fulfills all the subjective parameters or multiple dimensions. The company can use the A_2 raw material supplier if the A_1 raw material supplier fails to perform. The A_3 raw material supplier will be the worst choice for the company. The results obtained by MOORA and Full Multification Form with GREY are depicted by the line chart in Figures 5.6 and 5.7.

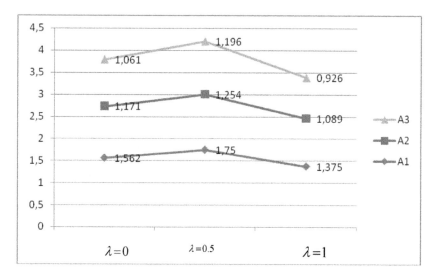

FIGURE 5.6   Performance score by GREY MOORA.

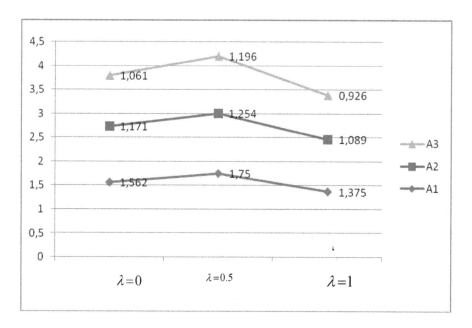

FIGURE 5.7   Performance score by GREY Full Multification Form.

5.5 CONCLUSION

The purpose of the study is to conduct a thorough literature review and understand the past, present, and future trends in green procurement research. The list of papers was downloaded using Scopus (https://www.scopus.com), which is the largest abstract and citation database for academic reputed scientific research journals, books, and conference proceedings. Through strict screening, only journal papers were selected for conducting the review of literature. The extant literature is categorized according to multiple classification schemes. A bibliometric analysis of the reviewed literature was presented as well. The review findings show that a total of 87 papers have been published over a span of 20 years. The number of publishing went up from the year 2008 onward and significantly from the year 2015.

The review of prior literature is done to develop the initial and ongoing cost taxonomy for sustainable procurement programs. Here, six costs are categorized under initial costs and nine costs under ongoing costs. These costs are basically incurred by the company involved in undertaking sustainable procurement practices.

The initial costs and ongoing costs in green procurement are largely influenced through selection of green suppliers (Dubey et al. 2013). Supplier selection is the basic step in green procurement programs. This motivated the authors to identify the key factors influencing supplier selection in green procurement. Thorough review of extant literature reveals that nine criteria are playing an important role in selection of green suppliers such as: Supplier understanding of green policy; Supplier knowledge of environmental impact of products; Supplier firm senior management support for green initiatives; Supplier's clear strategic goals on green program; Supplier's internal environmental coordinator; Supplier's competent green operations professionals; Supplier flexibility; Supplier's collaboration with tier two and tier three suppliers and Supplier's environmental management systems. Further, the paper further presents an illustration on green supplier selection through application of GREY-based methodology and uses a case study to decide the best raw material supplier through comparison of those nine criteria.

5.5.1 MANAGERIAL IMPLICATIONS

The key takeaway for managers are as follows:

Managers must design green procurement programs by considering the key enablers, which have the potential to influence the sustainability aspects. Mangers must map the external and internal risks and prepare contingency plans to deal with these risks. Mangers must educate suppliers to enhance their understanding of green policy and knowledge of environmental impact of products. Regular face-to-face meetings and discussions with the supplier firm's senior management is essential to get their support for green initiatives. This will also help suppliers to develop clear strategic goals on green programs. Transparency in communication improves environmental coordination with tier 1 supplier firms and helps in collaboration activities with tier two and tier three suppliers. Such collaboration ultimately improves flexibility to accept customer orders with low volume and high mix and results into increased innovations.

5.5.2 Unique Contribution, Limitations, and Future Research Directions

The study showcases a very interesting topic for study and portrays various dimensions that provide food for thought to future researchers. The uniqueness of the study is not only reporting the review of relevant papers, which presented evidences of green procurement concept, but also identifying the nine key enablers that influence green procurement programs.

The study suffers from certain limitations, such as small sample size used for analysing the case study. Such limitations can be eradicated in future studies by conducting an empirical study considering a larger sample size. Future studies can also be conducted on developing other green procurement taxonomies and developing advanced level research framework based on the review findings.

LIST OF PAPERS

DOI NO
10.1016/j.jclepro.2017.03.055
10.1504/IJPM.2013.050607
10.1016/j.jclepro.2015.11.057
10.1016/j.jclepro.2014.11.077
10.1108/JCRE-03-2016-0017
10.1016/j.jclepro.2014.08.106
10.1108/ICT-07-2016-0043
10.1680/mpal.2009.162.2.75
10.1177/0972262917700990
10.1504/IJPM.2017.086403
10.1504/IJPM.2016.077702
10.1108/17554211111142194
10.1080/00207543.2013.825748
10.1080/14767333.2016.1215290
10.1016/j.jclepro.2017.01.141
10.1016/j.jclepro.2015.09.085
10.1108/01443571111119551
10.1016/j.indmarman.2013.08.003
10.1108/S2043-052320160000009001
10.1016/j.jclepro.2014.11.068
10.1016/j.jclepro.2017.06.027
10.1016/j.pursup.2012.01.002
10.1108/eb021038
10.3311/PPso.2151
10.1504/IJPM.2013.052469
10.1080/09540962.2012.703422
10.1016/j.jclepro.2016.07.098
10.1002/csr.72
10.1016/j.jclepro.2012.08.026
10.1556/SocEc.33.2011.1.13

(Continued)

DOI NO
10.1016/j.jclepro.2016.09.126
10.1080/00207543.2015.1079341
10.1111/dsji.12007
10.1016/j.jclepro.2016.02.102
10.1080/09540962.2015.1007706
10.1080/13511610.2015.1024639
10.1080/14697017.2013.851950
10.1108/01409171011030453
10.1002/csr.92
10.1007/s10603-015-9282-8
10.1016/j.ijpe.2014.09.012
10.1080/00207543.2011.582184
10.1108/ECAM-01-2012-0007
10.1108/09699980910988348
10.5130/ajceb.v15i1.4281
10.1016/j.ijproman.2012.09.014
10.1080/00207543.2013.774467
10.1016/j.techsoc.2010.07.007
10.1016/j.pursup.2015.06.001
10.1016/j.pursup.2014.12.002
10.1016/j.pursup.2014.02.005
10.1108/IJOPM-07-2014-0359
10.1016/j.pursup.2014.01.002
10.1002/bse.678
10.1016/j.jclepro.2012.08.018
10.1016/S0959-6526(02)00021-5
10.3926/jiem.1057
10.1504/IJPM.2016.077705
10.1016/j.indmarman.2012.04.004
10.1016/j.jclepro.2016.05.056
10.1007/s10551-016-3376-3
10.1108/13598540910954557
10.1680/jmapl.15.00006
10.1016/j.jclepro.2013.03.027
10.1016/j.jclepro.2016.12.014
10.1504/IJPM.2012.047170
10.1016/j.jclepro.2015.07.007
10.4018/IJEIS.2015070102
10.1016/j.jclepro.2016.02.021
10.1111/poms.12263
10.1007/s10799-014-0193-1
10.1016/j.jclepro.2015.07.065
10.1108/IJLM-12-2015-0234
10.1680/mpal.12.00022
10.1080/01446190110093560

(Continued)

DOI NO
10.1016/S1066-7938(02)00072-6
10.1016/j.jclepro.2013.11.004
10.1016/j.jclepro.2009.04.001
10.1016/j.ijpe.2012.01.008
10.1504/IJPM.2009.021729
10.1108/13598540910941993
10.1177/1052562908323190
10.1504/IJISD.2010.034557
10.1504/IJPM.2012.047198
10.1016/j.jclepro.2016.07.001
10.1080/15332845.2013.752709
10.1080/14719037.2015.1051575

Source: Downloaded from Scopus.

REFERENCES

Ahsan, K., & Rahman, S. (2017). Green public procurement implementation challenges in Australian public healthcare sector. *Journal of Cleaner Production, 152,* 181–197. doi:10.1016/j.jclepro.2017.03.055.

Akenroye, T. O., Oyegoke, A. S., & Eyo, A. B. (2013). Development of a framework for the implementation of green public procurement in Nigeria. *International Journal of Procurement Management, 6*(1), 1–23. doi:10.1504/IJPM.2013.050607.

Aktin, T., & Gergin, Z. (2016). Mathematical modelling of sustainable procurement strategies: Three case studies. *Journal of Cleaner Production, 113,* 767–780. doi:10.1016/j.jclepro.2015.11.057.

Amarah, B., & Langston, C. (2017). Development of a triple bottom line stakeholder satisfaction model. *Journal of Corporate Real Estate, 19*(1), 17–35. doi:10.1108/JCRE-03-2016-0017.

Appolloni, A., Sun, H., Jia, F., & Li, X. (2014). Green procurement in the private sector: A state of the art review between 1996 and 2013. *Journal of Cleaner Production, 85,* 122–133. doi:10.1016/j.jclepro.2014.08.106.

Aragão, C. G., & Jabbour, C. J. C. (2017). Green training for sustainable procurement? Insights from the Brazilian public sector. *Industrial and Commercial Training, 49*(1), 48–54. doi:10.1108/ICT-07-2016-0043.

Aritua, B., Male, S., & Bower, D. A. (2009). Defining the intelligent public sector construction procurement client. *Management, Procurement and Law, 162*(2), 75–82. doi:10.1680/mpal.2009.162.2.75.

Bag, S. (2016). Green strategy, supplier relationship building and supply chain performance: Total interpretive structural modelling approach. *International Journal of Procurement Management, 9*(4), 398–426. doi:10.1504/IJPM.2016.077702.

Bag, S. (2017a). Identification of green procurement drivers and their interrelationship using total interpretive structural modelling. *Vision, 21*(2), 129–142. doi:10.1177/0972262917700990.

Bag, S. (2017b). Comparison of green procurement framework using fuzzy TISM and fuzzy DEMATEL methods. *International Journal of Procurement Management, 10*(5), 600–638. doi:10.1504/IJPM.2017.086403.

Bag, S., & Gupta, S. (2017). Antecedents of sustainable innovation in supplier networks: A South African experience. *Global Journal of Flexible Systems Management*, 18(3), 231–250.

B-BBBEE Procurement, Transformation and Verification, Retrieved from http://www.dti.gov. za/economic_empowerment/bee_veri.jsp on 18.8.18

Beer, S., & Lemmer, C. (2011). A critical review of "green" procurement: Life cycle analysis of food products within the supply chain. *Worldwide Hospitality and Tourism Themes*, 3(3), 229–244. doi:10.1108/17554211111142194.

Blome, C., Hollos, D., & Paulraj, A. (2014). Green procurement and green supplier development: Antecedents and effects on supplier performance. *International Journal of Production Research*, 52(1), 32–49. doi:10.1080/00207543.2013.825748.

Boak, G., Watt, P., Gold, J., Devins, D., & Garvey, R. (2016). Procuring a sustainable future: An action learning approach to the development and modelling of ethical and sustainable procurement practices. *Action Learning: Research and Practice*, 13(3), 204–218. doi:10.1080/14767333.2016.1215290.

Bohari, A. A. M., Skitmore, M., Xia, B., & Teo, M. (2017). Green oriented procurement for building projects: Preliminary findings from Malaysia. *Journal of Cleaner Production*, 148, 690–700. doi:10.1016/j.jclepro.2017.01.141.

Bradley, P. (2016). Environmental impacts of food retail: A framework method and case application. *Journal of Cleaner Production*, 113, 153–166. doi:10.1016/j.jclepro.2015.09.085.

Brammer, S., & Walker, H. (2011). Sustainable procurement in the public sector: An international comparative study. *International Journal of Operations & Production Management*, 31(4), 452–476. doi:10.1108/01443571111119551.

Brauers, W. K. M., & Zavadskas, E. K. (2006). The MOORA method and its application to privatization in a transition economy. *Control and Cybernetics*, 35(2), 445–469.

Brauers, W. K. M., & Zavadskas, E. K. (2009). Robustness of the multi-objective MOORA method with a test for the facilities sector. *Technological and Economic Development of Economy*, 15(2), 352–375.

Brindley, C., & Oxborrow, L. (2014). Aligning the sustainable supply chain to green marketing needs: A case study. *Industrial Marketing Management*, 43(1), 45–55. doi:10.1016/j.indmarman.2013.08.003.

Broomes, V. (2016). Organisational governance and strategic CSR to strengthen local supply chains–Navigating the maze. *Accountability and Social Responsibility: International Perspectives*, 9, 3–21. doi:10.1108/S2043-052320160000009001.

Butt, A. A., Toller, S., & Birgisson, B. (2015). Life cycle assessment for the green procurement of roads: A way forward. *Journal of Cleaner Production*, 90, 163–170. doi:10.1016/j.jclepro.2014.11.068.

Chiarini, A., Opoku, A., & Vagnoni, E. (2017). Public healthcare practices and criteria for a sustainable procurement: A comparative study between UK and Italy. *Journal of Cleaner Production*, 162, 391–399. doi:10.1016/j.jclepro.2017.06.027.

Crespin-Mazet, F., & Dontenwill, E. (2012). Sustainable procurement: Building legitimacy in the supply network. *Journal of Purchasing and Supply Management*, 18(4), 207–217. doi:10.1016/j.pursup.2012.01.002.

Dalgliesh, C. D., Bowen, P. A., & Hill, R. C. (1997). Environmental sustainability in the delivery of affordable housing in South Africa. *Engineering, Construction and Architectural Management*, 4(1), 23–39. doi:10.1108/eb021038.

Diófási, O., & Valkó, L. (2014). Step by step towards mandatory green public procurement. *Periodica Polytechnica. Social and Management Sciences*, 22(1), 21. doi:10.3311/PPso.2151.

Dubey, R., Bag, S., Ali, S. S., & Venkatesh, V. G. (2013). Green purchasing is key to superior performance: An empirical study. *International Journal of Procurement Management*, 6(2), 187–210. doi:10.1504/IJPM.2013.052469.

Dubey, R., Gunasekaran, A., Childe, S. J., Papadopoulos, T., Fosso-Wamba, S., & Song, M. (2016). Towards a theory of sustainable consumption and production: Constructs and measurement. *Resources, Conservation and Recycling, 106*, 78–89.

Erridge, A., & Hennigan, S. (2012). Sustainable procurement in health and social care in Northern Ireland. *Public Money & Management, 32*(5), 363–370. doi:10.1080/095409 62.2012.703422.

Esfahbodi, A., Zhang, Y., Watson, G., & Zhang, T. (2017). Governance pressures and performance outcomes of sustainable supply chain management: An empirical analysis of UK manufacturing industry. *Journal of Cleaner Production, 155*, 66–78. doi:10.1016/j.jclepro.2016.07.098.

Faith-Ell, C. (2005). The introduction of environmental requirements for trucks and construction vehicles used in road maintenance contracts in Sweden. *Corporate Social Responsibility and Environmental Management, 12*(2), 62–72. doi:10.1002/csr.72.

Fernández-Viñé, M. B., Gómez-Navarro, T., & Capuz-Rizo, S. F. (2013). Assessment of the public administration tools for the improvement of the eco-efficiency of small and medium sized enterprises. *Journal of Cleaner Production, 47*, 265–273. doi:10.1016/j.jclepro.2012.08.026.

Fet, A., Michelsen, O., & Boer, L. (2011). Green public procurement in practice: The case of Norway. *Society and Economy, 33*(1), 183–198. doi:10.1556/SocEc.33.2011.1.13.

Gan, V. J., Cheng, J. C., Lo, I. M., & Chan, C. M. (2017). Developing a CO_2-e accounting method for quantification and analysis of embodied carbon in high-rise buildings. *Journal of Cleaner Production, 141*, 825–836. doi:10.1016/j.jclepro.2016.09.126.

Ghadimi, P., Azadnia, A. H., Heavey, C., Dolgui, A., & Can, B. (2016). A review on the buyer–supplier dyad relationships in sustainable procurement context: Past, present and future. *International Journal of Production Research, 54*(5), 1443–1462. doi:10.1080/00207543.2015.1079341.

Goldschmidt, K., Harrison, T., Holtry, M., & Reeh, J. (2013). Sustainable procurement: Integrating classroom learning with university sustainability programs. *Decision Sciences Journal of Innovative Education, 11*(3), 279–294. doi:10.1111/dsji.12007.

Grandia, J. (2015). The role of change agents in sustainable public procurement projects. *Public Money & Management, 35*(2), 119–126. doi:10.1080/09540962.2015.1007706.

Grandia, J. (2016). Finding the missing link: Examining the mediating role of sustainable public procurement behaviour. *Journal of Cleaner Production, 124*, 183–190. doi:10.1016/j.jclepro.2016.02.102.

Grandia, J., Steijn, B., & Kuipers, B. (2015). It is not easy being green: Increasing sustainable public procurement behaviour. *Innovation: The European Journal of Social Science Research, 28*(3), 243–260. doi:10.1080/13511610.2015.1024639.

Guenther, E., Hueske, A. K., Stechemesser, K., & Buscher, L. (2013). The 'why not': Perspective of green purchasing—A multilevel case study analysis. *Journal of Change Management, 13*(4), 407–423. doi:10.1080/14697017.2013.851950.

Guenther, E., Scheibe, L., & Greschner Farkavcová, V. (2010). "The Hurdles Analysis" as an instrument for improving sustainable stewardship. *Management Research Review, 33*(4), 340–356. doi:10.1108/01409171011030453.

Günther, E., & Scheibe, L. (2006). The hurdle analysis: A self-evaluation tool for municipalities to identify, analyse and overcome hurdles to green procurement. *Corporate Social Responsibility and Environmental Management, 13*(2), 61–77. doi:10.1002/csr.92.

Hall, P., Löfgren, K., & Peters, G. (2016). Greening the street-level procurer: Challenges in the strongly decentralized Swedish system. *Journal of Consumer Policy, 39*(4), 467–483. doi:10.1007/s10603-015-9282-8.

Hayami, H., Nakamura, M., & Nakamura, A. O. (2015). Economic performance and supply chains: The impact of upstream firms' waste output on downstream firms' performance in Japan. *International Journal of Production Economics, 160*, 47–65. doi:10.1016/j.ijpe.2014.09.012.

Hollos, D., Blome, C., & Foerstl, K. (2012). Does sustainable supplier co-operation affect performance? Examining implications for the triple bottom line. *International Journal of Production Research, 50*(11), 2968–2986. doi:10.1080/00207543.2011.582184.

Jefferies, M., John Brewer, G., & Gajendran, T. (2014). Using a case study approach to identify critical success factors for alliance contracting. *Engineering, Construction and Architectural Management, 21*(5), 465–480. doi:10.1108/ECAM-01-2012-0007.

Jefferies, M., & McGeorge, W. D. (2009). Using public-private partnerships (PPPs) to procure social infrastructure in Australia. *Engineering, Construction and Architectural Management, 16*(5), 415–437. doi:10.1108/09699980910988348.

Jelodar, M. B., Yiu, T. W., & Wilkinson, S. (2015). Systematic representation of relationship quality in conflict and dispute: For construction projects. *Construction Economics and Building, 15*(1), 89–103. doi:10.5130/ajceb.v15i1.4281.

Lenferink, S., Tillema, T., & Arts, J. (2013). Towards sustainable infrastructure development through integrated contracts: Experiences with inclusiveness in Dutch infrastructure projects. *International Journal of Project Management, 31*(4), 615–627. doi:10.1016/j.ijproman.2012.09.014.

Li, C. (2013). An integrated approach to evaluating the production system in closed-loop supply chains. *International Journal of Production Research, 51*(13), 4045–4069. doi:10.1080/00207543.2013.774467.

Lo, S. F. (2010). Global warming action of Taiwan's semiconductor/TFT-LCD industries: How does voluntary agreement work in the IT industry? *Technology in Society, 32*(3), 249–254. doi:10.1016/j.techsoc.2010.07.007.

Mani, V., Gunasekaran, A., Papadopoulos, T., Hazen, B., & Dubey, R. (2016). Supply chain social sustainability for developing nations: Evidence from India. *Resources, Conservation and Recycling, 111*, 42–52.

Mansi, M. (2015). Sustainable procurement disclosure practices in central public sector enterprises: Evidence from India. *Journal of Purchasing and Supply Management, 21*(2), 125–137. doi:10.1016/j.pursup.2014.12.002.

Mansi, M., & Pandey, R. (2016). Impact of demographic characteristics of procurement professionals on sustainable procurement practices: Evidence from Australia. *Journal of Purchasing and Supply Management, 22*(1), 31–40. doi:10.1016/j.pursup.2015.06.001.

McMurray, A. J., Islam, M. M., Siwar, C., & Fien, J. (2014). Sustainable procurement in Malaysian organizations: Practices, barriers and opportunities. *Journal of Purchasing and Supply Management, 20*(3), 195–207. doi:10.1016/j.pursup.2014.02.005.

Meehan, J., & Bryde, D. (2011). Sustainable procurement practice. *Business Strategy and the Environment, 20*(2), 94–106. doi:10.1002/bse.678.

Meehan, J., & Bryde, D. J. (2014). Procuring sustainably in social housing: The role of social capital. *Journal of Purchasing and Supply Management, 20*(2), 74–81. doi:10.1016/j.pursup.2014.01.002.

Meehan, J., & Bryde, D. J. (2015). A field-level examination of the adoption of sustainable procurement in the social housing sector. *International Journal of Operations & Production Management, 35*(7), 982–1004. doi:10.1108/IJOPM-07-2014-0359.

Mosgaard, M., Riisgaard, H., & Huulgaard, R. D. (2013). Greening non-product-related procurement: When policy meets reality. *Journal of Cleaner Production, 39*, 137–145. doi:10.1016/j.jclepro.2012.08.018.

Mosgaard, M. A. (2015). Improving the practices of green procurement of minor items. *Journal of Cleaner Production, 90*, 264–274. doi:10.1016/j.jclepro.2014.11.077.

Nagel, M. H. (2003). Managing the environmental performance of production facilities in the electronics industry: More than application of the concept of cleaner production. *Journal of Cleaner Production, 11*(1), 11–26. doi:10.1016/S0959-6526(02)00021-5.

Odeyale, S. O. (2014). Performance appraisal for green/environmental friendliness of a supply chain department. *Journal of Industrial Engineering and Management, 7*(5), 1316–1333. doi:10.3926/jiem.1057.

Offei, I., Kissi, E., & Badu, E. (2016). Public procurement policies and strategies for capacity building of SME construction firms in Ghana. *International Journal of Procurement Management, 9*(4), 455–472. doi:10.1504/IJPM.2016.077705.

Oruezabala, G., & Rico, J. C. (2012). The impact of sustainable public procurement on supplier management: The case of French public hospitals. *Industrial Marketing Management, 41*(4), 573–580. doi:10.1016/j.indmarman.2012.04.004.

Pacheco-Blanco, B., & Bastante-Ceca, M. J. (2016). Green public procurement as an initiative for sustainable consumption: An exploratory study of Spanish public universities. *Journal of Cleaner Production, 133*, 648–656. doi:10.1016/j.jclepro.2016.05.056.

Perera, C., Auger, P., & Klein, J. (2016). Green consumption practices among young environmentalists: A practice theory perspective. *Journal of Business Ethics*, 1–22. doi:10.1007/s10551-016-3376-3.

Preuss, L. (2009). Addressing sustainable development through public procurement: The case of local government. *Supply Chain Management: An International Journal, 14*(3), 213–223. doi:10.1108/13598540910954557.

Rietbergen, M. G., & Blok, K. (2013). Assessing the potential impact of the CO_2 performance ladder on the reduction of carbon dioxide emissions in the Netherlands. *Journal of Cleaner Production, 52*, 33–45. doi:10.1016/j.jclepro.2013.03.027.

Roman, A. V. (2017). Institutionalizing sustainability: A structural equation model of sustainable procurement in US public agencies. *Journal of Cleaner production, 143*, 1048–1059. doi:10.1016/j.jclepro.2016.12.014.

Routroy, S., & Pradhan, S. K. (2012). Framework for green procurement: A case study. *International Journal of Procurement Management, 5*(3), 316–336. doi:10.1504/IJPM.2012.047170.

Ruparathna, R., & Hewage, K. (2015). Sustainable procurement in the Canadian construction industry: Current practices, drivers and opportunities. *Journal of Cleaner Production, 109*, 305–314. doi:10.1016/j.jclepro.2015.07.007.

Sakuragi, Y. (2002). A new partnership model for Japan: Promoting a circular flow society. *Corporate Environmental Strategy, 9*(3), 292–296. doi:10.1016/S1066-7938(02)00072-6.

Sambhanthan, A., & Potdar, V. (2015). Green business practices for software development companies: An explorative text analysis of business sustainability reports. *International Journal of Enterprise Information Systems, 11*(3), 13–26. doi:10.4018/IJEIS.2015070102.

Shen, L., Zhang, Z., & Zhang, X. (2017). Key factors affecting green procurement in real estate development: A China study. *Journal of Cleaner Production, 153*, 372–383. doi:10.1016/j.jclepro.2016.02.021.

Sheu, J. B. (2016). Buyer behavior in quality-dominated multi-sourcing recyclable-material procurement of green supply chains. *Production and Operations Management, 25*(3), 477–497. doi:10.1111/poms.12263.

Shi, P., Yan, B., Shi, S., & Ke, C. (2015). A decision support system to select suppliers for a sustainable supply chain based on a systematic DEA approach. *Information Technology and Management, 16*(1), 39–49. doi:10.1007/s10799-014-0193-1.

Smith, J., Andersson, G., Gourlay, R., Karner, S., Mikkelsen, B. E., Sonnino, R., & Barling, D. (2016). Balancing competing policy demands: The case of sustainable public sector food procurement. *Journal of Cleaner Production, 112*, 249–256. doi:10.1016/j.jclepro.2015.07.065.

Song, H., Yu, K., & Zhang, S. (2017). Green procurement, stakeholder satisfaction and operational performance. *The International Journal of Logistics Management, 28*(4), 1054–1077. doi:10.1108/IJLM-12-2015-0234.

Sourani, A., & Sohail, K. (2013). Enabling sustainable construction in UK public procurement. *Proceedings of Institution of Civil Engineers: Management, Procurement and Law, 166*(6), 297–312. doi:10.1680/mpal.12.00022.

Stanujkic, D., Magdalinovic, N., Jovanovic, R., & Stojanovic, S. (2012). An objective Multi-measures approach to optimization using MOORA method and interval grey numbers. *Technological and Economic Development of Economy, 18*(2), 331–363.

Sterner, E. (2002). 'Green procurement' of buildings: A study of Swedish clients' considerations. *Construction Management & Economics, 20*(1), 21–30. doi:10.1080/01446190110093560.

Suresh, S., Renukappa, S., Akintoye, A., & Egbu, C. (2016). Sustainable procurement strategies for competitive advantage: An empirical study. *Proceedings of the Institution of Civil Engineers-Management, Procurement and Law, 169*(1), 16–25. doi:10.1680/jmapl.15.00006.

Townsend, J., & Barrett, J. (2015). Exploring the applications of carbon footprinting towards sustainability at a UK university: Reporting and decision making. *Journal of Cleaner Production, 107*, 164–176. doi:10.1016/j.jclepro.2013.11.004.

Tseng, M. L., Lim, M., Wu, K. J., Zhou, L., & Bui, D. T. D. (2017). A novel approach for enhancing green supply chain management using converged interval-valued triangular fuzzy numbers-grey relation analysis. *Resources, Conservation and Recycling.* doi:10.1016/j.resconrec.2017.01.007.

Varnäs, A., Balfors, B., & Faith-Ell, C. (2009). Environmental consideration in procurement of construction contracts: Current practice, problems and opportunities in green procurement in the Swedish construction industry. *Journal of Cleaner Production, 17*(13), 1214–1222. doi:10.1016/j.jclepro.2009.04.001.

Walker, H., & Brammer, S. (2009). Sustainable procurement in the United Kingdom public sector. *Supply Chain Management: An International Journal, 14*(2), 128–137. doi:10.1108/13598540910941993.

Walker, H., & Brammer, S. (2012). The relationship between sustainable procurement and e-procurement in the public sector. *International Journal of Production Economics, 140*(1), 256–268. doi:10.1016/j.ijpe.2012.01.008.

Walker, H., & Phillips, W. (2009). Sustainable procurement: Emerging issues. *International Journal of Procurement Management, 2*(1), 41–61. doi:10.1504/IJPM.2009.021729.

Walker, H. L., Gough, S., Bakker, E. F., Knight, L. A., & McBain, D. (2009). Greening operations management: An online sustainable procurement course for practitioners. *Journal of Management Education, 33*(3), 348–371. doi:10.1177/1052562908323190.

Whatling, D. R., Hedges, P., Brown, R., & Fermor, P. (2010). Corporate responsibility reporting of biodiversity in the supply chain. *International Journal of Innovation and Sustainable Development, 5*(1), 51–64. doi:10.1504/IJISD.2010.034557.

Whitelock, V. G. (2012). Alignment between green supply chain management strategy and business strategy. *International Journal of Procurement Management, 5*(4), 430–451. doi:10.1504/IJPM.2012.047198.

Wong, J. K. W., San Chan, J. K., & Wadu, M. J. (2016). Facilitating effective green procurement in construction projects: An empirical study of the enablers. *Journal of Cleaner Production, 135*, 859–871. doi:10.1016/j.jclepro.2016.07.001.

Yen, C. H., Chen, C. Y., & Teng, H. Y. (2013). Perceptions of environmental management and employee job attitudes in hotel firms. *Journal of Human Resources in Hospitality & Tourism, 12*(2), 155–174. doi:10.1080/15332845.2013.752709.

Young, S., Nagpal, S., & Adams, C. A. (2016). Sustainable procurement in Australian and UK universities. *Public Management Review, 18*(7), 993–1016. doi:10.1080/14719037.2015.1051575.

6 New Model for Sustainable Supplier Selection Using Fuzzy DEMATEL

Oussama El Mariouli and
Abdellah Abouabdellah

CONTENTS

6.1   INTRODUCTION

The competitive edge of today's markets and the consideration of environmental and social issues in companies have led companies to attach great importance to the process of selecting the best suppliers who respect the three dimensions of economies, environments, and socials of the sustainable development (SD).

Any company that aims to remain competitive and meet the demands of the pressures of different stakeholders (shareholders, unions, employees, suppliers, local communities, governments, competitors, the media, other interest groups, customers, consumer associations, environmental groups, and the natural environment are the list of stakeholders) on economic prosperity, environmental protection, and the achievement of social justice with sensitivity and balance in all business processes is asking to adopt SD approach. This approach begins with the selection of the best supplier who respects the three pillars of SD.

The new supplier selection (SS) process requires new criteria for purchasing decisions instead of the standard criteria (quality, cost, and lead time), and has become a multi-criteria decision that is based on several qualitative and quantitative criteria.

In the literature, there are several studies that deal with the problems of SS. Most of this research does not integrate the three dimensions at the same time of SD in the selection, instead they focus on the environmental dimension and neglect the social dimension (see Zimmer et al., 2015, Vahidi et al., 2017). This research presents different methods and techniques of SS (see Shabanpour et al., 2016, El Mariouli and Aboubadellah, 2018, Sureeyatanapas et al., 2018).

In this chapter, we present our new mathematical model that selects the most important suppliers that respect the rules of SD. Our model uses the Fuzzy DEMATEL (Fuzzy Decision-Making Trial and Evaluation Laboratory) approach.

This model is articulated around four phases of realization, which we will detail below:

- The first phase: The identification of the SD criteria used to measure supplier performance;
- The second phase: For the calculation of the weight of the criteria for the three dimensions of SD, we used the method of Fuzzy DEMATEL;
- The third phase: The choice of a scenario to determine the coefficients of the economic, environmental, and social dimensions;
- The last phase: The presentation of our mathematical model that takes into account the SD criteria in the evaluation and the SS.

6.2 CRITERIA FOR MEASURING SUSTAINABLE SUPPLIER PERFORMANCE

In this section, we have made a broad synthetic analysis of an area that has experienced rapid growth in knowledge (see Zsidisin and Ancarani, 2016).

The literature review follows the systematic methodology of Briner and Denyer (see Briner and Denyer 2012) to reduce the number of articles published and focus on the major newspapers that publish research on the sustainable supply chain.

In our initial library search on relevant documents, we did an analysis of the different databases, including ScienceDirect, Springer, Emerald, Taylor & Francis, Inderscience, Google Scholar, by using the combination of keywords: "procurement, supply chain, sustainable development, performance, indicators, supplier's selection, review of the literature, criteria, economic, social, environment." These keywords were carefully chosen to ensure that as many relevant articles as possible would be included. We have also analyzed several international standards such as SCOR, GRI, OECD, ISO 26000, and so on.

We have identified that the first list of criteria that selects suppliers was made by Dickson in 1966. He surveyed 274 companies in Canada and the United States who were members of the National Association of Purchasing Managers. He came up

TABLE 6.1
SS Criteria in Literature

Economic Criteria	Environmental Criteria	Social Criteria
Innovation capacity (C_1)	Waste (C_{13})	Human rights (C_{22})
Production capacity (C_2)	Emissions (C_{14})	Jobs and wealth (C_{23})
Technical and technological capacity (C_3)	Environmental label (C_{15})	Training, support, and education (C_{24})
Cost (C_4)	Pollution (C_{16})	Health and security at work (C_{25})
Deadlines (C_5)	Program (C_{17})	Condition of work (C_{26})
Reliability (C_6)	Recycling (C_{18})	
Financial (C_7)	The respect of the rules Ethical Environmental (C_{19})	
Flexibility (C_8)	Use of Resources (C_{20})	
Delivery (C_9)	Toxic or dangerous substances (C_{21})	
Quality (C_{10})		
Reactivity (C_{11})		
Customer references (C_{12})		

Source: El Mariouli, O., and Abouabdellah, A., Model for assessing the economic, environmental and social performance of the supplier, *4th IEEE International Conference on Logistics Operations Management* (GOL'2018), Lehavre, France, 2018.

with 23 criteria used by these companies in the 1960s for SS (see Dickson, 1966). We followed the articles found in the literature that are published until the year 2016.

Our process of analysis of the scientific works and the international standards gave like result (see Table 6.1).

6.3 WEIGHTS OF THE CRITERIA

In this part, we present a Fuzzy DEMATEL hybrid approach to calculate the weight of each criterion for the three dimensions of SD.

The Fuzzy DEMATEL approach is the combination of the DEMATEL method presented by Geneva Research Center of the Battelle Memorial Institute (see Fontela and Gabus, 1972) and the fuzzy set theory developed by Zadeh (see Set and Zadeh, 1965). This new hybrid approach solves the problems of uncertainties related to human judgments. The Fuzzy DEMATEL method helps to build the interrelationships between criteria and factors (see Fontela and Gabus, 1974) based on the expert's point of view and to visualize in the IRM diagram (Impact Relation Map) the cause-effect relationships between the criteria.

6.3.1 THE FUZZY DEMATEL

The Fuzzy DEMATEL process (see Orji and Wei, 2014) develops according to the following steps:

Step 1: Linguistic scale direct-relation matrix
  In this step, experts are asked to complete a questionnaire to indicate the degree of connection between criterion i and j by five linguistic variables (see Table 6.2). The decision of the experts will be easy by the use of the theory of linguistic variables. The Linguistic scale direct-relation matrix $Z^k = [z_{ij}^k]$ is formed by Linguistic scale variables, and after these variable are transformed to fuzzy number according to the Table 6.2, Z^k is a non-negative matrix of dimension $n \times n$ and diagonal $z_{ij}^k = 0$, z_{ij}^k represents the effect of criteria i on criteria j.

$$Z^k = \left[z_{ij}^k\right] = \begin{bmatrix} 0 & z_{12} & \cdots & z_{1n} \\ z_{21} & 0 & \cdots & z_{2n} \\ . & : & : & : \\ z_{n1} & z_{n2} & \cdots & 0 \end{bmatrix} \tag{6.1}$$

Step 2: Direct-relation matrix
  The initial direct-relation matrix H^k is calculated according to the CFCS (Converting Fuzzy data into Crisps Scores) process (see Opricovic and Tzeng, 2003).
  Let $z_{ij} = (a_{ij}^n, b_{ij}^n, c_{ij}^n)$ the impact of i on j and $n = (1,2,3,..., p)$ questionnaire. The CFCS algorithm is presented as follows:

- **Standardization:**

$$xc_{ij}^n = (c_{ij}^n - \min a_{ij}^n) / \Delta_{\min}^{\max} \tag{6.2}$$

$$xb_{ij}^n = (b_{ij}^n - \min a_{ij}^n) / \Delta_{\min}^{\max} \tag{6.3}$$

$$xa_{ij}^n = (a_{ij}^n - \min a_{ij}^n) / \Delta_{\min}^{\max} \tag{6.4}$$

TABLE 6.2
Linguistic Variables

Linguistic Term Criteria	Fuzzy Value
No influence	(0, 0, 1/4)
Very low influence	(0, 1/4, 1/2)
Low influence	(1/4, 1/2, 3/4)
High influence	(1/2, 3/4, 1)
Very high influence	(3/4, 1, 1)

Source: Li, R.J., *Comput. Math. Appl.*, 38, 91–101, 1999.

- **Calculates normalized values right (cs) and left (as)**

$$xcs_{ij}^n = xc_{ij}^n / (1 + xc_{ij}^n - xb_{ij}^n)$$ (6.5)

$$xas_{ij}^n = xb_{ij}^n / (1 + xb_{ij}^n - xa_{ij}^n)$$ (6.6)

- **Calculate total normalized crips values**

$$x_{ij}^n = \left[xas_{ij}^n(1 - xas_{ij}^n) + xcs_{ij}^n \times xcs_{ij}^n \right] / (1 - xas_{ij}^n + xcs_{ij}^n)$$ (6.7)

- **Calculate crips values**

$$h_{ij}^n = \min a_{ij}^n + x_{ij}^n \times \Delta_{\min}^{\max}$$ (6.8)

- **Final crips values**

$$h_{ij} = \frac{1}{p}(h_{ij}^1 + h_{ij}^2 + h_{ij}^3 + ... + h_{ij}^p)$$ (6.9)

- **The initial direct-relation matrix is**

$$H^k = \begin{bmatrix} 0 & h_{12} & ... & h_{1n} \\ h_{21} & 0 & ... & h_{2n} \\ . & : & : & : \\ h_{n1} & h_{n2} & ... & 0 \end{bmatrix}$$ (6.10)

Step 3: Normalized fuzzy direct-relational matrix

In this step, we calculate the matrix normalized fuzzy direct-relation N.

$$N = \frac{H^k}{\max_{1 \le i \le n} \sum_{j=1}^n h_{ij}} \quad i, j=(1,2,3,..., n)$$ (6.11)

Step 4: Total-relation matrix

The total-relation matrix T is presented as follows:

$$T = N(I - N)^{-1}$$ (6.12)

I represent the identical matrix ($n \times n$).

Step 5: Causality-effect graph

To obtain the causality-effect graph we calculate ($r_i + c_j, r_i - c_j$):

r_i: sum of the ith row and

c_j: sum of the ith column.

$r_i + c_j$: represent horizontal axis
$r_i - c_j$: represent vertical axis

$$T = \left[t_{ij} \right]_{n \times n} \quad i, j = (1,2,3,\dots, n) \tag{6.13}$$

$$r_i = \sum_{1 \leq j \leq n}^{n} t_{ij} \forall i \tag{6.14}$$

$$c_j = \sum_{1 \leq i \leq n}^{n} t_{ij} \forall j \tag{6.15}$$

If $r_i - c_j > 0$, the criterion represents the cause in the group.
If $r_i - c_j < 0$, the criterion represents the effect in the group.

Step 6: Weights of criteria
The weight of the criteria is obtained by:

$$w_i = \sqrt{(r_i + c_j)^2 + (r_i - c_j)^2} \tag{6.16}$$

$$W_i = \frac{w_i}{\sum_{i=1}^{n} w_i} \tag{6.17}$$

6.3.2 DATA ANALYSIS

In enterprise X, we asked the three decision makers to evaluate the influence between the criteria (Table 6.2) with scores, where the scores 0, 1, 2, 3, and 4 represent: (no influence), (very low influence), (low influence), (high influence), and (very high influence), respectively.

Step 1: Linguistic scale direct-relation matrix
Every decision maker completed three direct-relation matrix for each dimension SD: economic matrix (12 × 12) (Table 6.6), environmental matrix (9 × 9) (Table 6.7), and social matrix (5 × 5) (Table 6.8).

For example, the direct-relation matrix obtained from the decision maker (DM) questionnaire 1 for each dimension of SD is shown in Tables 6.3 through 6.5):

Step 2: Direct-relation matrix
The direct-relation matrix is obtained by developing the CFCS process (Eqs. 6.2 through 6.9) of each dimension of SD for each DM, this matrix is shown in Table 6.6.

Step 3: Normalized fuzzy direct-relational matrix
The values of normalized fuzzy direct-relational matrix of economic dimension are bounded between 0 and 1, this matrix is shown in Table 6.7.

Step 4: Total-relation matrix
The total-relation matrix is acquired using Eq. 6.12. This matrix is shown in Table 6.8:

TABLE 6.3

Direct-Relation Matrix of Economic Dimension from DM Number 1

	C_1	C_2	C_3	C_4	C_5	C_6	C_7	C_8	C_9	C_{10}	C_{11}	C_{12}
C_1	0	VH	H	H	VH	H	H	H	L	H	VH	L
C_2	NO	0	NO	H	VH	H	H	L	NO	VH	VH	H
C_3	NO	VH	0	VH	H	H	L	VH	NO	VH	VH	H
C_4	NO	VL	VH	0	H	VL	H	NO	L	VH	L	L
C_5	NO	NO	NO	VH	0	VH	NO	VL	H	VL	VH	L
C_6	NO	NO	NO	VH	VH	0	NO	NO	NO	VH	H	VH
C_7	VH	NO	L	L	NO	NO	0	NO	NO	NO	NO	NO
C_8	NO	NO	NO	NO	VL	NO	NO	0	NO	VL	VH	L
C_9	NO	NO	NO	VH	NO	VH	NO	NO	0	NO	VL	VH
C_{10}	NO	NO	L	VH	VL	VH	NO	NO	NO	0	NO	VH
C_{11}	NO	NO	L	L	H	NO	NO	VL	NO	NO	0	H
C_{12}	NO	NO	NO	H	VL	H	NO	NO	NO	NO	NO	0

TABLE 6.4

Direct-Relation Matrix of Environmental Dimension from DM Number 1

	C_{13}	C_{14}	C_{15}	C_{16}	C_{17}	C_{18}	C_{19}	C_{20}	C_{21}
C_{13}	0	NO	NO	VH	NO	VH	VH	NO	NO
C_{14}	NO	0	NO	VH	NO	NO	VH	NO	NO
C_{15}	VH	VH	0	VH	VH	VH	H	VH	H
C_{16}	H	VH	NO	0	NO	NO	H	NO	NO
C_{17}	VH	VH	VH	VH	0	VH	H	VH	VH
C_{18}	H	NO	NO	NO	NO	0	L	H	H
C_{19}	VH	VH	H	H	H	L	0	VH	VH
C_{20}	VH	VH	NO	VH	NO	NO	NO	0	NO
C_{21}	VH	VH	NO	VH	NO	NO	NO	NO	0

TABLE 6.5

Direct-Relation Matrix of Dimension Social from DM Number 1

	C_{22}	C_{23}	C_{24}	C_{25}	C_{26}
C_{22}	0	NO	NO	VH	VH
C_{23}	NO	0	NO	NO	VH
C_{24}	VH	VH	0	VH	VH
C_{25}	H	NO	NO	0	VH
C_{26}	VH	VH	NO	VH	0

TABLE 6.6
The Direct-Relation Matrix of Economic Dimension

	C_1	C_2	C_3	C_4	C_5	C_6	C_7	C_8	C_9	C_{10}	C_{11}	C_{12}
C_1	0,000	2,670	2,430	1,500	2,670	1,730	2,670	2,670	1,270	1,030	2,670	0,800
C_2	0,090	0,000	0,090	1,960	2,670	2,190	1,960	1,500	0,330	2,430	2,430	1,730
C_3	0,090	2,670	0,000	2,670	1,730	2,430	1,730	2,430	0,800	2,910	2,430	2,190
C_4	0,330	1,040	2,430	0,000	1,730	0,330	2,430	0,330	1,030	2,670	1,260	1,260
C_5	0,330	0,090	1,030	2,670	0,000	2,430	0,090	1,500	2,190	1,040	2,430	1,040
C_6	0,330	0,090	0,330	2,430	2,670	0,000	0,570	0,090	0,330	2,430	2,190	2,430
C_7	2,670	0,560	1,500	1,500	0,090	0,570	0,000	0,090	0,090	0,330	0,090	0,090
C_8	0,090	0,570	0,800	0,330	0,800	0,090	0,090	0,090	0,330	1,730	2,430	0,560
C_9	0,090	0,090	0,570	2,430	0,090	2,670	0,330	0,090	0,000	0,570	0,330	1,970
C_{10}	0,090	0,090	0,800	2,670	0,330	2,430	0,090	0,330	0,090	0,000	0,090	2,670
C_{11}	0,570	0,330	0,800	0,560	2,430	0,800	0,090	0,570	0,090	0,090	0,000	1,260
C_{12}	0,330	0,090	0,330	2,190	0,330	2,430	0,090	0,330	0,090	0,090	0,090	0,000

TABLE 6.7
Normalized Fuzzy Direct-Relational Matrix of Economic Dimension

	C_1	C_2	C_3	C_4	C_5	C_6	C_7	C_8	C_9	C_{10}	C_{11}	C_{12}
C_1	0,000	0,128	0,116	0,072	0,128	0,083	0,128	0,128	0,061	0,049	0,128	0,038
C_2	0,004	0,000	0,004	0,094	0,128	0,105	0,094	0,072	0,016	0,116	0,116	0,083
C_3	0,004	0,128	0,000	0,128	0,083	0,116	0,083	0,116	0,038	0,139	0,116	0,105
C_4	0,016	0,050	0,116	0,000	0,083	0,016	0,116	0,016	0,049	0,128	0,060	0,060
C_5	0,016	0,004	0,049	0,128	0,000	0,116	0,004	0,072	0,105	0,050	0,116	0,050
C_6	0,016	0,004	0,016	0,116	0,128	0,000	0,027	0,004	0,016	0,116	0,105	0,116
C_7	0,128	0,027	0,072	0,072	0,004	0,027	0,000	0,004	0,004	0,016	0,004	0,004
C_8	0,004	0,027	0,038	0,016	0,038	0,004	0,004	0,000	0,016	0,083	0,116	0,027
C_9	0,004	0,004	0,027	0,116	0,004	0,128	0,016	0,004	0,000	0,027	0,016	0,094
C_{10}	0,004	0,004	0,038	0,128	0,016	0,116	0,004	0,016	0,004	0,000	0,004	0,128
C_{11}	0,027	0,016	0,038	0,027	0,116	0,038	0,004	0,027	0,004	0,004	0,000	0,060
C_{12}	0,016	0,004	0,016	0,105	0,016	0,116	0,004	0,016	0,004	0,004	0,004	0,000

TABLE 6.8

Total-Relation Matrix of Economic Dimension

	C_1	C_2	C_3	C_4	C_5	C_6	C_7	C_8	C_9	C_{10}	C_{11}	C_{12}
C_1	0,056	0,199	0,224	0,280	0,285	0,255	0,219	0,217	0,131	0,213	0,291	0,205
C_2	0,046	0,045	0,091	0,246	0,235	0,226	0,151	0,127	0,069	0,222	0,222	0,203
C_3	0,053	0,183	0,107	0,322	0,231	0,275	0,167	0,188	0,100	0,285	0,258	0,263
C_4	0,056	0,102	0,188	0,154	0,179	0,146	0,175	0,080	0,094	0,228	0,159	0,176
C_5	0,044	0,050	0,123	0,257	0,107	0,222	0,066	0,120	0,145	0,155	0,211	0,165
C_6	0,046	0,043	0,089	0,242	0,212	0,112	0,080	0,055	0,062	0,199	0,186	0,213
C_7	0,147	0,077	0,130	0,162	0,086	0,105	0,061	0,058	0,040	0,095	0,086	0,079
C_8	0,019	0,049	0,073	0,088	0,093	0,069	0,032	0,031	0,038	0,129	0,161	0,088
C_9	0,025	0,033	0,076	0,203	0,074	0,193	0,060	0,035	0,028	0,100	0,079	0,165
C_{10}	0,025	0,036	0,089	0,219	0,088	0,187	0,051	0,049	0,035	0,079	0,072	0,200
C_{11}	0,043	0,043	0,079	0,109	0,169	0,106	0,038	0,064	0,037	0,064	0,066	0,118
C_{12}	0,031	0,028	0,056	0,168	0,073	0,161	0,041	0,041	0,029	0,066	0,059	0,058

Step 5: Causality-effect graph

The sum of the rows and the sum of the columns are separately denoted as R and C within the total relation matrix (see Table 6.8).

The weight and the degree of the influence for each Supplier selection economic criteria is shown in Table 6.9.

Figure 6.1 represents the diagram of cause-effect relationships between economic criteria.

TABLE 6.9
Weights of Economic Criteria

	R	C	R + C	Rank	R − C	Rank	Group	Weight
C_1	2,575	0,591	3,166	5	1,984	1	Cause	0,101
C_2	1,883	0,887	2,769	8	0,996	3	Cause	0,079
C_3	2,432	1,325	3,757	2	1,108	2	Cause	0,106
C_4	1,737	2,451	4,188	1	−0,715	10	Effect	0,115
C_5	1,664	1,833	3,497	4	−0,169	6	Effect	0,095
C_6	1,541	2,058	3,599	3	−0,518	8	Effect	0,098
C_7	1,124	1,140	2,264	10	−0,016	5	Effect	0,061
C_8	0,870	1,066	1,935	11	−0,196	7	Effect	0,053
C_9	1,072	0,807	1,879	12	0,265	4	Cause	0,051
C_{10}	1,130	1,836	2,966	6	−0,706	9	Effect	0,082
C_{11}	0,937	1,849	2,786	7	−0,912	11	Effect	0,079
C_{12}	0,811	1,933	2,744	9	−1,121	12	Effect	0,080

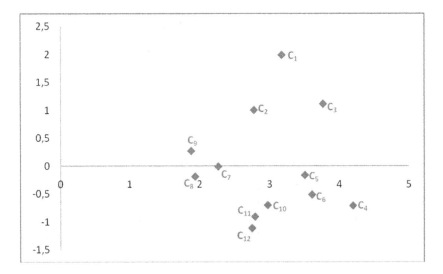

FIGURE 6.1   The causal diagram of economic dimension.

The importance of the economic criteria: $C_4 > C_3 > C_6 > C_5 > C_1 > C_{10} > C_{11} > C_2 > C_{12} > C_7 > C_8 > C_9$ according to the degree of importance of $(R + C)$. Contrary to the importance of the individual criteria, C_1, C_2, C_3, and C_9 are causes according to the value of the difference $(R - C)$.

If we follow the same steps in the economic dimension, we obtain (Tables 6.10 and 6.11):

Figure 6.2 shows the diagram of cause-effect relationships between environmental criteria.

The importance of the economic criteria: $C_{19} > C_{16} > C_{15} > C_{17} > C_{13} > C_{14} > C_{20} > C_{18} > C_{21}$ according to the degree of importance of $(R + C)$. In contrast to the importance of the individual criteria, C_{15}, C_{17}, and C_{19} are causes according to the value of the difference $(R - C)$.

Figure 6.3 represents the diagram of cause-effect relationships between social criteria.

TABLE 6.10
Weights of Environment Criteria

	R	C	R + C	Rank	R − C	Rank	Group	Weight
C_{13}	1,114	2,280	3,395	5	−1,166	7	Effect	0,112
C_{14}	0,942	2,452	3,394	6	−1,510	8	Effect	0,115
C_{15}	2,942	0,742	3,684	3	2,199	2	Cause	0,133
C_{16}	1,205	2,874	4,079	2	−1,669	9	Effect	0,137
C_{17}	2,876	0,670	3,546	4	2,206	1	Cause	0,130
C_{18}	0,917	1,424	2,342	8	−0,507	6	Effect	0,074
C_{19}	2,843	2,035	4,879	1	0,808	3	Cause	0,154
C_{20}	1,071	1,285	2,356	7	−0,214	5	Effect	0,074
C_{21}	1,072	1,221	2,293	9	−0,148	4	Effect	0,071

TABLE 6.11
Weights of Social Criteria

	R	C	R + C	Rank	R − C	Rank	Group
C_{22}	1,958	1,403	3,361	2	0,555	2	Cause
C_{23}	0,756	1,107	1,863	5	−0,351	3	Effect
C_{24}	2,024	0,366	2,391	4	1,658	1	Cause
C_{25}	1,138	1,632	2,771	3	−0,494	4	Effect
C_{26}	1,503	2,116	3,619	1	−0,612	5	Effect

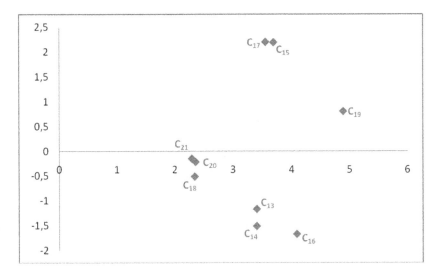

FIGURE 6.2 The causal diagram of environment dimension.

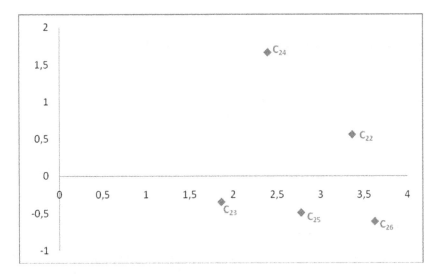

FIGURE 6.3 The causal diagram of social dimension.

The importance of the economic criteria: $C_{19} > C_{16} > C_{15} > C_{17} > C_{13} > C_{14} > C_{20} > C_{18} > C_{21}$ according to the degree of importance of $(D + R)$. In contrast to the importance of the individual criteria, C_{15}, C_{17}, and C_{19} are causes according to the value of the difference $(D - R)$.

Step 6: Weights of criteria

The weight of the criteria is obtained by Eq. 6.17 (see Table 6.12):

TABLE 6.12

Weights of Criteria

Economic Criteria		Environmental Criteria		Social Criteria	
Criteria	Weight	Criteria	Weight	Criteria	Weight
C_1	0,101	C_{13}	0,112	C_{22}	0,232
C_2	0,079	C_{14}	0,115	C_{23}	0,129
C_3	0,106	C_{15}	0,133	C_{24}	0,198
C_4	0,115	C_{16}	0,137	C_{25}	0,191
C_5	0,095	C_{17}	0,130	C_{26}	0,250
C_6	0,098	C_{18}	0,074		
C_7	0,061	C_{19}	0,154		
C_8	0,053	C_{20}	0,074		
C_9	0,051	C_{21}	0,071		
C_{10}	0,082				
C_{11}	0,079				
C_{12}	0,080				

6.4   CHOICE OF SUSTAINABILITY SCENARIO

In this step we choose a scenario for determining the weights of the three pillars of SD in such a way that:

$$W_{eco} + W_{env} + W_{soc} = 1 \quad W_{eco} > 0, W_{env} > 0, W_{soc} > 0 \tag{6.18}$$

For example, $W_{eco} = W_{env} = W_{soc} = \dfrac{1}{3}$

$$W_{eco} = \frac{1}{2} \text{ and } W_{env} = W_{soc} = \frac{1}{4}$$

6.5   PROPOSED MATHEMATICAL MODEL

In this part, we present our mathematical model that will allow decision makers to choose the most sustainable supplier to maximize sustainability. The objective function should include the three dimensions of SD: environmental, economic and social.

Each dimension index of SD is calculated by set of criteria and is banded between 0 and 1.

6.5.1   MATHEMATICAL MODEL DEVELOPMENT

- **Decision variables**
  I_{eco}: Economic dimension index
  I_{env}: Environmental dimension index
  I_{soc}: Social dimension index
  W_i: Weight associated to the criterion C_i

W_{eco}: Weight associated to the economic dimension
W_{env}: Weight associated to the environmental dimension
W_{soc}: Weight associated to the social dimension

- **The objective function**

  The objective function of the proposed model is obtained by calculating the economic, environmental and social indices:

$$I_{eco} = \frac{\sum_{i=1}^{12} W_i C_i}{12} \quad i = (1,2,\ldots,12) \tag{6.19}$$

$$I_{env} = \frac{\sum_{i=13}^{21} W_i C_i}{9} \quad i = (13,14,\ldots,21) \tag{6.20}$$

$$I_{soc} = \frac{\sum_{i=22}^{26} W_i C_i}{5} \quad i = (22,23,\ldots,26) \tag{6.21}$$

The sustainable performance of the supplier is given by the *Max P* function, which must be maximum ($0 \leq MaxP \leq 1$):

$$Max\ P = \frac{W_{eco} I_{eco} + W_{env} I_{env} + W_{soc} I_{soc}}{3} \tag{6.22}$$

- **Constraints:**

$$\sum_{i=1}^{12} W_i = 1 \tag{6.23}$$

$$\sum_{i=13}^{21} W_i = 1 \tag{6.24}$$

$$\sum_{i=22}^{26} W_i = 1$$

$$W_{eco} = W_{env} = W_{soc} = 1 \tag{6.25}$$

$$0 \leq C_i \leq 1, \forall i = \{1,2,\ldots,26\} \tag{6.26}$$

6.5.2 DATA ANALYSIS

The criteria value collected from supplier in case study. After all criteria value under each of the three pillars are multiplied by their respective weights (see Table 6.12) and aggregated as indicated by Eqs. (6.19 through 6.21). The result is three index

TABLE 6.13

Index of Environmental, Economic, and Social Dimensions

Dimension	Criteria	Weight	Supplier1	Supplier2	Supplier3	Supplier4	Supplier5
Economic	C_1	0,101	0,500	0,600	0,700	0,800	0,900
	C_2	0,079	0,200	0,400	0,600	0,800	1,000
	C_3	0,106	0,500	0,600	0,700	0,800	0,900
	C_4	0,115	0,910	0,920	0,930	0,940	0,950
	C_5	0,095	0,167	0,200	0,250	0,333	0,500
	C_6	0,098	0,500	0,600	0,700	0,800	0,900
	C_7	0,061	0,917	0,929	0,941	0,950	0,980
	C_8	0,053	0,333	0,500	0,500	1,000	1,000
	C_9	0,051	0,005	0,006	0,007	0,010	0,011
	C_{10}	0,082	0,710	0,770	0,790	0,870	0,920
	C_{11}	0,079	0,500	0,500	1,000	1,000	1,000
	C_{12}	0,080	0,850	0,880	0,900	0,980	1,000
	I_{eco}		0,528376	0,595915	0,691638	0,701738	0,858031
Environment	C_{13}	0,112	0,333	0,500	0,500	1,000	1,000
	C_{14}	0,115	0,333	0,333	0,500	0,500	1,000
	C_{15}	0,133	0,000	0,000	1,000	1,000	1,000
	C_{16}	0,137	0,333	0,333	0,500	0,500	1,000
	C_{17}	0,130	0,333	0,333	0,500	0,500	1,000
	C_{18}	0,074	0,000	0,000	1,000	1,000	1,000
	C_{19}	0,154	0,333	0,333	0,500	0,500	0,500
	C_{20}	0, 074	0,333	0,333	0,333	0,333	0,500
	C_{21}	0, 071	0	0	0	1,000	1,000
	I_{env}		0,240426	0,25913	0,555642	0,682642	0,886
Social	C_{22}	0,232	0,333	0,333	0,500	0,500	1,000
	C_{23}	0,129	0,333	0,333	0,500	0,500	1,000
	C_{24}	0,198	0,333	0,333	0,500	0,500	1,000
	C_{25}	0,191	0,200	0,240	0,250	0,290	0,300
	C_{26}	0,250	0,5	0,5	1,000	1,000	1,000
	I_{soc}		0,349347	0,356987	0,57725	0,58489	0,8663

values representing the environmental, economic and social dimensions, respectively (see Table 6.13).

In this case the enterprise X choice the scenario of weigh dimension like as:

$$W_{eco} = W_{env} = W_{soc} = 1$$

The final sustainable performance indices of suppliers were used to rank the suppliers in the X enterprise (see Table 6.14), supplier is identified as the best sustainable supplier with $Max\ P = 0,29$ (see Figure 6.4).

TABLE 6.14
The Sustainable Performance Indices of the Supplier

	I_{eco}	I_{env}	I_{soc}	Sustainable Performance	Rank
Supplier1	0,528376	0,240426	0,349347	0,12423878	5
Supplier2	0,595915	0,25913	0,356987	0,13467022	4
Supplier3	0,691638	0,555642	0,57725	0,20272556	3
Supplier4	0,701738	0,682642	0,58489	0,21880778	2
Supplier5	0,858031	0,886	0,8663	0,29003678	1

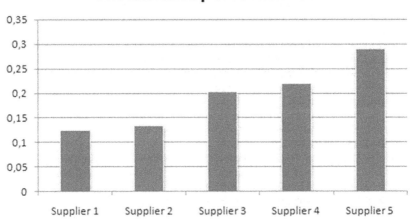

FIGURE 6.4   Sustainable performance of the supplier.

6.6   CONCLUSION

In this chapter, we have developed a new mathematical model that will help decision makers and managers to choose the best supplier who respects the three pillars of sustainable development. The selection of the most sustainable suppliers is the first step for companies that want to commit to a strategy of sustainable development as the SS process is the first link in the supply chain.

Our methodology starts with the selection of the most relevant criteria in the literature and international standards; we found 12 criteria for the economic dimension, 9 for the environmental dimension, and 5 for the social dimension. Then we used the Fuzzy DEMATEL hybrid method to classify and calculate the weight of the selected criteria. After we proposed a technique to choose a sustainability scenario, the scenario will be chosen by the managers of the company so that the sum of the weights assigned to the three dimensions of sustainable development is equal to one.

At the end, we presented a mathematical model that calculates the sustainability index of each supplier.

This model can be used easily in any company helping the decision makers to choose the best supplier who respects the three pillars of sustainable development and makes the supply chain sustainable. The current research can be extended by integrating other criteria.

REFERENCES

Briner, R. & Denyer. D. (2012). Systematic review and evidence synthesis as a practice and scholarship tool. In Rousseau, D. (Ed.), *Oxford Handbook of Evidence-Based Management.* Oxford University Press.

Dickson, G. W. (1966). An analysis of vendor selection systems and decisions. *Journal of Purchasing*, 2(1), 28–41.

El Mariouli, O. & Abouabdellah, A. (2018). Model for assessing the economic, environmental and social performance of the supplier. *4th IEEE International Conference on Logistics Operations Management* (GOL'2018), Lehavre, France.

Fontela, E. & Gabus, A. (1972). *World Problems, an Invitation to Further Thought within the Framework of DEMATEL.* Battelle Geneva Research Centre, Geneva, Switzerland.

Fontela, E. & Gabus. A. (1974). DEMATEL, innovative methods. Report No. 2, "Structural analysis of the world problematique (methods)," Battelle Geneva Research Institute, Switzerland.

Li, R. J. (1999). Fuzzy method in group decision making. *Computers & Mathematics with Application*, 38, 91–101.

Orji, I. J. & Wei, S. (2014). A decision support tool for sustainable supplier selection in manufacturing firms. *Journal of Industrial Engineering and Management JIEM*, 7(5), 1293–1315. doi:10.3926/jiem.1203.

Opricovic, S. & Tzeng, G. H. (2003). Defuzzification within a multicriteria decision model. *International Journal of Uncertainty, Fuzziness and Knowledge-Based Systems*, 11, 635–652. doi:10.1142/S0218488503002387.

Set, F. & Zadeh, L. A. (1965). Fuzzy sets. *Inform Control*, 8, 338–353.

Shabanpour, H., Yousefi, S., & Saen, R. F. (2016). Future planning for benchmarking and ranking sustainable suppliers using goal programming and robust double frontiers DEA. *Transportation Research Part D: Transport and Environment*, 50, 129–143. doi:10.1016/j.trd.2016.10.022.

Sureeyatanapas, P., Sriwattananusart, K., Niyamosothath, T., Setsomboon, W., & Arunyanart, S. (2018). Supplier selection towards uncertain and unavailable information: An extension of TOPSIS method. *Operations ResearchPerspectives*. doi:10.1016/j.orp.2018.01.005.

Vahidi, F., Torabi, S. A., & Ramezankhani, M. J. (2017). Sustainable supplier selection and order allocation under operational and disruption risks. *Journal of Cleaner Production.* doi:10.1016/j.jclepro.2017.11.012.

Zimmer, K., Fröhling, M., & Schultmann, F. (2015). Sustainable supplier management: A review of models supporting sustainable supplier selection, monitoring and development. *International Journal of Production Research.* doi:10.1080/00207546.2015.1079340.

Zsidisin, G. A. & Ancarani, A. (2016). Positioning the journal of purchasing and supply management in supply chain research: Perspectives from the outgoing editors. *Journal of Purchasing & Supply Management*, 22, 1–6.

7 Supplier Selection for Protective Relays of Power Transmission Network with the Fuzzy Approach

Ali Jahan and Alireza Panahande

CONTENTS

7.1 INTRODUCTION

Various products should be provided for customer demands in the global field of competition. An increase in customer demands provides an opportunity for the supplier to fulfill the biggest part of the job. In this case, the company will be more dependent on its suppliers. This dependence requires more coordination between the company and the suppliers, which highlights the importance of supply chain management (SCM). SCM is used with the aim of creating more efficient processes and, at the same time, decreasing the risk of chain disconnection (Christopher et al., 2011). SCM includes different kinds of procedures, such as integrated decision-making in the supply chain, which are affected by every member of the chain from raw material supplies to final consumers. This process has three kinds of supply, production, and distribution decisions, which are made in three strategic, tactical, and operational levels. Purchased materials and services make 80% of the total costs in many high-tech industries. Therefore, suppliers have a significant role in the production and distribution of service products with competitive quality (Gülen, 2007). In other words, for each kind of business service, selecting the right upstream supplier plays a key role in its success. It significantly reduces purchase costs, increases customer satisfaction, and improves competitiveness (Liao & Kao, 2010). Selecting the right supplier is an important factor in decision-making, which involves qualitative and quantitative factors with the highest potential of fulfilling the demands of the company at a reasonable cost. Basically, selecting a supplier is a decision made with the aim of limiting the collection of potential suppliers in order to become closer to the final selection (Wu & Barnes, 2011). Nowadays, selecting a supplier is not limited to some price lists and identifying the best option based on its financial dimension. Instead, selecting the right supplier should be carried out in a wide range of criteria: product quality, price, product reliability, product variety, timely delivery, profile and performance, financial state, commercial brand, guarantee, environmental performance, etc. All of these criteria are most effective in selecting a supplier. These factors are weighted based on demands, prioritizations, and strategies of the company. The significance of each criterion is different in each industry. Firms should adopt a strategic approach to supplier relationship management and avoid a *one-size-fits-all* strategy (Dyer et al., 1998; Gürler, 2007; Sagar & Singh, 2012; Yilmaz et al., 2011). Rapid growth and high competitions among companies require a proper supply chain management for higher profits. Today, intense competition among companies is replaced by supply chain competition. For an effective dominance of the market, production companies should reduce their costs and shorten their product chain. Therefore, many companies outsource many of their components, raw materials, and products in order to focus on their main objective. Since supplier performance directly affects the success of a supply chain, selecting the right supplier is a vital issue for overcoming organizational difficulties (Yu & Wong, 2015). Selection and evaluation of the suppliers is a significant strategy for supply chain management (Pang & Bai, 2013). The final objectives of this selection process are a reduction of purchase risks, maximizing the over value for the customer, and creating long-term relationships between the customer and the supplier (Chen et al., 2006). Notably, more than one option can reduce the risks and increase the supplier's success more efficiently.

This study aims at presenting a multi-criteria decision-making method for selecting the right supplier in fuzzy environment. For this purpose, the suggested methodologies are quality function deployment (QFD) and modified technique for order preference by similarity to ideal solution (M-TOPSIS). These two methods are used for the first time in the field of supply chain management and supplier selection. Moreover, protective relays of power transmission networks are used as a case study. This method optimizes energy management and power transmission network reliability and minimizes extensive power outages in power transmission systems. The remainder of this paper is organized as follows: Section 7.2 presents a literature review of studies on the selection of supplier chain management. Section 7.3 presents different multivariate decision-makings in the framework of a case study. Section 7.4 discusses the results of the paper and sensitivity analysis for further evaluation and validation of the results. Finally, Section 7.5 presents a conclusion and further recommendations.

7.2 LITERATURE REVIEW OF SUPPLIER SELECTION

SCM and selecting the right supplier is of great importance to universities and industries. The main objectives of SCM include reducing supply chain risks, reducing production costs, more income, better customer services, optimization of stock level along with commercial processes and cycle times, and increasing the level of satisfaction and profitability (Boran et al., 2009). De Boer et al. (2001) divided the supply chain into four sections: (1) statement of the problem, (2) formulation of selection criteria, (3) pre-selection of candidates, and (4) final selection. The second, third, and fourth sections use available tools for decision-making management.

7.2.1 BRIEF REVIEW OF SUPPLIER SELECTION METHODS

Soukup (1987) and Willis et al. (1993) estimated the limitations of classic techniques in evaluating the suppliers. Chai et al. (2013) divided these techniques into three categories: (1) Multi-criteria decision-making techniques (MCDM); (2) Math planning techniques (MP); and (3) artificial intelligence techniques (AI).

The most popular and most known MCDM techniques in this regard include the analytic hierarchy process (AHP) (Saaty, 2001) and analytical network process (ANP), which are recommended by many researchers (Liou et al., 2014). Both of these methods use paired comparison and judgment of the experts for estimating the evaluation criteria. Since discrete scales are not able to represent human thinking process, which is based on inaccuracy and unreliability, AHP method is considered classic and, generally, MCDM methods cannot represent the significance of qualitative criterion accurately. For this reason, fuzzy AHP (FAHP) was first introduced by Laarhoven and Pedrycz (1983). It is another format of classic AHP that transforms linguistic judgments (qualitative) into fuzzy values in order to create fuzzy paired comparison matrixes. These matrixes estimate relative weight of criteria and, consequently, provide a ranking for the existing options (Calabrese et al., 2013). Since then, several FAHP methods have been presented with the aim of solving some problems (Buckley, 1985; Chang, 1996; Herrera-Viedma et al., 2004; Kahraman et al., 2004; Zeng et al., 2007). Some authors used other methods for selecting the best supplier such as ANP (Lin et al., 2010), fuzzy ANP (Vinodh et al., 2011), fuzzy ELECTRE (Montazer et al., 2009; Sevkli, 2010), fuzzy PROMETHEE (Chai et al., 2012; Chen et al., 2011), TOPSIS (Saen, 2010), fuzzy TOPSIS (Wang et al., 2009), fuzzy VIKOR (Wu et al., 2009), fuzzy DEMATEL (Chang et al., 2011), fuzzy SMART (Chou & Chang, 2008), Gray Theory (Golmohammadi & Mellat-Parast, 2012), QFD (Ansari & Batoul, 2006), and fuzzy QFD (Bevilacqua et al., 2006; Lima-Junior & Carpinetti, 2016). Data covering analysis (DEA) is widely used in different investigations for selecting the suitable supplier (Azadeh & Alem, 2010; Azadi et al., 2015; Falagario et al., 2012; Wu & Blackhurst, 2009). Cooper et al. (2007) believe that when we use DEA, the number of alternatives should be three times more than the number of inputs and outputs (criteria). Therefore, the existing criteria and alternatives are limited. Except DEA, linear planning (Lin, 2012; Ozkok & Tiryaki, 2011; Ustun & Demi, 2008; Wang & Yang, 2009), nonlinear planning (Hsu et al., 2010; Rezaei & Davoodi, 2012; Yeh & Chuang, 2011), multi-objective planning (MOP) (Amin & Zhang, 2012; Feng et al., 2011; Haleh & Hamidi, 2011; Shankar et al., 2013; Shaw et al., 2012; Tsai & Hung, 2009; Yu et al., 2012), and random programming (Kara, 2011; Li & Zabinsky, 2011) are other important MP methods that are utilized for the selecting the best supplier. Artificial intelligence (AI) is used for studying and designing intelligent agents. An intelligent agent is a system that understands its surroundings and maximizes its success by some special actions (Russell et al., 2003). Genetic algorithm (GA) (Sadeghieh et al., 2012), artificial neural network (ANN) (GüNeri et al., 2011), Rough Set Theory (Bai & Sarkis, 2010), and GREY Theory (Li et al., 2007) are the main applied AI methods found through the literature review.

Yong Deng et al. (2018) combined *DEMETEL* and *game theory* for supplier selection. Abdel-Basset et al. (2018) employed the neutrosophic set for decision-making and evaluation method (DEMATEL) in order to analyze and determine the factors affecting the selection of suppliers. Torabi et al. (2018) suggested a hybrid SWOT-QFD systematic framework for choosing the most influential sustainability criteria. Awasthi et al. (2018) applied the fuzzy AHP to generate criteria weights for sustainable global supplier selection and then used fuzzy VIKOR to rate supplier performances against the evaluation criteria. Babbar and Amin (2018) used a novel mathematical model to select a set of suppliers and assign the order quantity. Because of the importance of environmental concerns, both qualitative and quantitative environmental criteria have been taken into account in this research. The proposed model comprises two phases namely a two-stage QFD, and a stochastic multi-objective mathematical model. Dobos and Vörösmarty (2018) used Data Envelopment Analysis (DEA) as supplier selection method, where green factors served as the output variables of a DEA model, and management variables were the inputs.

7.2.2 CASE STUDIES IN SUPPLIER SELECTION

A wide range of studies has been carried out on supplier selection with different objectives. Sambana and Kilaparthi (2018) used Kano integrated fuzzy analysis and VIKOR technique for selecting the best supplier of carpet washing company. For classification and prioritization of supplier selection criteria, fuzzy Kano analysis was used. The problem of unreliability in supplier selection was solved by VIKOR technique. Jain et al. (2018) analyzed supplier selection of an Indian automotive company by TOPSIS and AHP methods. The selection process was carried out by conducting some interviews with industry experts and the identification of the criteria that were used in the literature review. Criterion weights were estimated by the AHP method and ranking of suppliers was done by TOPSIS. Shi et al. (2018) used an integrated approach that consisted of interval-valued intuitionistic uncertain linguistic sets (IVIULSs), GREY relational analysis (GRA), and TOPSIS to evaluate and select the best green suppliers of the agricultural food industry. Assellaou et al. (2018) analyzed the problem of supplier selection in a known refining company in

Africa. In this analysis, they used an integrated model by DEMETEL-ANP-TOPSIS in order to select a supplier with the most satisfactory criteria. Stevic et al. (2017) applied the supplier selection process in a construction company based on a new multicriteria approach. Fazlollahtabar et al. (2017) used linear multi-objective fuzzy planning (FMOLP) for selecting the best supplier and allocation of desirable orders. Moreover, they used FAHP in order to estimate the significance of each criterion and estimated practical application of the recommended method in the steel industry. Gencer and Gurpinar (2007) used ANP for supplier selection in an electronic company. For this purpose, they evaluated supplier selection criteria. Scott et al. (2013) analyzed the supplier selection process in Britain for the bioenergy industry. They applied AHP and QFD methods on a list of evaluation criteria that could be utilized in the selection process of the potential supplier of the bioenergy industry in Britain.

7.2.3 STATEMENT OF THE PROBLEM

In this project, we tried to use a multi-criteria decision-making method in order to select and purchase the best protective relays of power transmission networks from the Territory of Semnan Regional Power Company. Since the mentioned methods are not able to connect customer values to the decision-making process, we used the QFD method that, unlike other methods, is able to hear the customer's voice and consider their needs and demands. Moreover, compared with other methods, such as AHP, QFD has fewer complex calculations and is more reliable and faster than MCDM. QFD/M-TOPSIS integrated method provides an opportunity for decision makers to consider customer demands along with technical criteria for selecting the best supplier. In fact, QFD increases the efficiency of supplier selection and connects the customers to suppliers.

7.3   METHODOLOGY

This study aims to present a multivariate decision-making method for selecting the right supplier in fuzzy environment. The methodology of this study is a combination of QFD and modified TOPSIS (M-TOPSIS). In the first step of the methodology, a team of decision makers is selected, including three experts for selection and evaluation of the criteria. In the second step, criteria are weighted by FQFD method. Finally, in the third step, a ranking of existing options is presented by M-TOPSIS method. The recommended methodology presented an efficient performance in Semnan Regional Power Company. This methodology was performed with the aim of selecting the best protective relays of power transmission networks in order to increase network reliability. The suggested algorithm is as follows.

7.3.1   FUZZY QUALITY FUNCTION DEPLOYMENT MODEL (FQFD)

QFD model is a quality control method. It is considered to be a linear and structured strategy for fulfilling customer demands. This fulfillment is done by increasing the special properties of products and presenting new services. This model has four

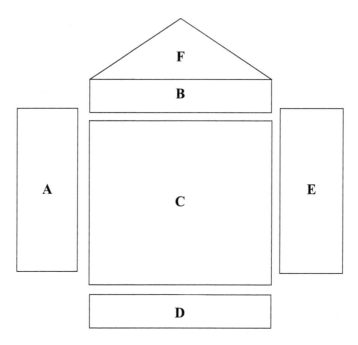

FIGURE 7.1 House of quality structure: (A) Represents customer needs or voice of customers or WHATs; (B) represents Technical requirement or HOWs; (C) represents the relationship matrix between WHATs and HOWs; (D) represents importance weight of expectation; (E) represents customers benchmarking; and (F) represents interrelationships between engineering characteristics. (From Babbar, C., and Amin, S.H., *Expert Syst. Appl.*, 92, 27–38, 2018.)

matrixes or so-called houses (Bevilacqua et al., 2006). The first one is a house of the quality matrix (HOQ) shown in Figure 7.1 (Hauser & Clausing, 1988).

The main steps of the QFD model (Karsak & Dursun, 2014; Yazdani et al., 2017) are as follows:

Step 1: Selection of Criteria
At first, customer requirements (CR) and related technical requirements (TRs) that affect the supply chain are determined.

Step 2: In this step, the significance of customer requirements is determined by linguistic triangular fizzy variables and mean of fuzzy weights along with the correlation of CR and technical demands. A fuzzy set is created by a membership function by which each element reaches its membership degree in distance [0, 1]. If the allocated value is located in the specified space, the elements will be provided with a specific membership degree (it partially belongs to the fuzzy set). Figure 7.2 presents the triangular fuzzy values used in this project (Tseng et al., 2017).

Step 3: Computation of fuzzy QFD weights for supply chain technical requirements.

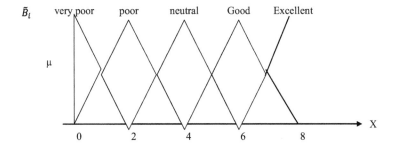

FIGURE 7.2   Fuzzy linguistic terms for house of quality evaluation.

A normalization rule is used to deliver normalized weights of main decision criteria for the final selection process. The recommended approach, i.e., QFD model, is a house of the quality matrix that is used for final weighting. The matrix is also used for connecting customer demands with technical requirements of suppliers. The details of the QFD model are presented in (Germani et al., 2012; Tavana et al., 2017). Fuzzy QFD model (Zaim et al., 2014) is created in a decision-making group in order to connect customer requirements with designing steps (technical requirements). It is of note that technical requirements provide a tool for customer satisfaction.

Step 4: A survey is done about customer understandings and final users in order to estimate their satisfaction level and determine the significance of customer's point of view in technical requirements. Linguistic fuzzy values determine the significance of customer's point of view and technical requirements. Afterward, these values are converted to triangular fuzzy values used in the calculations. Using the variable that converts linguistic values to fuzzy values can help decision makers in their coping with ambiguities and decision-making procedures. We selected 5 linguistic labels to facilitate the participation of decision makers. Decision makers have a significant role in determining linguistic variables. In this way, they can make a reliable and efficient decision. Linguistic terminologies of Table 7.1 (Lee et al., 2015)— *Very poor, poor, Neutral, Good,* and *Excellent*—represent the satisfaction level and the connection of customer demands with technical supply chain

TABLE 7.1

Linguistic Preference and the Corresponding Fuzzy Numbers

Linguistic Terms	Fuzzy Number
Very poor (VP)	(0, 0, 2)
Poor (P)	(0, 2, 4)
Neutral (N)	(2, 4, 6)
Good (G)	(4, 6, 8)
Excellent (E)	(6, 8, 8)

TABLE 7.2

Linguistic Values for Weight of CRs

Linguistic Terms	Fuzzy Number
Very low (VL)	(0, 2, 4)
Low (L)	(2, 4, 5)
Medium (M)	(4, 6, 8)
High (H)	(6, 8, 10)
Very high (VH)	(8, 9, 10)

variable. For weighting process of CRs, we used linguistic variables shown in Table 7.2; i.e., *Very low, Low, Medium, High,* and *Very high* (Kannan et al., 2014).

Step 5: Based on the participation of decision-making groups, the mean of triangular fuzzy values should be calculated by q fuzzy values. Moreover, the significance of customer requirements is estimated by q individual members according to their specialties. q_j member denotes the weight of customer demands, $w_{\tilde{c}_{ij}} = (w_{c_1}^{ij}, w_{c_2}^{ij}, w_{c_3}^{ij})$. By using these weights and estimation of the correlation of customer requirements and technical requirements, the final weights of each technical factor will be calculated as follows (Lee et al., 2015):

$$\tilde{w}_{\tilde{c}_{ij}} = \frac{\sum_{j=1}^{q} w_{c_{ij}}}{q} = \frac{\left(\sum_{j=1}^{q} w_{c_1}^{ij}, \sum_{j=1}^{q} w_{c_2}^{ij}, \sum_{j=1}^{q} w_{c_3}^{ij}\right)}{q} = \left(w_{c_1}^{i}, w_{c_2}^{i}, w_{c_3}^{i}\right) \quad (7.1)$$

where q stands for the number of decision makers and $\tilde{w}_{\tilde{c}_{ij}}$ stands for the weight of customer demands. The duty of QFD members is to present their decisions and judgments by creating a correlation between CRs and TRs; i.e., $\tilde{c}_{ij} = (c_{ij1}, c_{ij2}, c_{ij3})$.

$$\tilde{c}_{ij} = \frac{\sum_{k=1}^{q} c_{ij}^{k}}{q} = \frac{\left(\sum_{j=1}^{q} c_{ij1}^{k}, \sum_{j=1}^{q} c_{ij2}^{k}, \sum_{j=1}^{q} c_{ij3}^{k}\right)}{q} = \left(c_{ij1}, c_{ij2}, c_{ij3}\right) \quad (7.2)$$

Moreover, Eq. 7.3 is used for defuzzification performance calculation of triangular fuzzy values (Cheng, 1999).

$$x = \frac{(b_1 + 2b_2 + b_3)}{4} \quad (7.3)$$

In this step, Eq. 7.4 is used for weighting of technical requirements and, at the same time, considering customer satisfaction ($w_{\tilde{c}s}$).

$$w_{\tilde{c}s} = \frac{\sum_{i=1}^{m} \tilde{w}_{\tilde{c}_{ij}} \tilde{c}_{ij}}{\sum_{i=1}^{m} \tilde{w}_{\tilde{c}_i}} \quad (7.4)$$

7.3.2 THE M-TOPSIS METHOD

TOPSIS, originally introduced by Hwang and Yoon (1981), attempts to find the most preferred alternative that is as close as possible to the positive ideal solution (PIS) and as far as possible to the negative ideal solution (NIS) (Hwang & Yoon, 1981). Contrary to NIS, PIS is a solution that maximizes the benefit criteria and minimizes the cost criteria. Among many modifications and improvements of TOPSIS introduced so far (Antuchevičiene et al., 2010; Chamodrakas et al., 2011; Marković, 2016; Ren et al., 2007), M-TOPSIS (Hatami-Marbini & Kangi, 2017; Pinter & Pšunder, 2013) is the most suitable method for our purpose. The procedure for M-TOPSIS can be expressed in the following steps:

Step 1: Creating the Decision Matrix

$$D = \begin{matrix} & X_1 & \cdots & X_1 & \cdots & X_1 \\ A_1 \\ \vdots \\ A_i \\ \vdots \\ A_m \end{matrix} \begin{bmatrix} X_{11} & \cdots & X_{1J} & \cdots & X_{1n} \\ \vdots & & \vdots & & \vdots \\ X_{j1} & \cdots & X_{ij} & \cdots & X_{in} \\ \vdots & & \vdots & & \vdots \\ X_{m1} & \cdots & X_{mj} & \cdots & X_{mn} \end{bmatrix} \quad (7.5)$$

$$W = (w_1, \ldots, w_j, \ldots, w_n)$$

where X_{mn} is the significance level of each A_m option by considering X_n criterion, which is presented as a linguistic triangular fuzzy value. Each decision maker (expert) can evaluate the options based on different criteria and the calculated weights of house of the quality matrix.

To create an aggregated matrix, it is assumed that in a decision-making committee with k members, a fuzzy ranking of each decision maker can be presented with a triangular fuzzy value $\tilde{R}_k (k = 1, 2, \ldots, k)$ and a membership function. Therefore, the aggregated fuzzy ranking can be estimated as follows:

$$\tilde{R} = (a, b, c) \text{ and } k = 1, 2, \ldots, k \text{ where } a = Min\{c_k\}$$

$$b = \frac{1}{k} \sum_{k=1}^{k} b_k \text{ and } c = Max\{c_k\} \quad (7.6)$$

Step 2: Converting an Existing Decision-Making Matrix to an *Unscaled Matrix*

This step performs the normalization process of the aggregated fuzzy decision matrix. The data in the aggregated fuzzy decision matrix are normalized to unify different measurement scales. The normalized aggregated fuzzy decision matrix is presented as: $\tilde{R} = [\tilde{r}_{ij}]_{m+n}$.

Therefore, normalized values of cost and profit criteria are calculated as follows:

$$\tilde{r}_{ij} = \left(\frac{a_{ij}}{c_j^*}, \frac{b_{ij}}{c_j^*}, \frac{c_{ij}}{c_j^*} \right), j \in B \tag{7.7}$$

$$c_j^* = maxc_{ij}, \ j \in B$$

$$\tilde{r}_{ij} = \left(\frac{a_j^-}{c_{ij}}, \frac{a_j^-}{b_{ij}}, \frac{a_j^-}{a_{ij}} \right), j \in C \tag{7.8}$$

$$a_j^- = mina_{ij}, \ j \in C$$

Step 3: Creating a weighted *unscaled* matrix with assuming w vector (algorithm input).

The elements of the weighted normalized matrix $(\tilde{v}_{ij})$ are calculated as follows:

$$\tilde{v}_{ij} = \tilde{r}_{ij}.\tilde{w}_{ij} \tag{7.9}$$

Step 4: Identifying the positive ideal solution and a negative ideal solution:

$$A^* = \left[V_1^*, \ldots, V_j^*, \ldots, V_n^* \right]; \ V_j^* = max\{V_{ij}\} \tag{7.10}$$

$$A^- = \left[V_1^-, \ldots, V_j^-, \ldots, V_n^- \right]; \ V_j^- = min\{V_{ij}\} \tag{7.11}$$

Step 5: Calculating the distance from positive and negative ideal.

The distance of each option from fuzzy positive and negative ideals are calculated by:

$$d_i^* = \sum_{j=1}^n d\left(\tilde{v}_{ij}, \tilde{v}_j^* \right), \ i = 1,2,\ldots,m \tag{7.12}$$

$$d_i^- = \sum_{j=1}^n d\left(\tilde{v}_{ij}, \tilde{v}_j^- \right), \ i = 1,2,\ldots,m \tag{7.13}$$

For triangular fuzzy values, the distance of two triangular values (a_1,b_1,c_1) and (a_2,b_2,c_2) can be calculated as follows:

$$d\left(\tilde{M}_1, \tilde{M}_2 \right) = \sqrt{\left(\frac{1}{3} \right) \left[(a_1 - a_2)^2 + (b_1 - b_2)^2 + (c_1 - c_2)^2 \right]} \tag{7.14}$$

Step 7: Define the optimized ideal reference point (D) as:

$$D = \left(S^L, S^R \right) = \left(min\left(d_i^* \right), max\left(d_i^- \right) \right), \ i = 1,2,\ldots,m \tag{7.15}$$

Step 8: Calculate the distance between the point (d_i^*, d_i^-) for each alternative and point D as follows:

$$CC_i = \sqrt{(d_i^* - S^L)^2 + (d_i^- - S^R)^2}, \; i = 1, 2, \ldots, m \qquad (7.16)$$

Step 9: Rank the alternatives based on CC_i in an increasing order. The alternative A_i is closer to A^* and farther from A^- as CC_i approaches 0.

7.4   RANKING OF PROTECTIVE RELAYS OF POWER TRANSMISSION NETWORKS

In electrical installations, such as power transmission networks, generators, transformers, and other electrical equipment, there will be a power outage resulted from insulation defects, weakness of electric-dynamic strength against unexpected voltages, and excessive temperature increase. These defects may represent themselves in the form of a short circuit, earthing system, tear and cut in electrical conductors, and the breaking of insulators. The equipment having such kind of defects will be removed from the network in order to prevent expansion of defect and outage of other parts of the network. Therefore, the network should be designed with an efficient stability. For this end, a relay should be used. A relay detects network defects and prepares the warning system. If necessary, it disconnects the electrical circuit itself. A relay used for protecting electrical equipment is called a protective relay. The aim of a protective relay is to prevent further damages and to protect undamaged parts (Blackburn & Domin, 2006).

The aim of this study is to select a suitable supplier for protective relays of power transmission networks of Semnan Regional Power Company. This supplier optimizes properties of the purchased relays, increases network reliability, and decreases outage costs. Therefore, the mentioned company confirmed the process of supplier selection in fuzzy MCDM environment. Consequently, the decisions are made by FQFD and FTOPSIS methods. In this methodology, the weight of each criterion is determined by FQFD method and option rankings are done with FTOPSIS method. Figure 7.3 illustrates the flowchart of this study.

7.4.1   FIRST PHASE: PRIMARY PHASE

At first, a decision-making team including three managers was selected from the mentioned company (DM_1, DM_2, and DM_3). Then, by studying the literature review of this research (Table 7.3) and interviewing the decision-making team, evaluation criteria were determined and limited. Accordingly, the decision-making team determined nine technical requirements (TRs) and seven customer requirement (CRs) criteria. Moreover, five alternatives were selected as suppliers (S), which are selected for a successful screening procedure and are considered to be suitable for purchasing the desired product. Decision alternatives are Lian Vision Alborz Company (S_1), Artnos Company (S_2), Modje Niroo Company (S_3), Energy Company of Spadana (S_4), and Parin Control Company (S_5). Customer requirement

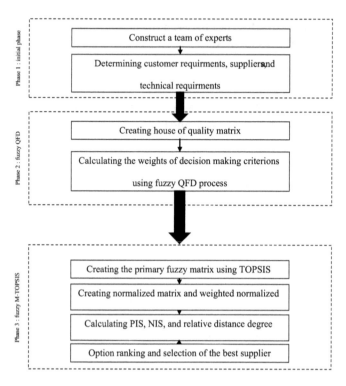

FIGURE 7.3 The flowchart of the presented methodology.

TABLE 7.3
Utilized Criterions for Supplier Evaluation in Different Articles

Year and References	Supplier Selection Criterions
Bevilacqua et al. (2006)	Product quality, suitable price (competition price), delivery reliability, ability of corrective action, customer support, and delivery planning
Mendoza and Ventura (2008)	Flexibility, quality, price, services, timely delivery
Juan et al. (2009)	Quality, cost, information, schedule, integration, service, performance, punctuality, management, response, communication, assurance, empathy
Ho et al. (2010)	Quality, timely delivery, production ability, price, management services, research and improvement, financial ability, flexibility, reputation
Kannan et al. (2013)	Cost, quality, delivery reliability, technology capability, environmental competency
Dweiri et al. (2016)	Price, quality, delivery, services
Lima-Junior and Carpinetti (2016)	Cost, delivery reliability, financial situation, management practices, product development, quality, relationship
Çakır (2017)	Quality, services, delivery, production capability, financial position
Pramanik et al. (2017)	Quality, delivery, reliability, processing time, profit margin
Banaeian et al. (2018)	Service level, quality, price, EMS (environmental prerequisite, planning, and certificates)

criteria are confidence in the safe and durable product (CR$_1$), availability (CR$_2$), quality assurance system (CR$_3$), affordable price (CR$_4$), commitment to the health of employees (CR$_5$), flexibility (CR$_6$), and assessing previous customers (CR$_7$). On the other hand, technical requirement criteria are financial stability (TR$_1$), product quality (TR$_2$), geographical location (TR$_3$), creativity and innovation (TR$_4$), experience (TR$_5$), delivery (TR$_6$), services (TR$_7$), production and supply capability (TR$_8$), and variety of the products (TR$_9$).

7.4.2 Phase 2: Fuzzy QFD Phase

Decision makers (DMs) use fuzzy linguistic variables (Tables 7.1 and 7.2), quality house matrix, and some questionnaires for evaluating the suppliers and express their priority. After creating the house of the quality matrix, three decision makers are asked to express their opinions on the correlation between customer factors and decision-making criteria. This information is indicated in Tables 7.3 through 7.6. Moreover, they are asked to estimate the weight of each CR using linguistic variables presented in the last column of the mentioned table.

TABLE 7.4

House of Quality Matrix Process for Decision Makers Number 1 (DM$_1$)

TRs CRs	TR$_1$	TR$_2$	TR$_3$	TR$_4$	TR$_5$	TR$_6$	TR$_7$	TR$_8$	TR$_9$	Weight of CRs
CR$_1$	—	G	—	—	N	—	G	—	G	VH
CR$_2$	N	—	G	—	N	E	G	G	—	H
CR$_3$	—	G	—	P	—	N	N	N	N	H
CR$_4$	G	G	N	—	N	N	N	G	G	VH
CR$_5$	N	N	—	—	N	—	P	N	N	M
CR$_6$	—	N	—	G	G	N	N	N	N	VH
CR$_7$	N	N	P	P	N	N	N	P	N	H

TABLE 7.5

House of Quality Matrix Process for Decision Makers Number 2 (DM$_2$)

TRs CRs	TR$_1$	TR$_2$	TR$_3$	TR$_4$	TR$_5$	TR$_6$	TR$_7$	TR$_8$	TR$_9$	Weight of CRs
CR$_1$	N	E	VP	G	E	VP	G	P	N	VH
CR$_2$	P	VP	G	—	—	G	G	—	—	H
CR$_3$	G	E	—	G	—	—	N	—	—	H
CR$_4$	E	E	N	E	N	N	N	G	G	VH
CR$_5$	E	—	—	—	—	—	—	N	N	L
CR$_6$	—	G	N	N	P	N	E	G	G	H
CR$_7$	—	E	—	N	—	N	N	P	P	M

TABLE 7.6

House of Quality Matrix Process for Decision Makers Number 3 (DM₃)

TRs CRs	TR₁	TR₂	TR₃	TR₄	TR₅	TR₆	TR₇	TR₈	TR₉	Weight of CRs
CR₁	G	E	—	G	G	P	E	G	G	VH
CR₂	P	N	G	N	N	G	G	G	G	M
CR₃	N	E	—	E	G	N	E	N	N	M
CR₄	E	E	G	E	N	N	G	N	G	H
CR₅	G	E	—	G	G	G	G	G	—	M
CR₆	G	G	N	G	G	N	E	P	G	H
CR₇	G	E	N	E	G	G	E	N	G	H

In this step, linguistic forms are converted to fuzzy triangular values and HOQ aggregated matrix is created based on Eqs. 7.1 and 7.2 along with experts' opinions (Table 7.7). Triangular fuzzy values are converted to defuzzification values by Eq. (7.3). For example, decision maker number 1 assumes the significance of "CR₁" as a very high criterion according to fuzzy values (8, 9, 10). Defuzzification values are calculated as follows:

$$x = \frac{(b_1 + 2b_2 + b_3)}{4} = \frac{8 + 2(9) + 10}{4} = 9$$

According to Eq. (7.4), after relative weighting, the scales are normalized. Then, all the final weights of technical requirements are presented in Table 7.8. Based on this table, *product quality* criterion is proved to be more significant in the experts' opinions. These weights can be used as M-TOPSIS inputs.

7.4.3 PHASE 3: FUZZY M-TOPSIS PHASE

In the TOPSIS method, the selected option, and the best option should have the shortest geometric distance from positive ideal solution, and the furthest distance from the negative ideal solution. This method compares a set of options according to their criterion weights, normalized weights, and the geometric distance between each option and the ideal option, which is considered to have the best score among criteria (Vipul Jain et al., 2016).

For evaluating supplier options, decision makers are asked to fill out another questionnaire. They should present an experiment-based judgment and evaluate the suppliers by linguistic fuzzy variables according to each criterion. All the collected data are presented in Table 7.9.

Aggregated fuzzy matrix is created from Eq. 7.6 and is presented in Table 7.10. For example, decision makers assume that the significances of *product quality* for

TABLE 7.7

Aggregated HOQ Matrix

	TR₁	TR₂	TR₃	TR₄	TR₅	TR₆	TR₇	TR₈	TR₉	Weight of CRs
CR₁	(2,3.33,4.67)	(5.33,7,33,8)	(0,0,0.67)	(2.67,4,5.33)	(4,6,7.33)	(0,0.67,2)	(4.67,6.67,8)	(1.33,2,67,4)	(3.33,5.33,7.33)	(8,9,10)
CR₂	(0.67,2.67,4.67)	(0.67,1.33,2.67)	(4,6,8)	(0.67,1.33,2)	(1.33,2.67,4)	(4.67,6.67,8)	(4,6,8)	(2.67,4,5.33)	(1.33,2,2.67)	(5.33,7.33,9.33)
CR₃	(2,3.33,4067)	(5.33,7,33,8)	(0,0,0)	(3.33,5.33,6.67)	(1.33,2,2.67)	(1.33,2,67,4)	(3.33,5.33,6.67)	(1.33,2,67,4)	(1.33,2,67,4)	(5.33,7,33,9.33)
CR₄	(5.33,7,33,8)	(5.33,7,33,8)	(2.67,4.67,6.67)	(4,5.33,5.33)	(2,4,6)	(2,4,6)	(2.67,4.67,6.67)	(2.67,4.67,6.67)	(4,6,8)	(7.33,8.67,10)
CR₅	(4,6,7.33)	(2.67,4,4.67)	(0,0,0)	(1.33,2,2.67)	(2,3.33,4.67)	(2,4,6)	(1.33,2,67,4)	(2.67,4.67,6.67)	(1.33,2,67,4)	(3.33,5,33,7)
CR₆	(1.33,2,2.67)	(3.33,5.33,7.33)	(1.33,2,67,4)	(3.33,5.33,7.33)	(2.67,4.67,6.67)	(4.67,6,67,7.33)	(4.67,6.67,7.33)	(2,4,6)	(3.33,5.33,7.33)	(6.67,8.33,10)
CR₇	(2,3.33,4.67)	(4.67,6.67,7.33)	(0.67,2,3.33)	(2.67,4.67,6)	(2,3.33,4.67)	(2.67,4.67,6.67)	(3.33,5.33,6.67)	(0.67,2.67,4.67)	(2,4,6)	(5.33,7.33,9.33)

TABLE 7.8

Calculating the Final Weight Using QFD Method

TRs / CRs	TR₁	TR₂	TR₃	TR₄	TR₅	TR₆	TR₇	TR₈	TR₉	Weight of CRs
CR₁	3.333	7	0.166	4	5.833	0.833	6.5	2.666	5.333	9
CR₂	2.666	1.5	6	1.333	2.666	6.5	6	4	2	7.333
CR₃	3.333	7	0	5.166	2	2.666	5.166	2.666	2.666	7.333
CR₄	7	7	4.666	5	4	4	4.666	5.333	6	8.666
CR₅	5.833	3.833	0	2	3.333	2	2.666	4.666	2.666	5.25
CR₆	2	5.333	2.666	5.333	4.666	4	6.333	4	5.333	8.333
CR₇	3.333	6.333	2	4.5	3.333	4.666	5.166	2.666	4	7.333
Relative weights	206.402	297.013	122.833	214.944	202.222	187.444	285.5	196.5	222	1712.861
Normalized & aggregatedweights	0.1205	0.1734	0/0717	0.1255	0.1181	0.1094	0.1667	0.1147	0.1296	

TABLE 7.9

Suppliers Evaluation by Experts (DMs)

Suppliers	DMs	TR₁	TR₂	TR₃	TR₄	TR₅	TR₆	TR₇	TR₈	TR₉
S₁	DM₁	G	G	G	N	E	E	G	G	N
	DM₂	G	E	G	G	E	E	E	N	G
	DM₃	G	G	G	P	G	G	G	G	N
S₂	DM₁	G	G	G	N	G	G	G	N	N
	DM₂	N	E	G	G	G	N	N	N	N
	DM₃	N	G	G	P	G	N	N	G	N
S₃	DM₁	G	G	G	N	N	G	G	N	N
	DM₂	G	E	G	G	G	G	G	N	P
	DM₃	G	G	G	VP	G	P	G	N	N
S₄	DM₁	G	G	N	N	G	N	N	N	N
	DM₂	G	G	N	G	N	N	N	N	G
	DM₃	G	G	G	P	N	N	G	N	N
S₅	DM₁	N	G	N	N	G	N	N	N	N
	DM₂	N	G	G	N	N	N	P	P	G
	DM₃	N	G	G	P	N	N	N	N	N

TABLE 7.10

Aggregated Fuzzy Decision-Making Matrix

Suppliers	TR₁	TR₂	TR₃	TR₄	TR₅	TR₆	TR₇	TR₈	TR₉
S₁	(4,6,8)	(4,6.7,8)	(4,6,8)	(0,4,8)	(4,7.3,8)	(4,7.3,8)	(4,6.7,8)	(2,5.3,8)	(2,4.7,8)
S₂	(2,4.7,8)	(4,6.7,8)	(4,6,8)	(0,4,8)	(4,6,8)	(2,4.7,8)	(2,4.7,8)	(2,4.7,8)	(2,4,6)
S₃	(4,6,8)	(4,6.7,8)	(4,6,8)	(0,3.3,8)	(2,5.3,8)	(0,4.7,8)	(4,6,8)	(2,4,6)	(0,3.3,6)
S₄	(4,6,8)	(4,6,8)	(2,4.7,8)	(0,4,8)	(2,4.7,8)	(2,4,6)	(2,4.7,8)	(2,4,6)	(2,4.7,8)
S₅	(2,4,6)	(4,6,8)	(2,5.3,8)	(0,3.3,6)	(2,4.7,8)	(2,4,6)	(0,3.3,6)	(0,3.3,6)	(2,4.7,8)

the first supplier (S₁) are (4, 6, 8) and (6, 8, 8) in fuzzy numbers. Defuzzification value is calculated as follows:

$$\tilde{R} = (a,b,c) = \left(Min\{c_k\}, \frac{1}{k}\sum_{k=1}^{k} b_k, Max\{c_k\} \right) = (4,6.7,8)$$

In the next step, the normalized matrix is calculated using Eqs. 7.7 and 7.8. Table 7.11 indicates the normalized fuzzy matrix. Then, Eq. 7.9 is used for creating a weighted normalized matrix, presented in Table 7.12. In this table, positive (PIS) and negative (NIS) ideals are extracted and presented. These two PIS and NIS criteria are calculated by Eqs. 7.10 and 7.11.

Then, distances from PIS and NIS are calculated by Eq. 7.14, which are presented in Tables 7.13 and 7.14. As can be seen, the distance from PIS is shorter,

TABLE 7.11
Normal Fuzzy Matrix

Suppliers	TR₁	TR₂	TR₃	TR₄	TR₅	TR₆	TR₇	TR₈	TR₉
S_1	(0.5,0.75,1)	(0.5,0.83,1)	(0.5,0.75,1)	(0,0.5,1)	(0.5,0.92,1)	(0.5,0.92,1)	(0.5,0.83,1)	(0.25,0.67,1)	(0.25,0.58,1)
S_2	(0.25,0.58,1)	(0.5,0.83,1)	(0.5,0.75,1)	(0,0.5,1)	(0.5,0.75,1)	(0.25,0.58,1)	(0.25,0.58,1)	(0.25,0.58,1)	(0.33,0.67,1)
S_3	(0.5,0.75,1)	(0.5,0.83,1)	(0.5,0.75,1)	(0,0.42,1)	(0.25,0.67,1)	(0,0.58,1)	(0.5,0.75,1)	(0.33,0.67,1)	(0,0.56,1)
S_4	(0.5,0.75,1)	(0.5,0.75,1)	(0.25,0.58,1)	(0,0.5,1)	(0.25,0.58,1)	(0.33,0.67,1)	(0.25,0.58,1)	(0.33,0.67,1)	(0.25,0.58,1)
S_5	(0.33,0.67,1)	(0.5,0.75,1)	(0.25,0.67,1)	(0,0.56,1)	(0.25,0.58,1)	(0.33,0.67,1)	(0,0.56,1)	(0,0.56,1)	(0.25,0.58,1)

TABLE 7.12
Weighted Normal Matrix

Suppliers	TR₁	TR₂	TR₃	TR₄	TR₅	TR₆	TR₇	TR₈	TR₉
S_1	(0.6,0.09,0.12)	(0.09,0.14,0.17)	(0.04,0.05,0.07)	(0,0.06,0.13)	(0.06,0.11,0.12)	(0.05,0.1,0.11)	(0.08,0.14,0.17)	(0.03,0.08,0.11)	(0.03,0.08,0.13)
S_2	(0.03,0.07,0.12)	(0.09,0.14,0.17)	(0.04,0.05,0.07)	(0,0.06,0.13)	(0.06,0.09,0.12)	(0.03,0.06,0.11)	(0.04,0.1,0.17)	(0.03,0.07,0.11)	(0.4,0.9,0.13)
S_3	(0.06,0.09,0.12)	(0.09,0.14,0.17)	(0.04,0.05,0.07)	(0,0.05,0.13)	(0.03,0.08,0.12)	(0.0,0.06,0.11)	(0.08,0.13,0.17)	(0.04,0.08,0.11)	(0.0,0.07,0.13)
S_4	(0.06,0.09,0.12)	(0.09,0.13,0.17)	(0.02,0.04,0.07)	(0,0.06,0.13)	(0.03,0.07,0.12)	(0.04,0.07,0.11)	(0.04,0.1,0.17)	(0.04,0.08,0.11)	(0.03,0.08,0.13)
S_5	(0.04,0.08,0.12)	(0.09,0.13,0.17)	(0.02,0.05,0.07)	(0,0.07,0.13)	(0.03,0.07,0.12)	(0.03,0.07,0.12)	(0.0,0.09,0.17)	(0.0,0.06,0.11)	(0.03,0.08,0.13)
PIS(v^+)	(0.12,0.12,0.12)	(0.17,0.17,0.17)	(0.07,0.07,0.07)	(0.13,0.13,0.13)	(0.12,0.12,0.12)	(0.11,0.11,0.11)	(0.17,0.17,0.17)	(0.11,0.11,0.11)	(0.13,0.13,0.13)
NIS(v^-)	(0.03,0.03,0.03)	(0.09,0.09,0.09)	(0.02,0.02,0.02)	(0,0,0)	(0.03,0.03,0.03)	(0,0,0)	(0,0,0)	(0,0,0)	(0,0,0)

TABLE 7.13

Distance from PIS

Suppliers	TR$_1$	TR$_2$	TR$_3$	TR$_4$	TR$_5$	TR$_6$	TR$_7$	TR$_8$	TR$_9$	d_i^+
S$_1$	0/038504	0/050334	0/021846	0/084545	0/035869	0/032406	0/053192	0/050855	0/064509	0/657313
S$_2$	0/059297	0/050334	0/021846	0/084545	0/039625	0/054654	0/085196	0/053201	0/056079	0/710476
S$_3$	0/038504	0/050334	0/021846	0/087483	0/057435	0/068875	0/056406	0/045809	0/082186	0/713357
S$_4$	0/038504	0/053373	0/034195	0/084545	0/060016	0/047533	0/085196	0/045809	0/064509	0/716715
S$_5$	0/051468	0/053373	0/032695	0/082774	0/060016	0/047533	0/107861	0/068951	0/064509	0/754440
S^L										0/657313

TABLE 7.14

Distance from NIS

Suppliers	TR$_1$	TR$_2$	TR$_3$	TR$_4$	TR$_5$	TR$_6$	TR$_7$	TR$_8$	TR$_9$	d_i^-
S$_1$	0/065195	0/057554	0/036819	0/081002	0/070039	0/091346	0/134190	0/081307	0/088627	0/840285
S$_2$	0/057196	0/057554	0/036819	0/081002	0/063312	0/074831	0/113977	0/078447	0/093328	0/810226
S$_3$	0/065195	0/057554	0/036819	0/078488	0/058102	0/073145	0/129557	0/082608	0/085601	0/816743
S$_4$	0/065195	0/053451	0/032430	0/081002	0/055575	0/078801	0/113977	0/082608	0/088627	0/807258
S$_5$	0/060076	0/053451	0/033920	0/082881	0/055575	0/078801	0/110086	0/075769	0/088627	0/799491
S^R										0/840285

TABLE 7.15
Final Ranking of Suppliers

Suppliers	CC_i	Ranking
S_1	0	1
S_2	0/061073	3
S_3	0/060788	2
S_4	0/067966	4
S_5	0/105347	5

suggesting that the selection of all the suppliers is possible. However, M-TOPSIS method selects the best option. According to M-TOPSIS method, after calculating d_i^* and d_i^-, the shortest distance from positive ideal (S^L) and furthest distance from negative ideal (S^R) are identified in the bottom of each d_i column. Therefore, by applying Eq. 7.16, the distance between each point (d_i^*, d_i^-) and D is calculated for each M-TOPSIS option and the ranking of suppliers is provided by this method. This ranking is presented in Table 7.15. According to these information, Lian Vision Alborz Company is the best supplier of protective relays of power transmission networks with CC_i = 0. However, the weakest supplier is Parin Control Company with CC_i = 0.1053. It is of note that in the mentioned, if the value of CC_i is closer to zero, it will have a higher priority. In the final ranking of Semnan Regional Power Company of protective relays of power transmission networks, done by fuzzy QFD-TOPSIS method, are:

Parin Control > Energy Company of Spadana > Artnos Company > Modjeniroo Company > Lian Vision Alborz.

7.4.4 Sensitivity Analysis

Sensitivity analysis is a method for estimating the effects of parametric changes on different elements. This estimation is usually done by substituting the weight of each criterion (Yazdani et al., 2017). According to the results of the ranking section, it was found that the difference between the first and second selection is so significant such that any change in the weights of these two options will not change their priority. However, the difference between the third and fourth options is not significant. Therefore, changing their weights will result in a change of their priorities. Difference between *financial stability*, *experience*, and *geographic location* is s significant. The utilized hypothesis and the weights of their variations are indicated in Table 7.16 as 10 sensitivity analysis tests. These tests are:

1. Weight of *geographical location* criterion is reduced down to 10%.
2. Weight of *geographical location* criterion is reduced to down 10% and weight of *financial stability* criterion is increased up to 10%.

TABLE 7.16
Sensitivity Analysis Tests

Theories	W_1	W_2	W_3	W_4	W_5	W_6	W_7	W_8	W_9
Test 1	0/1205	0/1734	0/0645	0/1255	0/1181	0/1094	0/1667	0/1147	0/1296
Test 2	0/1326	0/1734	0/0645	0/1255	0/1181	0/1094	0/1667	0/1147	0/1296
Test 3	0/1326	0/1734	0/0645	0/1255	0/1063	0/1094	0/1667	0/1147	0/1296
Test 4	0/1446	0/1734	0/0645	0/1255	0/1181	0/1094	0/1667	0/1147	0/1296
Test 5	0/1385	0/1734	0/0609	0/1255	0/1181	0/1094	0/1667	0/1147	0/1296
Test 6	0/1446	0/1734	0/0645	0/1255	0/1063	0/1094	0/1667	0/1147	0/1296
Test 7	0/1024	0/1734	0/0609	0/1255	0/1358	0/1094	0/1667	0/1147	0/1296
Test 8	0/1326	0/1734	0/0573	0/1255	0/1181	0/1094	0/1667	0/1147	0/1296
Test 9	0/1446	0/1734	0/0573	0/1255	0/1181	0/1094	0/1667	0/1147	0/1296
Test 10	0/1385	0/1734	0/0573	0/1255	0/1358	0/1094	0/1667	0/1147	0/1296

3. Weight of *geographical location* and *experience* criteria are reduced down to 10% and *financial stability* criterion is increased up to 10%.
4. Weight of *geographical location* is reduced down to 10% and the weight of *financial stability* criterion is increased up to 20%.
5. Weight of *geographical location* criterion is reduced down to 15% and *financial stability* criterion is increased up to 15%.
6. Weight of *geographical location* and *experience* criteria are reduced down to 10% and *financial stability* is increased up to 20%.
7. Weight of *geographical location* and *financial stability* criteria are reduced down to 15% and *experience* is increased up to 15%.
8. Weight of *geographical location* criterion is reduced down to 20% and *financial stability* criterion is increased up to 10%.
9. Weight of *geographical location* criterion is reduced down to 20% and *financial stability* criterion is increased up to 20%.
10. Weight of *geographical location* criterion is reduced down to 20% and *financial stability* and *experience* criteria are increased up to 15%.

The strategies for sensitivity analysis of this project were divided into 10 tests in order to present different rankings of the suppliers. Therefore, the required changes for sensitivity analysis were done on criterion weights shown in Figure 7.4. This figure indicates supplier-ranking changes for each test, which indeed are similar to each other. Similar rankings are observed in tests 1–10 except 7. According to Table 7.16 of tests and the presented diagram of the first and second supplier's stability in tests, it can be claimed that no change in weights will affect the final ranking of suppliers and the confirmed stability of the presented model.

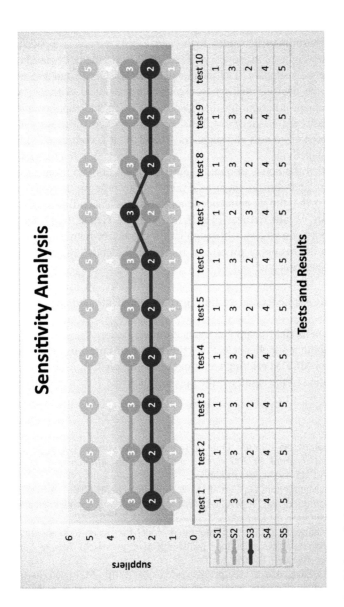

FIGURE 7.4   Sensitivity analysis.

7.5 CONCLUSION AND RECOMMENDATIONS FOR FURTHER STUDY

To enhance the efficiency of the obtained results, a combinational method was used in this study, including QFD and M-TOPSIS methods in fuzzy environments for identification and ranking of Semnan Regional Power Company suppliers of protective relays of power transmission networks. To this end, a QFD method is applied to estimate the final criterion weights of the suppliers. Then, M-TOPSIS method is used for ranking and selecting the best supplier. The integrated QFD/M-TOPSIS method provides an opportunity for the decision makers to use customer demand criterion, along with technical requirement criterion, for their decision-making and supplier selections. In fact, QFD increases the efficiency of decisions and supplier selections and creates a connection between customers and suppliers. Therefore, it is considered to be the best supplier selection method.

It is of note that the house of quality matrix helps users and project members to attract customers and convert them to a multivariate process. Sensitivity analysis strategies indicate that the recommended method is efficient and stable. This method is not affected by weight changes of different criteria. Hence, it is confirmed that Lian Vision Alborz can be a suitable supplier for protective relays.

According to the results of this study, some recommendations are presented for future supplier selection studies.

- Conducting a study on the relationship between mentioned factors and present model
- Designing a model for supplier evaluation
- Analyzing other supplier selection criteria and rankings in various sections
- Using other MCDM ranking techniques such as VICOR, ELECTERE, etc. for evaluation of the results of the present study

REFERENCES

Ansari, A., & Batoul, M. (2006). Quality function deployment: The role of suppliers. *International Journal of Purchasing and Materials Management*, *30*(3), 27–35. doi:10.1111/j.1745-493X.1994.tb00271.x.

Abdel-Basset, M., Manogaran, G., Gamal, A., & Smarandache, F. (2018). A hybrid approach of neutrosophic sets and DEMATEL method for developing supplier selection criteria. *Design Automation for Embedded Systems*, *22*(3), 257–278.

Amin, S. H., & Zhang, G. (2012). An integrated model for closed-loop supply chain configuration and supplier selection: Multi-objective approach. *Expert Systems with Applications*, *39*(8), 6782–6791.

Antuchevičiene, J., Zavadskas, E. K., & Zakarevičius, A. (2010). Multiple criteria construction management decisions considering relations between criteria. *Technological and Economic Development of Economy*, *16*(1), 109–125.

Assellaou, H., Ouhbi, B., & Frikh, B. (2018). A hybrid MCDM approach for supplier selection with a case study. In *Operations Research/Computer Science Interfaces Series* (Vol. 62). doi:10.1007/978-3-319-58253-5_12.

Awasthi, A., Govindan, K., & Gold, S. (2018). Multi-tier sustainable global supplier selection using a fuzzy AHP-VIKOR based approach. *International Journal of Production Economics*, *195*, 106–117. doi:10.1016/j.ijpe.2017.10.013.

Azadeh, A., & Alem, S. M. (2010). A flexible deterministic, stochastic and fuzzy Data Envelopment Analysis approach for supply chain risk and vendor selection problem: Simulation analysis. *Expert Systems with Applications*, *37*(12), 7438–7448.

Azadi, M., Jafarian, M., Saen, R. F., & Mirhedayatian, S. M. (2015). A new fuzzy DEA model for evaluation of efficiency and effectiveness of suppliers in sustainable supply chain management context. *Computers & Operations Research*, *54*, 274–285.

Babbar, C., & Amin, S. H. (2018). A multi-objective mathematical model integrating environmental concerns for supplier selection and order allocation based on fuzzy QFD in beverages industry. *Expert Systems with Applications*, *92*, 27–38.

Bai, C., & Sarkis, J. (2010). Integrating sustainability into supplier selection with grey system and rough set methodologies. *International Journal of Production Economics*, *124*(1), 252–264.

Banaeian, N., Mobli, H., Fahimnia, B., Nielsen, I. E., & Omid, M. (2018). Green supplier selection using fuzzy group decision making methods: A case study from the agri-food industry. *Computers & Operations Research*, *89*, 337–347.

Bevilacqua, M., Ciarapica, F. E., & Giacchetta, G. (2006). A fuzzy-QFD approach to supplier selection. *Journal of Purchasing and Supply Management*, *12*(1), 14–27.

Blackburn, J. L., & Domin, T. J. (2006). *Protective Relaying: Principles and Applications*. Boca Raton, FL: CRC Press.

Boran, F. E., Genç, S., Kurt, M., & Akay, D. (2009). A multi-criteria intuitionistic fuzzy group decision making for supplier selection with TOPSIS method. *Expert Systems with Applications*, *36*(8), 11363–11368. doi:10.1016/j.eswa.2009.03.039.

Buckley, J. J. (1985). Fuzzy hierarchical analysis. *Fuzzy Sets and Systems*, *17*(3), 233–247.

Çakır, S. (2017). Selecting the best supplier at a steel-producing company under fuzzy environment. *International Journal of Advanced Manufacturing Technology*, *88*(5–8), 1345–1361. doi:10.1007/s00170-016-8861-4.

Calabrese, A., Costa, R., & Menichini, T. (2013). Using fuzzy AHP to manage intellectual capital assets: An application to the ICT service industry. *Expert Systems with Applications*, *40*(9), 3747–3755.

Chai, J., Liu, J. N. K., & Ngai, E. W. T. (2013). Application of decision-making techniques in supplier selection: A systematic review of literature. *Expert Systems with Applications*, *40*(10), 3872–3885.

Chai, J., Liu, J. N. K., & Xu, Z. (2012). A new rule-based SIR approach to supplier selection under intuitionistic fuzzy environments. *International Journal of Uncertainty, Fuzziness and Knowledge-Based Systems*, *20*(03), 451–471.

Chamodrakas, I., Leftheriotis, I., & Martakos, D. (2011). In-depth analysis and simulation study of an innovative fuzzy approach for ranking alternatives in multiple attribute decision making problems based on TOPSIS. *Applied Soft Computing*, *11*(1), 900–907.

Chang, B., Chang, C.-W., & Wu, C.-H. (2011). Fuzzy DEMATEL method for developing supplier selection criteria. *Expert Systems with Applications*, *38*(3), 1850–1858.

Chang, D.-Y. (1996). Applications of the extent analysis method on fuzzy AHP. *European Journal of Operational Research*, *95*(3), 649–655.

Chen, C.-T., Lin, C.-T., & Huang, S.-F. (2006). A fuzzy approach for supplier evaluation and selection in supply chain management. *International Journal of Production Economics*, *102*(2), 289–301.

Chen, T.-Y., Wang, H.-P., & Lu, Y.-Y. (2011). A multicriteria group decision-making approach based on interval-valued intuitionistic fuzzy sets: A comparative perspective. *Expert Systems with Applications*, *38*(6), 7647–7658.

Cheng, C.-H. (1999). Evaluating weapon systems using ranking fuzzy numbers. *Fuzzy Sets and Systems, 107*(1), 25–35.

Chou, S.-Y., & Chang, Y.-H. (2008). A decision support system for supplier selection based on a strategy-aligned fuzzy SMART approach. *Expert Systems with Applications, 34*(4), 2241–2253.

Christopher, M., Mena, C., Khan, O., & Yurt, O. (2011). Approaches to managing global sourcing risk. *Supply Chain Management: An International Journal, 16*(2), 67–81.

Cooper, W. W., Seiford, L. M., & Tone, K. (2007). *Data Envelopment Analysis: A Comprehensive Text with Models, Applications, References and DEA-Solver Software.* Springer US. Retrieved from https://books.google.com/books?id=bJMRBwAAQBAJ

De Boer, L., Labro, E., & Morlacchi, P. (2001). A review of methods supporting supplier selection. *European Journal of Purchasing & Supply Management, 7*(2), 75–89.

Dobos, I., & Vörösmarty, G. (2018). Inventory-related costs in green supplier selection problems with Data Envelopment Analysis (DEA). *International Journal of Production Economics.* doi:10.1016/J.IJPE.2018.03.022.

Dweiri, F., Kumar, S., Khan, S. A., & Jain, V. (2016). Designing an integrated AHP based decision support system for supplier selection in automotive industry. *Expert Systems with Applications, 62*, 273–283.

Dyer, J. H., Cho, D. S., & Chu, W. (1998). Strategic supplier segmentation: The next "best practice" in supply chain management. *California Management Review, 40*(2), 57–77.

Falagario, M., Sciancalepore, F., Costantino, N., & Pietroforte, R. (2012). Using a DEA-cross efficiency approach in public procurement tenders. *European Journal of Operational Research, 218*(2), 523–529.

Feng, B., Fan, Z.-P., & Li, Y. (2011). A decision method for supplier selection in multi-service outsourcing. *International Journal of Production Economics, 132*(2), 240–250.

Gencer, C., & Gürpinar, D. (2007). Analytic network process in supplier selection: A case study in an electronic firm. *Applied Mathematical Modelling, 31*(11), 2475–2486. doi:10.1016/J.APM.2006.10.002.

Germani, M., Mengoni, M., & Peruzzini, M. (2012). A QFD-based method to support SMEs in benchmarking co-design tools. *Computers in Industry, 63*(1), 12–29.

Golmohammadi, D., & Mellat-Parast, M. (2012). Developing a grey-based decision-making model for supplier selection. *International Journal of Production Economics, 137*(2), 191–200.

Gülen, K. G. (2007). Supplier selection and outsourcing strategies in supply chain management. *Journal of Aeronautics and Space Technologies, 3*(2), 1–6.

Güneri, A. F., Ertay, T., & YüCel, A. (2011). An approach based on ANFIS input selection and modeling for supplier selection problem. *Expert Systems with Applications, 38*(12), 14907–14917.

Gürler, A. G. İ. (2007). Supplier selection criteria of Turkish automotive industry. *Journal of Yaşar University, 2*(6), 555–569.

Haleh, H., & Hamidi, A. (2011). A fuzzy MCDM model for allocating orders to suppliers in a supply chain under uncertainty over a multi-period time horizon. *Expert Systems with Applications, 38*(8), 9076–9083.

Hatami-Marbini, A., & Kangi, F. (2017). An extension of fuzzy TOPSIS for a group decision making with an application to tehran stock exchange. *Applied Soft Computing, 52*, 1084–1097.

Hauser, J. R., & Clausing, D. (1988). The house of quality. *Harvard Business Review, 66*(3), 63–73.

Herrera-Viedma, E., Herrera, F., Chiclana, F., & Luque, M. (2004). Some issues on consistency of fuzzy preference relations. *European Journal of Operational Research, 154*(1), 98–109.

Ho, W., Xu, X., & Dey, P. K. (2010). Multi-criteria decision making approaches for supplier evaluation and selection: A literature review. *European Journal of Operational Research, 202*(1), 16–24.

Hsu, B.-M., Chiang, C.-Y., & Shu, M.-H. (2010). Supplier selection using fuzzy quality data and their applications to touch screen. *Expert Systems with Applications, 37*(9), 6192–6200.

Hwang, C.-L., & Yoon, K. (1981). Methods for multiple attribute decision making. In *Multiple Attribute Decision Making* (pp. 58–191). New York: Springer.

Jain, V., Sangaiah, A. K., Sakhuja, S., Thoduka, N., & Aggarwal, R. (2016). Supplier selection using fuzzy AHP and TOPSIS: A case study in the Indian automotive industry. *Neural Computing and Applications, 29*(7), 555–564.

Jain, V., Sangaiah, A. K., Sakhuja, S., Thoduka, N., & Aggarwal, R. (2018). Supplier selection using fuzzy AHP and TOPSIS: A case study in the Indian automotive industry. *Neural Computing and Applications, 29*(7), 555–564. doi:10.1007/s00521-016-2533-z.

Juan, Y.-K., Perng, Y.-H., Castro-Lacouture, D., & Lu, K.-S. (2009). Housing refurbishment contractors selection based on a hybrid fuzzy-QFD approach. *Automation in Construction, 18*(2), 139–144.

Kahraman, C., Cebeci, U., & Ruan, D. (2004). Multi-attribute comparison of catering service companies using fuzzy AHP: The case of Turkey. *International Journal of Production Economics, 87*(2), 171–184.

Kannan, D., de Sousa Jabbour, A. B. L., & Jabbour, C. J. C. (2014). Selecting green suppliers based on GSCM practices: Using fuzzy TOPSIS applied to a Brazilian electronics company. *European Journal of Operational Research, 233*(2), 432–447.

Kannan, D., Khodaverdi, R., Olfat, L., Jafarian, A., & Diabat, A. (2013). Integrated fuzzy multi criteria decision making method and multi-objective programming approach for supplier selection and order allocation in a green supply chain. *Journal of Cleaner Production, 47*, 355–367.

Kara, S. S. (2011). Supplier selection with an integrated methodology in unknown environment. *Expert Systems with Applications, 38*(3), 2133–2139.

Karsak, E. E., & Dursun, M. (2014). An integrated supplier selection methodology incorporating QFD and DEA with imprecise data. *Expert Systems with Applications, 41*(16), 6995–7004.

Katiraee, N., Shirazi, B., & Fazlollahtabar, H. (2017). Fuzzy multi-objective supplier selection considering production requirements in resilient supply chain: Case study in steel industry. *International Journal of Information Systems and Supply Chain Management, 10*(3), 65–83. doi:10.4018/IJISSCM.2017070104.

Kilaparthi, S., & Sambana, N. (2018). Fuzzy kano: VIKOR integrated approach for supplier selection—A case study. *International Journal of Mechanical and Production Engineering Research and Development, 8*(2), 337–348. doi:10.24247/ijmperdapr201837.

Lee, C. K. M., Ru, C. T. Y., Yeung, C. L., Choy, K. L., & Ip, W. H. (2015). Analyze the healthcare service requirement using fuzzy QFD. *Computers in Industry, 74*, 1–15.

Li, G.-D., Yamaguchi, D., & Nagai, M. (2007). A grey-based decision-making approach to the supplier selection problem. *Mathematical and Computer Modelling, 46*(3–4), 573–581.

Li, L., & Zabinsky, Z. B. (2011). Incorporating uncertainty into a supplier selection problem. *International Journal of Production Economics, 134*(2), 344–356.

Liao, C.-N., & Kao, H.-P. (2010). Supplier selection model using Taguchi loss function, analytical hierarchy process and multi-choice goal programming. *Computers & Industrial Engineering, 58*(4), 571–577.

Lima-Junior, F. R., & Carpinetti, L. C. R. (2016). A multicriteria approach based on fuzzy QFD for choosing criteria for supplier selection. *Computers & Industrial Engineering, 101*, 269–285.

Lin, R.-H. (2012). An integrated model for supplier selection under a fuzzy situation. *International Journal of Production Economics, 138*(1), 55–61.

Lin, Y.-T., Lin, C.-L., Yu, H.-C., & Tzeng, G.-H. (2010). A novel hybrid MCDM approach for outsourcing vendor selection: A case study for a semiconductor company in Taiwan. *Expert Systems with Applications, 37*(7), 4796–4804.

Liou, J. J. H., Chuang, Y.-C., & Tzeng, G.-H. (2014). A fuzzy integral-based model for supplier evaluation and improvement. *Information Sciences, 266*, 199–217.

Liu, T., Deng, Y., & Chan, F. (2018). Evidential supplier selection based on DEMATEL and game theory. *International Journal of Fuzzy Systems, 20*(4), 1321–1333.

Marković, Z. (2016). Modification of TOPSIS method for solving of multicriteria tasks. *Yugoslav Journal of Operations Research, 20*(1), 117–143.

Mendoza, A., & Ventura, J. A. (2008). An effective method to supplier selection and order quantity allocation. *International Journal of Business and Systems Research, 2*(1), 1–15.

Montazer, G. A., Saremi, H. Q., & Ramezani, M. (2009). Design a new mixed expert decision aiding system using fuzzy ELECTRE III method for vendor selection. *Expert Systems with Applications, 36*(8), 10837–10847.

Ozkok, B. A., & Tiryaki, F. (2011). A compensatory fuzzy approach to multi-objective linear supplier selection problem with multiple-item. *Expert Systems with Applications, 38*(9), 11363–11368.

Pang, B., & Bai, S. (2013). An integrated fuzzy synthetic evaluation approach for supplier selection based on analytic network process. *Journal of Intelligent Manufacturing, 24*(1), 163–174.

Pinter, U., & Pšunder, I. (2013). Evaluating construction project success with use of the M-TOPSIS method. *Journal of Civil Engineering and Management, 19*(1), 16–23.

Pramanik, D., Haldar, A., Mondal, S. C., Naskar, S. K., & Ray, A. (2017). Resilient supplier selection using AHP-TOPSIS-QFD under a fuzzy environment. *International Journal of Management Science and Engineering Management, 12*(1), 45–54.

Ren, L., Zhang, Y., Wang, Y., & Sun, Z. (2007). Comparative analysis of a novel M-TOPSIS method and topsis. *Applied Mathematics Research eXpress, 2007.* doi:10.1093/amrx/abm005.

Rezaei, J., & Davoodi, M. (2012). A joint pricing, lot-sizing, and supplier selection model. *International Journal of Production Research, 50*(16), 4524–4542.

Russell, S. J., Norvig, P., Canny, J. F., Malik, J. M., & Edwards, D. D. (2003). *Artificial Intelligence: A Modern Approach* (Vol. 2). Upper Saddle River, NJ: Prentice Hall.

Saaty, T. L. (2001). *Decision Making with Dependence and Feedback: The Analytic Network Process* (Vol. 7). Pittsburgh, PA: RWS Publications, 557–570.

Sadeghieh, A., Dehghanbaghi, M., Dabbaghi, A., & Barak, S. (2012). A genetic algorithm based grey goal programming (G3) approach for parts supplier evaluation and selection. *International Journal of Production Research, 50*(16), 4612–4630.

Saen, R. F. (2010). Developing a new data envelopment analysis methodology for supplier selection in the presence of both undesirable outputs and imprecise data. *The International Journal of Advanced Manufacturing Technology, 51*(9–12), 1243–1250.

Sagar, M. K., & Singh, D. (2012). Supplier selection criteria: Study of automobile sector in India. *International Journal of Engineering Research and Development, 4*(4), 34–39.

Scott, J. A., Ho, W., & Dey, P. K. (2013). Strategic sourcing in the UK bioenergy industry. *International Journal of Production Economics, 146*(2), 478–490.

Sevkli, M. (2010). An application of the fuzzy ELECTRE method for supplier selection. *International Journal of Production Research, 48*(12), 3393–3405.

Shankar, B. L., Basavarajappa, S., Kadadevaramath, R. S., & Chen, J. C. H. (2013). A bi-objective optimization of supply chain design and distribution operations using non-dominated sorting algorithm: A case study. *Expert Systems with Applications, 40*(14), 5730–5739.

Shaw, K., Shankar, R., Yadav, S. S., & Thakur, L. S. (2012). Supplier selection using fuzzy AHP and fuzzy multi-objective linear programming for developing low carbon supply chain. *Expert Systems with Applications, 39*(9), 8182–8192.

Shi, H., Quan, M.-Y., Liu, H.-C., & Duan, C.-Y. (2018). A novel integrated approach for green supplier selection with interval-valued intuitionistic uncertain linguistic information: A case study in the agri-food industry. *Sustainability (Switzerland), 10*(3). doi:10.3390/su10030733.

Soukup, W. R. (1987). Supplier selection strategies. *Journal of Supply Chain Management, 23*(2), 7–12.

Stević, Ž., Pamučar, D., Vasiljević, M., Stojić, G., & Korica, S. (2017). Novel integrated multi-criteria model for supplier selection: Case study construction company. *Symmetry, 9*(11). doi:10.3390/sym9110279.

Tavana, M., Yazdani, M., & Di Caprio, D. (2017). An application of an integrated ANP–QFD framework for sustainable supplier selection. *International Journal of Logistics Research and Applications, 20*(3), 254–275.

Tsai, W.-H., & Hung, S.-J. (2009). A fuzzy goal programming approach for green supply chain optimisation under activity-based costing and performance evaluation with a value-chain structure. *International Journal of Production Research, 47*(18), 4991–5017.

Tseng, M.-L., Lim, M., Wu, K.-J., Zhou, L., & Bui, D. T. D. (2017). A novel approach for enhancing green supply chain management using converged interval-valued triangular fuzzy numbers-grey relation analysis. *Resources, Conservation and Recycling, 128,* 122–133.

Ustun, O., & Demi, E. A. (2008). An integrated multi-objective decision-making process for multi-period lot-sizing with supplier selection. *Omega, 36*(4), 509–521.

Vahidi, F., Torabi, S. A., & Ramezankhani, M. J. (2018). Sustainable supplier selection and order allocation under operational and disruption risks. *Journal of Cleaner Production, 174,* 1351–1365.

Van Laarhoven, P. J. M., & Pedrycz, W. (1983). A fuzzy extension of Saaty's priority theory. *Fuzzy Sets and Systems, 11*(1–3), 229–241.

Vinodh, S., Ramiya, R. A., & Gautham, S. G. (2011). Application of fuzzy analytic network process for supplier selection in a manufacturing organisation. *Expert Systems with Applications, 38*(1), 272–280.

Wang, J.-W., Cheng, C.-H., & Huang, K.-C. (2009). Fuzzy hierarchical TOPSIS for supplier selection. *Applied Soft Computing, 9*(1), 377–386.

Wang, T.-Y., & Yang, Y.-H. (2009). A fuzzy model for supplier selection in quantity discount environments. *Expert Systems with Applications, 36*(10), 12179–12187.

Willis, T. H., Huston, C. R., & Pohlkamp, F. (1993). Evaluation measures of just-in-time supplier performance. *Production and Inventory Management Journal, 34*(2), 1.

Wu, C., & Barnes, D. (2011). A literature review of decision-making models and approaches for partner selection in agile supply chains. *Journal of Purchasing and Supply Management, 17*(4), 256–274.

Wu, H.-Y., Tzeng, G.-H., & Chen, Y.-H. (2009). A fuzzy MCDM approach for evaluating banking performance based on Balanced Scorecard. *Expert Systems with Applications, 36*(6), 10135–10147.

Wu, T., & Blackhurst, J. (2009). Supplier evaluation and selection: An augmented DEA approach. *International Journal of Production Research, 47*(16), 4593–4608.

Yazdani, M., Zarate, P., Coulibaly, A., & Zavadskas, E. K. (2017). A group decision making support system in logistics and supply chain management. *Expert Systems with Applications, 88,* 376–392.

Yeh, W.-C., & Chuang, M.-C. (2011). Using multi-objective genetic algorithm for partner selection in green supply chain problems. *Expert Systems with Applications, 38*(4), 4244–4253.

Yilmaz, O., Gulsun, B., Guneri, A. F., & Ozgurler, S. (2011). Supplier selection of a textile company with ANP. TMT.

Yu, C., & Wong, T. N. (2015). An agent-based negotiation model for supplier selection of multiple products with synergy effect. *Expert Systems with Applications*, *42*(1), 223–237.

Yu, M.-C., Goh, M., & Lin, H.-C. (2012). Fuzzy multi-objective vendor selection under lean procurement. *European Journal of Operational Research*, *219*(2), 305–311.

Zaim, S., Sevkli, M., Camgöz-Akdağ, H., Demirel, O. F., Yayla, A. Y., & Delen, D. (2014). Use of ANP weighted crisp and fuzzy QFD for product development. *Expert Systems with Applications*, *41*(9), 4464–4474.

Zeng, J., An, M., & Smith, N. J. (2007). Application of a fuzzy-based decision-making methodology to construction project risk assessment. *International Journal of Project Management*, *25*(6), 589–600.

8 Multiple Objective Decision-Making Methods in Supplier Selection

Nurullah Umarusman

CONTENTS

8.1 INTRODUCTION

Businesses in the twenty-first century focus on their supply chain management (SCM) to increase their competitive power and, therefore, market share. The correct handling of the complex process of supply chain from raw materials to marketing and distribution of final product improves productivity and adds to quality. In this regard, supply chain is an organization consisting of all the processes starting from the raw material supplier to delivering the finished product to consumers. Information flows through the organizations connecting the supply chain activities, and it is used by the components among the supply chain.

A supply chain is a facility structure that fulfills the processes of gathering raw materials, transforming materials into intermediate products, and finally distributing final products to customers. There are uncertainties among the supply chain that stem from sources such as demand, process, and supply. Firms widely use inventories to protect the supply chain from such problems. Therefore, SCM is the administration of materials and information in and among sellers, manufacturing facilities, assembly lines, and distribution facilities (Rota et al. 1998). Supply chain is made up of a number of activities in terms of planning, coordination, and control of materials, components, and the final product from supplier to customer. The contribution of purchasing function to supply chain profitability has increased lately. Choosing the correct supplier is one of the most significant functions of purchasing (Ravindran and Wadhwa 2009). One important feature of the purchasing function is to choose the supplier to purchase the required material, equipment, and service for any business venture. The purchasing function is naturally a fundamental piece of business management. It is impossible in the current competitive environment to manufacture low-cost and high-quality products successfully with no satisfying seller. Therefore, one of the most vital purchasing decisions is to choose a competent group of suppliers and continue working with them (Weber et al. 1991). In order to stand up to the competition in customer-centered economies, the suppliers, manufacturers, contracted manufacturer, distributers, retailers, which make up the rings of the supply chain, have to participate together in the activities of design, production, and distribution. All members of this chain have to unite and be managed effectively to contribute to the supply chain profitability. It is because the members of a supply chain are the vital components of a supply chain formation. Therefore, partner selection is one of the significant steps in forming a supply chain (Chen et al. 2005).

The concept of sustainability is accepted today as proposed by Elkington (1998), whose approach was termed as triple bottom line with economic, environmental, and social dimensions. Each of these dimensions can be analyzed under three subdimensions. An improvement in any of the dimensions without a negative impact on other dimensions will result in more sustainable supply chains (Çetinkaya et al. 2011). The establishment of an effective supplier selection model based on the aim of sustainability is one of the most significant topics in creating a strategical partnership. In view of the increasing uncertainty and complexity of socioeconomic world, it is hard to suppose that the assumptions in a complex model would be totally realized with perfect information. Additionally, the decisions for supplier selection are usually taken under the pressure of time, and decision makers might have restricted expertise, focus, or information processing abilities about the problem (Büyüközkan and Çifçi 2011). Sustainability primarily focuses on social and environmental effects of sustainability, their interaction with one another, and conformity with economic factors. Many sustainable SCM models rely on the elements of eco-efficiency and environmental factors but disregard the social dimension. Holistic models that include all dimensions of sustainability have attracted interest recently. Ecological and social dimensions are usually based on the use of generic factors; however, more definite measures are also utilized (Brandenburg et al. 2014). Therefore, social, economic, and environmental dimensions should be all realized simultaneously and increase performance to achieve sustainability.

Numerous authors have documented the development of SCM over years. These authors take a slightly different perspective to the development of supply chain thinking, research, and practice, but a common theme of consolidation and integration flows through the timeline (Templar et al. 2016). While Monczka et al. (2009) categorized the evolution of purchasing and SCM into seven periods, Jain et al. (2010) argue that the evolution of SCM can be plotted against six eras.

Supplier management is critical to enhance how a company can benefit at the functional, operational, economic, and financial levels. Supplier evaluation is the usual step in managerial operations as new products, parts, or materials must be assessed to increase performance of the current supplier (Sepulveda and Derpich 2015). The quality of the goods supplier is very subjective as it differs from company to company. It also depends on the type of sourcing decisions. Sourcing decisions are classified into six groups: consumable supplies, production materials and components, intellectual property, capital purchases, services, and subcontractors (Mukherjee 2017).

In the literature, many ways of classifying the supplier base have been proposed. The most important way to develop different purchasing strategies is to classify the supplier base in a comprehensive manner (Lilliecreutz and Ydreskog 1999). Kraljic (1983) defined categories of suppliers based on the economic impact and the risks for suppliers. A matrix of routine, leverage, bottleneck, and critical quadrants define these categories. The purchasing portfolio matrix groups them in two dimensions: profit impact and supply risk. Monczka et al. (2009) created a classification as "Critical, Routine, Leverage, and Bottleneck". Sepulveda and Derpich (2015) categorized suppliers into three groups: transactional suppliers, collaborative suppliers, and integrated suppliers. These classifications are unique for the company in question; at the same time, the method can also be applied in similar situations.

8.2 PROCESS AND CRITERIA FOR SUPPLIER SELECTION

Traditional methods for selecting and evaluating suppliers focus only on the businesses' demands, but they do not account for the supply chain as a whole. The successful management of the link between the suppliers and customers in a supply chain requires active cooperation. Businesses usually tend to work in a close manner with fewer or one trusted supplier to have a high supply chain performance and maintain it. Outsourcing is inevitable to increase the competitiveness of businesses, and it is a critical success factor to decide whether to use outsourcing or not in appropriate supplier selection. Supplier selection is a MCDM problem that requires a combination of various quantitative and qualitative factors to determine the best supplier. For example, a supplier with low unit price could also offer the lowest quality. Therefore, it possesses a complex structure as conflicting criteria must be reviewed during such decision-making problems (Ravindran and Warsing 2013).

The Supplier Selection Process offers a structure that guides and directs companies. Selecting the right supplier is like choosing the right business partner; if you make the right decision, you form a potentially permanent relation. Selecting the wrong supplier can damage or ruin a possibly good project (Power et al. 2006). Today, supply management means building a long-term cooperation with a few reliable suppliers. Choosing the right supplier depends on various factors that include both quantitative and qualitative subjects (Ho et al. 2010). It is because the members of a supply chain are the crucial determinants of a supply chain formation. Therefore, partner selection is one of the substantial steps in forming a supply chain (Chen et al. 2005). Selecting the right supplier is the key in supplying process to lower costs. However, choosing the incorrect supplier would bring operational and financial problems (Weber et al. 1991).

8.2.1 AIM OF SUPPLIER SELECTION

The decision of supplier selection is a vital and crucial for buyers and enterprises. The complexity increases when different supplier performances and related factors are taken into consideration, which effects the selection decision. Managers should analyze and report the significance of a number of factors to convert the indicators into precise empirical measures (Bai and Sarkis 2010). The goal of supplier selection is to determine the suppliers with high potential and sustainability, which would satisfy business demands in the long term with acceptable costs. A correct selection of suppliers lowers purchasing costs and increases customer satisfaction, thus competition power. Buyers and sellers have to consider each and every factor to increase their profit. According to de Boer et al. (2001), there are four steps to supplier selection: the determination of the aim of supplier selection, definition of decision criteria, pre-selection of possible suppliers, and final decision. The most vital element is supplier selection is the criteria upon which the selection would be made. In this regard, the selection of the supplier based on appropriate criteria will result in the right decision. Supplier selection traditionally depends on the analyses of objectives, which include quality, price/ cost, and after-sale services. Additionally, suppliers' attributes such as their financial position, reputation, facility capacity, communication, and location create the previous information about them and present the importance of criteria (Johnson et al. 2005).

As the criteria for supplier selection are conflicting, supplier selection process is complex and difficult. Therefore, it is regarded as a MCDM problem. That's why different qualitative and quantitative factors directly affect selection process to determine the best suppliers. For example, a supplier with high-quality products might have low percentage of after-sale services. Consequently, the selection process to determine the most appropriate supplier(s) becomes so complex due to the existence of more than one criterion. Apart from choosing the most suitable supplier(s), it is also crucial in supplier selection process to determine how much product would be acquired from which supplier.

8.2.2 MISTAKES IN SUPPLIER SELECTION

In the decision-making process, it is extremely easy to make supplier selection depending on a criterion. The process of deciding on supplier selection involves multicriteria including both quantitative and qualitative prospects in real-world problems. When there are numerous criteria, equal or different weighting factor are assigned the criteria in decision-making (Thiruchelvam and Tookey 2011). Although supplier selection is considered to be easy, increase of the supplier numbers involves the increasing number of properties on which they must be evaluated, and the difficulties of selecting and stating supplier selection parameters (Altinoz et al. 2010). As the criteria for supplier selection are conflicting, the process itself is complex and difficult. Therefore, supplier selection is regarded as a MCDM problem. The reason is that different qualitative and quantitative factors directly affect the selection process while determining the best suppliers. For example, a supplier with high-quality products might have low percentage of after-sale services. Consequently, the selection process to determine the most appropriate supplier(s) becomes so complex due to the existence of more than one criterion. Apart from selecting the most suitable supplier(s), it is also crucial in supplier selection process to determine how much product would be purchased from which supplier. Supplier selection is a part of the supply chain, and mistakes should be avoided as it may be impossible to compensate for them (Power et al. 2006).

A good integration of a supply chain by a business, cooperation between members of the chain, establishment of communication and loyalty decrease costs and increase profits and market share. Deficiencies and problems in information transfer and sharing along the supply chains effect the chain negatively resulting in further problems. It causes information degradation from the first player to the last in the chain and results in failure due to mistakes in supplier selection.

8.2.3 CRITERIA FOR SUPPLIER SELECTION

Criteria for supplier selection vary depending on the characteristics of businesses. In addition to their needs, each and every business may have different principles and policies. Companies determine their own Supplier Selection Criteria (SSC) based on their methods and workflows. Dickson (1966) was the first to define 23 criteria to be used in supplier selection. Afterwards, Dempsey (1978), Roa and Kiser (1980), and Bache et al. (1987) defined criteria with various numbers. While purchasing managers determine their suppliers based on traditional evaluation and selection criteria, they rarely consider social and environmental criteria. As sustainability is the important driving force behind supplier chain, businesses should pay attention to the other dimensions of sustainability in addition to the traditional economic criteria (Özçelik and Öztürk 2014). Sustainable-Supplier Selection Problem (S-SSP) includes the definition of sustainability factors or valid attributes for supplier selection and decision-making based on evaluation criteria of supplier performance. In addition, it is inevitable that data and information might be inaccurate, expert opinions are uncertain, and such inaccuracy and uncertainties are also present in sustainability factors to precisely solve S-SSP in Sustainable SCM. There are complex relations among the criteria required by S-SSP to evaluate supplier performances. Therefore, the evaluation approach must be able to define such complex relations (Rabbania et al. 2017).

Ghoushchi et al. (2017) defined the criteria for sustainable supplier selection process from the articles they analyzed. They classified them in terms of the economic, social, and environmental dimensions which were called as triple bottom line by Elkington (1998). All the sustainable economic selection criteria directly or indirectly comply with Dickson's criteria defined as *very important* average value. It means that traditional supplier selection mostly utilizes economic criteria. Technological developments, decreasing natural resources with increasing population, damage to the ecosystem due to environmental pollution, and socio-cultural changes brought the criteria belonging to the concept of sustainability in supplier selection and evaluation to a more significant level in the last 20 years.

8.2.4 MATHEMATICAL METHODS USED IN SUPPLIER SELECTION

The decision of supplier selection is a part of managing a supply chain and plays a significant role in the success of a business. SSP, which aims to determine the most appropriate suppliers in terms of the demands of a business, is a MCDM problem with many conflicting qualitative and quantitative criteria. MCDM methods are used extensively in supplier selection. These methods enable choosing the best supplier by ranking among the suppliers from the best efficient to the least and calculating the amount of material to be purchased from suppliers. Multiple Objective Decision Making (MODM) methods such as Goal Programming (GP), Multiple Objective Linear Programming (MOLP), Compromise Programming, and their fuzzy models can be used to determine the amount to be purchased from suppliers considering objective functions based on basic criteria such as transportation price, cost, quality, and after-sale service coverage. Multiple Attribute Decision Making (MADM) methods such as ANP, AHP, ELECTRE, TOPSIS, VIKOR, and their fuzzy environment approaches can be used to select the supplier with the highest efficiency. Additionally, hybrid models were developed based on the models stated above.

Methods to solve in supplier selection problems are classified in various ways by the researchers. A number of such methods can be stated below. Ghodsypour and O'Brien (1998) classified supplier selection problems as single-source and multiple-source based on the number of sources. The classification of the methods of Ding et al. (2003) for solving selection problems is as elimination methods, optimization methods, and probabilistic method. Pal et al. (2013) classified supplier selection problems as methods for pre-qualification of suppliers, MADM techniques, Mathematical Programming (MP) methods, AI models, fuzzy approach, and hybrid methods. Mukherjee et al. (2013) and Mukherjee (2017) included *methods for single* and *integrated model* in his classification.

The literature study reviewed 45 articles for MCDM methods in Supplier Selection. This study concluded that the said methods can be analyzed in three groups: The first group is based on the MADM method used in the solution, the second is according to the MODM method employed in the process, and the third is the group that includes a combination of both techniques. Additionally, the fuzzy models of the methods in the classification are effectively used in supplier selection. The literature study was conveyed based on the three groups to clarify this classification.

MADM techniques carry out the selection process among a limited number of very well-defined alternatives. Therefore, a limited number of suppliers are evaluated in terms of a limited number of criteria to make the selection in the problems where MADM methods are used. The literature summary for the articles analyzed in this section can be given as follows:

Chou and Chang (2008) present a strategy-aligned Fuzzy Smart approach to solve the SSP from the point of strategic management of the supply chain. Lee (2009) developed a model by using Fuzzy Analytic Hierarchy Process in fuzzy decision environment for SSP. Aydin and Kahraman (2010) used the F-AHP in supplier selection for an air conditioner firm. Chang et al. (2011) used Fuzzy DEMATEL method to define the significant factors selection suppliers in SCM. Yücenur et al.

(2011) based their model on linguistic variable weight to choose a global supplier with the help of Analytical Hierarchy and Network Processes (http://link.springer.com/search?facet-author=%22Qiansheng+Zhang%22). Zhang and Huang (2012) used Fuzzy TOPSIS method on supplier selection in IT service outsourcing problems. Wu and Liu (2013) made use of VIKOR and Fuzzy TOPSIS in supplier selection and demonstrated the effectiveness of each algorithm in an illustrative example. Focusing on socially sustainable supplier selection. Mani et al. (2014) demonstrated the development of social sustainability indicators and described how those indicators may be utilized in prioritizing alternatives to make decisions with AHP. Hruška et al. (2014) described his design using AHP with an application. Yu and Tseng (2014) illustrated their proposed Social Compliance Program criteria to demonstrate if SCP is able to evaluate and select sustainable global suppliers from the social compliance perspective. They used a multicriteria decision framework to measure the SSP within S-SCM by adopting AHP methodology. Tosun and Akyüz (2015) proposed a new F-TODIM for SSP and provided an application. Keskin (2015) suggested a new model to increase the quality of supplier selection and evaluation, which uses Fuzzy DEMATEL and Fuzzy c-means algorithms. Singh et al. (2016) combined VIKOR and AHP in an IVF environment to propose a hierarchal MCDM method to rank sustainable manufacturing strategies. Girubha et al. (2016) compared ISM–ANP–ELECTRE and ISM–ANP–VIKOR to select sustainable suppliers. They studied the results from these two hybrid MCDM methods on which they based their supplier selection. Büyüközkan and Çifçi (2017) created a hybrid Fuzzy MCDM, which integrates ANP, TOPSIS, and DEMATEL methods in a fuzzy context. Polat et al. (2017) suggested an integrated F-MCGDM approach, which employs fuzzy TOPSIS and fuzzy AHP together. Shu-ping et al. (2017) created a new hybrid method, which integrates TL-ANP and IT-ELECTRE II, and worked on a kind of MCDM problems.

MODM methods are used for goal problems, which include unlimited number of alternatives defined with mathematical constraints. Known as mathematical programming, these methods determine how many products would be purchased from which supplier among a limited number of suppliers. The evaluation of literature analysis in this regard is given below:

Gaballa (1974) is the first to use Mixed-Integer Programming (MIP) to decrease the total discounted price of allocated items to the suppliers. Additionally, Weber and Ellram (1992) demonstrated the benefits of applying a multiple objective mathematical programming model in a decision support context. Weber and Current (1993) presented a multiple objective approach to systematically study the inert balance involved in multicriteria Vendor Selection Problem (VSP). Ghodsypour and O'Brien (1998) presented integrated AHP and LP. Karpak et al. (1999) suggest using visual interactive GP for SSP. Kumar et al. (2004) aimed to solve an SSP by applying a Fuzzy GP approach with fuzzy parameters. Amid et al. (2006) studied supplier selection with asymmetrical fuzzy decision-making technique by assigning different weights in accordance to needs of the decision maker. Kumar et al. (2006) suggested a fuzzy Multiple Objective Integer Programming vendor selection problem' formulation. Famuyiwa et al. (2008) used a fuzzy logic-based GP approach to present a model for decision makers for evaluating suppliers' suitability. Amid et al. (2009) developed a fuzzy multi-objective model for the SSP under price breaks. Elahi et al. (2011) suggested MOLP and utilized fuzzy Compromise Programming to turn it into a single objective model and mix the weights of objectives through various decision makers' opinions. Nazari-Shirkouhi et al. (2013) offered interactive FMOLP model to solve an SSP under multiple product and multiple price level. Arikan (2013) regarded an SSP with objectives such as maximum quality, minimum costs, and maximum on-time delivery as MOLP, proposed fuzzy solution approach as the solution, and demonstrated his approach with an example. Jadidi et al. (2014) considered an SSP with the objectives of minimization of lead-time, price, and rejects as a MOP and conducted the solution with seven methods, providing a comparative analysis. Jadidi et al. (2015) suggested a new Multi-Choice GP approach for the solution of SSP that he modelled as a MOOP. Yousefi et al. (2017) transformed supply chain visibility, reduction of the number of defective or delayed parts, and minimizing supply chain cost objectives into one function by using Global Criterion Method. Umarusman and Hacıvelioğulları (2018) dealt with the SSP of a business in farming industry restructuring it based on MOLP and solving it according to Global Criterion Method. Ghoushchi et al. (2018) presented a new Data Envelopment Analysis model based on GP with both cardinal and ordinal data to evaluate performance and to select sustainable suppliers.

It is possible to see articles that use MADM and MODM together in supplier selection. The most striking case in these articles is that the relative weight factors and significance levels that could be used in MODM techniques were defined with MADM methods: Ghodsypour and O'Brien (1998) presented an integrated AHP and LP for SSP. Kasirian et al. (2010) used Lexicographic GP to solve an SSP with two goals and defined the priority ranking among the goals with both AHP and ANP. Liao and Kao (2010) offered a method by combining TLF, AHP, and Multi-Choice GP. Shaw et al. (2012) present an integrated method to choose the suitable supplier in the SSC using fuzzy Analytic Hierarchy Process and fuzzy MOLP. Perić et al. (2013) proposed a model to select vendor and determine supply quotas by using MOLP and Analytic Hierarchy Process model solved by Fuzzy Linear Programming. Huang and Hu (2013) developed a systematic process for supplier selection using Fuzzy ANP-GP and DNP. Azadnia et al. (2015) suggested an integrated approach of rule-based weighted fuzzy method, F-AHP and multiple objective MP. It aims for SSC and order allocation that is merged with a multiple period, multiple product lot-sizing problem. Singh (2016) presented a hybrid algorithm that allocates the customer demand among the suppliers based on order preference by Fuzzy Logic, GP, MILP, and TOPSIS approaches. Kumar et al. (2017) used F-Analytic Hierarchy Process and FMOLP to develop a model to select suppliers and optimize orders. The developed model was applied in a car firm in India. Luthra et al. (2017) suggested a framework to investigate Sustainable SS by using an integrated AHP, VIKOR, a multiple criteria optimization, and compromise solution.

8.2.4.1 Formulation of Multiple Objective Supplier Selection Problem

In terms of MCDM, SSP can be regarded as a special case of MOLP (Amid et al. 2011). Mathematical programming for SSP was first used by Gaballa (1974) who utilized mixed-integer programming. The multiple objective mathematical model for SSP is defined as follows (Weber and Current 1993) and (Amid et al. 2011):

$$Min \ W_1 = \sum_{i=1}^{n} P_i(x)$$

$$Min \ W_2 = \sum_{i=1}^{n} F_i(x)$$

$$Max \ Z_1 = \sum_{i=1}^{n} S_i(x) \qquad (8.1)$$

Subject to

$$\sum_{i=1}^{n} x_i \geq D$$

$$x_i \leq C_i$$

$$x \geq 0 \text{ and integer}$$

where:

   x_i: The product amount to be received from ith supplier,
   n: The number of suppliers,
   D: Periodical demand amount,
   C_i: Capacity of supplier i,
   P_i: Cost of the product to be received from the supplier i,
   F_i: The percentage of return to ith supplier,
   S_i: The percentage of service success of ith supplier.

Multiple Objective SSP-(1) includes the objectives of minimum cost, minimum return percentage, and maximum service success objective and the constraints of product amount to be supplied from supplier and supplier capacity. It is possible to see a different definition of the objective and constraint functions in some problems. Therefore, the organizations for (8.1) can be used for the objectives defined by the reader as well. Positive ideal solution and negative ideal solution values should be determined for each objective function in (8.1). Positive ideal solution of (8.1) can be obtained by solving each objective function. Negative ideal solution of (8.1) can be obtained by minimizing the objective function $Z_1(x)$ and by maximizing the objective function $W_1(x)$ and $W_2(x)$. Positive ideal solutions are defined as $I^* = \left\{ Z_1^*; W_1^*; W_2^* \right\}$ and negative ideal solutions are shown as $I^- = \left\{ Z_1^-; W_1^-; W_2^- \right\}$.

Depending on these solutions, there are many methods used in MODM classification to minimize the distance from the positive ideal solution. These methods solve the problem with different viewpoints based on the *preference information* provided from the decision maker. Hwang and Masud (1979), Romero (1985), and Romero (1991), Lai and Hwang (1994), and Gal et al. (1999) can be referred as the fundamental sources about these methods. Among the methods to be used in solving supplier selection problems, Global Criterion Method, Compromise Programming, Minmax GP, Step Method are analyzed here. Their theoretical processes are briefly described, and they are reorganized for SSP based on ideal solutions.

8.2.4.1.1   Global Criterion Method

Global Criterion Method is located in the classification of MODM methods where *preference information* is not used, and it reaches the solution without requiring any information from the decision maker. The preliminary studies on Global Criterion Method were initiated by Yu (1973) and Zeleny (1973). The most important feature of this method is that it is a scaler function that transforms more than one objective function mathematically into one single objective function (Marler and Arora 2004). The vector that minimizes at most one collective criterion, which is composed of multiple criteria based on a specific rule, is the optimal vector. In other words, it transforms k number of objective functions into one single objective problem (Tabucannon 1988). Each objective function in the global function is stated as a ratio. Nondimensionalization is important as objective functions have different dimensions (units). The best solution for the problem varies on the selected p-value. While these deviations are weighted proportionally to the largest deviation with the greatest weight based on $p = 2$ proposed by Salukvadze (1974), the same significance is given to all deviations based on $p = 1$ proposed by Boychuk and Ovchinnikov (1973). In case of $p > 2$, the weight of deviations will be greater. As the normalization is carried out based on the positive ideal solutions of objectives, the global objective value will be within the range of $[0;1]$. Global Criteria Method is provided mathematically below.

$$Min \sum_{i=1}^{i} \left(\frac{f_i\left(x^*\right) - f_i(x)}{f_i\left(x^*\right)} \right)^p$$

Subject to   (8.2)

$$Ax \leq b$$

$$x \geq 0.$$

where, $f_i(x^*)$ is the positive ideal point belonging to ith objective function. Umarusman and Türkmen (2013) generalized the Global Criteria Method for problems with maximization and minimization directed objectives. Below is the Global Criterion Methods, which are generalized using Equations 8.3 and 8.4.

$$\sum_{k=1}^{l} \left[\frac{Z_k^* - Z_k(x)}{Z_k^*} \right]^p \tag{8.3}$$

for maximization-type and minimization-type objectives to positive ideal solutions

$$\sum_{s=1}^{r} \left[\frac{W_s(x) - W_s^*}{W_s^*} \right]^p \tag{8.4}$$

to solve Multiple Objective Linear Programming problems

$$Min \ G = \sum_{k=1}^{l} \left[\frac{Z_k^* - Z_k(x)}{Z_k^*} \right]^p + \sum_{s=1}^{r} \left[\frac{W_s(x) - W_s^*}{W_s^*} \right]^p$$

Subject to $\tag{8.5}$

$$Ax \le b$$

$$x \ge 0.$$

Here;

$Z_k(x)$: kth objective function to be maximized,
Z_k^*: Positive ideal solution value for kth objective,
$W_s(x)$: sth objective function to be minimized,
W_s^*: Positive ideal solution value for sth objective,
$p : (1 \le p \le \infty)$.

The reorganized structure of Global Criterion Method for Multiple Objective SSP (8.1) is given below. Global objective function (8.6) is acquired using (8.3) and (8.4) of Multiple Objective SSP.

$$Min \ G = \left[\frac{Z_1^* - \sum_{i=1}^{n} S_i(x)}{Z_1^*} \right]^p + \left[\frac{\sum_{i=1}^{n} P_i(x) - W_1^*}{W_1^*} \right]^p + \sum_{s=1}^{r} \left[\frac{\sum_{i=1}^{n} F_i(x) - W_2^*}{W_2^*} \right]^p \tag{8.6}$$

Can be stated rather briefly as

$$Min \ G = \left[1 - \frac{\sum_{i=1}^{n} S_i(x)}{Z_1^*} \right]^p + \left[\frac{\sum_{i=1}^{n} P_i(x)}{W_1^*} - 1 \right]^p + \sum_{s=1}^{r} \left[\frac{\sum_{i=1}^{n} F_i(x)}{W_2^*} - 1 \right]^p \tag{8.7}$$

or

$$Min\ G = \left[1 - \frac{Z_1(x)}{Z_1^*}\right]^p + \left[\frac{W_1(x)}{W_1^*} - 1\right]^p + \sum_{s=1}^{r}\left[\frac{W_2(x)}{W_2^*} - 1\right]^p \qquad (8.8)$$

Here, $p = 1$ is used to provide the same amount of importance to all the objectives.

$$Min\ G = \left[\left(-\frac{\sum_{i=1}^{n}S_i(x)}{Z_1^*} + \frac{\sum_{i=1}^{n}P_i(x)}{W_1^*} + \frac{\sum_{i=1}^{n}F_i(x)}{W_2^*}\right) - 1\right] \qquad (8.9)$$

$$\left(-\frac{\sum_{i=1}^{n}S_i(x)}{Z_1^*} + \frac{\sum_{i=1}^{n}P_i(x)}{W_1^*} + \frac{\sum_{i=1}^{n}F_i(x)}{W_2^*}\right)$$ should be minimized in order to mini-

mize (8.9). Using this information;

$$Min\ F = \left(-\frac{\sum_{i=1}^{n}S_i(x)}{Z_1^*} + \frac{\sum_{i=1}^{n}P_i(x)}{W_1^*} + \frac{\sum_{i=1}^{n}F_i(x)}{W_2^*}\right)$$

Subject to

$$\sum_{i=1}^{n}x_i \geq D \qquad (8.10)$$

$$x_i \leq C_i$$

$$x \geq 0 \text{ and integer.}$$

acquired. Here;

x_i: The product amount to be received from ith supplier,
D: Periodical demand amount,
C_i: Capacity of supplier i,
n: The number of suppliers,
P_i: Cost of the product to be received from the supplier i,
F_i: The percentage of return to ith supplier,
S_i: The percentage of service success of ith supplier.

Z_1^*: Positive ideal solution value for cost objective,
W_1^*: Positive ideal solution for service success percentage objective,
W_2^*: Positive ideal solution value for return percentage objective.

8.2.4.1.2 Compromise Programming

Compromise Programming (Zeleny 1973 and Yu 1973) is a method that investigates the compromise solution, which primarily uses preference information, and its main goal is to minimize the distances of functions from their positive ideal points. In this method, which aims to minimize the distance from the ideal point, it is a behavior form to prefer the alternative, which is the closest to the ideal point, also known as Zeleny's (1973) preference action, and it is a rational phenomenon of human choices to prefer the point closest to the ideal. If there is no ideal solution considering many and conflicting objectives, the second best should be targeted (Zeleny 2011). Compromise Programming enables the selection from alternatives by considering the ideal solutions, which belong to each objective function also called as reference point (Malczewski 1999). Compromise Programming is given mathematically below.

$$Min\ d_\infty$$

Subject to

$$\alpha_k^p \left[\frac{Z_k^* - Z_k(x)}{Z_k^* - Z_k^-} \right] \le d_\infty$$

$$\alpha_s^p \left[\frac{W_s(x) - W_s^*}{W_s^- - W_s^*} \right] \le d_\infty$$

(8.11)

$$x \in X$$

The result (8.11) for $p = \infty$ is called the compromise solution in the non-superior solutions cluster. The result of Compromise Programming solution is within the range of $0 \le d \le 1$. The fact that d-value is "0" means that the positive ideal solutions of objective functions are acquired. The fact that d-value equals "1" means objective functions equal negative ideal solutions. Considering these two cases, the normalized degree of distance shows the success percentage of objective functions based on the positive ideal and negative ideal solutions. Multiple Objective SSP (8.1) is organized in terms of Compromise Programming as seen below.

$$Min\ d_\infty$$

Subject to

$$\alpha_1^p \left[\frac{\sum_{i=1}^{n} P_i(x) - W_1^*}{W_1^- - W_1^*} \right] \leq d_\infty$$

$$\alpha_2^p \left[\frac{\sum_{i=1}^{n} F_i(x) - W_2^*}{W_2^- - W_2^*} \right] \leq d_\infty \qquad (8.12)$$

$$\alpha_1^p \left[\frac{Z_1^* - \sum_{i=1}^{n} S_i(x)}{Z_1^* - Z_1^-} \right] \leq d_\infty$$

$$\sum_{i=1}^{n} x_i \geq D$$

$$x_i \leq C_i$$

$x \geq 0$ and integer.

Here;

x_i: Is the product amount to be received from ith supplier,

D: Periodical demand amount,

n: The number of suppliers,

P_i: Cost of the product to be received from the supplier i,

W_1^*: Positive ideal solution for service success percentage objective,

W_1^-: Negative ideal solution for service success percentage objective,

F_i: The percentage of return to ith supplier,

W_2^*: Positive ideal solution value for return percentage objective,

W_2^-: Negative ideal solution value for return percentage objective,

S_i: The percentage of service success of ith supplier.

Z_1^*: Positive ideal solution value for cost objective,

Z_1^-: Negative ideal solution value for cost objective.

8.2.4.1.3 Minmax Goal Programming

GP is one of MODM methods where preference information is used primarily. It is considered as an extension of classical LP and aims to reach the target values determined for each goal instead of maximizing or minimizing goal functions.

GP studies were initiated by Charnes et al. (1955), and Charnes and Cooper (1961) were the first to use Goal Programming as a term. The minimization process in GP approach can be applied in various ways. The basic methods of GP are Archimedean GP (Charnes and Cooper 1977), Lexicographic GP (Ijiri 1965, Lee 1972), and Minmax GP (Flavell 1976).

In this chapter, the Minmax GP proposed by Flavell (1976) is reorganized for the SSP. In this kind of GP, maximum deviation is minimized instead of minimization of the sum of deviating variables (Ignizio and Cavalier 1994). Flavell (1976) was the first to propose this method. It is defined mathematically as follows:

$$Min\ d_\infty$$

Subject to

$$\left(w_i^- \frac{d_i^-}{t_i} + w_i^+ \frac{d_i^+}{t_i} \right) \le d_\infty$$

$$\sum_{j=1}^{n} a_{ij} x_j - b_i + d_i^- - d_i^+ = b_i \qquad (8.13)$$

$$d_i^+ . d_i^- = 0$$

$$x_j \ge 0,\ i = 1,2,\ldots, m \text{ ve } j = 1,2,\ldots, n$$

where:

d_∞: Maximum deviation amount whose minimum value is investigated,
w_i^+: Weight of the positively deviating variable,
w_i^-: Weight of the negatively deviating variable.

Minmax GP is also named as Chebyshev Goal Programming. Instead of d_1 metric, which is used by Archimedean and Lexicographic GP models, d_∞ metric is used in this method. If the positive ideal solution of minimum cost objective (W_1^*) is added to the objective function as an equation,

$$\sum_{i=1}^{n} P_i(x) + d_1^- - d_1^+ = W_1^* \qquad (8.14)$$

is acquired. If positive ideal solutions are used for minimization-type objectives, d_i^- should be "0."

$$\sum_{i=1}^{n} P_i(x) - d_1^+ = W_1^* \qquad (8.15)$$

If a similar reorganization is made for return percentage objective,

$$\sum_{i=1}^{n} F_i(x) - d_2^+ = W_2^*$$ (8.16)

is acquired. Service success percentage objective, which is maximization-type objective, is transformed into a goal.

$$\sum_{i=1}^{n} S_i(x) + d_3^- - d_3^+ = Z_1^*$$ (8.17)

As this goal cannot exceed (Z_1^*), d_i^+ will be "0." Accordingly,

$$\sum_{i=1}^{n} S_i(x) + d_1^- = Z_1^*$$ (8.18)

The goal is acquired. On the other hand, normalization process is required as the units of the goals are transformed into targets are different from each other. Positive and negative ideal solutions of each goal function are used for the normalization process.

$$Min \; d_\infty$$

Subject to

$$\sum_{i=1}^{n} P_i(x) - d_1^+ = W_1^*$$

$$\left(\frac{d_1^+}{W_1^- - W_1^*} \right) - d_\infty \le 0$$

$$\sum_{i=1}^{n} F_i(x) + d_2^- - d_2^+ = W_2^*$$

$$\left(\frac{d_2^+}{W_2^- - W_2^*} \right) - d_\infty \le 0$$

$$\sum_{i=1}^{n} S_i(x) + d_1^- = Z_1^*$$ (8.19)

$$\left(\frac{d_1^-}{Z_1^* - Z_1^-} \right) - d_\infty \le 0$$

$$\sum_{i=1}^{n} x_i \ge D$$

$$x_i \le C_i$$

$$x_j \ge 0, \; i = 1,2,\ldots,m, \text{ and integer.}$$

Here;

  x_i: The product amount to be received from ith supplier,
  D: Periodical demand amount,
  n : The Number of Suppliers,
  P_i: Cost of the product to be received from the supplier i,
  W_1^* : Positive ideal solution for service success percentage objective,
  W_1^- : Negative ideal solution for service success percentage objective,
  F_i: The percentage of return to ith supplier,
  W_2^* : Positive ideal solution value for return percentage objective,
  W_2^- : Negative ideal solution value for return percentage objective,
  S_i: The percentage of service success of ith supplier.
  Z_1^* : Positive ideal solution value for cost objective,
  Z_1^- : negative ideal solution value for cost objective.

8.2.4.1.4   Step Method

The Step Method (or Stem), the progressive orientation procedure, and the method of constraints are used to solve MOLP problems (Benayoun et al. 1971). This method involves consecutive investigation of solutions guided partially by a decision maker who answers the questions set by the algorithm. Thus, each iteration consists of a calculation phase and a decision phase (Tabucanon 1988). Step Method is one of the early interactive MODM methods. It can be basically applied to both a linear problem and non-linear problem with integer and/or continuous variables. The approach first solves single-objective problems to form a *payoff table*, which gives an optimal solution and corresponding p objective function value for each p single objective problem solved. Following this initial step, an iterative sequence of a single objective problem is solved. Each problem leads to a nondominated solution, which minimizes the maximum weighted distance to the ideal solution found in the initial step of the algorithm. In addition to the initial problem constraints, the decision maker is allowed to impose additional constraints on the problem's objective function. The weights imposed at this step are determined automatically and do not necessarily relate to the decision maker's preferences (Evans 2016). Step Method suggests that the compromising solution has minimum combined deviation from positive ideal solutions. According to the Step Method, an objective function is minimized by being transformed into maximization by an appropriate sign change. The Step Method process follows the aforementioned steps (Hwang and Masud 1979). The flow-chart of Step is given in Figure 8.1. The flow-chart stages for model (8.1) are summarized below:

Step 0: Construction of a pay-off table
The first and second objective functions in (8.1) are of the minimization type.
   Both objective functions should be converted into maximization type by multiplying by (−1). This conversion is as follows:

$$Min \ W_1(x) = Max - 1 \qquad (8.20)$$

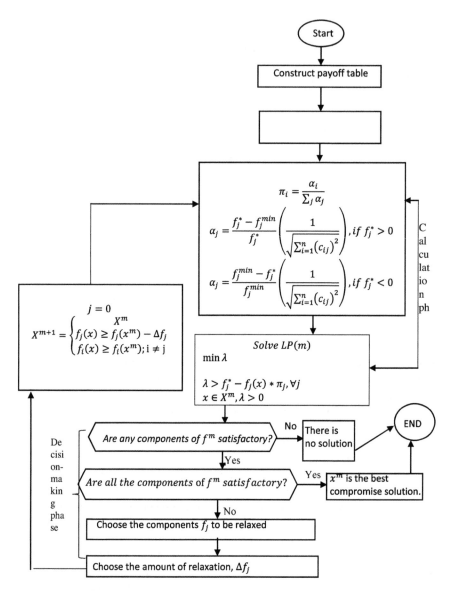

FIGURE 8.1 The flow-chart of Step Method by Hwang and Masud. (From *Multiple Objective Decision Making-Methods and Applications: A State-of-The-Art Survey*, Springer-Verlag, Berlin, Germany, 1979.)

$$Min\ W_2\left(x\right) = Max - 1 \qquad (8.21)$$

After these transformations, positive ideal solutions are determined through having a solution for each objective function in (8.1). The trade-off matrix, which consists of positive ideal solutions is illustrated in Table 8.1.

TABLE 8.1
Trade-Off Matrix

Decision Variables	$-W_1(x)$	$-W_2(x)$	$Z_1(x)$
x^{1*}	$-W_1(x^{1*}) = -W_1^*$	$-W_2(x^{1*})$	$Z_1(x^{1*})$
x^{2*}	$-W_1(x^{2*})$	$-W_2(x^{2*}) = -W_2^*$	$Z_1(x^{2*})$
x^{3*}	$-W_1(x^{3*})$	$-W_2(x^{3*})$	$Z_1(x^{3*}) = Z_1^*$

The positive ideal solution set are acquired from (8.1) is $I^* = \{-W_1^*, -W_2^*, Z_1^*\}$.

Step 1: Calculation phase

At the mth cycle, the feasible solution to MOLP of (8.21) sought is the nearest, in minimax sense, to the positive ideal solution of each objective function. π_1, π_2, and π_3 give the relative significance of distance to the optima, but it should be stated that they are effective only locally.

$$Min\,\lambda$$

Subject to

$$\lambda \geq \left(W_1(x) - W_1^*\right)\pi_1$$

$$\lambda \geq \left(W_2(x) - W_2^*\right)\pi_2$$

$$\lambda \geq \left(Z_1^* - Z_1(x)\right)\pi_3 \qquad (8.22)$$

$$\sum_{i=1}^{n} x_i \geq D$$

$$x_i \leq C_i$$

$$x \geq 0 \text{ and integer.}$$

where;

$$\pi_1 = \frac{\alpha_1}{\alpha_1 + \alpha_2 + \alpha_3} \qquad (8.23)$$

$$\pi_2 = \frac{\alpha_2}{\alpha_1 + \alpha_2 + \alpha_3} \qquad (8.24)$$

and

$$\pi_3 = \frac{\alpha_3}{\alpha_1 + \alpha_2 + \alpha_3} \qquad (8.25)$$

$$\alpha_1 = \frac{W_1^{min} - W_1^*}{W_1^{min}} \left(\frac{1}{\sqrt{\sum_{j=1}^{n} (c_{ij})^2}} \right) \qquad (8.26)$$

$$\alpha_2 = \frac{W_2^{min} - W_2^*}{W_2^{min}} \left(\frac{1}{\sqrt{\sum_{j=1}^{n} (c_{ij})^2}} \right) \qquad (8.27)$$

and

$$\alpha_3 = \frac{Z_1^* - Z_1^{min}}{Z_1^*} \left(\frac{1}{\sqrt{\sum_{j=1}^{n} (c_{ij})^2}} \right) \qquad (8.28)$$

The α_1, α_2 and α_3 are used to define the weights π_1, π_2 and π_3 such a way that $\pi_1 + \pi_2 + \pi_3 = 1$.

Step 2: Decision phase

Here, the decision maker compares the objective function values determined from Step 2 with the positive ideal solutions determined in Step 0. The compromise solution X^m is presented to the decision maker, who compares its objective vector $f^m = \{-W_1(x^{m*}), -W_2(x^{m*}), Z_1(x^{m*})\}$ with $\{-W_1^*, -W_2^*, Z_1^*\}$ the ideal one. If the decision maker is satisfied with the current solution, it means that the best compromise solution is reached. If there are no satisfactory objectives, it means the best compromise solution cannot be reached with this method. If some objectives are satisfactory, the decision maker should chance the opportunity loss for some other objectives to develop the unsatisfactory objective. This situation is illustrated below for any objective selected from (8.1). Supposing that satisfaction is not achieved from the objective in (8.1) for $-W_1(x^{m*})$ and $-W_2(x^{m*})$. It is assumed that $Z_1(x^{m*})$ will be risked with the opportunity loss in order to raise the satisfaction for these objectives. Accordingly, the following function is created;

$$X^{m+1} = \begin{cases} X^m \\ Z_1(x) \geq Z_1(x^{m*}) - \Delta Z_1 \\ -W_1(x) \geq -W_1(x^{m*}) \, and - W_2(x) \geq -W_2(x^{m*}) \end{cases} \qquad (8.29)$$

At this stage, the weight of $Z_1(x)$ should be "0." For other objectives, weights can be determined by the operations in step 1. Then the above operations are repeated and the $(m + 1)$ iteration is completed.

8.3　APPLICATION

A business that manufactures on-vehicle equipment desires to procure oil-coolers to be used in manufacturing concrete mixers. Among the possible suppliers of oil-coolers, one business is in Turkey, and the other three are located in three different countries across Europe. During the interviews with the business management, they stated that their aim was to increase their competitive power and to purchase based on Quality (%), Guarantee and Compensation (%), and Product Unit Price ($) criteria without relying solely on one single supplier. On the other hand, they wished to do a ranking among their requested supplies according to their previous demand information and present manufacturing capacity. Table 8.5 gives the supplier capacities. It is aimed to determine how many units of products are to be supplied from which supplier based on the aforementioned information in order to ensure the maximum benefit for the business. Each concrete mixer requires one oil-cooler. The information on the oil-coolers to be supplied from the suppliers are given in Table 8.2 based on the criteria of Quality (%), Guarantee and Compensation (%), and Product Unit Price ($).

- The business plans to manufacture a minimum of 1200 and maximum of 1500 concrete mixers and allocates a budget of $315,000.00 for the oil-coolers required for the aforementioned products.
- The table provides the maximum and minimum number of units to be purchased from each supplier.

Using the above information, the MOLP model for the Multiple Objective SSP is given below:

$$Z_1(x) = 0.92x_1 + 0.93x_2 + 0.91x_3 + 0.89x_4$$
$$Z_2(x) = 0.96x_1 + 0.9x_2 + 0.95x_3 + 0.92x_4$$
$$W_1(x) = 247x_1 + 239x_2 + 235x_3 + 224x_4$$

Subject to

$$x_1 + x_2 + x_3 + x_4 \le 1500$$
$$x_1 + x_2 + x_3 + x_4 \ge 1200$$
$$250x_1 + 260x_2 + 255x_3 + 243x_4 \le 315000$$
$$x_1 \ge 270$$
$$x_1 \le 420$$
$$x_2 \ge 150$$
$$x_2 \le 285$$
$$x_3 \ge 230$$
$$x_3 \le 365$$
$$x_4 \ge 320$$
$$x_4 \le 380$$
$$x_1, x_2, x_3, x_4 \ge \text{and integer.}$$

(8.30)

TABLE 8.2

Supplier Information

	Quality (%)	Guarantee and Compensation (%)	Product Unit Price ($)	Cost ($)	Minimum Capacity	Maximum Capacity
Supplier 1	0.92	0.96	247	250	270	420
Supplier 2	0.93	0.9	239	260	150	285
Supplier 3	0.91	0.95	235	255	230	365
Supplier 4	0.89	0.92	224	243	320	380

The first step of solving MOLP problems is determining positive and negative ideal solutions. Then, the solution is completed after the most appropriate method is selected, based on the preference information obtained from the decision maker. Four of the MODM techniques were chosen for the solution of (8.30). These are Global Criterion Method, Compromise Programming, Minmax GP, and Step Method.

The positive ideal solutions and negative ideal solutions of (8.30) are $I^* = \{1144.97; 1180.7; 281800\}$ and $I^- = \{1090.45; 1117.05; 297200\}$, respectively. Table 8.3 presents the variable values and objective function values acquired from the solution of (8.30) as MOLP is based on each objective function.

According to Table 8.3, each objective function's solution appears on different values for (8.30), an optimum solution cannot be found; which is an expected situation. The methods used for solving (8.30) and their solution phases are given below.

8.3.1 GLOBAL SOLUTION

According to the Global Criterion Method, to solve (8.30), it is necessary to use positive ideal solutions. The model that will provide a Global solution for (8.30) is shown below:

First, the global objective function is established.

$$MinG = \left[\frac{1144.97 - (Z_1(x))}{1144.97}\right] + \left[\frac{1180.7 - (Z_2(x))}{1180.7}\right] + \left[\frac{W_1(x) - 281800}{297200 - 281800}\right]$$

or

$$MinG = \left[1 - \left(\frac{Z_1(x)}{1144.97} + \frac{Z_2(x)}{1180.7} - \frac{W_1(x)}{281800}\right)\right]$$

TABLE 8.3

Trade-Off Matrix

Decision Variables	$Z_1(x)$	$Z_2(x)$	$W_1(x)$
$x^{1*} = \{420; 225; 232; 380\}$	1144.97	1175.7	297155
$x^{2*} = \{420; 150; 318; 370\}$	1144.58	1180.7	297200
$x^{3*} = \{270; 185; 365; 380\}$	1090.8	1122.05	281800

So that G can occur at the minimum level.

$$\left(\frac{Z_1(x)}{1144.97} + \frac{Z_2(x)}{1180.7} - \frac{W_1(x)}{281800} \right)$$

should be maximum. Moving from these functions, the Global model of (8.30) becomes as follows:

$$Max F = \left(\frac{Z_1(x)}{1144.97} + \frac{Z_2(x)}{1180.7} - \frac{W_1(x)}{281800} \right)$$

or

$$Max F = \left(0.00074x_1 + 0.000726x_2 + 0.000765x_3 + 0.000762x_4 \right)$$

Subject to

$$x_1 + x_2 + x_3 + x_4 \leq 1500$$
$$x_1 + x_2 + x_3 + x_4 \geq 1200$$
$$250x_1 + 260x_2 + 255x_3 + 243x_4 \leq 315000$$
$$x_1 \geq 270$$
$$x_1 \leq 420$$
$$x_2 \geq 150$$
$$x_2 \leq 285$$
$$x_3 \geq 230$$
$$x_3 \leq 365$$
$$x_4 \geq 320$$
$$x_4 \leq 380$$
$$x_1, x_2, x_3, x_4 \geq \text{ and integer.}$$

(8.31)

Table 8.4 provides the results acquired from the solution of (8.31).

Using the decision variable values in Table 8.4, $Max F = 0.945$ is determined. Then $Min G = \left[1 - (0.945) \right] = 0.055$ is obtained. In addition to these results, the distance degree to the positive ideal solution of each objective function can also be calculated:

$$\text{The distance degree of } Z_1(x) \left[\frac{1144.97 - (1142.89)}{1144.97} \right] = d_1 \Rightarrow d_1 = 0.0018$$

$$\text{The distance degree of } Z_2(x) \left[\frac{1180.7 - (1178.87)}{1180.7} \right] = d_1 \Rightarrow d_1 = 0.00155$$

$$\text{The distance degree of } W(x) \left[\frac{(296159) - 281800}{281800} \right] = d_\infty \Rightarrow d_1 = 0.051$$

TABLE 8.4
Global Solution

Decision	$Z_1(x)$	$Z_2(x)$	$W_1(x)$
x_1	362	362	362
x_2	150	150	150
x_3	365	365	365
x_4	380	380	380
Objective Function Values	1142.89	1178.87	296159

When these values are examined, each objective function is realized at a value very close to its own ideal solution. (8.30) with respect to the global solution can be interpreted as follows:

The amounts purchased from suppliers are 362, 150, 365, and 380, respectively. The total cost is \$296,159.00. Also, the mean percentage values for Z_1 and Z_2 are calculated as follows:

$$Z_1(\%) = \frac{1142.89}{362 + 150 + 365 + 380} = 0.909$$

$$Z_2(\%) = \frac{1178.87}{362 + 150 + 365 + 380} = 0.938$$

The mean of objective functions is 90.9% and 90.38% respectively.

8.3.2 COMPROMISE SOLUTION

Compromise Programming Model, based on positive and negative ideal solutions of (8.30), is given below:

Minimum d_∞

$$\alpha_1 \left[\frac{1144.97 - (0.92x_1 + 0.93x_2 + 0.91x_3 + 0.89x_4)}{1144.97 - 1090.45} \right] \le d_\infty$$

$$\alpha_2 \left[\frac{1180.7 - (0.96x_1 + 0.9x_2 + 0.95x_3 + 0.92x_4)}{1180.7 - 1117.05} \right] \le d_\infty$$

$$\beta_1 \left[\frac{247x_1 + 239x_2 + 0235x_3 + 224x_4 - 281800}{297200 - 281800} \right] \le d_\infty$$

$$x_1 + x_2 + x_3 + x_4 \le 1500$$

$$x_1 + x_2 + x_3 + x_4 \ge 1200$$

$$250x_1 + 260x_2 + 255x_3 + 243x_4 \le 315000$$

$$x_1 \geq 270$$
$$x_1 \leq 420$$
$$x_2 \geq 150$$
$$x_2 \leq 285$$
$$x_3 \geq 230$$
$$x_3 \leq 365$$
$$x_4 \geq 320$$
$$x_4 \leq 380$$
$$x_1, x_2, x_3, x_4 \geq \text{and integer.}$$
$$(\alpha_1 + \alpha_2 + \beta_3 = 1) \tag{8.32}$$

Different weights can be used for each objective function for the solution of the Compromise Programming problem (8.32). The solutions based on different weights are indicated in Table 8.5. Weights in Compromise solution-2 were determined using AHP (solution steps are not given for weights determined using AHP). The weights in the other solutions are assigned by the decision maker.

Decision variables are determined according to different weights of objectives in Table 8.5 and objective function values are determined depending on the determined decision variables. These decision variable values indicate the number of products to be supplied from suppliers. On the other hand, since the unit of the first two objectives is (%), the (%) values of these objectives are determined as follows. It is assumed that compromise solution-2 is chosen for this calculation:

$$Z_1 = \frac{1113.83}{270 + 285 + 293 + 375} \cong 0.91$$

TABLE 8.5
Compromise Solution

	Compromise Solution-1	Compromise Solution-2	Compromise Solution-3	Compromise Solution-4
β_1	0.33	0.45	0.3	0.2
α_1	0.33	0.3	0.25	0.2
α_2	0.33	0.25	0.45	0.6
x_1	289	270	270	270
x_2	202	258	285	225
x_3	365	330	293	344
x_4	374	380	375	376
Z_1	1118.75	1128.84	1113.83	1105.33
Z_2	1150.07	1154.5	1139.05	1134.42
W_1	289212	291022	287660	285529
d_∞	0.160272	0.1497078	0.1713499	0.1454203

$$Z_2 = \frac{1139.05}{270 + 285 + 293 + 375} \cong 0.93$$

According to these results, the firm will purchase 270 units from the first supplier, 285 units from the second supplier, 293 units from the third supplier, and 375 units from the fourth supplier. In terms of these figures, the products have an average of 91% quality, average 93% guarantee and compensation value, and the cost has been calculated as $287,660.00 Additionally, normalized distances to positive ideal solutions are given in the last line of the table. The most important situation in these results is that the heavier objectives are closer to their own positive ideal solutions. Now, let's set the distance of each objective function to its own positive ideal solutions according to the objective function values in the selected compromise solution-3.

$$0.3 \left[\frac{1144.97 - (1113.83)}{1144.97 - 1090.45} \right] = d_\infty \Rightarrow d_\infty = 0.17135$$

$$0.25 \left[\frac{1180.7 - (1139.05)}{1180.7 - 1117.05} \right] = d_\infty \Rightarrow d_\infty = 0.16359$$

$$0.45 \left[\frac{(287660) - 281800}{297200 - 281800} \right] = d_\infty \Rightarrow d_\infty = 0.17134$$

According to these results, the compromise solution is $Min\ d = 0.1713499$; in fact, it is the distance degree of the function to its positive ideal solution. As a result, the maximum value of each objective function in its distance from its positive ideal solutions will give the degree of compromise solution.

8.3.3 MINMAX SOLUTION

The model according to Minmax Goal Programming is given below by using the positive ideal solution set and the negative ideal solution set of (8.30). In this model, the normalization constant for targets converted into targets has been determined in terms of positive and negative ideal solutions. The positive ideal solution set of (8.30) is as $I^* = \{1144.97; 1180.7; 281800\}$ and the negative ideal solution set is $I^- = \{1090.45; 1117.05; 297200\}$:

$$H_1 : 0.92x_1 + 0.93x_2 + 0.91x_3 + 0.89x_4 + d_1^- - d_1^+ = 1144.97$$

$$H_2 : 0.96x_1 + 0.9x_2 + 0.95x_3 + 0.92x_4 + d_2^- - d_2^+ = 1180.7$$

$$H_3 : 247x_1 + 239x_2 + 0235x_3 + 224x_4 + d_3^- - d_3^+ = 281800$$

Because H_1 and H_2 of these targets are maximization oriented goals, the target shall never exceed the positive ideal solution; therefore, it will be $d_1^+ = 0$ and $d_2^+ = 0$. H_3 will not be smaller than its own positive ideal solution because it belongs to the minimization-oriented goal. Therefore, it will be $d_3^- = 0$. According to this information, the Minmax Goal Programming model of (8.30) appears as follows;

$$Minimum \ d_\infty$$

$$Subject \ to$$

$$H_1 : 0.92x_1 + 0.93x_2 + 0.93x_3 + 0.89x_4 + d_1^- = 1144.97$$

$$\left(w_1^- \frac{d_1^-}{1144.97 - 1090.45} \right) \le d_\infty$$

$$H_2 : 0.96x_1 + 0.9x_2 + 0.95x_3 + 0.92x_4 + d_2^- = 1180.7$$

$$\left(w_2^- \frac{d_2^-}{1180.7 - 1117.05} \right) \le d_\infty$$

$$H_3 : 247x_1 + 239x_2 + 0235x_3 + 224x_4 - d_3^+ = 281800$$

$$\left(w_3^+ \frac{d_i^+}{297200 - 281800} \right) \le d_\infty$$

$$x_1 + x_2 + x_3 + x_4 \le 1500$$

$$x_1 + x_2 + x_3 + x_4 \ge 1200$$

$$250x_1 + 260x_2 + 255x_3 + 243x_4 \le 315000 \qquad (8.33)$$

$$x_1 \ge 270$$

$$x_1 \le 420$$

$$x_2 \ge 150$$

$$x_2 \le 285$$

$$x_3 \ge 230$$

$$x_3 \le 365$$

$$x_4 \ge 320$$

$$x_4 \le 380$$

$$x_1, x_2, x_3, x_4 \ge and \ integer.$$

Table 8.6 lists the values determined from the solution of (8.33). The weights of each target in P4 are values determined previously using AHP and these are respectively $w_1 = 0.45$, $w_2 = 0.3$, $w_3 = 0.25$.

According to the results given in Table 8.6, 270 units from the first supplier, 285 units from the second supplier, 330 units from the third supplier, and 380 units from the fourth supplier should be purchased. The values of $d_1^- = 18.1399$, $d_2^- = 26.2$, $d_3^+ = 9222$ indicate the deviations from positive ideal solutions of the objective functions, which are converted to the targets. By setting different weights for the targets, new solutions can be determined. An important feature emerges with this solution; the solution of MOLP problem through

TABLE 8.6

Minmax Solution

Decision Variable	$Z_1(x)$	$Z_2(x)$	$w_1(x)$
x_1	270	270	270
x_2	258	258	258
x_3	330	330	330
x_4	380	380	380
d_1^-	18.1399	0	0
d_2^-	0	26.2	0
d_3^+	0	0	9222
Objective Function Values	1128.84	1154.5	291022

Minmax GP and the Compromise Programming, which are made using the same weights, are the same.

8.3.4 STEP METHOD SOLUTION

The objective in minimization type in (8.30) is converted to the type of maximization by multiplying by (−1).

$$Max\, Z_1(x) = 0.92x_1 + 0.93x_2 + 0.93x_3 + 0.89x_4$$

$$Max\, Z_2(x) = 0.96x_1 + 0.9x_2 + 0.95x_3 + 0.92x_4$$

$$Max\, -W_1(x) = -247x_1 - 239x_2 - 0235x_3 - 224x_4$$

Subject to

$$x_1 + x_2 + x_3 + x_4 \le 1500$$

$$x_1 + x_2 + x_3 + x_4 \ge 1200$$

$$250x_1 + 260x_2 + 255x_3 + 243x_4 \le 315000$$

$$x_1 \ge 270$$

$$x_1 \le 420$$

$$x_2 \ge 150$$

$$x_2 \le 285$$ (8.34)

$$x_3 \ge 230$$

$$x_3 \le 365$$

$$x_4 \ge 320$$

$$x_4 \le 380$$

$$x_1, x_2, x_3, x_4 \ge \text{and integer.}$$

TABLE 8.7

Trade-Off Matrix

Decision Variable	$Z_1(x)$	$Z_2(x)$	$-W_1(x)$
$x^{1*} = \{420;225;232;380\}$	1144.97	1175.7	−297155
$x^{2*} = \{420;150;318;370\}$	1144.58	1180.7	−297200
$x^{3*} = \{270;185;365;380\}$	1090.8	1122.05	−281800

Step 0: Construction of a pay-off table

The decision variable values obtained for each objective function through the solution of (8.34) and the positive ideal solutions are illustrated in Table 8.7.

Step1: Calculation phase

$$Z_1(x);$$

$$\alpha_1 = \frac{1144.97 - 1090.8}{1144.97} \left(\frac{1}{\sqrt{(0.92)^2 + (0.93)^2 + (0.91)^2 + (0.94)^2}} \right) = 0.02592$$

$$Z_2(x);$$

$$\alpha_2 = \frac{1180.7 - 1122.05}{1180.7} \left(\frac{1}{\sqrt{(0.96)^2 + (0.9)^2 + (0.95)^2 + (0.92)^2}} \right) = 0.02662$$

$$-W_1(x);$$

$$\alpha_3 = \frac{-297200 - (-281800)}{-297200} \left(\frac{1}{\sqrt{(-247)^2 + (-239)^2 + (-235)^2 + (-224)^2}} \right)$$

$$\alpha_3 = 0.0001096$$

The weights of the objective functions in (8.34) are obtained as $\pi_1 = 0.49226$, $\pi_2 = 0.505659$ and $\pi_3 = 0.002081$, respectively. From this information, (8.34) is arranged as follows;

$$Min \, \lambda$$

Subject to

$$[1144.97 - (0.92x_1 + 0.93x_2 + 0.93x_3 + 0.89x_4)].(0.49226) \leq \lambda$$

$$[1180.7 - (0.96x_1 + 0.9x_2 + 0.95x_3 + 0.92x_4)].(0.505659) \leq \lambda$$

$$[(-281800) - (-247x_1 - 239x_2 - 0235x_3 - 224x_4)].(0.002081) \leq \lambda$$

$$x_1 + x_2 + x_3 + x_4 \leq 1500$$

$$x_1 + x_2 + x_3 + x_4 \geq 1200$$

$$250x_1 + 260x_2 + 255x_3 + 243x_4 \leq 315000$$

$$x_1 \geq 270$$

$$x_1 \leq 420$$

$$x_2 \geq 150$$

$$x_2 \leq 285$$

$$x_3 \geq 230 \tag{8.35}$$

$$x_3 \leq 365$$

$$x_4 \geq 320$$

$$x_4 \leq 380$$

$$x_1, x_2, x_3, x_4 \geq \text{ and integer.}$$

$$x^1 = \{334; 150; 364; 380\} \text{ and } f^1 = \{1116.22; 1151.04; -289008\}.$$

Step 2: Decision phase

Now, when $f^1 = \{1116.22; 1151.04; -289008\}$ and $I^* = \{1144.97; 1180.7; -281800\}$ are compared, decision maker is satisfied with the third objective but not satisfied with the first two objectives. For this reason, the decision maker aims to increase the satisfaction level of the other two objectives by taking chance of the cost increase of \$500.00 from the third objective.

$$X^{m+1} = \begin{cases} X^m \\ -W_1(x) \geq -W_1(x^m) - \Delta W_1 = -289008 - 500 \\ Z_1(x) \geq -1116.22 \text{ and } -Z_2(x) \geq 1151.04 \end{cases}$$

Assuming that $\Delta W_1 = 500$, the iterations of the *Step* method are repeated. The calculations in this step, which is in the second iteration, are left to the reader. The solution has been made for the ΔW_1 values, which is supposed to come from the decision maker (or determined by the decision maker) and calculation results are illustrated in Table 8.8.

From Table 8.8, the DM can see the trade-off between objectives $Z_1(x)$, $Z_2(x)$, and $W_1(x)$ and decide which set of solutions is the best compromised one. From the information in this table, it is seen that the changes in ΔW_1 (cost increases) increase the values of the other two objectives.

TABLE 8.8
Sensitivity Analysis for (P5)

	$Z_1(x)$	$Z_2(x)$	$W_1(x)$	x_1	x_2	x_3	x_4	λ
$\Delta f_3 = 0$	1116.22	1151.04	289008	334	150	264	380	15
$\Delta f_3^1 = 500$	1118.08	1152.96	289508	336	150	364	380	14.05
$\Delta f_3^2 = 1000$	1119,92	1154.91	290008	338	150	365	379	13.069
$\Delta f_3^3 = 1500$	1121.76	1156.83	290508	340	150	365	379	12.096
$\Delta f_3^4 = 2000$	1123.61	1158.76	291008	343	150	364	379	11.117
$\Delta f_3^5 = 3000$	1127.31	1162.63	292008	347	150	365	378	9.156
	$Z_1^* = 1144.97$	$Z_2^* = 1180.7$	$W_1^* = 281800$					

8.4 CONCLUSION

The selection and evaluation of the supplier is one of the vital actions in the competitive business environment. Incorrect supplier selection can cause significant financial, operational, and managerial losses for businesses as well as environmental and social disadvantages. Therefore, the potential supplier should try to understand the objectives, aims, and goals of the business. In addition to these qualifications, the supplier must possess technological innovations, customer satisfaction, and qualities that will enhance product quality. In the case of a Sustainable SSP, decisions taken by a business in order to provide the continuity of the system should not only depend on economic criteria, but also make the best choice taking social and environmental criteria into account. Therefore, the most significant element in the SSP is to determine the supplier criteria that will provide the selection. Although the criteria used in the literature are generalized, the sector of the business, geographical location, knowledge of the business managers, and experience on the selection process can actually lead to different criteria. The literature survey on SSP shows that a large number of evaluation criteria must be taken into consideration in order to ensure sustainability in supplier selection. In this process, the important question is which of the existing criteria should be chosen. To find the answer to this question, it is obvious that managers will be directed to MADM methods. It is possible in MADM methods to prioritize, or weight, the existing criteria. It should not be forgotten that they are only selected from a limited number of alternatives. For this reason, alternatives (criteria to be evaluated) must be defined carefully. On the other hand, the amount to be purchased from the pre-determined suppliers can be calculated according to a number of criteria by using MODM methods. Objective functions in supplier selection problems that are solved by MODM methods consist of selection criteria. When looking at these two classifications in terms of MCDM, the most important element in the supplier selection problems is the determination of SSC. From this perspective, the success of the SSP is directly proportional to the determination of SSC.

REFERENCES

Altinoz, C., Kilduff, P., and Winchester Jr., S.C. (2010). Current issues and methods in supplier selection. *Journal of the Textile Institute 92*(2), 128–141.

Amid, A., Ghodsypour, S.H., and O'Brien, O. (2006). Fuzzy multiobjective linear model for supplier selection in a supply chain. *International Journal of Production Economics 104*(2), 394–407.

Amid, A., Ghodsypour, S.H., and O'Brien, C. (2009). A weighted additive fuzzy multiobjective model for the supplier selection problem under price breaks in a supply chain. *International Journal of Production Economics 121*(2), 323–332.

Amid, A., Ghodsypour, S.H., and O'Brien, C. (2011). A weighted max–min model for fuzzy multi-objective supplier selection in a supply chain. *International Journal of Production Economics 131*(1), 139–145.

Arikan, F. (2013). A fuzzy solution approach for multi objective supplier selection. *Expert Systems with Applications 40*(3), 947–952.

Aydin, S. and Kahraman, C. (2010). Multiattribute supplier selection using fuzzy analytic hierarchy process. *International Journal of Computational Intelligence Systems 3*(5), 553–565.

Azadnia, A.H., Zameri, M., Saman, M., and Wong, K.Y. (2015). Sustainable supplier selection and order lot-sizing: An integrated multi-objective decision-making process. *International Journal of Production Research 53*(2), 383–408.

Bache, J., Carr, R., Parnaby, J., and Tobias, A.M. (1987). Supplier development systems. *International Journal of Technology Management 2*(2), 219–228.

Bai, C. and Sarkis, J. (2010). Integrating sustainability into supplier selection with Grey system and rough set methodologies. *International Journal of Production Economics 124*(1), 252–264.

Benayoun, R., de Montgolfier, J., Tergny, J., and Larichev, O. (1971). Linear programming with multiple objective functions: Step method (STEM). *Mathematical Programming 1*(3), 366–375.

Boychuk, L. and Ovchinnikov, V. (1973). Principal methods of solution of multicriterial optimization problems. *Soviet Automatic Control 6*, 1–4.

Brandenburgad, M., Govindan, K., Sarkis, J., and Seuring, S. (2014). Quantitative models for sustainable supply chain management: Developments and directions. *European Journal of Operational Research 233*(2), 299–312.

Büyüközkan, G. and Çifçi, G. (2011). A novel fuzzy multi-criteria decision framework for sustainable supplier selection with incomplete information. *Computers in Industry 62*(2), 164–174.

Büyüközkan, G. and Çifçi, G. (2017). A novel hybrid MCDM approach based on fuzzy DEMATEL, fuzzy ANP and fuzzy TOPSIS to evaluate green suppliers. *Expert Systems with Applications 39*(3), 3000–3011.

Chang, B., Chang, C., and Wu, C. (2011). Fuzzy DEMATEL method for developing supplier selection criteria. *Expert Systems with Applications 38*(3), 1850–1858.

Charnes, A. and Cooper, W.W. (1961). *Management Models and Industrial Applications of Linear Programming*, Wiley, New York.

Charnes, A. and Cooper, W.W. (1977). Goal programming and multiple objective optimizations. *European Journal of Operational Research 1*(1), 39–54.

Charnes, A., Cooper, W.W., and Ferguson, R. (1955). Optimal estimation of executive compensation by linear programming. *Management Science 1*(2), 138–151.

Chen, K.L., Chen, K.S., and Li, R.K. (2005). Suppliers capability and price analysis chart. *International Journal of Production Economics 98*(3), 315–327.

Chou, S.-Y. and Chang, Y.-H. (2008). A decision support system for supplier selection based on a strategy-aligned fuzzy smart approach. *Expert Systems with Applications 34*(4), 2241–2253.

Çetinkaya, B., Cuthbertson, R., Ewer, G., Klaas-Wissing, T., Piotrowicz, T., and Tyssen, C. (2011). *Sustainable Supply Chain Management*, Springer, Heidelberg, Germany.

de Boer, L., Labro, E., and Morlacchi, P. (2001). A review of methods supporting supplier selection. *European Journal of Purchasing & Supply Management 7*, 75–89.

Dempsey, W.A. (1978). Vendor selection and the buying process. *Industrial Marketing Management 7*, 257–267.

Dickson, G.W. (1966). An analysis of vendor selection systems and decisions. *Journal of Purchasing 2*(1), 5–17.

Ding, H., Benyoucef, L., and Xie, X. (2003). A simulation-optimization approach using genetic search for supplier selection, in: Chick, S., Sanchez, P.J., Ferrin, D., and Murrice, D.J. (Eds.), *Proceedings of the 2003 Winter Simulation Conference*.

Elahi, B., Hosseini, S.M., and Makui, A. (2011). A fuzzy compromise programming solution for supplier selection in quantity discounts situation. *International Journal of Industrial Engineering & Production Research 22*(1), 107–114.

Elkington, J. (1998). *Cannibals with Forks: The Triple Bottom Line of the 21st Century Business*, 2nd ed., Capstone, Oxford, UK.

Evans, G.W. (2016). *Multiple Criteria Decision Analysis for Industrial Engineering: Methodology and Applications*, CRC Press, Boca Raton, FL.

Famuyiwa, O., Monplaisir, L., and Nepal, B. (2008). An integrated fuzzy-goal-programming-based framework for selecting suppliers in strategic alliance formation. *International Journal of Production Economics 113*(2), 862–875.

Flavell, R.B. (1976). A new goal programming formulation. *Omega, The International Journal of Management Science 4*, 731–732.

Gaballa, A.A. (1974). Minimum cost allocation of tenders. O*perational Research Quarterly 25*(3), 389–398.

Gal, T., Stewart, T., and Hanne, T. (1999). *Multicriteria Decision Making: Advances in MCDM Models, Algorithms, Theory, and Application*, Springer, New York.

Ghodsypour, S.H. and O'Brien, C. (1998). A decision support system for supplier selection using an integrated analytic hierarchy process and linear programming. *International Journal of Production Economics 56*–57, 199–212.

Ghoushchi, S.J., Milan, M.D., and Rezaee, M.J. (2017). Evaluation and selection of sustainable suppliers in supply chain using new GP-DEA model with imprecise data. *Journal of Industrial Engineering International 14*(3), 613–625.

Girubha, J., Vinodh, S., and Kek, V. (2016). Application of interpretative structural modelling integrated multi criteria decision making methods for sustainable supplier selection. *Journal of Modelling in Management 11*(2), 358–388.

Ho, W., Xu, X. and Dey, P.K. (2010). Multi-criteria decision making approaches for supplier evaluation and selection: A literature review. *European Journal of Operational Research 202*(1), 16–24.

Hruška, R., Průša, P., and Babić, D. (2014). The use of AHP method for selection of supplier. *Transport 29*(2), 195–203.

Huang, J.-D. and Hu, M.H. (2013). Two-stage solution approach for supplier selection: A case study in a Taiwan automotive industry. *International Journal of Computer Integrated Manufacturing 26*(3), 237–225.

Hwang, C.L. and Masud, A.S. (1979). *Multiple Objective Decision Making-Methods and Applications: A State-of-The-Art Survey*, Springer-Verlag, Berlin, Germany.

Ignizio, J.P. and ve Cavalier, T.M. (1994), *Linear Programming*, Prentice Hall, Englewood Cliffs, NJ.

Ijiri, Y. (1965). *Management Goals and Accounting for Control*, North-Holland, Amsterdam, the Netherlands.

Jadidi, O., Cavalieri, S., and Zolfaghari, S. (2015). An improved multi-choice goal programming approach for supplier selection problems. *Applied Mathematical Modelling 39*(14), 4213–4222.

Jadidi, O., Zolfaghari, S., and Cavalieri, S. (2014). A new normalized goal programming model for multi-objective problems: A case of supplier selection and order allocation. *International Journal of Production Economics 148*, 158–165.

Jain, J., Dangayach, G.S., Agarwal, G., and Banerjee, S. (2010). Supply chain management: Literature review and some issues. *Journal of Studies on Manufacturing 1*(1), 11–25.

Johnson, P.F., Leenders, M.R., and Flynn, A.E. (2005). *Purchasing and Supply Management*, 14th ed., McGraw-Hill, New York.

Karpak, B., Kasuganti, R.R., and Kumcu, E. (1999). Multi-objective decision-making in supplier selection: An application of visual interactive goal-programming. *The Journal of Applied Business Research 15*(2), 57–71.

Kasirian, M.N., Rosnah, M.Y., and Ismail, M.Y. (2010). Application of AHP and ANP in supplier selection process: A case in an automotive company. *International Journal of Management Science and Engineering Management 5*(2), 125–135.

Keskin, G.A. (2015). Using integrated fuzzy DEMATEL and fuzzy C: Means algorithm for supplier evaluation and selection. *International Journal of Production Research 53*(12), 3586–3602.

Kraljic, P. (1983). Purchasing must become supply management. *Harvard Business Review 61*(5), 109–117.

Kumar, D. Rahman, Z., and Chan, F.T.S. (2017). A fuzzy AHP and fuzzy multi-objective linear programming model for order allocation in a sustainable supply chain: A case study. *International Journal of Computer Integrated Manufacturing 30*(6), 535–551.

Kumar, M., Vrat, P., and Shankar, R. (2004). A fuzzy goal programming approach for vendor selection problem in a supply chain. *Computers & Industrial Engineering 46*(1), 69–85.

Kumar, M., Vrat, P., and Shankar, R. (2006). A fuzzy programming approach for vendor selection problem in a supply chain. *International Journal of Production Economics 101*(2), 273–285.

Lai, Y.-J. and Hwang, C.-L. (1994). *Fuzzy Multiple Objective Decision Making: Methods and Applications (Lecture Notes in Economics and Mathematical Systems)*, Springer, New York.

Lee, A.H.I. (2009). A fuzzy supplier selection model with the consideration of benefits, opportunities, costs and risks. *Expert Systems with Applications 36*(2), 2879–2893.

Lee, S.M. (1972). *Goal Programming for Decision Analysis*, Auerbach, Philadelphia, PA.

Liao, C.N. and Kao, H.P. (2010). Supplier selection model using Taguchi loss function, analytical hierarchy process and multi-choice goal programming. *Computers & Industrial Engineering 58*(4), 571–577.

Lilliecreutz, J. and Ydreskog, L., (1999). Supplier classification as an enabler for a differentiated purchasing strategy. *Global Purchasing & Supply Chain Management 11*, 66–74.

Luthra, S., Govindan, K., Kannan, D., Mangla, S.K., and Garg, C.P. (2017). An integrated framework for sustainable supplier selection and evaluation in supply chains. *Journal of Cleaner Production 140*(3), 1686–1698.

Malczewski, J. (1999). *GIS and Multicriteria Decision Analysis*, John Wiley & Sons, New York.

Mani, V., Agrawal, R., and Sharma, V. (2014). Supplier selection using social sustainability: AHP based approach in India. *International Strategic Management Review 2*(2), 98–112.

Marler, R.T. and Arora, J.S. (2004). Survey of multi-objective optimization methods for engineering. *Structural Multidisciplinary Optimization 26*, 369–395.

Monczka, R.M., Handfield, R.B., Giunipero, L.C., and Patterson, J.L. (2009). *Purchasing and Supply Chain Management, South-Western Cengage Learning*, 4th ed., South-Western, Mason, OH.

Mukherjee, K. (2017). *Supplier Selection: An MCDA-Based Approach*, Springer (India) Pvt. Ltd., New Delhi, India.

Mukherjee, K., Sarkar, B., and Bhattacharyya, A. (2013). Supplier selection by f-compromise method: A case study of cement industry of NE India. *International Journal of Computational Systems Engineering 1*(3), 162–174.

Nazari-Shirkouhi, S., Shakouri, H., Javadi, B., and Keramati, A. (2013). Supplier selection and order allocation problem using a two-phase fuzzy multi-objective linear programming. *Applied Mathematical Modelling 37*(22), 9308–9323.

Özçelik, F. and Avcı Öztürk, B. (2014). A research on barriers to sustainable supply chain management and sustainable supplier selection. *Dokuz Eylül Üniversitesi Sosyal Bilimler Enstitüsü Dergisi 16*(2), 259–279.

Pal, O., Gupta, A.K., and Garg, R.K. (2013). Supplier selection criteria and methods in supply chains: A review. *International Journal of Social, Education, Economics and Management Engineering 7*(10), 1395–1401.

Perić, T., Babić, Z., and Veža, I. (2013). Vendor selection and supply quantities determination in a bakery by AHP and fuzzy multi-criteria programming. *International Journal of Computer Integrated Manufacturing 26*(9), 816–829.

Polat, G., Eray, E., and Bingol, N.B. (2017). An integrated fuzzy mcgdm approach for supplier selection problem. *Journal of Civil Engineering and Management 23*(7), 926–942.

Power, M.J., De Souza, K., and Bonifazi, C. (2006). *The Outsourcing Handbook: How to Implement a Successful Outsourcing Process*, 107–109, 1st ed., Kogan Page, London, UK.

Rabbania, M., Foroozesha, N., Mousavib, S.M., and Farrokhi-Asla, H. (2017). Sustainable supplier selection by a new decision model based on interval-valued fuzzy sets and possibilistic statistical reference point systems under uncertainty. *International Journal of Systems Science: Operations & Logistics* 1–17. doi:10.1080/23302674.2017.1376232.

Ravindran, A.R. and Wadhwa, V. (2009). Multiple criteria optimization models for supplier selection, in: *Handbook of Military Industrial Engineering*, Badiru, A.B. and Thomas, M.U. (Eds.), Chapter 4-1. CRC Press, Boca Raton, FL.

Ravindran, A.R. and Warsing Jr, D.P. (2013). *Supply Chain Engineering Models and Applications*, CRC Press, Boca Raton, FL.

Roa, C.P. and Kiser, G.E. (1980). Educational buyers perceptions of vendor attributes. *Journal of Purchasing and Materials Management 16*, 25–30.

Romero, C. (1985). Multi-objective and goal programming as a distance function model. *JORS 36*(3), 249–251.

Romero, C. (1991). *Handbook of Critical Issues in Goal Programming*, Perganom Press, New York.

Rota, K., Thierry, C., and Bel, G. (1998). Supply chain management: A supplier perspective. *Production Planning & Control 13*(4), 370–380.

Salukvadze, M. (1974). On the existence of solution in problems of optimization under vector valued criteria. *Journal of Optimization Theory and Applications 12*(2), 203–217.

Sepulveda, J.M. and Derpich, I.S. (2015). Multicriteria supplier classification for dss: Comparative analysis of two methods. *International Journal of Computers Communications & Control 10*(2), 238–247.

Shaw, K., Shankar, R., Yadav, S.S., and Thakur, L.S. (2012). Supplier selection using fuzzy AHP and fuzzy multi-objective linear programming for developing low carbon supply chain. *Expert Systems with Applications 39*(9), 8182–8192.

Singh, A. (2016). A goal programming approach for supplier evaluation and demand allocation among suppliers. *International Journal of Integrated Supply Management 10*(1), 38–62.

Singh, S., Olugu, E.U., Musa, S.N., Mahat, A.B., and Wong, K.Y. (2016). Strategy selection for sustainable manufacturing with integrated AHP-VIKOR method under interval-valued fuzzy environment. *The International Journal of Advanced Manufacturing Technology 84*(1–4), 547–563.

Tabucannon, M. (1988). *Multiple Criteria Decision Making in Industry*, Elsevier, Amsterdam, the Netherlands.

Templar, S., Findlay, C., and Hofman, E. (2016). *Financing the End-to-End Supply Chain: A Reference Guide to Supply Chain Finance*, Kogan Page, London, UK.

Thiruchelvam, S. and Tookey, J.E. (2011). Evolving trends of supplier selection criteria and methods. *International Journal of Automotive and Mechanical Engineering (IJAME)* 4, 437–454.

Tosun, Ö. and Akyüz, G. (2015). A fuzzy TODIM approach for the supplier selection problem. *International Journal of Computational Intelligence Systems* 8(2), 317–329.

Umarusman, N. and Ahmet, T. (2013). Building optimum production settings using de novo programming with global criterion method. *International Journal of Computer Applications* 82(18), 75–87.

Umarusman, N. and Hacıvelioğulları, T. (2018). Solution proposal for supplier selection problem: An application in agricultural machinery sector with global criterion method. *Dokuz Eylul University Faculty of Economics and Administrative Sciences Journal* 33(1), 353–368.

Wan, S.P., Xu, G.L., and Dong, J.Y. (2017). Supplier selection using ANP and ELECTRE II in interval 2-tuple linguistic environment. *Information Sciences 385–386*, 19–38.

Weber, C.A., Current, J.R., and Benton, W.C. (1991). Vendor selection criteria and methods. *European Journal of Operational Research 50*, 2–18.

Weber, C.A. and Current, J.R. (1993). A multiobjective approach to vendor selection. *European Journal of Operational Research 68*(2), 173–184.

Weber, C.A. and Ellram, L.M. (1992). Supplier selection using multi-objective programming: A decision support system approach. *International Journal of Physical Distribution & Logistics Management 23*(2), 3–14.

Wu, M. and Liu, Z. (2011). The supplier selection application based on two methods: VIKOR algorithm with entropy method and Fuzzy TOPSIS with vague sets method. *International Journal of Management Science and Engineering Management 6*(2), 110–116.

Yousefi, S., Mahmoudzadeh, H., and Jahangoshai Rezaee, M. (2016). Using supply chain visibility and cost for supplier selection: A mathematical model. *International Journal of Management Science and Engineering Management 12*(3), 196–205.

Yu, P.L. (1973). A class of solutions for group decision problems. *Management Science 19*(8), 936–946.

Yu, V.F. and Tseng, L.-C. (2014). Measuring social compliance performance in the global sustainable supply chain: An AHP approach. *Journal of Information and Optimization Sciences 35*(1), 47–72.

Yücenur, G.N., Vayvay, Ö., and Çetin-Demirel, N. (2011). Supplier selection problem in global supply chains by AHP and ANP approaches under fuzzy environment. *The International Journal of Advanced Manufacturing Technology 56*(5–8), 823–833.

Zeleny, M. (1973). Compromise programming, in: Cochrane, J.L. and Zeleny, M. (Eds.), *Multiple Criteria Decision Making*, pp. 262–301, University of South Carolina, Columbia, SC.

Zeleny, M. (2011). Multiple Criteria Decision Making (MCDM): From paradigm lost to paradigm regained? *Journal of Multi-Criteria Decision Analysis 18*, 77–89.

Zhang, Q. and Huang, Y. (2012). Intuitionistic fuzzy decision method for supplier selection in information technology service outsourcing, in: Lei, J., Wang, F.L., Deng, H. and Miao, D. (Eds.), *Emerging Research in Artificial Intelligence and Computational Intelligence: International Conference, AICI 2012, Chengdu, China, October 26–28. Proceedings* (pp. 432–439). Berlin, Germany: Springer.

Section III

Enablers/Barriers for Sustainable Procurement Operations

9 Assessing the Drivers to Gain Sustainable Competitive Advantage

An Interpretive Structural Modeling Approach

Surya Prakash, Satydev, Kaliyan Mathiyazhagan, Maheshwar Dwivedy, and Sanjiv Narula

CONTENTS

9.1 INTRODUCTION

9.1.1 BACKGROUND

Globalization happens to be responsible for bringing more complexities to all industrial processes (Reuter et al., 2010). Moreover, the increasing pressure of accomplishing sustainable development at the organizational level is affecting many other dimensions such as society, economy, and environment (Beske et al., 2008). This situation leads to an intensified competition among the firms to adopt out-of-the-box philosophies for sustainability and innovations (Hansen et al., 2009). In his study, Oliver (1997) said that the process of selecting resources impacts the sustainable competitive advantage of firms. He also argued that the organization should manage sustainable advantage by strengthening the institutional context of its resource decisions. Recently, Dubey et al. (2016) explored the importance of diversity, delivery, equity, safety, and other socio-economic issues in a supply chain network context to adopt sustainability. In order to attain sustainability in business, adaptive strategic planning is required that can be derived from a number of sustainability factors. Contemporary managers must consider some factors like the distinction of the value aided process, while aiming for attaining a competitive advantage (Priyanto et al., 2012; Vinayan et al., 2012). Traditionally, the existent sources of sustainable competitive advantage (SCA) for any company are human and social capital. Hence, the managers should develop and sustain these sources carefully (Mahdi and Almsafir, 2014; Mappigau, 2012). In today's overexcited competitive times, the strategic advertisement and feedback may become a success driver for the manufacturing sector (Maheshwari and Seth, 2017). We can safely say that the organizations of the modern era are facing fierce competition to gain a competitive advantage (Lin et al., 2016; Priyanto et al., 2012; Verhulst and Lambrechts, 2015). Additionally, the indispensable advent of information technology also plays a significant role in the development of sustainability characteristics in modern business (Priyanto et al., 2012). However, such isolated thoughts derived from the studies fail to analyze the critical issue of dealing with this subject comprehensively. This leaves a gap to carry out a comprehensive analysis of primary factors that impact the SCA adoption from literature as well as from an industry perspective. The contemplation of factors that have a positive impact on achieving SCA is necessary to make a business viable and to place it on the track to long-term success. Moreover, the concept of SCA is not limited to manufacturing but also holds true for services (Bharadwaj et al., 1993).

To start with, the SCA may be conceived in the form of cost advantage, value advantage, and focus advantage for the firms. Let's understand the difference in these perspectives, the cost advantage shows the competency of business on price, while the value advantage indicates differentiation of offers that provide the superior value to the firm and customer. The focus advantage concentrates on a market niche with modified offering design for a particular sector of the market. In this context, the leading global organizations or market leaders master the art of achieving the SCA and increases the values discussed above. There is a wide range of factors that can drive the process of achieving SCA, and it also depends on the manner of the factor's implementation (Dubey et al., 2018). For many leading firms, maintaining the advantageous position is always embedded in their vision and mission statements. However, it is the mid-size firms that face the music, particularly the small-to-medium enterprises (SMEs), and unorganized sectors of the economy to be sustainable. The SMEs are large in numbers, particularly in developing countries, and their integration has a major impact on the business environment, the financial system of other firms, as well as society (Freeman, 2011). For these firms, it is always very tough to maintain a wide range of critical aspects of performance to maintain the SCA, for example, customer loyalty, the location of facilities, distribution network, information systems, unique products and services, trusted vendors, adequate customer service level, etc. All such factors work as driving force for getting most benefits from SCA. Therefore, it is required to plan meticulously to achieve SCA by considering various aspects such as understanding the segments of the markets, products, customers, and product quality and price.

It is essential to identify the relationship among the driving forces, success factors, and procedures to follow to achieve the SCA in long terms. Moreover, the driving factors for achieving the SCAs can follow the precedence as well. In literature, there are some decision making approaches available to achieve a hieratical model. Interpretive structural modeling (ISM) is one of the effective methods that can be employed for this purpose. The positive aspect of ISM is that it is an interactive process to structuring a set of elements in an entirely logical manner when these elements are different in nature and have a relation to each other, directly or indirectly (Jayant and Azhar, 2014; Prakash et al., 2018). This study demonstrates a methodology by employing ISM to gain SCA in the competitive market scenario. The following are the steps in this proposed process:

- Identification of valuable drivers to gain SCA
- Finding contextual relationships between drivers identified and established their hierarchical levels
- Developing and analysis of hierarchical relationship model (HRM) of these drivers

The chapter is organized in the following sections: In Section 9.2, state of the art is explored to get insight into the sustainable competitive advantage keeping aim to identify the drivers of SCA. The complete methodology proposed to fulfill the research gap identified is provided in Section 9.3. Section 9.4 of this chapter represents the outcome of the analysis, and finally, the last section concludes the research findings.

9.2 SUSTAINABLE COMPETITIVE ADVANTAGES AND DRIVERS

9.2.1 SUSTAINABLE COMPETITIVE ADVANTAGE (SCA)

Traditionally, the SCA is perceived by firms as the phenomenon of getting a lead over competitors either by providing more value or services at very competitive prices. Sustainable advantage indicates that the process of achieving these advantages also contributes to attaining goals of economic, environmental, and social aspects. The literature defines the sustainable development as the means through which an organization proposes to gain a state of affairs that recognized as goals of sustainability (Droege and Dess, 2002). Hence, on the same lines, sustainability aims to alter physical, social, as well as economic situations and motivated action to attain the goals (Freeman, 2011; Vinayan et al., 2012).

The methods and approaches employed to achieve the competitive advantage are termed as competitive strategies. From literature, it was observed that cost management, differentiation, and focus are three widely used competitive strategies (Freeman, 2011). Thus, the SCA is the intended integration to social, economic, and ecological aspects in attaining the business lead by firms (Freeman, 2011; Mappigau, 2012; Vinayan et al., 2012). The operations strategies like the capability of operation development play a vital role in the industrial environment. These strategies depend on competitors and the idea to create a roadmap for competitive advantage (Sansone et al., 2017). Many developed and developing counties have their own competitive advantages (CA) because of the determinants and outside factors (such as government policies) (Kathuria et al., 2010). For them, the capability of competitive advantage can be generated through knowledge management, which is termed as a determinant for SCA (Sook-ling, 2015).

The companies may enable CA by utilizing the opportunities and counteracting the threats. The internal (links of the company) and external environments (competitors) influence the process of competitive advantage. Therefore, the analysis of linking factors to the firm such as resources becomes more beneficial (Bakri, 2017). Strategic management of capabilities enables the firm to sustainable competitive advantage (Munir et al., 2012). To endure in a highly global environment, the organization is required to focus on many factors simultaneously. These factors may include differential product strategies, creating core competencies, the growth of intellectual property, and technologies differentiation (Srivastava et al., 2013). Broadly, the SCA can be understood as a competitive advantage, which has an intention of maintaining social, economic, and ecological constancy for long-term sustainable development (Dubey et al., 2016). These SCAs may have various categories and scope, which must be identified carefully for a given context. We have identified generic SCA factors from literature in the next section, which further narrowed down our search for relevant factors for valid SCA elements.

9.2.2 Identification of Drivers for Sustainable Competitive Advantage

We have adopted a methodology that is a mix of literature review, benchmarking, and brainstorming to identify the critical factor for achieving SCA. In the first step, the SCA-related literature is collected and investigated. A team of two experts from the academic domain worked on this aspect. This team has a proven scholarly record of conducting empirical research. The team collected the important factors from the literature. The identification of drivers was carried out from research papers assessed from the databases of Science Direct (www.sciencedirect.com), Emerald (www. emeraldinsight.com), Ecopus (www.scopus.com) and Springer (www.springer.com) and Taylor & Francis (www.tandfonline.com). The systematic search process is adopted using the keywords to locate the relevant studies in the above databases and includes *sustainable competitive advantage, sustainability, drivers, sustainable strategies, sustainable factors, competition*, and their combination with AND/OR operator. Along with these databases, some other open resources were also explored including Google Scholar and open industry white papers.

Though, a scuttle of factors found can contribute to a competitive advantage for Indian manufacturing firms. We have come out with the 13 most important factors for the SCA analysis in this study. The drivers selected in this study are based on their potential to provide an advantage in a competitive environment. Additionally, the SCA factors are validated through the benchmarking process to test their validity and usefulness for analysis. Multiple brainstorming sessions were conducted with two industry experts who have the knowledge and expertise for managing Indian SMEs for more than ten years in supply chain and operations domain at engineer or manager level. The definition, scope and description of these identified drivers are given below in the following section:

1. *Innovative product*: The customers always search for the differentiation of the existing products. The innovative ideas to produce a new product

may lead to high demand and customer's satisfaction, which further leads to competitive advantage (Janet et al., 2015).

2. *Managerial capacity*: SCA is the intended integration to social, economic and ecological facts, therefore, SCA can maintain through an intention toward managing constancy of these facts for long-term sustainable development, thus empowered management is inevitable (Dubey et al., 2016; Mnjala, 2014).

3. *Environmental education and training*: The ecological development depends on the knowledge of environmental factors, and it may drive to one of the sustainable factor (ecological factor), which helps to gain environment related competitive advantage in terms of carbon credits, water footprints, greenhouse gas emissions, etc. (Freeman, 2011; Somsuk, 2014).

4. *Manpower involvement*: The personnel participation plays a vital role in the proper management, smooth operations, competitive strategies and to find the challenges for the business (Mahdi and Almsafir, 2014). This fact will derive the human resource in the organization to achieve the tangible goals.

5. *Tracking the development of directives*: The SCA is a benefit over competitors gained by following the SCA drivers, but it is necessary to know about the strategies of competitors. The understanding of tracking the development of directives is required to command on competitors (Somsuk, 2014).

6. *Cleaner and sustainable production*: Manufacturing and production are the core activities of the economy of a nation. It can be one of the most influential aspects to incorporate the sustainable practices. The need of the hour is to work on adopting cleaner technologies. The global trade participation is also a significant feature for the country for inclusive and sustainable development (Mappigau, 2012).

7. *Green design and purchasing*: The concept of green is not the same as sustainable, but it impacts the factors of sustainability in a significant manner. The green design and purchasing tend to implement many procedures in a sustainable direction, thus become a driver to gain SCA (Somsuk, 2014).

8. *Information system*: It is the era of information, and most of the organizations do information sharing for managing supply capacity and delivery lead times (Badea et al., 2014). The use of robust information sharing strategies between the business entities plays a vital role in SCA (Janet et al., 2015; Mappigau, 2012).

9. *Collaborative R&D with suppliers*: The supplier is one of the most important links of any organization to gain SCA. The supplier can identify the customer's thinking and requirements of the market. If the determination of a firm is based on the requirements of the market, it may lead easy management of processes (Somsuk, 2014).

10. *Product dynamicity*: The variability in the products is necessary to make an attraction of consumer by implementing the innovative ideas so that company can survive the competition (Mappigau, 2012).

11. *Branding and enduring*: The production of durable and qualitative products can establish a brand, which leads to increased demand in the market. If the

product once endured in the market, people trust to it and refer to follow the same (Janet et al., 2015). The factor leads the company to catch the leading position in its domain of operations.

12. *Real intellectual property*: The SCA is the advantage over the competitors which gain through the intellectual support of the management team. The real intellectual property imparts many characters associated with the personal and actual property. The managers are free to license, sell, and buy these properties (Mnjala, 2014).

13. *Strong focus and differentiation*: As we know, the success depends on the intensity of determination for the aim. If the determination is stronger, the result will be positive and vice versa. The changeability in the products and processes is necessary to cover demand versatility (Janet et al., 2015; Stinnett and Gibson, 2016).

Although the literature provides a great insight to sustainability and the competition among organizations, to gain a competitive advantage no concentration is acknowledged for evaluation and modeling the drivers of SCA for proposing hieratical relations so that the priorities of the process can be fixed to achieve the goals and market leadership. In today's competitive and rapidly changing environment, the manufacturing firms seeking the advantages from these drivers may adopt them in a manner (Cinzia et al., 2016; Lew et al., 2013; Ravi et al., 2009; Yugal et al., 2011). All of the above research articles pointed out that the factors related to cost, product features, workforce, and management support are potential factors that suggest an organization to implement SCA drivers. Manufacturing firms have numerous factors in the form of cost and product quality, which force adopting competitive strategies in traditional systems. In this research, the above-identified SCA drivers are used, and ISM methodology is deployed to fulfill this gap. Total 13 drivers were identified in the literature survey. The open-ended questions on the identified factors are asked for industry persons and captured in the reachability matrix. The results of the above exercise yielded the potential SCA drivers related to manufacturing firms and the most significant drivers selected for a further analysis using ISM method.

9.3 INTERPRETIVE STRUCTURAL MODELING (ISM) METHOD

ISM methodology was recommended by Prof. J. Warfield to investigate the comprehensive systems (Sage, 1977; Warfield, 1974). ISM is a well established and popular method to identify and summarize the relationships between specific factors. This method transforms vague and weakly articulated intellectual models into visible hierarchal models to take further actions. The ISM method is useful approach to analyze the qualitative as well as quantitative data. Therefore, ISM methodology found as best suitable for the analysis of SCA drivers. In this study, the most of the data is quantitative data and require its conversion in decision-making values. The decision science domain is quite rich in using the ISM to solve a variety of problems including analysis of decisions and assessment of worth in a large system (Faisal et al., 2007; Mishra et al., 2012; Raut et al., 2018; Vanita et al., 2009). The objective of this study is to explore the application side of the ISM methodology, we are leaving readers to read about this method in detail in given studies here. The exhaustive review of ISM and its implementation modalities is provided in Luthra et al. (2011), Dubey et al. (2016), Haleem et al. (2012), Kumar et al. (2013), Mangla et al. (2013), Ansari et al. (2013), Mangla et al. (2012), and Chong et al. (2012). The critical steps used in this methodology are described as follows:

Step 1: List the elements considered for the study/analysis. In this study, SCA drivers have been identified as elements through the literature review.

Step 2: Construction of structural self-interaction matrix (SSIM) that shows contextual relations between the components of the adapted system.

Step 3: Operate SSIM to develop the primary/initial reachability matrix (IRM). After that, the verification of transitivity observed to construct final reachability matrix (FRM). The meaning of transitivity is if a component "A" is related to component "B" and "B" is related to "C," the component "A" is inevitably related to component "C."

Step 4: Partitioning of different hierarchical levels with the help of the reachability matrix and draw the digraph using the contextual relationship identified in the reachability matrix. Figure 9.1 shows the flow diagram summarizes all these steps of methodology.

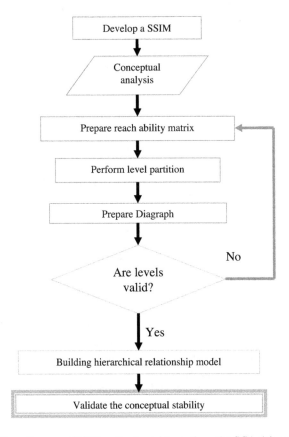

FIGURE 9.1   Flow diagram of ISM method used to analyze the SCA drivers.

9.4   IMPLEMENTATION OF INTERPRETIVE STRUCTURAL MODELING (ISM)

The excellent relationship between the drivers promotes the strategic planning for effective use of the drivers, which gives the drive to gain SCA. According to Bolanos et al. (2005), some form of contextual relationship exists among pairwise variables include definitive (shows genesis, is accessible from), comparative (is comparatively more critical), influence (effects on different activities), and temporal (must follow).

In our work, first, the set of questions was asked of industry experts to set up the contextual relations among drivers, and the inputs were converted in SSIM matrix (See Table 9.1). In order to analyze the drivers to gain SCA, the *leads to* contextual relationship were applied to assess the relationship dynamics between each pair of drivers i and j; (i) for columns and (j) for rows. The primary meaning of *leads to* relationship is the one driver (i) leads to other drivers (j) of SCAs. Four symbols (V_i, A_j, X_{ij}, and O) are used to symbolize the influence way, and the category of relationship that exists between any two SCA drivers.

The meaning of V_i is the effectiveness of driver "i" leading to the performance of driver "j" in one direction. The term A_j is used when effectiveness of "j" leads to the performance of "i" in one direction. X_{ij} represent the situation when the effectiveness of "i" and "j" leads to the performance of each other. The symbol O is used when there is no relation between the effectiveness of drivers "i" and "j."

The next step is to establish the direct reachability matrix (DRM), which is constructed by converting SSIM into a binary matrix. Table 9.2 represents the DRM and depicts the replacement of the symbols V_i, A_j, X_{ij}, and O with 1 and 0 in SSIM. The rules for the substitution followed are, if "V_i" shows the relationship (i, j) in SSIM, then the similar relationship shows 1 in place of (i, j) and 0 in place of (j, i). If "A_j" shows the relationship (i, j) in SSIM, then the similar relationship is 1 for (j, i) and is 0 for (i, j). If "X_{ij}" shows the relationship (i, j) in SSIM, then the analogous relationship shows 1 in place of (i, j) and (j, i), whereas "O" shows the relationship (i, j) in SSIM, then the analogous relationship is 0 replace to (i, j) and (j, i).

The subsequent step after creating DRM is to construct the final reachability matrix (FRM) was also checked for its impact on transitivity. The preparation of FRM is also intentioned the effect of transitivity. The transitivity is indicated if driver 1 leads to driver 2, and driver 2 leads to driver 3, therefore. As a result, driver 1 must also lead to driver 3. Table 9.3 shows the final reachability matrix also intentioned the effect of transitivity and relationships among drivers SCA.

The driving power and dependence power are evaluated by the FRM for all SCA drivers. The driving power for each driver indicates the total number of drivers, which may help to gain SCA. On the other hand, dependence power for each driver sums up the number of drivers, which may strengthen the approach to SCA. The ranking of drivers defined by values of driving power and dependence power.

TABLE 9.1
Relations of Effectiveness Interaction of SCA Drivers

Drivers	Driver1	Driver 2	Driver 3	Driver 4	Driver 5	Driver 6	Driver 7	Driver 8	Driver 9	Driver 10	Driver 11	Driver 12	Driver 13
Driver 13													—
Driver 12												—	A_j
Driver 11											—	V_i	A_j
Driver 10										—	A_j	X_{ij}	A_j
Driver 9									—	O	O	O	O
Driver 8								—	O	O	O	O	O
Driver 7							—	X_{ij}	A_j	O	O	O	O
Driver 6						—	A_j	O	O	O	O	V_i	O
Driver 5					—	O	X_{ij}	A_j	X_{ij}	O	O	O	O
Driver 4				—	A_j	O	O	O	O	O	O	O	O
Driver 3			—	A_j	A_j	O	A_j	A_j	A_j	O	O	O	O
Driver 2		—	O	O	O	O	O	O	O	A_j	O	A_j	A_j
Driver 1	—	O	O	A_j	A_j	O	O	A_j	A_j	O	O	O	O

TABLE 9.2

Direct Reachability Matrix for Drivers

Drivers	Driver1	Driver 2	Driver 3	Driver 4	Driver 5	Driver 6	Driver 7	Driver 8	Driver 9	Driver 10	Driver 11	Driver 12	Driver 13
Driver 13	0	1	0	0	0	0	0	0	0	1	1	1	1
Driver 12	0	1	0	0	0	0	0	0	0	1	0	1	1
Driver 11	0	0	0	0	0	0	0	0	0	1	1	1	0
Driver 10	0	1	0	0	0	0	0	0	0	1	0	1	0
Driver 9	1	0	1	0	1	0	1	0	1	0	0	0	0
Driver 8	1	0	1	0	1	1	1	1	0	0	0	0	0
Driver 7	0	0	1	0	1	1	1	1	0	0	0	0	0
Driver 6	0	0	0	0	0	0	0	0	0	0	0	1	0
Driver 5	1	0	1	1	1	0	1	0	1	0	0	0	0
Driver 4	1	0	1	1	0	0	0	0	0	0	0	0	0
Driver 3	0	0	1	0	0	0	0	0	0	0	0	0	0
Driver 2	0	1	0	0	0	0	0	0	0	0	0	0	0
Driver 1	1	0	0	0	0	0	0	0	0	0	0	0	0

TABLE 9.3
Final Reachability Matrix with Driving and Dependence Power of SCA Drivers

Drivers	Driver 1	Driver 2	Driver 3	Driver 4	Driver 5	Driver 6	Driver 7	Driver 8	Driver 9	Driver 10	Driver 11	Driver 12	Driver 13	Driving power
Driver 13	0	0	0	0	0	0	0	0	0	0	0	0	1	1.0
Driver 12	0	0	0	0	0	0	0	0	0	0	0	1	0	1.0
Driver 11	0	1[a]	0	0	0	0	0	0	0	0	1	1	0	1.0
Driver 10	0	1	0	0	0	0	0	0	0	1	1	1	0	3.0
Driver 9	1	1[a]	1[a]	1[a]	1	1[a]	1[a]	1[a]	1	1[a]	0	1[a]	0	11.0
Driver 8	1	1[a]	1[a]	1[a]	1[a]	1[a]	1	1	1[a]	1[a]	0	1[a]	0	4.0
Driver 7	1[a]	1[a]	1[a]	1[a]	1	1	1	1	1[a]	1[a]	0	1[a]	0	11.0
Driver 6	0	1[a]	0	0	1[a]	1	0	0	1[a]	1[a]	0	1	0	11.0
Driver 5	1	1[a]	1	1	1	1[a]	1[a]	1[a]	1[a]	1[a]	0	1[a]	0	11.0
Driver 4	1	0	1	1	0	0	0	0	0	0	0	0	0	3.0
Driver 3	0	0	1	0	0	0	0	0	0	0	0	0	0	4.0
Driver 2	0	1	0	0	0	0	0	0	0	1	0	0	0	3.0
Driver 1	1	0	0	0	0	0	0	0	0	0	0	0	0	5.0
Dependence power	6.0	10.0	6.0	5.0	4.0	5.0	4.0	4.0	4.0	9.0	2.0	9.0	1.0	69/69

[a] Shows the transivity among factors.

9.4.1   DEVELOPMENT OF AN ISM-BASED MODEL

By the final reachability matrix, the set of reachability, antecedent, and intersection was obtained for all drivers to implement the strategies for SCA. The SCA drivers having the equal importance of reachability and intersection deposit was specified level 1 and was given as the highest location in the hierarchical structure of drivers in executing SCA strategies (Luthra et al., 2014). The same iteration repeated until the hierarchical height of every driver was found. The final hierarchical positions for drivers in executing for a competitive advantage over competitors are shown in Table 9.4.

The structural model constructed by the final reachability matrix as shown in Figure 9.2. In this model, the drivers are positioned in a hierarchy from the top (level 1) to bottom (level 6). The variables "Tracking the development of directives," "Green design and purchasing," "Information system," and "Collaborative R&D with suppliers" having level VI (highest level) and are placed at the bottom of the hierarchy. They have a potential to drive all other variables of all levels. The driving power for each driver indicates the total number of drivers, which may help to gain SCA. On the other hand, dependence power for each driver sums up the number of drivers, which may strengthen the approach to goal of SCA.

TABLE 9.4
Hierarchical Values of SCA Drivers to Find the Levels

Driver code	Reachability Set	Antecedent Set	Intersection Set	Levels
Dr1	[1]	[1], [4], [5], [7], [8], [9]	[1]	I
Dr2	[2]	[2], [5], [6], [7], [8], [9], [10], [11], [12], [13]	[2]	I
Dr3	[3]	[3], [4], [5], [7], [8], [9]	[3]	I
Dr4	[1], [3], [4]	[4], [5], [7], [8], [9]	[4]	II
Dr5	[1], [2], [3], [4], [5], [6], [7], [8], [9], [10], [12]	[5], [7], [8], [9]	[5], [7], [8], [9]	VI
Dr6	[2], [6], [10], [12]	[5], [6], [7], [8], [9]	[6]	V
Dr7	[1], [2], [3], [4], [5], [6], [7], [8], [9], [10], [12]	[5], [7], [8], [9]	[5], [7], [8], [9]	VI
Dr8	[1], [2], [3], [4], [5], [6], [7], [8], [9], [10], [12]	[5], [7], [8], [9]	[5], [7], [8], [9]	VI
Dr9	[1], [2], [3], [4], [5], [6], [7], [8], [9], [10], [12]	[5], [7], [8], [9]	[5], [7], [8], [9]	VI
Dr10	[2], [10], [12]	[5], [6], [7], [8], [9], [10], [11], [12], [13]	[10], [12]	III
Dr11	[2], [10], [11], [12]	[11], [13]	[11]	IV
Dr12	[2], [10], [12]	[5], [6], [7], [8], [9], [10], [11], [12], [13]	[10], [12]	III
Dr13	[2], [10], [11], [12], [13]	[13]	[13]	V

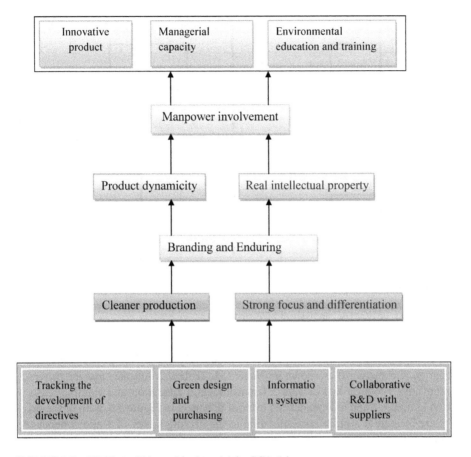

FIGURE 9.2 ISM-based hierarchical model for SCA drivers.

9.4.2 MICMAC ANALYSIS

The main aim of MICMAC (Matriced' Impacts Croise's Multiplication Appliquée a UN Classement) analysis is to analyze the driving power and dependence of the factors (Mandal and Deshmukh, 1994). The outcomes of ISM analysis works as the input to the MICMAC analysis so that driving and dependence power can be identified (Gorane and Kant, 2013). In our case, the SCA drivers described earlier categorized into four clusters (Figure 9.3). The first cluster indicates *independent variables* that have less driving power and less dependence. The second cluster is made for *dependent variables* that have less driving power but more dependence. The *linkage variables* constitute the third cluster that has strong driving power and dependence, and the *independent variables* constitute the fourth cluster that has strong driving power but poor dependence. This analysis is similar to that given in (Kumar et al., 2013; Luthra et al., 2011).

The driving power and dependence of these variables are provided in Table 9.3. In this table, the entry "1" along the column indicates the dependence power and along

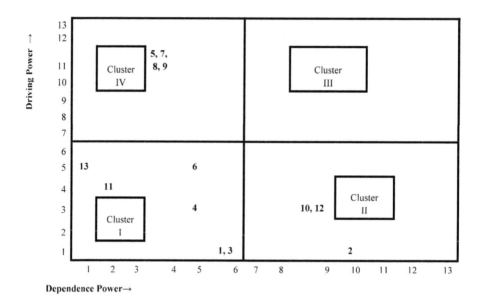

FIGURE 9.3 MICMAC analysis of SCA drivers.

the rows indicates the driving power. Consequently, the driver power-dependence diagram is constructed as shown in Figure 9.3. Table 9.2 indicates that the drivers 1 (innovative product) and drivers 3 (environmental education and training) have a driving power 1 and a dependence power of 6; hence, in Figure 9.3, both of the drivers are located at a place corresponding to the driving power of 1 and dependence power of 6.

The cluster I (autonomous region) consists of innovative product (1), environmental education and training (3), manpower involvement (4), cleaner production (6), branding and enduring (11), and strong focus and differentiation (13) drivers, which have less driving power as well as dependence. Managerial capacity (2), product dynamicity (10), and real intellectual property (12) have been identified as dependent drivers of SCA.

The third cluster (linkage region) does not have any driver. This indicates that no drivers have any more dependence along with the driving power. In our study, the fourth cluster (independent or driver region) consists of tracking the development of directives (5), green design and purchasing (7), information system (8), and collaborative R&D with suppliers (9) as elements of driving region; and placed at the lower level in the hierarchical structure.

9.5 CONCLUSION

The successful process to gain SCA over the competitors depends upon the capabilities of the drivers of the industry. We have identified 13 driving factors that can enable the organization to sustain in a competitive environment. These 13 drivers have a direct influence on the complete organizational systems to impact the competitiveness sustainably. The consequences of these drivers to gain SCA may be mild depending on its impact on management yield and service quality. In order to develop successful business strategies, the primary opening point would be to recognize and investigate SCA drivers. Our results show that innovative product (1), managerial capacity (2), and environmental education and training (3) have the highest impact for achieving SCA goals. The driving factors such as tracking the development of directives (5), green design and purchasing (7), information system (8), and collaborative R&D with suppliers (9) are of more independent nature.

An ISM-based model developed to define the relationship among all SCA drivers. To know the impact of the interlinking of enabling factors, the MICMAC analysis that was carried out helped to validated the model developed and analysis of drivers on the basis of driving and dependence power. The drivers have the highest driving power and low dependence power, i.e., tracking the development of directives, green design and purchasing, information systems, and collaborative R&D with suppliers will play a significant role to gain a competitive advantage over competitors. This analysis reveals that decision makers must formulate the strategies that have a focus on such areas. The study set the basis for future work, such as cross-validation of the drivers in various industries (service industry, as well), empirical investigations, case studies, etc. This study is limited to the factor identification from literature, which can be further improved by applying confirmatory factor analysis (CFA), structural equation modeling (SEM) etc. The study can also be extended by applying other suitable MADM/MCDM approaches.

REFERENCES

Ahuja, V., Yang, J., and Shankar, R. (2009). Benefits of collaborative ICT adoption for building project management. *Construction Innovation*, 9(3), 323–340.

Ansari, M. F., Kharb, R. K., Luthra, S., Shimmi, S. L., and Chatterji, S. (2013). Analysis of barriers to implement solar power installations in India using interpretive structural modeling technique. *Renewable and Sustainable Energy Reviews*, 27, 163–174.

Badea, A., Prostean, G., Goncalves, G., and Allaoui, H. (2014). Assessing risk factors in collaborative supply chain with the analytic hierarchy process (AHP). *Procedia: Social and Behavioral Sciences*, 124, 114–123.

Beske, P., Koplin, J., and Seuring, S. (2008). The use of environmental and social standards by German first-tier suppliers of the Volkswagen AG. *Corporate Social Responsibility and Environmental Management*, 15(2), 63–75.

Bharadwaj, S. G., Varadarajan, P. R., and Fahy, J. (1993). Sustainable competitive advantage in service industries: A conceptual model and research propositions. *The Journal of Marketing*, 83–99.

Bolaños, R., Fontela, E., Nenclares, A., and Pastor, P. (2005). Using interpretive structural modelling in strategic decision-making groups. *Management Decision*, 43(6), 877–895.

Cheah, A. C. H., Wong, W. P., and Deng, Q. (2012). Challenges of lean manufacturing implementation: A hierarchical model. In *Proceedings of the 2012 International Conference on Industrial Engineering and Operations Management*. Istanbul, Turkey (pp. 2091–2099).

Dubey, R., Altay, N., Gunasekaran, A., Blome, C., Papadopoulos, T., and Childe, S. J. (2018). Supply chain agility, adaptability and alignment Empirical evidence from the Indian auto components industry. *International Journal of Operations and Production Management*, 38(1), 129–148.

Dubey, R., Gunasekaran, A., Papadopoulos, T., Stephen, J., Shibin, K. T., and Wamba, S. F. (2016). Sustainable supply chain management: Framework and further research directions. *Journal of Cleaner Production*, 142, 1119–1130.

Faisal, M. N., Banwet, D. K. and Shankar, R. (2007). Supply chain agility: Analysing the enablers. *International Journal of Agile Systems and Management*, 2(1), 76–91.

Farris, D. R. and Sage, A. P. (1975). On the use of interpretive structural modeling for worth assessment. *Computers and Electrical Engineering*, 2, 149–174.

Freeman, E. M. (2011). *Restaurant Industry Sustainability: Barriers and Solutions to Sustainable Practice Indicators*, Arizona State University.

Gorane, S. J. and Kant, R. (2013). Modelling the SCM enablers: An integrated ISM-fuzzy MICMAC approach. *Asia Pacific Journal of Marketing and Logistics*, 25(2), 263–286.

Haleem, A., Sushil, Qadri, M. A., and Kumar, S. (2012). Analysis of critical success factors of world-class manufacturing practices: An application of interpretative structural modelling and interpretative ranking process. *Production Planning and Control*, 23(10–11), 722–734.

Hansen, E. G., Grosse-Dunker, F., and Reichwald, R. (2009). Sustainability innovation cube: A framework to evaluate sustainability-oriented innovations. *International Journal of Innovation Management*, 13(4), 683–713.

Janet, M., Wilbrodah, M. M., and Mbithi, M. S. (2015). Factors influencing competitive advantage among supermarkets in Kenya: A case of nakumatt holdings limited. *International Journal of Novel Research in Humanity and Social Sciences*, 2(3), 63–77.

Jayant, A. and Azhar, M. (2014). Analysis of the barriers for implementing green supply chain management (GSCM) practices: An Interpretive Structural Modeling (ISM) Approach. *Procedia Engineering*, 97, 2157–2166.

Jumadi, R. and Bakri, S. (2017). Strategic resources for sustainable competitive advantage E. *International Journal of Advanced Research*, 5(3), 237–241. doi:10.21474/IJAR01/3507.

Kathuria, R., Porth, S. J., Kathuria, N. N., and Kohli, T. K. (2010). Competitive priorities and strategic consensus in emerging economies: Evidence from India. *International Journal of Operations and Production Management*, 30(8), 879–896.

Kumar, S., Luthra, S., and Haleem, A. (2013). Customer involvement in greening the supply chain: An interpretive structural modeling methodology. *Journal of Industrial Engineering International*, 9(1), 6.

Lin, M. H., Hu, J., Tseng, M. L., Chiu, A. S. F., and Lin, C. (2016). Sustainable development in technological and vocational higher education: Balanced scorecard measures with uncertainty. *Journal of Cleaner Production*, 120, 1–12.

Lumpkin, G. T., Droege, S. B., and Dess, G. G. (2002). E-commerce strategies: Achieving sustainable competitive advantage and avoiding pitfalls. *Organizational Dynamics*, 30(4), 325–340.

Luthra, S., Kumar, S., Kharb, R., Ansari, M. F., and Shimmi, S. L. (2014). Adoption of smart grid technologies: An analysis of interactions among barriers. *Renewable and Sustainable Energy Reviews*, 33, 554–565.

Luthra, S., Kumar, V., Kumar, S., and Haleem, A. (2011). Barriers to implement green supply chain management in automobile industry using interpretive structural modeling technique-an Indian perspective. *Journal of Industrial Engineering and Management*, 4(2), 231–257.

Mahdi, O. R. and Almsafir, M. K. (2014). The role of strategic leadership in building sustainable competitive advantage in the academic environment. *Social and Behavioral Sciences*, 129, 289–296.

Maheshwari, P., Seth, N., and Gupta, A. K. (2018). An interpretive structural modeling approach to advertisement effectiveness in the Indian mobile phone industry. *Journal of Modelling in Management*, 13(1), 190–210.

Mandal, A. and Deshmukh, S. G. (1994). Vendor selection using Interpretive Structural Modelling (ISM). *International Journal of Operations and Production Management*, 14(6), 52–59.

Mangla, S., Madaan, J., and Chan, F. T. S. (2012). Analysis of performance focused variables for multi-objective flexible decision modeling approach of product recovery systems. *Global Journal of Flexible Systems Management*, 13(2), 77–86.

Mangla, S., Madaan, J., and Chan, F. T. S. (2013). Analysis of flexible decision strategies for sustainability-focused green product recovery system. *International Journal of Production Research*, 51(11), 1–15.

Mappigau, P. (2012). Core competence and sustainable competitive adventage of Small Silk Weaving Industries (SIs) In Wajo District, South Sulawesi. *Procedia Economics and Finance*, 4, 160–167.

Mishra, S., Datta, S., and Mahapatra, S. S. (2012). Interrelationship of drivers for agile manufacturing: An Indian experience. *International Journal of Services and Operations Management*, 11(1), 35–48.

Mnjala, D. M. (2014). The challenges of creating sustainable competitive advantage in the banking industry in Kenya. *IOSR Journal of Business and Management*, 16(4), 82–87.

Munir, A., Lim, M. K., and Knight, L. (2012). Social and sustaining competitive advantage in SMEs. *Procedia-Social and Behavioral Sciences*, 25, 408–412. doi:10.1016/j.sbspro.2012.02.052.

Oliver, C. (1997). Sustainable competitive advantage: Combining institutional and resource-based views. *Strategic Management Journal*, 18(9), 697–713.

Prakash, S., Dwivedy, M., Poudel, S. S., and Shrestha, D. R. (2018). Modelling the barriers for mass adoption of electric vehicles in Indian automotive sector: An Interpretive Structural Modeling (ISM) approach. In *IEEE 2018 5th International Conference on Industrial Engineering and Applications* (ICIEA), (pp. 458–462).

Prakash, S., Soni, G., Rathore, A. P., and Singh, S. (2017). Risk analysis and mitigation for perishable food supply chain: A case of dairy industry. *Benchmarking: An International Journal*, 24(1), 2–23.

Priyanto, A., Aslichati, L., and Kuncoro, S. (2012). The custom made strategy of "Satu Kayu Desain Enterprise" in efforting to achieve sustainable competitive advantage. *Procedia Economics and Finance*, 4, 54–58.

Raut, R., Narkhede, B. E., Gardas, B. B., and Luong, H. T. (2018). An ISM approach for the barrier analysis in implementing sustainable practices: The Indian oil and gas sector. *Benchmarking: An International Journal*, 25(4), 1245–1271.

Reuter, C., Foerstl, K., Hartmann, E., and Blome, C. (2010). Sustainable global supplier management: The role of dynamic capabilities in achieving competitive advantage. *Journal of Supply Chain Management*, 46(2), 45–63.

Sage, A. P. (1977). *Interpretive Structural Modeling: Methodology for Large-Scale Systems*. New York, McGraw-Hill.

Sansone, C., Hilletofth, P., and Eriksson, D. (2017). Critical operations capabilities for competitive manufacturing: A systematic review. *Industrial Management and Data Systems*, 117(5), 801–837. doi:10.1108/IMDS-02-2016-0066.

Somsuk, N. (2014). Prioritizing drivers of sustainable competitive advantages in green supply chain management based on fuzzy AHP. *Journal of Medical and Bioengineering*, 3(4).

Sook-ling, L. (2015). Information infrastructure capability and organisational framework. *International Journal of Operations and Production Management*, 35(7), 1032–1055. doi:10.1108/IJOPM-12-2013-0553.

Srivastava, M., Franklin, A., and Martinette, L. (2013). Building a sustainable competitive advantage. *Journal of Technology Management & Innovation*, 8(2), 47–60.

Stinnett, B. and Gibson, F. (2016). Sustainable facility development: Perceived benefits and challenges. *International Journal of Sustainability in Higher Education*, 17(5), 601–612.

Verhulst, E. and Lambrechts, W. (2015). Fostering the incorporation of sustainable development in higher education. Lessons learned from a change management perspective. *Journal of Cleaner Production*, 106, 189–204.

Vinayan, G., Jayashree, S., and Marthandan, G. (2012). Critical success factors of sustainable competitive advantage: A study in Malaysian manufacturing industries. *International Journal of Business and Management*, 7(22), 29–45.

Warfield, J. N. (1974). Developing subsystem matrices in structural modeling. IEEE transactions on systems. *Man and Cybernetics*, SMC-4(1), 74–80.

10 Analyzing the Drivers of Sustainable Procurement Using the DEMATEL Approach

K. E. K. Vimal, P. Sasikumar, Kaliyan
Mathiyazhagan, M. Nishal, and K. Sivakumar

CONTENTS

10.1 INTRODUCTION

Sustainable Procurement is a follow-through process of sustainable development objective where it involves procuring process that is consistent and aligns strongly with sustainable development principles (Allen, 2006). Procurement has a big part in sustainable development, and it needs to extend to incorporate the supply chain. The principle of sustainability inspires procurement to make decisions that considers the environmental, economic, and society. The sustainable development started with understanding the ways to recycle, reuse, remanufacture and to increase the life expectancy of the product (Bjurling, 2007). Now, it appears that there is a growing interest in implementing sustainable concepts, specifically to procurement as it contributes its share to numerous fields. Currently, procurement has taken the direction of addressing different dimensions such as society, the environment, and economy. With each dimension, the level of focus towards different attributes is studied theoretically and practiced (Wallace, 2006).

Green supply chain management includes reduction of waste, packaging, assessment of the vendors' environmental performance, carbon emissions reduction related to goods transport, and development of products in an eco-friendly manner (Mont et al., 2009). Existing research findings show that improving the environmental supply leads to cost reductions and improves organizational performance, which would also enhance the firm's reputation. Government regulations also drive the organization to move ahead to implement sustainability procurement (Jennings et al., 1995). In recent years, procurement became the vital function in sustainable supply chain management, which the organization should focus on (Chen, 2005). Despite the growing importance, there is no proper understanding or demonstration in the existing literature about the motivating factors of sustainable procurement. In this regard, to create the awareness and add knowledge to the existing literature, this chapter attempts to analyze the drivers of sustainable procurement. This chapter aims to collect the common drivers of sustainable procurement from the literature review and in consultation with the subject expert, the drivers are categories into internal and external drivers to reveal better information.

Interestingly, many methods are available in the literature to analyze the interdependence of factors. The widely seen techniques are Graph Theory, DEMATEL, Interpretive structural modeling (ISM), and Structural equation modeling (Mangla et al., 2018). Graph theory is used to understand the inheritance and interdependent strengths among the factors. DEMATEL helps to understand the causal interrelationships based on the cause and effect group. Structural equation modeling can be used to establish theory building with the support of a large sample size. On the other hand, ISM helps to develop the relationships based on the driving and dependencies of the factors. Thus, the unique characteristics of DEMATEL, over the other methods, are flexibility, ability to distinguish the relationships with wide variation, ability to establish multi-directional relationships, and ability to produce sound analysis with limited data. The above discussion reveals that the DEMATEL technique is comparatively sound in establishing the interrelationships among the factors with limited data. Thus, the interrelationships between the identified drivers of sustainable procurement are developed in this chapter by applying the DEMATEL technique.

10.2   DRIVERS OF SUSTAINABLE PROCUREMENT

The common drivers motivating the adoption of sustainable procurement are identified from the literature. The common drivers are identified from the literature with the keywords search, which includes *sustainable procurement, green procurement, drivers of sustainable procurement,* and *motivating factors of sustainable procurement.* The combination of keywords was also used in various journal databases namely: Google Scholar, Scopus, Springer, Emerald, Taylor & Francis, and

Inderscience. With the above search, the relevant papers were identified, and with the collective effort of the authors, the common drivers are listed. The identified drivers are presented to the formulated subject expert for their views. Based on the opinion of the subject expert, 18 drivers are identified and categorized into internal and external drivers. Internal drivers include the organizational factors, and external drivers comprise regulatory policies, customers, competitiveness, society, and suppliers, which is shown in Table 10.1.

10.2.1 ORGANIZATIONAL DRIVERS

The organizational factors are drivers from within the organization could reduce the cost through pollution, which is reflected as a hidden cost in the form of misused resources and effort. The investor's pressure the system to adopt sustainable procurement, as it would pose the vision of the organization and concern in the value and ethics of the organization (Benn et al., 2014). The intermediate management are associated positively with support for green procurement, and most of the policymakers for procurement are placed at this level (Walker et al., 2008). The operational and environmental purchasing is more so related to employee involvement. These initiatives were taken to reduce the cost, removal of waste, and enhancement of quality; thus, in this way, environmental performance become a motivation for achieving the higher quality (Brammer et al., 2012).

TABLE 10.1
Drivers of Sustainable Procurement

S.No	Drivers	Aspects	Internal/ External
1	Skillful policy makers (Walker et al., 2008)	Organizational	Internal
2	Company value (Mont et al., 2009)		
3	Employee involvement (Häkkinen et al., 2011)		
4	Manage economic risk (Häkkinen et al., 2011)		
5	Legislative and regulatory compliance (Brammer and Walker, 2011)	Regulatory	External
6	Pre-regulation (Grob et al., 2014)		
7	ISO 14000 Certification (Walker and Brammer, 2009)		
8	Public knowledge (Brammer and Walker, 2011)	Customer	
9	Environment and e-logistic (Meehan et al., 2011)		
10	Marketing pressure (Wallace, 2006)		
11	Firm performance improvement (Rao et al., 2005)	Competitiveness	
12	Brand value (Kumar et al., 2014)		
13	Stakeholders encouragement for environment concern (Walker et al., 2008)	Society	
14	Public pressure (Zhu and Sarkis, 2007)		
15	Criticism by public (Zhu and Sarkis, 2007)		
16	Potential for publicity (Amran et al., 2011)		
17	Supplier collaboration (Grob et al., 2014)	Suppliers	
18	Integration supply (Walker et al., 2008)		

10.2.2 Regulatory Drivers

The government regulation and legislation are one of the vital drivers for companies' adaption of sustainability (Grob et al., 2014). As said, green purchasing has value for the buying company as it is related to the compliance of environmental legislation, but it does not improve the environmental performance. Companies which comply with the legislation does not seem to share or integrate their ecological concern with their value chain process (Brammer et al., 2012). Thus, regulations become a stimulus to diminish the effect of impact on the environment.

10.2.3 Customer-Based Drivers

The public has developed knowledge about the environment practice over the products they buy. The marketing is also tends to highlight the environment aspects involved within the system (Grob et al., 2014).

10.2.4 Competitiveness-Oriented Drivers

The firm can improve their competitive advantage with the help of proactive environmental strategy through the development of capabilities of its supply chain management. It mostly reflects the company value over the market and environment purchasing, which may be advantageous in improving the financial performance system (Grob et al., 2014).

10.2.5 Societal Drivers

The current scenario shows the deterioration of the environment; thus, people are demanding products to be more environment concerned relatively showing the public awareness of environmental issues (Amran et al., 2011). The ecological supply practice of the firms is assessed because of pressure from the public and shareholders. This awareness generates the opportunity for firms to gain new customers through developing the ways to deal with ecological practices (Häkkinen et al., 2011).

10.2.6 Suppliers-Based Drivers

Environmental management benefits from the collaboratively integrated supply chain in its operations (Chen, 2005). There is an increase in the involvement of environmental association with suppliers when the base of supply is reduced (Rao et al., 2005). This also enables the suppliers to adopt and enable sustainable practices.

10.3 RESEARCH METHOD

The data collected procedure and the steps of DEMATEL method is presented in the following subsection.

10.3.1 Data Collection

The experts were identified based on the subject and domain of expertise. In total, 15 experts were identified (for details, refer to Table 10.2), and the responses were collected from each expert individually. The DEMATEL technique is adopted to analyze the interrelationships among the drivers.

10.3.2 DEMATEL Method

The DEMATEL was developed to identify cause-effect relationships in a complex system (Sivakumar et al., 2018). The DEMATEL method is relatively flexible and has wide variations among relationships between elements as compared to interpretive structural modeling. DEMATEL provides multiple directional relationships as compared to an analytical hierarchal process, which is unidirectional in a relationship (Girubha et al., 2016). Because of these methodological advantages, the DEMATEL approach is used for this study. The widely followed steps of DEMATEL are as follows.

Step: 1
Generate the initial relationship matrix (A_k) for each expert using Equation 10.1. The initial relation matrix depicts the pairwise comparison among the drivers with scale shown in Table 10.3. Each element in the matrix A_k represents the influence on driver i over j given by a kth expert.

TABLE 10.2
Subject Expert Details

S.No	Designation	Member Expertise	Years of Experience
1	Procurement Manager	Auto Components	20
2	Operations Management	Auto Components	5
3	Production Engineering	Auto Components	4
4	Research and Development	Auto Components	10
5	Procurement Manager	Electrical/Electronics	15
6	Research and Development	Electrical/Electronics	6
7	Procurement Manager	Plastic Part Manufacturing	12
8	Research and Development	Plastic Part Manufacturing	17
9	Operations Management	Plastic Part Manufacturing	6
10	Procurement Manager	Paper Production	4
11	Operations Management	Paper Production	6
12	Research and Development	Paper Production	20
13	Supply Chain Management and Reverse logistics	Independent (Academic)	20
14	Environmental Conscious manufacturing	Independent (Academic)	8
15	World Class Manufacturing	Independent (Academic)	8

TABLE 10.3

Interrelationship Scores

Score	Meaning
0	No influence
1	Very low influence
2	Low influence
3	Medium influence
4	High influence
5	Very high influence

$$A_k = \begin{bmatrix} 0 & a_{12k} & a_{13k} & \cdots & a_{1(n-1)k} & a_{1nk} \\ a_{21k} & 0 & a_{23k} & \cdots & & a_{2nk} \\ \cdots & \cdots & \cdots & \cdots & \cdots & \cdots \\ \cdots & \cdots & \cdots & \cdots & \cdots & \cdots \\ a_{(n-1)1k} & a_{(n-1)2k} & a_{(n-1)3k} & \cdots & 0 & a_{(n-1)nk} \\ a_{n1k} & a_{n2k} & a_{n3k} & \cdots & a_{n(n-1)k} & 0 \end{bmatrix} \tag{10.1}$$

Step: 2

Using Equation 10.2, the opinion of all the experts were integrated using Equation 10.2 to the overall direct relationship matrix A.

$$A = \frac{[a_{ij}]}{k} \tag{10.2}$$

$$a_{ij} = \text{average of } a_{ijk}$$

Step: 3

Using Equation 10.3, the initial direct relationship matrix B is calculated.

$$B = [b_{ij}]_{nxn} = \frac{A}{\max\limits_{1 \le i \le n} \sum\limits_{j=1}^{n} a_{ij}}, \text{ where } 0 \le b_{ij} \le 1 \tag{10.3}$$

B = normalized overall direct relationship matrix
Step: 4

Using Equation 10.4, the total relationship matrix C is computed. The sum of rows (R) and the sum of columns (C) for the total relationship matrix C is computed using Equations 10.5 and 10.6, respectively.

$$C = [C_{ij}]_{n*n} = B[I - B]^{-1} \tag{10.4}$$

$$R = \left[r_{ij} \right]_{n \times 1} = \left[\sum_{j=1}^{n} c_{ij} \right]_{n \times 1} \quad (10.5)$$

$$C = \left[c_{ij} \right]_{1 \times n} = \left[\sum_{i=1}^{n} d_{ij} \right]_{1 \times n1} \quad (10.6)$$

I = Identity matrix
R = Sum of rows
C = Sum of columns

Step: 5

The threshold value (α) is computed using Equation 10.7 for identifying the critical relationships.

$$\alpha = \frac{\sum_{j=1}^{n} \sum_{i=1}^{n} c_{ij}}{n^2} \quad (10.7)$$

n = The number of elements in the matrix C.

Step: 6

The causal diagram is drawn by taking ($R_i + C_j$, $R_i - C_j$) as the coordinates for each driver in the graph were the horizontal axis refers to cause/effect and the vertical axis refers to prominence.

Step: 7

The element c_{ij} in matrix C, which meets or exceeds the threshold value (α), are considered to have a interrelationship. The interrelationships are visualized through a directed graph.

10.4 APPLICATION OF DEMATEL FOR ANALYZING THE DRIVERS

The initial interaction matrix for aspects and drivers are obtained from 15 experts through a face-to-face interview. With the obtained responses and by applying Equation 10.3, the initial normalized matrix A_k is computed. The total relationship matrix is computed using the Equation 10.4 (Table 10.4 for aspects). The threshold value to identify the relationship is calculated as 0.075 using Equation 10.7. With the identified significant relationship, a digraph is developed as shown in Figure 10.1. Similarly, the threshold value to identify the significant relationship between the drivers is computed as 0.026. The digraph depicting the significant relationship among the drivers is presented in Figure 10.2. Equations 10.5 and 10.6 is used to compute the prominence level and causal effect, and the corresponding scores are shown in Table 10.4.

Similarly, the digraph depicting the interaction among the drivers are developing using the Table 10.5 is shown in Figure 10.2.

TABLE 10.4

Total Relationship Matrix for Aspects

	A_1	A_2	A_3	A_4	A_5	A_6	R_i	C_j	$R_i + C_j$	$R_i - C_j$
Organizational (A_1)	0	0	0	0	0	0.33	0.33	1.17	1.51	−0.843
Regulatory (A_2)	0.31	0	0	0.25	0	0.416	0.97	0.16	1.145	0.8
Customer (A_3)	0.36	0.166	0	0.291	0	0.402	1.22	0	1.22	1.225
Competitiveness (A_4)	0.25	0	0	0	0	0.33	0.58	0.54	1.125	0.041
Society (A_5)	0.25	0	0	0	0	0.083	0.33	0	0.333	0.333
Suppliers (A_6)	0	0	0	0	0	0	0	1.56	1.56	−1.56

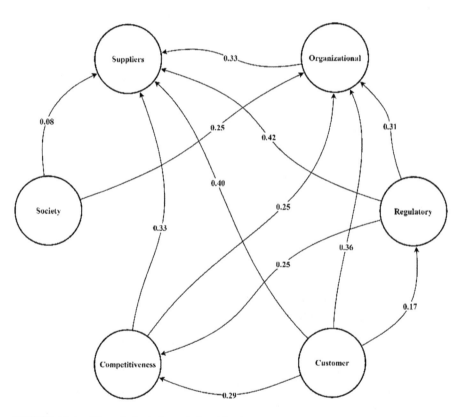

FIGURE 10.1   Digraph depicting relationships between aspects.

TABLE 10.5

Drivers with $R_i + C_j$ and $R_i - C_j$ Values

Driver	$R_i + C_j$	$R_i - C_j$
Organizational (A₁)		
Skillful policymakers (D₁)	1.38	1.387
Company value (D₂)	1.21	−0.08
Employee involvement (D₃)	0.68	0.135
Manage economic risk (D₄)	0.85	−0.63
Regulatory (A₂)		
Legislative and regulatory compliance (D₅)	1.46	0.917
Pre-regulation (D₆)	0.784	−0.08
ISO 14000 Certification (D₇)	1.53	−0.447
Customer (A₃)		
Public knowledge (D₈)	0.61	0.32
Environment and e-logistic (D₉)	0.57	−0.37
Marketing pressure (D₁₀)	0.58	−0.133
Competitiveness (A₄)		
Firm performance improvement (D₁₁)	1.11	−0.028
Brand value (D₁₂)	1.21	−0.44
Society (A₅)		
Stakeholders encouragement for environment concern (D₁₃)	0.646306	−0.28
Public pressure (D₁₄)	0.836183	0.020
Criticism by public (D₁₅)	0.832939	−0.068
Potential for publicity (D₁₆)	0.731424	−0.15
Suppliers (A₆)		
Supplier collaboration (D₁₇)	0.86	0.007
Integration supply (D₁₈)	0.80	−0.056

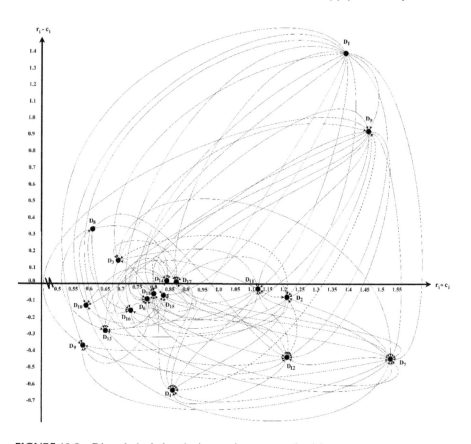

FIGURE 10.2   Digraph depicting the interactions among the drivers.

10.5   DISCUSSIONS ON RESULTS

The results about aspects and criteria are discussed below.

10.5.1   RESULTS ON ASPECTS

Concerning Figure 10.1, the customer (A_3) is the most critical driver influencing almost all aspects, except supplier (A_6). On the basis of ($R_i + C_j$) values, the importance of an aspect is prioritized, i.e., $A_6 > A_1 > A_3 > A_2 > A_4 > A_5$. On the basis of ($R_i - C_j$) values, the aspects are divided into either the cause or effect group (Table 10.4):

1. If the value of ($R_i - C_j$) is positive, then the aspects will be categorized in the cause group. On comparing the $R_i - C_j$ value, the highest positive ($R_i - C_j$) factors will be having a highest direct effect on the others. This study consists of Regulatory (A_2), Customer (A_3), Competitiveness (A_4), and Society (A_5) are categorized in the Cause group, which is having the

$(R_i - C_j)$ values of 0.81, 1.22, 0.041, and 0.33, respectively. A_3 (Customer) was having the highest effect on the others.

2. If the value of $(R_i - C_j)$ is negative, then the aspects will be categorized in the effect group. In this study, Organization (A_1) and Suppliers (A_6) were classified in the effect group, with the $(R_i - C_j)$ values of −0.84 and −1.56, respectively. Moreover, A_6 (suppliers) was the most affected by the other factors (A_1), (A_2), (A_3), (A_4), (A_5).

10.5.2 RESULTS ON DRIVERS

The cause and effect level of the drivers are seen in the prominence and causal diagram (Figure 10.2). Figure 10.2 shows that D_1 is the highly interrelated driver, which is presented in the right corner, and similarly D_9, is the least correlated driver, which is presented in the left corner. From Figure 10.2, it is understandable that D_4 and D_9 have a single relationship (cause) with the other drivers on the contrary the maximum correlated drivers had found to be D_1 followed by D_5. Similarly, D_1 has a significant relationship with all the other drivers apart from D_8 and D_{14}. These two drivers are related to the public, which does not have any role in skillful policymaking. Skillful policymaking can manage economic risks, which are confirmed with the relationship value of 0.113. Table 10.5 indicates that under organizational (A_1), this study finds out that the skillful policymakers (D_1) and company value were having the highest $(R_i + C_j)$ values of 1.387 and 1.212, respectively. The skillful policymakers (D_1) and employee involvement (D_3) has the highest value in $(R_i - C_j)$ of 1.387 and 0.1358, and that is why they are in the cause group. Company value (D_2) and manage economic risk (D_4) has values of −0.084 and −0.63, which is why they are in effect group. A skillful policy maker (D_1) is one of the most important drivers as it influences the other drivers directly. Table 10.5 indicates that under regulatory (A_2), the study finds out that the legislative and regulatory compliance (D_5) has the highest value of 1.46 in $(R_i + C_j)$ and also has a positive value 0.91 in $(R_i - C_j)$, which let it be in the cause group. Pre-regulation (D_6) and ISO 14000 certification (D_7) has the value of −0.085 and −0.44, which is leading them to effect group. Legislative and regulatory compliance is an essential driver as it influences the other drivers directly.

Table 10.5 indicates that under customer (A_3), the study finds out that the public knowledge (D_8) has the highest value of 0.61 in $(R_i + C_j)$ and also it has a positive value in the $(R_i - C_j)$ of 0.32 leads it to be in the cause group. Environment and e logistics (D_9) and marketing pressure (D_{10}) has a negative value of −0.37 and −0.133 that is why they are in the effect group. Public knowledge (D_8) is an essential driver as it has significant influences. Table 10.5 indicates that under competitiveness (A_4), the study finds out that brand value (D_{12}) has the highest value in the $(R_i + C_j)$ of 1.21. The aspects do not have any positive value in $(R_i - C_j)$ leading to cause group. Firm performance improvement (D_{11}) and brand value (D_{12}) has a negative value in $(R_i - C_j)$ of −0.028 and −0.443 leading them to effect group. Firm performance improvement (D_{11}) is an essential driver with key influence.

Table 10.5 indicates that under society (A_5), the study finds out that public pressure (D_{14}) and criticism by the public (D_{15}) has the highest value of 0.83 and 0.83 in $(R_i + C_j)$. Public pressure (D_{14}) has a positive value of 0.02012 in $(R_i - C_j)$ leading to cause group. Stakeholder's encouragement for environment concern (D_{13}), criticism by the public (D_{15}), and potential for publicity (D_{16}) has a negative value of -0.28, -0.068, and -0.157 in ($R_i - C_j$) that it falls in effect group. Public pressure (D_{14}) has a key influence over the other drivers. Table 10.5 indicates that under suppliers (A_6), the study finds out that supplier collaboration (D_{17}) has the highest value of 0.866 in ($R_i + C_j$). Supplier collaboration (D_{17}) has a positive value of 0.0073 in ($R_i - C_j$) leading to cause group. Integration supply (D_{18}) has a negative value of -0.0569 in ($R_i - C_j$) that it rests upon the effect group. Supplier collaboration (D_{17}) and integration supply (D_{18}) are critical factors influencing suppliers (A_6).

10.6 CONCLUSIONS

This study applied the DEMATEL method to find the relationship among the drivers of sustainability procurement. Through literature along with experts' opinion, 18 drivers have been identified that are categorized into 6 aspects. The three important aspects in the cause group that are skillful policymakers, employee involvement, legislative and regulatory compliance, public knowledge, public pressure, and supplier collaboration were given more importance, and more research and innovative ways should be introduced to improve them. In the whole process of arranging the value of driver under the six fundamental aspects, it was found that the skillful policymakers (D_1), company value (D_2), legislative and regulatory compliance (D_5), ISO 14000 certification (D_7), firm performance improvement (D_{11}), and brand value (D_{12}) were the most critical drivers. Thus, the companies should start with implementing the critical drivers to start moving ahead in implementing sustainability in the procurement sector.

The study presented in this chapter presents rich insights to the managers and practicing engineers. The findings will help the managers understand the significant driver, least significant driver, and interrelationship among the drivers. This understanding will help the managers move towards a better adaptation of sustainable procurement practices. The drivers considered in this chapter are generic; thus, the managers of similar organizations can utilize the results of the study. Further, the procedure can be utilized to extend the study with inclusion or exclusion of a few drivers. Thus, the demonstrated study is expected to provide a roadmap for implementing sustainable procurement in their organization successfully. The major limitation of the current study is the DEMATEL analysis has been done with a minimum number of expert opinions. In the near future, the study can be done by considering the opinion of the wider population.

REFERENCES

Allen, B. (2006). In pursuit of responsible procurement. *Summit*, 9(4), 7.
Amran, A., & Haniffa, R. (2011). Evidence in development of sustainability reporting: A case of a developing country. *Business Strategy and the Environment*, 20(3), 141–156.
Benn, S., Edwards, M., & Williams, T. (2014). *Organizational Change for Corporate Sustainability*. London, UK: Routledge.

245

Bjurling, K. (2007). Vita rockar och vassa saxar. *En rapport om landstingens brist på etiska inköp, SwedWatch, Fair Trade Center and Rena Kläder, 86.*

Brammer, S., & Walker, H. (2011). Sustainable procurement in the public sector: An international comparative study. *International Journal of Operations & Production Management, 31*(4), 452–476.

Brammer, S., Jackson, G., & Matten, D. (2012). Corporate social responsibility and institutional theory: New perspectives on private governance. *Socio-economic Review, 10*(1), 3–28.

Chen, C. C. (2005). Incorporating green purchasing into the frame of ISO 14000. *Journal of Cleaner Production, 13*(9), 927–933.

Girubha, J., Vinodh, S., & Kek, V. (2016). Application of interpretative structural modelling integrated multi criteria decision making methods for sustainable supplier selection. *Journal of Modelling in Management, 11*(2), 358–388.

Grob, S., & Benn, S. (2014). Conceptualising the adoption of sustainable procurement: An institutional theory perspective. *Australasian Journal of Environmental Management, 21*(1), 11–21.

Häkkinen, T., & Belloni, K. (2011). Barriers and drivers for sustainable building. *Building Research & Information, 39*(3), 239–255.

Jennings, P. D., & Zandbergen, P. A. (1995). Ecologically sustainable organizations: An institutional approach. *Academy of Management Review, 20*(4), 1015–1052.

Kumar, V., & Christodoulopoulou, A. (2014). Sustainability and branding: An integrated perspective. *Industrial Marketing Management, 43*(1), 6–15.

Mangla, S. K., Luthra, S., Mishra, N., Singh, A., Rana, N. P., Dora, M., & Dwivedi, Y. (2018). Barriers to effective circular supply chain management in a developing country context. *Production Planning & Control, 29*(6), 551–569.

Meehan, J., & Bryde, D. (2011). Sustainable procurement practice. *Business Strategy and the Environment, 20*(2), 94–106.

Mont, O., & Leire, C. (2009). Socially responsible purchasing in supply chains: Drivers and barriers in Sweden. *Social Responsibility Journal, 5*(3), 388–407.

Rao, P., & Holt, D. (2005). Do green supply chains lead to competitiveness and economic performance? *International Journal of Operations & Production Management, 25*(9), 898–916.

Sivakumar, K., Jeyapaul, R., Vimal, K. E. K., & Ravi, P. (2018). A DEMATEL approach for evaluating barriers for sustainable end-of-life practices. *Journal of Manufacturing Technology Management, 29*(6), 1065–1091.

Walker, H., & Brammer, S. (2009). Sustainable procurement in the United Kingdom public sector. *Supply Chain Management: An International Journal, 14*(2), 128–137.

Walker, H., Di Sisto, L., & McBain, D. (2008). Drivers and barriers to environmental supply chain management practices: Lessons from the public and private sectors. *Journal of Purchasing and Supply Management, 14*(1), 69–85.

Wallace, W. (2006). Marketing sustainable products in the retail sector: The potential integration of sustainability marketing. PhD dissertation, Lund University, Lund, Sweden.

Zhu, Q., & Sarkis, J. (2007). The moderating effects of institutional pressures on emergent green supply chain practices and performance. *International Journal of Production Research, 45*(18–19), 4333–4355.

11 Barriers on Sustainable Supply Chain Management Implementation for Turkish SME Suppliers
A Fuzzy Analytic Hierarchy Process Approach

Bahar Türk and Ali Kemal Çelik

CONTENTS

11.1   INTRODUCTION

The widely-adopted perception in modern business argues that competition is not amongst companies but their supply chains in the globalized era. Therefore, along with overwhelmingly increasing competition levels, competitive companies must efficiently manage their underlying chains by considering customer expectations and the efficient use of natural resources in terms of social responsibility (Vachon and Klassen, 2006). One convenient way to maintain the underlying efficiency is to adopt sustainable supply chain operations. The primary understanding of sustainable supply chain operations is to avoid environmental issues that can be caused by outcomes occurred before and after the production process (New and Westbrook, 2004; Dubey et al., 2017a). Additionally, such a process of Small- and Medium-sized Enterprises (SMEs) and their suppliers deserves considerable attention, since SMEs constitute a significant number of economic units in both Turkey and worldwide, and they account for more than half of the total employment and production.

SMEs' adaption of their sustainable understanding to supply chains has a substantial potential to overcome environmental issues. However, SME suppliers may principally encounter several barriers on the implementation of sustainable supply chain management (SSCM). In this regard, the corresponding potential barriers have to be carefully recognized in order to maintain a SSCM. Moreover, better understanding the order of importance of these barriers for enterprises and/or suppliers may create more efficient future solution strategies and to have more accurate resource transfer planning. For this purpose, following Govindan et al. (2014), five main criteria, finance, involvement and support, technology, knowledge and outsourcing, were taken into consideration and utilized from a fuzzy AHP approach. Besides, the ultimate consumers' level of environmental sensitivity for all these dynamics is adopted to reflect environmentally-friendly policies and supplier selection of SMEs.

When the relevant operational flow is reversely considered, suppliers who can successfully perform SSCM are primarily chosen by SMEs; hence, the underlying SMEs are chosen by customers with an important competitive advantage. In other words, this interactive process can be assessed as profitable for all parties since the sustainability phenomenon can be accomplished.

11.2   SUSTAINABLE SUPPLY CHAIN MANAGEMENT

In recent years, it has been essential for businesses to re-consider their supply chain management processes due to ascending competition among businesses, globalization, increased complicacy of supply chain networks and the number of members, and decreasing product life cycle periods (Beamon, 1999; Dubey et al., 2017b; Bazan and Jaber, 2018). Along with challenging competition conditions, customer expectations have also increased simultaneously. Hence, businesses tend to act more socially responsible to meet such expectations and improve their supply chain management. In that sense, businesses are obliged to manage their supply chains with an understanding of sustainability by properly taking the fair use of natural resources into consideration (Vachon and Klassen, 2006; Mathiyazhagan et al., 2018).

More efficient management of the supply chains both requires simultaneous control of each member in the chain and reverse logistics and sustainable management approaches. These requirements have led to a transition from traditional to SSCM process by expanding the structure of the traditional supply chain management. At that point, the concept of sustainability takes place at every phase of the supply chain from stock management and logistics functions to the ultimate consumer (Simpson and Power, 2005; Büyüközkan and Vardaloğlu, 2008; Carvalho et al., 2017).

Traditional supply chain includes two or more organizations legally separated though they are connected with stock, information, and financial flow (Stadtler and Kilger, 2002; Fredendall and Hill, 2016; Reefke and Sundaram, 2017). The traditional supply chain can also be concluded as a process that encompasses all operational flows from raw material to the ultimate customer (Chopra and Meindl, 2001; Ballou, 2004; Christopher, 2016). The underlying process involves various units—that meet different needs in line with common goals and objectives, that constitute an entire system by complementing each other, and that are dependent on different levels—such as producer, distributor, supplier, retailer, logistics service provider, and the ultimate customer.

Supply chain management refers to the management of stock and information and financial flows in a relevant network composed of suppliers, producers, distributors, and customers. In other words, supply chain management is described as the act of coordination that provides the most efficient mix to be introduced to the market through several units in the supply chain including production, transportation, and inventory (Fredendall and Hill, 2016; Hugos, 2018). The successful management of the underlying flows implies the success of supply chain management (Bowersox et al., 2002; Christopher, 2016). At the same time, the supply chain management requires the relevant coordination of businesses that intend to improve their strategic positions and operational efficiency (Murphy et al., 2004; Jiao et al., 2017; Chen, 2018) and, not surprisingly, a successful supply chain management and successful businesses are significantly correlated. Businesses that are aiming to take a larger share from the market, concentrating on the ultimate consumer, and increasing their profitability in this way, have undertaken a competition within supply chain management dimensions. In that context, supply chains that are managed more effectively, and efficiently have become as one of the most influential competition tools (Lee and Billington, 1992; Levi et al., 2003; Hugos, 2018).

SSCM can be characterized as a broad-based innovation that can provide a *win-win* strategy by minimizing environmental risks and increasing ecological efficiency during the process of achieving profit and market share targets (Van Hoek, 1999; Govindan and Soleimani, 2017; Sarkis, 2018). Supply chain management can be briefly described as the combination of environmental concerns into the supply chain process (Zhu et al., 2008; Mumtaz et al., 2018). Likewise, the supply chain management process refers to acting with a sense of sustainability in all decision-making and choice phases of the underlying supply chain from production, distribution, and purchasing to marketing (Morana, 2013). Besides, the SSCM is considered as the success of economic goals, transparent integration, and strategies to improve the long-term performance of the businesses and supply chains. The underlying definition of SSCM is associated with the supporting facets of sustainability, such as triple bottom line (so-called as the milestone for sustainability), risk management, transparency, strategy, and culture (Carter and Rogers, 2008; Busse and Mollenkopf, 2017; Dubey et al., 2017a).

As a novel paradigm with adopted sustainability strategies, SSCM serves a variety of returns to its practitioner companies (Sarkis, 2003; Zhu and Sarkis, 2004; Simpson and Power, 2005). SSCM decreases unfavorable outcomes for both tactical and strategic decisions, whereas it strengthens control processes, provides the proper and efficient use of resources, and encourages recycling (Kumar et al., 2012). SSCM contributes to job satisfaction and social quality of life and, additionally, it plays an important role on creating values for companies by providing customer satisfaction through efficient asset use and increased service quality. Consequently, SSCM both succeeds to create value for supply chain of companies and explains the importance of environment (Zhu and Sarkis, 2004; Kumar et al., 2012; Bag et al., 2017; Mirghafoori et al., 2017).

11.3 SMEs AND THEIR SUPPLIERS WITHIN THE SCOPE OF SUSTAINABLE SUPPLY CHAIN MANAGEMENT

SSCM provides a competitive advantage for SMEs to enter international markets, particularly in developing countries, since SMEs commonly utilize labor-intensive production technologies. However, since SMEs have to compete with many small, medium, and large companies, potential minor issues to be experienced in both product/service production and procurement processes may lead to many major issues. Particularly, quality is one of the major issues in the procurement process of raw materials and main goods for SMEs, which produce upon regular ordering. In that context, finding and maintaining reliable suppliers becomes reasonably crucial. Moreover, building strong relationships with suppliers, improving these relations, and creating alternatives emerge as a supply chain management function than a logistics function (Kehoe and Boughton, 2001; Tao et al., 2017).

A successful supply chain management for SMEs can be accomplished by expanding communication and information channels all throughout the supply chain. The information and planning shared with suppliers increases the productivity of the supply chain and contributes to competition power of companies, as well (Kehoe and Boughton, 2001; Dweekat et al., 2017). Within the scope of sustainability, the interdependency among the members of supply chain implies that a decision by any member has an impact on other members of the supply chain. Since SMEs purpose to have particular standards on operations and decisions, in terms of sustainability, decisions of SMEs will reflect on their suppliers. Moreover, a number of SME suppliers, which exhibit relatively unsatisfactory performance, may be eliminated from the supply chain management system. Therefore, sustainable supply chain management implementation (SSCMI) of SME suppliers will have a significant impact on their relationship with SMEs and will create the reason of further decisions (Lyon and Maxwell, 2013; Adams et al., 2017).

In this contemporary era, companies are no longer perceived as corporations that strive to only gain profit, they also pay attention to several important concepts including quality and social issues. Accordingly, competitive companies also concentrate on the concept of sustainability for their future management strategies (Lyon and Maxwell, 2013). Particularly, the underlying significant transition has been initially introduced by companies that intend to be perceived as sustainable corporations, and it has indispensable impact on supply chains as well due to interdependency. In fact, the understanding of sustainability is perceived across all supply chains due to the relevant demand and pressure from customers on companies and from companies to their suppliers to conform to the requirements of sustainability (Lyon and Maxwell, 2008, 2013; Singla et al., 2017; Johnson et al., 2018).

In the existing literature, it can be recognized that the SSCM is generally reduced to *supplier selection*. Green et al. (1996) argue that an attempt to consider the industrial purchase in an environmental context can be regarded as environmental supply chain management. Gilbert (2001) describes an environmentalist supply chain management as the involvement of environmental facts and concerns to purchasing decisions and long-run relationships with suppliers. Zhu and his colleagues (2008) describe SSCM as the integration of suppliers to environmental management process. Similarly, Lee (2008) describes environmentalist attempts as the efforts for the transfer and the diffusion of advanced sustainable management implementation throughout the supply chain using the relationship between companies and/or SMEs with high volume purchasing capacity and their suppliers (Güner and Coşkun, 2013). In this context, companies, which tend to conform their operations to the environmental standards, consider their relationships with their suppliers again and probably make further decisions about their suppliers, which do not remain their operations within the lines of sustainability. Earlier research puts forward that the requirements for sustainability have an important influence on the existing relationships with their suppliers and their future decisions on supplier selection (Skjoett-Larsen, 2000; Luthra et al., 2017).

Nowadays, many companies and SMEs with high sanction from the market will demand from their suppliers to conform to a variety of rigorous regulations as a precondition for sustainability (Ho et al., 2009). They also enforce compliance with environmental standards and increasingly abandon suppliers that do not assure environmentalist criteria (Sroufe and Melnyk, 2017). Unfortunately, only a limited number of companies request an official environmental certificate from their suppliers at the present time. Nevertheless, customers' future attempts to request from suppliers to have environmental standards will probably enforce suppliers to comply with sustainability. Consequently, companies that tend to be sustainable in their operations should precisely take their suppliers into account and they should encourage and even enforce their suppliers to be environmentally-friendly (Heying and Sanzero, 2009; Song et al., 2017; Yun and Chuluunsukh, 2018).

11.4   METHODOLOGY

11.4.1   THE AIM AND THE IMPORTANCE OF THE STUDY

SMEs offer a major contribution to a country's economy, and their suppliers provide SMEs products and/or semi-products to assist SMEs to sustain quality and environmentally-friendly operations. This chapter purposes to determine the order of importance for the barriers that SMEs and their suppliers experience during sustainable chain management applications. In this way, one can also determine strategies that are more effective and efficient during the transition or improvement process of SSCMI and SMEs, which would perform more environmentally-friendly operations along with their more environmental suppliers.

As SMEs have a crucial role on many national and international operations, the impact of the operation capacity of SME suppliers on the environment deserves particular attention. However, several suppliers may avoid being an environmentally-friendly corporation even if SMEs cooperate to sustain to be environmentalist. Therefore, the present study may contribute to the existing literature on the determination of barriers that lead to the abovementioned unfavorable situations, and the proposal of convenient solution strategies to overcome the underlying barriers.

11.4.2   THE SCOPE AND THE LIMITATIONS

The present study utilized from a multi-criteria decision-making technique, namely a fuzzy AHP, to determine the order of importance for barriers encountered by SMEs and their suppliers. Following Govindan et al. (2014)'s classification on such barriers, five main criteria, including *finance, involvement and support, technology, knowledge, and outsourcing*, and an additional twenty-six sub-criteria were taken into account and a decision hierarchy was developed. The data used in this study were gathered from 10 SME supplier managers' opinions under the structure of the authorized institution for the development of Turkish SMEs, namely the KOSGEB. The corresponding experts' opinions were requested through a well-established written questionnaire. The underlying questionnaire involves comparative questions that may give information on weights/priorities of several barriers they encounter in their SSCM applications. Saaty (1990)'s fundamental AHP scale was used to carry out the relevant comparisons as shown in Table 11.1.

TABLE 11.1
Saaty (1990)'s Fundamental AHP Scale

Weight	Definition
1	Equal importance
3	Moderate importance of one over another
5	Essential or strong importance
7	Very strong importance
9	Extreme importance
2, 4, 6, 8 are intermediate values between the two adjacent judgments	

SSCMI is a comprehensive process affected by a respectable number of indicators. However, this paper is limited within the content of five criteria, such as finance, involvement and support, technology, knowledge, and outsourcing, and their subcriteria. Similarly, the data being obtained is also limited to SME managers' expert opinions.

11.4.3 ANALYTIC HIERARCHY PROCESS (AHP) AND FUZZY ANALYTIC HIERARCHY PROCESS (FUZZY AHP)

Analytic hierarchy process is a fundamental approach associated with decision-making among a number of alternatives by providing a comparison of among opinions and evaluating the structure of complex systems. The AHP utilizes from both subjective and objective evaluation criteria and, subsequently, it makes a consistency test about the underlying evaluation. Hence, this approach lends assistance to the decision maker on the priority among all possible alternatives (Chang, 1996). The AHP was developed by Saaty (1990) and it principally depends on assessing the problem in a hierarchical form. The corresponding hierarchical form facilitates the researcher to arrange the critical dimensions of the problem in a sequential framework. In that way, the AHP provides a simple comparison, arrangement, and interpretation of complex alternatives and guarantees to make the optimum decision and to exhibit the apparent rationale of this optimum decision (Saaty, 1990).

The AHP is widely adopted as an intuitive model as it compares both tangible and intangible criteria that may possibly affect the decision process by considering them numerically and determining the priorities relative to each other. In the underlying procedure, decision makers'/experts' experience and knowledge are utilized through convenient questionnaires or interviews in order to determine criteria on the problem being concerned before deciding. Namely, experts' judgements are taken into account in order to introduce criteria affecting future decisions and to determine their order of importance (Güngör and İşler, 2012). In the traditional AHP process, these individuals should be necessarily chosen from experts or they should have at least an average knowledge on the corresponding subject, as the results of the AHP are interpreted depending on these individuals' judgements through several comparison procedure. Hence, priority and pairwise comparison matrices are introduced with respect to the underlying judgements by transforming the judgements into numerical values. All essential calculations are made using numerical values (Saaty, 1990, 2000).

The steps relating to a traditional AHP can be explained briefly as follows:

Step 1: The decision criteria are determined as the hierarchy of the objectives. There exists a rigorous hierarchy from objectives to intermediate levels; criteria and their sub-criteria of related levels, and to the bottom level, namely, alternatives.

Step 2: The main criteria, sub-criteria and alternatives are adequately weighted as a function of their order of importance for related unit member in the upper level. In order to concentrate on only two factors simultaneously, simple pairwise comparisons are used to defined weights and classification.

Step 3: After the introduction of decision matrix, a priority or a normalized eigenvalue vector is introduced to assign weights to each element of the underlying matrix. The order of importance is conducted by determining the weights of criteria using calculated eigenvalue vector (Saaty and Vargas, 2012).

Step 4: A consistency analysis is performed to measure whether or not the calculated values are consistent. Here, a consistency refers to a rationale or a mathematical association among priority values as a result of pairwise comparisons (Göksu and Güngör, 2008). Accordingly, a pairwise comparison matrix is considered as consistent when the maximum eigenvalue (λ_{max}) is equal to the size of the matrix (n). Finally, the procedure is concluded by calculating consistency index and consistency ratio.

The traditional AHP provides a number of criteria simultaneously and the ability to involve both qualitative and quantitative criteria to the decision-making process with an elastic hierarchy regarding the unique structure of the problem (Mahmoodzade et al., 2007). However, a fuzzy AHP was proposed due to some concerns, including the traditional AHP might not reflect experts' authentic mentality. A fuzzy AHP generates a confidence interval compared to absolute values in the traditional AHP procedure. Additionally, the fuzzy AHP serves the researchers a variety of methods intended for determining the optimum alternative or arranging alternatives within the scope of multi-criteria decision-making using the concept of fuzzy sets and hierarchical framework (Şengül et al., 2013).

Van Laarhoven and Pedrcyz (1983) proposed a comparison of fuzzy ratios using triangular fuzzy numbers, while Buckley (1985) utilized from trapezoidal fuzzy numbers. Moreover, Chang (1996) proposed a novel approach using an extended fuzzy AHP algorithm (Kaptanoğlu and Özok, 2006). The main advantage of this approach is the ease of the calculation procedure and not to require additional procedures, except for traditional AHP steps, where the underlying approach concentrates on triangular fuzzy numbers only (Göksu and Güngör, 2008).

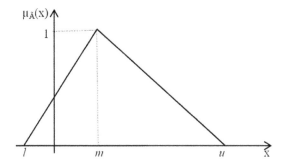

FIGURE 11.1   Triangular membership function.

Triangular membership function is principally defined using three parameters: Let l, m, u denote the corresponding three parameters, and the essential components of triangular membership function are calculated as the following (Figure 11.1):

$$\mu_{\tilde{A}}(x;l,m,u) = \begin{cases} 1 \le x \le m & \text{if} \quad \dfrac{x-l}{m-l} \\[2mm] m \le x \le u & \text{if} \quad \dfrac{u-x}{u-m} \\[2mm] x > u \text{ or } x < l & \text{if} \quad 0 \end{cases} \tag{11.1}$$

For $\mu_A(m) = 1$, m denotes the peak of a triangular fuzzy number, while this peak does not necessitate to be the mid-point of l and u (Baykal and Beyan, 2004).

11.4.4   EXTENDED FUZZY AHP ALGORITHM

Let $X = \{x_1, x_2, ..., x_n\}$ be an object set, and $U = \{u_1, u_2, ..., u_n\}$ be a goal set. According to Chang (1996)'s extent analysis method, for each g_i goal, an extent analysis is respectively performed to each object. Thus, m extent analysis values for each object are described as the following:

$$M_{g_i}^1, M_{g_i}^2, ..., M_{g_i}^m, \quad i = 1, 2, ..., n \tag{11.2}$$

In Equation 11.2, all $M_{g_i}^j (j = 1, 2, ..., m)$'s are triangular fuzzy numbers. The steps of Chang (1996)'s extent analysis are as follows:

1. For i-th object, a fuzzy synthetic extension value is derived as

$$S_i = \sum_{j=1}^{m} M_{g_i}^j \otimes \left[\sum_{i=1}^{n} \sum_{j=1}^{m} M_{g_i}^j \right] \tag{11.3}$$

2. A fuzzy summation procedure of m extent analysis values is conducted for a particular matrix to obtain $\sum_{j=1}^{m} M_{gi}^{j}$ value as follows:

$$\sum_{j=1}^{m} M_{gi}^{j} = \left(\sum_{j=1}^{m} l_i, \sum_{j=1}^{m} m_i, \sum_{j=1}^{m} u_i \right) \tag{11.4}$$

3. An additional fuzzy summation procedure is conducted for M_{gi}^{j} ($j = 1,2,...,m$) to obtain $\left[\sum_{i=1}^{n} \sum_{j=1}^{m} M_{gi}^{j} \right]^{-1}$ inverse matrix as follows:

$$\sum_{i=1}^{n} \sum_{j=1}^{m} M_{gi}^{j} = \left(\sum_{j=1}^{m} l_i, \sum_{j=1}^{m} m_i, \sum_{j=1}^{m} u_i \right) \tag{11.5}$$

4. The inverse of the vector in Step 3 is calculated as follows:

$$\left[\sum_{i=1}^{n} \sum_{j=1}^{m} M_{gi}^{j} \right]^{-1} = \left(\frac{1}{\sum_{i=1}^{n} u_i}, \frac{1}{\sum_{i=1}^{n} m_i}, \frac{1}{\sum_{i=1}^{n} l_i} \right) \tag{11.6}$$

After Step 4, the ordering of the calculated fuzzy numbers is conducted.

The ordering of the calculated fuzzy values are based on different features of fuzzy sets, such as the centre of gravity, the area under membership degree function, or some intersection points. As fuzzy numbers do not have a natural ordering compared to real numbers, several different methods are used for the ordering of fuzzy numbers, and additionally, different methods may give different ordering outcome (Kaptanoğlu and Özok, 2006). Some ordering methods are Chang (1996) Method, Liou and Wang (1992) Method, Abdel-Kader and Dagdale (2001) Method, Quadratic Mean Method, and Kwong and Bai (2003) Method (Şengül et al., 2013).

11.4.4.1 Chang (1996) Method

This method is principally based on comparing obtained synthesis values and determining weights with respect to the underlying comparison values (Chang, 1996). M_1 and M_2 are fuzzy numbers to be ordered, the degree of $M_2 = (l_2, m_2, u_2) \geq M_1 = (l_1, m_1, u_1)$ probability is described as:

$$V(M_2 \geq M_1) = \sup_{y \geq x} \left[\min \left(\mu_{M_1}(x), \mu_{M_2}(y) \right) \right] \tag{11.7}$$

Equation 11.7 can be analogously written as follows:

$$V(M_2 \geq M_1) = hgt(M_2 \cap M_1) = \mu_{M_2}(d) = \begin{cases} 1 & , \text{if } m_2 \geq m_1 \\ 0 & , \text{if } l_1 \geq u_2 \\ \dfrac{l_1 - u_2}{(m_2 - u_2) - (m_1 - l_1)} & , \text{otherwise} \end{cases} \tag{11.8}$$

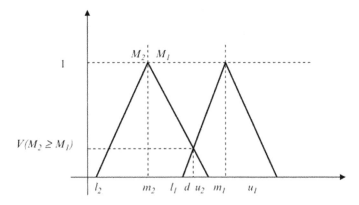

FIGURE 11.2   The intersection point between μ_{M_1} and μ_{M_2}.

In Equation 11.8, d denotes the ordinate of D, and D denotes the highest intersection point between μ_{M_1} and μ_{M_2}, as illustrated in Figure 11.2.

For $k \neq i$ and $k = 1, 2, ..., n$, the weight vector is given as $W' = (d'(A_1), d'(A_2), ..., d'(A_n))$, where $A_i (i = 1, 2, ..., n)$ are n elements. Through normalization, normalized weight vector is $W = (d(A_1), d(A_2), ..., d(A_n))$, where W denotes a nonfuzzy number (Chang, 1996).

11.4.4.2   Liou and Wang (1992) Method

This method is principally based on total integral values methodology for the ordering of triangular fuzzy numbers. In that methodology, the total integral value for $\tilde{A} = (l, m, u)$ triangular fuzzy number is calculated as follows (Liou and Wang, 1992):

$$I_T^\alpha(\tilde{A}) = \frac{1}{2}\alpha(m + u) + \frac{1}{2}(1 - \alpha)(l + m)$$

$$= \frac{1}{2}\left[a.u + m + (1 - a).l \right]$$

(11.9)

Equation 11.9 refers to an optimism index, which takes value within the range of $a \in [0, 1]$ closed interval. When the optimism index increases and decreases, the decision maker is assessed as optimist or pessimist, respectively.

11.4.4.3   Abdel-Kader and Dugdale (2001) Method

This method divides a triangular fuzzy number into three categories including full memberships, partial memberships located in the right-hand size, partial memberships located in the left-hand side. As other ordering methods reflect partial memberships located in the left-hand size, this method also proposes a novel methodology that involves all three sides of the triangular fuzzy number in the ordering procedure. The underlying method uses an a optimism index (Abdel-Kader and Dugdale, 2001):

$$\tilde{A}_1 = (l_1, m_1, u_1), ... \quad ... \quad ... \quad ..., \tilde{A}_k = (l_k, m_k, u_k)$$

(11.10)

For fuzzy numbers in Equation 11.10, $S = (l_1, m_1, u_1, \ldots \ldots \ldots, l_k, m_k, u_k)$ and $V(\tilde{A}_k)$ denotes the value of $\tilde{A}_k$ described as follows:

$$V(\tilde{A}_k) = m_k \left\{ (\alpha) \left[\frac{u_k - x_{min}}{x_{max} - x_{min} + u_k - m_k} \right] \right\} + (1-\alpha) \left[1 - \frac{x_{max} - l_k}{x_{max} - x_{min} m_k - l_k} \right] \quad (11.11)$$

$$x_{min} = \inf S, x_{max} = \sup S$$

11.4.4.4 Quadratic Mean Method

The Quadratic Mean Method is commonly used when there exist zero or negative numbers. In this way, an ordering becomes possible when one of the boundaries of the fuzzy number is zero and negative. For $\tilde{A}_k = (l, m, u)$ triangular fuzzy number, the quadratic mean is calculated as

$$K(\tilde{A}_k) = \sqrt{\frac{l^2 + m^2 + u^2}{3}} \quad (11.12)$$

and $K(\tilde{A}_k)$ values are ordered (Göksu and Güngör, 2008).

11.4.4.5 Kwong and Bai (2003) Method

For $\tilde{A}_k = (l, m, u)$ triangular fuzzy number, this method uses the following equation for the ordering of triangular fuzzy numbers (Kwong and Bai, 2003):

$$M_k = \frac{l + 4m + u}{6} \quad (11.13)$$

11.5 DATA ANALYSIS

In the present study, the data being utilized from SME experts were analyzed using a fuzzy AHP approach. Since several weights of selected criteria in the sum of weights vector for the fuzzy AHP approach were found as zero, synthesis values were obtained using the Chang (1996) Method. Additionally, the Quadratic Mean Method was performed to transform fuzzy numbers into real numbers. Table 11.2 presents fuzzy ordering-scale used in the present study following Chang (1996) Method:

The first stage in the fuzzy AHP was the construction of hierarchical framework as illustrated in Figure 11.3.

The pairwise comparison of five main criteria and their sub-criteria were performed with respect to the fuzzy AHP approach. The pairwise comparisons and related abbreviations used in the underlying hierarchy were as follows (Tüzemen and Özdağoğlu, 2007):

TABLE 11.2

The Fuzzy AHP Scale Used in this Study Following Chang (1996)

		Fuzzy AHP Scale	
Saaty (1990)'s Relative Importance Scale	Oral Importance	Triangular Fuzzy Scale	Triangular Fuzzy Opposite Scale
1	Equal importance	(1, 1, 1)	(1/1, 1/1, 1/1)
3	Moderate importance of one over another	(1, 3, 5)	(1/5, 1/3, 1/1)
5	Essential or strong importance	(3, 5, 7)	(1/7, 1/5, 1/3)
7	Very strong importance	(5, 7, 9)	(1/9, 1/7, 1/5)
9	Extreme importance	(7, 9, 9)	(1/9, 1/9, 1/7)

Source: Kaptanoğlu and Özok (2006).

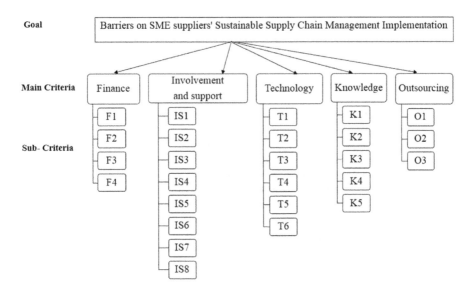

FIGURE 11.3 The hierarchical structure for barriers on SME suppliers' supply chain management implementation.

P_{ij}: The comparative order of the importance of i-th alternative or scale with respect to j-th alternative or scale

W_{ik}: The order of importance of i-th alternative with respect to k-th scale

C_k: k-th scale

WS_{ik}: Sum of weights of i-th alternative with respect to k-th scale

n: The number of alternatives being compared (Table 11.3)

TABLE 11.3

The Main Criteria and Sub-criteria Used in the Hierarchy Process

Main Criteria	Sub-criteria
Finance	**F1:** High cost of hazardous waste disposal
	F2: Financial constraints
	F3: Non-availability of bank loans to encourage green products/processes
	F4: High investments and less return-on-investments
Involvement and support	**IS1:** Lack of training courses/consultancy/institutions to train, monitor/mentor progress specific to each industry
	IS2: Lack of customer awareness and pressure about green supply chain system
	IS3: Lack of corporate social responsibility
	IS4: Lack of involvement of top management in adopting green supply chain management
	IS5: Restrictive company policies towards product/process stewardship
	IS6: Poor supplier commitment/unwilling to exchange information
	IS7: Lack of inter-departmental co-operation in communication
	IS8: Not much involvement in environmental related programs/meetings
Technology	**T1:** Lack of new technology, materials and processes
	T2: Complexity of design to reuse/recycle used products
	T3: Lack of technical expertise
	T4: Lack of human resources
	T5: Lack of effective environmental measures
	T6: Fear of failure
Knowledge	**K1:** Lack of green system exposure to professionals
	K2: Lack of environmental knowledge
	K3: Perception of *out-of-responsibility* zone
	K4: Disbelief about environmental benefits
	K5: Lack of awareness about reverse logistics adoption
Outsourcing	**O1:** Lack of government support to adopt environmentally friendly policies
	O2: Complexity in measuring and monitoring suppliers' environmental practices
	O3: Problem in maintaining environmental suppliers

Source: Govindan, K. et al., *Int. J. Prod. Econ.*, 147, 555–568, 2014.

The second stage involves the procedure of weighting the main criteria and determining the order of importance. Tables 11.4 through 11.6 exhibit the steps being followed for pairwise comparison matrices and the determination of weights.

Step 1: Pairwise comparison
Step 2: Normalization

As shown in Table 11.6, knowledge criterion (27.4%) was found as the most important criterion Turkish SME suppliers encounter during SSCMI process. The corresponding result may be principally interpreted as Turkish SME suppliers' lack of professionals who are experts on environmentalist systems and lack of

TABLE 11.4

Pairwise Comparison Matrix for Main Criteria

	Finance	Involvement and Support	Technology	Knowledge	Outsourcing
Finance	1	1.367	1.144	0.492	1.015
Involvement and support	0.682	1	1.178	0.692	1.529
Technology	0.823	0.979	1	1.510	3.251
Knowledge	2.154	1.585	0.777	1	2.686
Outsourcing	0.989	0.678	0.315	0.436	1
Total	5.648	5.609	4.414	4.130	9.481

TABLE 11.5

Normalization Matrix for Main Criteria

	Finance	Involvement and Support	Technology	Knowledge	Outsourcing	Weight (W)
Finance	0.171	0.240	0.255	0.117	0.106	0.177
Involvement and support	0.121	0.189	0.262	0.164	0.160	0.179
Technology	0.146	0.172	0.240	0.358	0.340	0.251
Knowledge	0.383	0.278	0.173	0.256	0.280	0.274
Outsourcing	0.176	0.119	0.070	0.103	0.112	0.116
Total						1.000

TABLE 11.6

Weights for Main Criteria

Rank	Main Criteria	Weight
1	Knowledge	0.274
2	Technology	0.251
3	Involvement and support	0.179
4	Finance	0.177
5	Outsourcing	0.116

environmental knowledge. Moreover, one can argue that the Turkish SME suppliers may necessitate higher responsibility to gain an understanding on SSCMas and they may not believe in the environmental benefits of the SSCM procedure. Additionally, there may exist a lack of reverse logistic awareness among the Turkish

SME suppliers as well. In that context, the crucial role of sustainability awareness revisits regardless of financial constraints, involvement, and support.

The results of the fuzzy AHP approach reveal the importance of technology on SSCM for the Turkish SME suppliers. Along with the knowledge criterion, technology criterion (25.1%) is another important criterion to move the Turkish SME suppliers one more step forward. The technology criterion cannot be solely considered within the scope of the concept of digitalization. The main concern in the SSCM procedure is the negative outcome of not being close to technological advances, lack of essential materials, a potential perception that reused and/or recycle products may create complexity, lack of personnel and human resources to manage technology, lack of efficient environmental precautions, and fear of failure.

The empirical results indicate that involvement and support (17.9%) and finance criteria (17.7%) have almost analogous importance on the Turkish SME suppliers' SSCM process. Involvement and support criterion may reflect several drawbacks including lack of industrial specific educational courses and consultancy companies, lack of customer awareness, customer pressure and executive management involvement in terms of environmentalist supply chain management, lack of corporate social responsibility, restrictive company policies, lack of connection of information sharing between suppliers, and lack of involvement in programs and/or meetings concentrated on environmental issues. On the other hand, the finance criterion encompasses several difficulties including the increasing costs of hazardous waste disposal, higher amount of bank loans for environmentally-friendly products or procedures, requirement of high investment for environmentalist supply chain, and the relatively small amount of investments, respectively. In a sense, this circumstance puts forward that the Turkish SME suppliers are analogously affected by both internal and external factors in the SSCM process.

The outsourcing criterion (11.6%) can be associated with the lack of governmental support to the adoption of environmentally-friendly policies, the complexity of measurement and monitoring of environmental efforts, and insufficient precautions to protect environmentalist suppliers. The reason of being the least important barrier for the outsourcing criterion can be interpreted as the awareness of the Turkish SME suppliers to improve and/or increase their intrinsic motivation initially in terms of sustainable supply chain.

11.5.1 Consistency Analysis

In the fuzzy AHP, the consistency of experts' judgements throughout the pairwise comparisons should be tested in terms of their characteristics (Taha, 2003). If the consistency degree lies within the range of recommended levels, the decision-making procedure proceeds. Otherwise, the decision maker should reconsider pairwise comparison judgements before proceeding the corresponding analysis (Taylor, 2002). Accordingly, when the consistency ratio is above the level of 0.10, then there exists a significant inconsistency about experts' judgements. When the consistency ratio is equal to or less than 0.10, then pairwise consistency level is considered as acceptable (Taylor, 2002).

There are three main calculations to be made in the consistency analysis: (i) the calculation of sum of weighted vector, (ii) the calculation of consistency index (CI), and (iii) the calculation of consistency ratio (CR = CI/RI where RI denotes a random index). The sum of weighted vector is calculated as follows:

$$W_{11} \begin{bmatrix} P_{11} \\ P_{21} \\ P_{31} \end{bmatrix} + W_{21} \begin{bmatrix} P_{12} \\ P_{22} \\ P_{32} \end{bmatrix} + W_{31} \begin{bmatrix} P_{13} \\ P_{23} \\ P_{33} \end{bmatrix} = \begin{bmatrix} WS_{11} \\ WS_{21} \\ WS_{31} \end{bmatrix} \tag{11.14}$$

Table 11.7 presents the outcome of the sum of weighted vector. After the sum of weighted vector is constructed, each element of the underlying vector is divided into its priority value. Later, their mean values are calculated and designated as λ_{max} in Equation 11.15.

$$\lambda_{max} = \frac{WS_{11}/W_{11} + WS_{21}/W_{21} + WS_{31}/W_{31}}{n} \tag{11.15}$$

The CI is calculated as the following:

$$CI = \frac{\lambda_{max} - n}{n - 1} \tag{11.16}$$

The CI for the main criteria was calculated as 0.05. Later the CR is calculated where RI is a randomly-generated consistency index and takes values with respect to the number of items as shown in Table 11.8.

TABLE 11.7
The Sum of Weighted Vector

	Finance	Involvement and Support	Technology	Knowledge	Outsourcing	Total
Finance	0.170	0.245	0.288	0.135	0.118	0.957
Involvement and support	0.121	0.193	0.296	0.190	0.178	0.979
Technology	0.146	0.175	0.272	0.414	0.378	1.387
Knowledge	0.383	0.284	0.195	0.296	0.312	1.472
Outsourcing	0.176	0.121	0.079	0.119	0.125	0.622

TABLE 11.8
Saaty (1990)'s RI Values

n	1	2	3	4	5	6	7	8	9	10
RI	0.00	0.00	0.58	0.90	1.12	1.24	1.32	1.41	1.45	1.49

For $n = 5$ and the RI $= 1.12$, the CR was calculated as 0.08 implying that the model was consistent (CR $= 0.08 \leq 0.10$). However, in order to make a robust assessment, sub-criteria should be weighted as well. Thus, a general fuzzy AHP ordering of all criteria would be accomplished.

11.5.2 WEIGHTING AND THE ORDER OF IMPORTANCE FOR SUB-CRITERIA OF FINANCE

Tables 11.9 through 11.11 present pairwise comparison, normalization matrices, and weighted ratios for four sub-criteria of finance, respectively. As seen in Table 11.11, financial constraints (F2, 32.2%) was found as the most significant barrier under the main finance criteria. The insufficient share of resources for environmental operations and not to allocate additional financial support may be the main cause of financial constraints. As results indicate, non-availability of bank loans to encourage green products/processes (F3, 29.1%) was found as the second most important sub-criteria for barriers on the Turkish SME suppliers' SSCMI. This result may be expected when the limited share of environmental operations on the existing banking system.

Among others, high investments and less investment returns (F4, 20.7%) and the high cost of hazardous waste disposal (F1, 18.1%) were found as third and fourth most important sub-criteria among financial factors. The consistency analysis for finance sub-criteria was performed and was found as consistent (CI $= 0.06$; RI $= 0.90$; CR $= 0.06$).

TABLE 11.9
Pairwise Comparison Matrix for Sub-criteria of Finance

	F1	F2	F3	F4
F1	1	0.621	0.740	0.710
F2	1.726	1	1.256	1.571
F3	1.442	0.830	1	1.844
F4	1.458	0.669	0.566	1
Total	5.626	3.120	3.562	5.125

TABLE 11.10
Normalization Matrix for Sub-criteria of Finance

	F1	F2	F3	F4	Weight (W)
F1	0.189	0.194	0.203	0.136	0.181
F2	0.302	0.337	0.344	0.301	0.322
F3	0.253	0.259	0.296	0.354	0.291
F4	0.256	0.209	0.155	0.207	0.207
Total					1.000

TABLE 11.11
Weights for Sub-criteria of Finance

Rank	Sub-criteria	Weight
1	F2	0.322
2	F3	0.291
3	F4	0.207
4	F1	0.181

11.5.3 WEIGHTING AND THE ORDER OF IMPORTANCE FOR SUB-CRITERIA OF INVOLVEMENT AND SUPPORT

Tables 11.12 through 11.14 present pairwise comparison, normalization matrices, and weight ratios for eight sub-criteria of involvement and support main criterion, respectively. As shown in Table 11.14, restrictive company protocol to product/process management (IS5, 16.6%) sub-criterion was found as the most important sub-criterion to explain the Turkish SME suppliers' barriers on SSCMI. This result may be interpreted as the reflection of suppliers' main problem cooperating with many different SMES on the restriction of environmentalist efforts due to company policies and intervention by executive managers.

The second most important sub-criteria of involvement and support was lack of customer awareness and pressure about green supply chain management (IS2, 159%). When the crucial role of customers on company policies is considered, their lack of awareness on environmental issues will directly make a significant impact on possible delays on the adoption of supply chain management operations. The third most important involvement and support sub-criteria was the lack of involvement of top management with adopting green supply chain management (IS4, 12.8%). When the first important sub-criteria and this sub-criterion were jointly considered, the crucial role of suppliers' management and managers on present and/or future environmentalist operations is displayed.

TABLE 11.12
Pairwise Comparison Matrix for Sub-criteria of Involvement and Support

	IS1	IS2	IS3	IS4	IS5	IS6	IS7	IS8
IS1	1	1.111	0.763	0.781	0.878	1.054	0.765	0.840
IS2	0.972	1	1.136	1.967	1.051	2.009	1.553	1.393
IS3	1.364	0.911	1	0.925	0.880	1.450	0.947	0.854
IS4	1.358	0.547	1.154	1	1.202	1.528	0.962	1.180
IS5	1.195	1.038	1.212	0.885	1	2.789	2.006	1.822
IS6	1.007	0.529	0.725	0.713	0.377	1	1.371	1.202
IS7	1.395	0.689	1.127	1.137	0.518	0.787	1	1.091
IS8	1.237	0.764	1.251	0.902	0.602	0.860	0.981	1
Total	9.528	6.589	8.368	8.310	6.508	11.477	9.585	9.382

TABLE 11.13
Normalization Matrix for Sub-criteria of Involvement and Support

	IS1	IS2	IS3	IS4	IS5	IS6	IS7	IS8	Weight (W)
IS1	0.112	0.167	0.090	0.0931	0.133	0.091	0.079	0.089	0.106
IS2	0.101	0.162	0.134	0.2344	0.16	0.174	0.161	0.147	0.159
IS3	0.142	0.137	0.128	0.1103	0.134	0.125	0.098	0.09	0.120
IS4	0.141	0.082	0.137	0.1287	0.183	0.132	0.099	0.125	0.128
IS5	0.124	0.156	0.143	0.1054	0.164	0.241	0.208	0.193	0.166
IS6	0.105	0.079	0.086	0.085	0.057	0.093	0.142	0.127	0.096
IS7	0.145	0.103	0.133	0.1355	0.079	0.068	0.112	0.115	0.111
IS8	0.129	0.115	0.148	0.1075	0.091	0.074	0.101	0.114	0.110
Total									1.000

TABLE 11.14
Weights for Sub-criteria of Involvement and Support

Rank	Sub-criteria	Weight
1	IS5	0.166
2	IS2	0.159
3	IS4	0.128
4	IS3	0.120
5	IS7	0.111
6	IS8	0.110
7	IS1	0.106
8	IS6	0.096

In terms of the order of importance, the abovementioned three sub-criteria were followed by the lack of corporate social responsibility (IS3, 12.1%), lack of inter-departmental co-operation in communication (IS7, 11.1%), not much involvement in environmental related programs/meetings (IS8, 11%), lack of training courses/consultancy/institutions to train, monitor/mentor progress specific to each industry (IS1, 10.6%), and finally poor supplier commitment/unwilling to exchange information. The involvement and support sub-criteria were found as consistent with the CI = 0.09, the RI = 1.41, and the CR = 0.06.

11.5.4 WEIGHTING AND THE ORDER OF IMPORTANCE FOR SUB-CRITERIA OF TECHNOLOGY

Tables 11.15 through 11.17 present pairwise comparison, normalization matrices, and weight ratios for technology sub-criteria, respectively. Particularly, the most important sub-criteria for technology main criterion was found as lack of human resources

(T4, 21.5%). In the contemporary era with increasing debates on post-digitalization, companies concentrated only on providing supply to the SMEs have certain difficulties in terms of human resources.

As shown in Table 11.17, the second most important sub-criterion for the technology main criterion was found as the lack of technical expertise (T3, 21.5%), which can be highly correlated with the lack of human resources. The third most important

TABLE 11.15
Pairwise Comparison Matrix for Sub-criteria of Technology

	T1	T2	T3	T4	T5	T6
T1	1	1.306	0.749	0.743	1.499	1.385
T2	0.809	1	0.671	0.781	0.786	0.647
T3	1.431	1.576	1	1.035	1.422	1.218
T4	1.429	1.332	1.015	1	1.481	2.061
T5	0.701	1.360	0.739	0.697	1	1.066
T6	0.765	1.647	0.880	0.528	1.005	1
Total	6.135	8.221	5.054	4.784	7.193	7.377

TABLE 11.16
Normalization Matrix for Sub-criteria of Technology

	T1	T2	T3	T4	T5	T6	Weight (W)
T1	0.174	0.157	0.146	0.153	0.206	0.186	0.170
T2	0.130	0.130	0.131	0.161	0.108	0.087	0.124
T3	0.230	0.190	0.210	0.213	0.196	0.163	0.200
T4	0.230	0.160	0.198	0.222	0.204	0.276	0.215
T5	0.113	0.164	0.144	0.143	0.149	0.143	0.143
T6	0.123	0.198	0.171	0.109	0.138	0.145	0.147
Total							1.000

TABLE 11.17
Weights for Sub-criteria of Technology

Rank	Sub-criteria	Weight (%)
1	T4	0.215
2	T3	0.200
3	T1	0.170
4	T6	0.147
5	T5	0.143
6	T2	0.124

sub-criterion was found as the lack of new technology, materials, and processes (T1, 17%). This result may be interpreted as the Turkish SME suppliers have still to encounter problems on following the advanced technology in terms of SSCM adoption and increasing technological efficiency. These sub-criteria were followed by the fear of failure (T6, 14.7%), the lack of effective environmental measures (T5, 14.3%), and finally complexity of design to reuse/recycle used products (T2, 12.4%), respectively. The consistency analysis for sub-criteria of technology confirms that the model is consistent (CI = 0.04; RI = 1.24; CR = 0.03).

11.5.5 WEIGHTING AND THE ORDER OF IMPORTANCE FOR SUB-CRITERIA OF KNOWLEDGE

Tables 11.18 through 11.20 present pairwise comparison, normalization matrices, and weight ratios for sub-criteria of knowledge main criterion. As shown in Table 11.20, the most important sub-criterion for the knowledge criterion was found as the lack of environmental knowledge (K2, 26.4%), while the second and the third most important sub-criterion was found as the lack of green system exposure to professionals (K1, 23.2%) and disbelief about environmental benefits (K4, 19.6%), respectively.

The abovementioned results show consistency with earlier research concentrated on environmental operations in the existing literature. Accordingly, individuals and

TABLE 11.18
Pairwise Comparison Matrix for Sub-criteria of Knowledge

	K1	K2	K3	K4	K5
K1	1	1.500	1.740	0.706	1.329
K2	0.706	1	2.370	2.165	1.360
K3	0.614	0.445	1	0.808	0.710
K4	1.500	0.492	1.310	1	1.076
K5	0.811	0.786	1.500	0.974	1
Total	4.631	4.223	7.920	5.653	5.475

TABLE 11.19
Normalization Matrix for Sub-criteria of Knowledge

	K1	K2	K3	K4	K5	Weight (W)
K1	0.229	0.349	0.220	0.123	0.239	0.232
K2	0.150	0.251	0.300	0.378	0.245	0.264
K3	0.130	0.103	0.130	0.141	0.128	0.127
K4	0.318	0.114	0.160	0.188	0.194	0.196
K5	0.172	0.183	0.190	0.170	0.194	0.181
Total						1.000

TABLE 11.20
Weights for Sub-criteria of Knowledge

Rank	Sub-criteria	Weight (%)
1	K2	0.264
2	K1	0.232
3	K4	0.196
4	K5	0.181
5	K3	0.127

corporations that have never behaved as environmentally friendly with lower environmental awareness and that do not believe in the reality of environmental issues seriously damage the sustainability process. Unfortunately, the current situation holds for a number of suppliers. The lack of knowledge can be considered as one of the most important barriers on SSCM adoption. The ordering of the remaining sub-criteria were found as the lack of reverse logistics awareness (K5, 18.1%) and the perception of *out-of-responsibility* zone (K3, 12.7%). The consistency of judgements about sub-criteria of knowledge was also found as consistent (CI = 0.08; RI = 1.12; CR = 0.07).

11.5.6 WEIGHTING AND THE ORDER OF IMPORTANCE FOR SUB-CRITERIA OF OUTSOURCING

Tables 11.21 through 11.23 show the output of pairwise comparison, normalization matrices, and weight ratios for sub-criteria of outsourcing. The empirical results reveal that the order of importance for outsourcing was found as the lack of government support to adopt environmentally friendly policies (O1, 48.9%), complexity in measuring and monitoring suppliers' environmental practices (O2, 28.9%), and problems in maintaining environmental suppliers (O3, 22.2%).

The abovementioned findings may exhibit that limited support to suppliers on environmentally friendly operations, as well as perceptions and approaches within the company, seriously damages SSCM process, environmentalist motivations, and further environmental related attempts. The consistency test implies that sub-criteria judgements for outsourcing criterion was found as consistent (CI = 0.01; RI = 0.58; CR = 0.01).

TABLE 11.21
Pairwise Comparison Matrix for Sub-criteria of Outsourcing

	O1	O2	O3
O1	1	1.524	2.549
O2	0.685	1	1.157
O3	0.399	0.893	1
Total	2.084	3.417	4.706

TABLE 11.22
Normalization Matrix for Sub-criteria of Outsourcing

	O1	O2	O3	Weight (W)
O1	0.499	0.436	0.533	0.489
O2	0.317	0.309	0.242	0.289
O3	0.184	0.255	0.226	0.222
Total				1.000

TABLE 11.23
Weights for Sub-criteria of Outsourcing

Rank	Sub-criterion	Weight
1	O1	0.489
2	O2	0.289
3	O3	0.222

TABLE 11.24
The Order of Importance for Main Criteria

Rank	Main Criteria	Weight
1	Knowledge	0.274
2	Technology	0.251
3	Involvement and support	0.179
4	Finance	0.177
5	Outsourcing	0.116

11.5.7 THE GENERAL FUZZY AHP ORDERING FOR ALL MAIN AND SUB-CRITERIA

This sub-section evaluates the main criteria and sub-criteria used in this study to obtain a general assessment on the Turkish SME suppliers' barriers being encountered on SSCMI. Tables 11.24 through 11.26 summarize the order of importance for main criteria and sub-criteria and general assessment of all criteria, respectively.

When all sub-criteria are considered, lack of environmental knowledge (K2, 0.072) was found as the most important barrier, followed by lack of green system exposure to professionals (K1, 0.064), financial constraints (F2, 0.057), and lack of government support to adopt environmentally friendly policies (O1, 0.057). These findings also confirm that the lack of knowledge was the Turkish SME suppliers' most important barrier on SSCMI.

TABLE 11.25
The Order of Importance for Sub-criteria

Finance		Involvement and Support		Technology		Knowledge		Outsourcing	
F1	0.181	IS1	0.106	T1	0.170	K1	0.232	O1	0.489
F2	0.322	IS2	0.159	T2	0.124	K2	0.264	O2	0.289
F3	0.291	IS3	0.120	T3	0.200	K3	0.127	O3	0.222
F4	0.207	IS4	0.128	T4	0.215	K4	0.196		
		IS5	0.166	T5	0.143	K5	0.181		
		IS6	0.096	T6	0.147				
		IS7	0.111						
		IS8	0.110						

TABLE 11.26
The General Order of Importance for All Criteria

Finance		Involvement and Support		Technology		Knowledge		Outsourcing	
F1	0.032	IS1	0.019	T1	0.043	K1	0.064	O1	0.057
F2	0.057	IS2	0.029	T2	0.031	K2	0.072	O2	0.034
F3	0.052	IS3	0.022	T3	0.050	K3	0.035	O3	0.026
F4	0.037	IS4	0.023	T4	0.054	K4	0.054		
		IS5	0.030	T5	0.036	K5	0.050		
		IS6	0.017	T6	0.037				
		IS7	0.020						
		IS8	0.020						

11.6 MANAGERIAL IMPLICATIONS

The eco-friendly discourse has frequently attracted attention in recent years and it has taken its respectable place on many businesses' daily and strategic plans. Both SMEs and large enterprises tend to design their supply chain operations not to cause serious environmental issues, and thus, they gain social, economic, and operational earnings. Companies from different industries involve in value creating efforts within an eco-friendly approach during product/service production, sales, and after-sales services. In other words, both economic and ecological values are created in all stages of product lifecycle, and the concept of sustainability comes into question.

Although emphasis is growing on the concept of environmental awareness as a general approach, there still exist several misunderstandings, lack of knowledge, and experience in Turkey. Accordingly, relatively more eco-friendly companies encounter a variety of barriers on the implementation of SSCM.

11.7 CONCLUSION

This investigation aimed at determining the order of importance for such barriers encountered by the Turkish SME suppliers using a fuzzy AHP approach.

The ordering among the main criteria reveals that lack of knowledge is the most important barrier encountered by the Turkish SME suppliers on sustainable supply chain implementation. At that point, the perception of *out-of-responsibility* zone has a negative impact on the entire sustainability process. Businesses that fail to concentrate on environmental awareness and to adopt the significant contributions of environmental operations on their future performance will not able to find a place in the future, as sustainable supply chains are one of the most important competition tools. On the other side, lack of sufficient professional knowledge by top managers and qualified personnel on environmental issues, though there exist some environmental concerns, will complicate the sustainability process. Other potential barriers that hinder the SSCM process are the lack of awareness on reverse logistics and lack of expertise on the management of environmental systems.

The analysis results indicate that the technology variable is another significant barrier on SSCM. The concept of digitalization, which manifests in all areas of life, emerges as a crucial contributor to supply chain management process. However, businesses that do not adopt technological advances and that experience lack of materials have certain difficulties on the transition to sustainable supply chain. While businesses have a tendency to adopt advanced technology, lack of human resources and technical expertise to manage technology appear to be another barrier on the sustainability of supply chain management.

The Turkish SME suppliers encounter several barriers in terms of involvement and support, including restrictive company policies to product/process management, lack of communication among suppliers, and lack of inter-departmental cooperation. On the other hand, lack of training courses and consultancy to train, monitor, and mentor progress specific to each industry is an important problem. As the significant impact of customer awareness and pressure on environmental responsibility of a business is well-known, lack of such customer pressure and lack of involvement by the top administration in the adoption of sustainability process are recognized as other important barriers on the SSCM process.

The finance criterion is considered as another important barrier for SMEs and their suppliers. Although many businesses that tend to adopt sustainability do not consider the finance criterion as the most important barrier, financial constraints lead to discouraging the process in many aspects. Businesses that have financial constraints and lack the support of bank loans to encourage green products/processes cannot experience the transition of the sustainable supply chain, as the sustainability requires higher amounts of investments. The outsourcing support has also emerged as one of the significant barriers of the Turkish SME suppliers for SSCMI. Within the framework of the SMEs, they highly necessitate governmental support to survive, and the lack of governmental incentives for the adoption of environmentally friendly efforts will significantly hinder the underlying process. Additionally, lack of encouragement of environmentalist suppliers increases the uncertainty for future environmental policies.

In terms of selected sub-criteria; lack of environmental knowledge, lack of green system exposure to professionals, financial constraints, and lack of government support to adopt environmentally friendly policies were found as the most important barriers of the Turkish SME suppliers for SSCMI. These findings generally show consistency with the outcome of the abovementioned main criteria. These results also suggest that the Turkish SME suppliers may concentrate on increasing intra-organizational motivations before considering problems out of the organization. In this circumstance, future efforts should concentrate on operations that tend to increase SMEs and their suppliers' environmental awareness, the level of intra-organizational environmental knowledge, and environmental responsibility.

The number of training courses, consultancy, and institutions to train and monitor progress specific to each industry should be increased. Similarly, further incentives to encourage the training of professionals with expertise on green systems and employment staff with a high level of technology skills may contribute to the transition of the SSCM. In that sense, one further incentive may provide these qualified personnel a relatively high salary. Fortunately, these individuals may also be a solution to the lack of technical expertise. Accordingly, new technology, materials, and processes demanded for the SSCM process should be provided. The budget allocation requirement for the underlying items would be possible by courtesy of suppliers' belief in the contribution of environmental efforts to further performance. On the other hand, involvement of personnel in environmental related programs and meetings will increase their motivation on environmental operations and benefits.

Another important subject is the governmental incentives for businesses that tend to adopt environmentally friendly policies. The governments and the authorized institutions should encourage environmentally friendly SMEs and their suppliers to maintain environmentalist operations. In the existing literature, the ultimate customers' environmentally friendly attitudes were found as to have a significant impact on environmentalist operations of SMEs. Not surprisingly, an increase on customers' environment sensitivity will encourage the adoption of environmental operations and the choice of environmentalist suppliers by the SMEs, as well. As the SMEs having customers with low environment sensitivity tend to move away from environmentally friendly operations and to concentrate on decreasing costs only, ultimately environmentally friendly customers play a crucial role in a successful SSCMI.

The positive impact of environmental operations on the Turkish economy by the courtesy of the Turkish SMEs and their suppliers will probably be more evident in the near future. Therefore, both Turkish SMEs and their suppliers should consider the SSCMI and should involve the concept of sustainability in their future strategies. Thus, the Turkish SMEs respond to future environmental standards that might be encountered and to adopt unforeseeable changes.

REFERENCES

Abdel-Kader, M. G., & Dugdale, D. (2001). Evaluating investments in advanced manufacturing technology: A fuzzy set theory approach. *The British Accounting Review, 33*(4), 455–489.

Adams, F. G., Gabler, C. B., & Landers, V. M. (2017). Green logistics competency: A resource hierarchy view of supply chain sustainability. In *Academy of Marketing Science World Marketing Congress* (pp. 31–40). Singapore: Springer.

Bag, S., Anand, N., & Pandey, K. K. (2017). Green supply chain management model for sustainable manufacturing practices. In *Green Supply Chain Management for Sustainable Business Practice* (pp. 153–189). Hersey, PA: IGI Global.

Ballou, R. H. (2004). *Business Logistics/Supply Chain Management, Planning, Organizing and Controlling the Supply Chain* (5. Edition). Upper Saddle River, NJ: Pearson Prentice Hall.

Baykal, N., & Beyan, T. (2004). *Bulanık mantık ilke ve temelleri* (1. Baskı). Ankara, Turkey: Bıçaklar Kitabevi.

Bazan, E., & Jaber, M. Y. (2018). The development and analysis of environmentally responsible supply chain models. In *Operations and Service Management: Concepts, Methodologies, Tools, and Applications* (pp. 1294–1317). Hersey, PA: IGI Global.

Beamon, B. M. (1999). Designing the green supply chain. *Logistics Information Management*, *12*(4), 332–342.

Bowersox, D. J., Closs, D. J., & Cooper, M. B. (2002). *Supply Chain Logistics Management* (1st ed.). New York: McGraw-Hill.

Buckley, J. J. (1985). Fuzzy hierarchical analysis. *Fuzzy Sets and Systems*, *17*(3), 233–247.

Busse, C., & Mollenkopf, D. A. (2017). Under the umbrella of sustainable supply chain management: Emergent solutions to real-world problems. *International Journal of Physical Distribution & Logistics Management*, *47*(5), 342–343.

Büyüközkan, G., & Vardaloğlu, Z. (2008). Yeşil tedarik zinciri yönetimi. *Lojistik Dergisi*, *8*, 66–73.

Carter, C. R., & Rogers, D. S. (2008). A framework of sustainable supply chain management: Moving toward new theory. *International Journal of Physical Distribution & Logistics Management*, *38*(5), 360–387.

Carvalho, H., Govindan, K., Azevedo, S. G., & Cruz-Machado, V. (2017). Modelling green and lean supply chains: An eco-efficiency perspective. *Resources, Conservation and Recycling*, *120*, 75–87.

Chang, D. (1996). Applications of the extent analysis method on fuzzy AHP. *European Journal of Operational Research*, *95*(3), 649–655.

Chen, P. (2018). Development of highway logistics in china. In *Contemporary Logistics in China* (pp. 95–114). Singapore: Springer.

Chopra, S., & Meindl, P. (2001). *Supply Chain Management: Strategy, Planning, Operation* (1st ed.). Upper Saddle River, NJ: Prentice Hall.

Christopher, M. (2016). *Logistics & Supply Chain Management* (1st ed.). London, UK: Pearson

Dubey, R., Gunasekaran, A., & Papadopoulos, T. (2017b). Green supply chain management: Theoretical framework and further research directions. *Benchmarking: An International Journal*, *24*(1), 184–218.

Dubey, R., Gunasekaran, A., Childe, S. J., Papadopoulos, T., & Fosso Wamba, S. (2017a). World class sustainable supply chain management: Critical review and further research directions. *The International Journal of Logistics Management*, *28*(2), 332–362.

Dweekat, A. J., Hwang, G., & Park, J. (2017). A supply chain performance measurement approach using the internet of things: Toward more practical SCPMS. *Industrial Management & Data Systems*, *117*(2), 267–286.

Fredendall, L. D., & Hill, E. (2016). *Basics of Supply Chain Management* (1st ed.). Boca Raton, FL: CRC Press.

Gilbert, S. (2001). *Greening Supply Chain: Enhancing Competitiveness through Green Productivity*. *16*, 1–6. Tokyo, Japan: Asian Productivity Organization.

Göksu, A., & Güngör, İ. (2008). Bulanık analitik hiyerarşik proses ve üniversite tercih sıralamasında uygulanması. *Süleyman Demirel Üniversitesi İktisadi ve İdari Bilimler Fakültesi Dergisi*, *13*(3), 195–204.

Govindan, K., & Soleimani, H. (2017). A review of reverse logistics and closed-loop supply chains: A journal of cleaner production focus. *Journal of Cleaner Production*, *142*, 371–384.

Govindan, K., Kaliyan, M., Kannan, D., & Haq, A. N. (2014). Barriers analysis for green supply chain management implementation in Indian industries using analytic hierarchy process. *International Journal of Production Economics, 147,* 555–568.

Green, K., Barbara, M., & Steve, N. (1996). Purchasing and environmental management: Interactions, policies and opportunities. *Business Strategy and the Environment, 5*(3), 188–197.

Güner, S., & Coskun, E. (2013). Küçük ve orta ölçekli işletmelerin çevre algıları ve alıcı-tedarikçi ilişkilerinin çevreci uygulamalar üzerindeki etkisi. *Ege Akademik Bakış, 13*(2), 151–167.

Güngör, İ., & İşler, D. B. (2012). Analitik hiyerarşi yaklaşımı ile otomobil seçimi. *Uluslararası Yönetim İktisat ve İşletme Dergisi, 1*(2), 21–33.

Heying, A., & Sanzero, W. (2009). A case study of walmart's "green" supply chain management. MGT 520.

Ho, J. C., Shalishali, M. K., Tseng, T., & Ang, D. S. (2009). Opportunities in green supply chain management. *The Coastal Business Journal, 8*(1), 18–31.

Hugos, M. H. (2018). *Essentials of Supply Chain Management* (4th ed.). Hoboken, NJ: John Wiley & Sons.

Jiao, Z. L., Lee, S. J., Wang, L., & Liu, B. L. (2017). *Contemporary Logistics in China.* Singapore: Springer.

Johnson, M., Redlbacher, F., & Schaltegger, S. (2018). Stakeholder engagement for corporate sustainability: A comparative analysis of b2c and b2b companies. *Corporate Social Responsibility and Environmental Management, 25*(4), 659–673.

Kaptanoğlu, D., & Özok, A. H. (2006). Akademik performans değerlendirilmesi için bir bulanık model. *İTÜ Dergisi, 5*(1), 193–204.

Kehoe, D., & Boughton, N. (2001). Internet based supply chain management: A classification of approaches to manufacturing planning and control. *International Journal of Operations & Production Management, 21*(4), 516–525.

Kumar, S., Teichman, S., & Timpernagel, T. (2012). A green supply chain is a requirement for profitability. *International Journal of Production Research, 50*(5), 1278–1296.

Kwong, C. K., & Bai, H. (2003). Determining the importance weights for the customer requirements in QFD using a fuzzy AHP with an extent analysis approach. *IIE Transactions, 35*(7), 619–626.

Lee, H. L., & Billington, C. (1992). Managing supply chain inventory: Pitfalls and opportunities. *Sloan Management Review, 33*(3), 65–74.

Lee, S. Y. (2008). Drivers for the participation of small and medium-sized suppliers in green supply chain initiatives. *Supply Chain Management: An International Journal, 13*(3), 185–198.

Levi, D. S., Kaminsky, P., & Levi, E. S. (2003). *Designing and Managing the Supply Chain: Concepts, Strategies, and Case Studies* (2nd ed.). New York: McGraw-Hill.

Liou, T. S., & Wang, M. J. J. (1992). Ranking fuzzy numbers with integral value. *Fuzzy Sets and Systems, 50*(3), 247–255.

Luthra, S., Govindan, K., Kannan, D., Mangla, S. K., & Garg, C. P. (2017). An integrated framework for sustainable supplier selection and evaluation in supply chains. *Journal of Cleaner Production, 140,* 1686–1698.

Lyon, T. P., & Maxwell, J. W. (2008). Corporate social responsibility and the environment: A theoretical perspective. *Review of Environmental Economics and Policy, 2*(2), 240–260.

Lyon, T., & Maxwell, J. W. (2013). On the profitability of corporate environmentalism. *The Oxford Handbook of Managerial Economics* (1st ed.). Oxford, UK: Oxford University Press.

Mahmoodzadeh, S., Shahrabi, J., Pariazar, M., & Zaeri, M. (2007). Project selection by using fuzzy AHP and TOPSIS technique. *International Journal of Human and Social Sciences, 1*(3), 135–140.

Mathiyazhagan, K., Datta, U., Singla, A., & Krishnamoorthi, S. (2018). Identification and prioritization of motivational factors for the green supply chain management adoption: Case from Indian construction industries. *Opsearch, 55*(1), 202–219.

Mirghafoori, S. H., Andalib, D., & Keshavarz, P. (2017). Developing green performance through supply chain agility in manufacturing industry: A case study approach. *Corporate Social Responsibility and Environmental Management, 24*(5), 368–381.

Morana, J. (2013). *Sustainable Supply Chain Management* (1st ed.). Hoboken, NJ: John Wiley & Sons.

Mumtaz, U., Ali, Y., & Petrillo, A. (2018). A linear regression approach to evaluate the green supply chain management impact on industrial organizational performance. *Science of the Total Environment, 624*, 162–169.

Murphy Jr., P. R., & Donald, F. W. (2004). *Contemporary Logistics* (8th ed.). Upper Saddle River, NJ: Pearson Prentice Hall.

New, S., & Westbrook, R. (Eds.). (2004). *Understanding Supply Chains: Concepts, Critiques, and Futures* (1st ed.). Oxford, UK: Oxford University Press.

Reefke, H., & Sundaram, D. (2017). Key themes and research opportunities in sustainable supply chain management: Identification and evaluation. *Omega, 66*, 195–211.

Saaty, T. L. (1990). How to make a decision: The analytic hierarchy process. *European Journal of Operations Research, 48*(3). 9–26.

Saaty, T. L. (2000). *Fundamentals of Decision Making and Priority Theory* (1st ed.). Pittsburg, PA: RWS Publications.

Saaty, T. L., & Vargas, L. G. (2012). *Models, Methods, Concepts & Applications of the Analytic Hierarchy Process* (2nd ed.). Singapore: Springer.

Sarkis, J. (2003). A strategic decision framework for green supply chain management. *Journal of Cleaner Production, 11*(4), 397–409.

Sarkis, J. (2018). Sustainable and green supply chains: Advancement through resources, conservation and recycling. *Resources, Conservation and Recycling, 134* (2018), A1–A3.

Şengül, Ü., Eren, M., Eslemian Shiraz, S., Gezder, V., & Şengül, A. B. (2013). Fuzzy TOPSIS method for ranking renewable energy supply systems in Turkey. *Renewable Energy, 75*, 617–625. doi:10.1016/j.renene.2014.10.045.

Şengül, Ü., Miraç, E., & Shıraz, S. E. (2012). Bulanık AHP ile belediyelerin toplu taşıma araç seçimi. *Erciyes Üniversitesi İktisadi ve İdari Bilimler Fakültesi Dergisi, 40*, 143–165.

Simpson, D. F., & Power, D. J. (2005). Use the supply relationship to develop lean and green suppliers. *Supply Chain Management: An International Journal, 10*(1), 60–68.

Singla, A., Ahuja, I. P. S., & Sethi, A. P. S. (2017). The effects of demand pull strategies on sustainable development in manufacturing industries. *International Journal of Innovations in Engineering and Technology, 8*(2), 27–34.

Skjoett-Larsen, T. (2000). European logistics beyond 2000. *International Journal of Physical Distribution & Logistics Management, 30*(5), 377–387.

Song, W., Chen, Z., Wang, X., Wang, Q., Shi, C., & Zhao, W. (2017). Environmentally friendly supplier selection using prospect theory. *Sustainability, 9*(3), 377.

Sroufe, R. P., & Melnyk, S. A. (2017). *Developing Sustainable Supply Chains to Drive Value* (1st ed.). New York: Business Expert Press LLC.

Stadtler, H., & Kilger, C. (2002). *Supply Chain Management and Advanced Planning* (2nd ed.). Singapore: Springer.

Taha, H. A. (2003). *Operations Research: An Introduction*. Fayetteville, NC: Prentice-Hall.

Tao, F., Cheng, Y., Zhang, L., & Nee, A. Y. (2017). Advanced manufacturing systems: Socialization characteristics and trends. *Journal of Intelligent Manufacturing, 28*(5), 1079–1094.

Taylor, B. W. (2002). *Introduction to Management Science* (1st ed.). Hoboken, NJ: Pearson Education Inc.

Tüzemen, A., & Özdağoğlu, A. (2007). Doktora öğrencilerinin eş seçiminde önem verdikleri kriterlerin analitik hiyerarşi süreci yöntemi ile belirlenmesi. *Atatürk Üniversitesi İktisadi ve İdari Bilimler Dergisi*, *21*(1), 215–232.

Vachon, S., & Klassen, R. D. (2006). Green project partnership in the supply chain: The case of the package printing industry. *Journal of Cleaner Production*, *14*(6–7), 661–671.

Van Hoek, R. I. (1999). From reversed logistics to green supply chains. *Supply Chain Management: An International Journal*, *4*(3), 129–135.

Van Laarhoven, P. J. M., & Pedrycz, W. (1983). A fuzzy extension of Saaty's priority theory. *Fuzzy Sets and Systems*, *11*(1–3), 229–241.

Yun, Y., & Chuluunsukh, A. (2018). Environmentally-friendly supply chain network with various transportation types. *Journal of Global Tourism Research*, *3*(1), 17–24.

Zhu, Q., & Sarkis, J. (2004). Relationships between operational practices and performance among early adopters of green supply chain management practices in Chinese manufacturing enterprises. *Journal of Operations Management*, *22*(3), 265–289.

Zhu, Q., Sarkis, J., Lai, K. H., & Geng, Y. (2008). The role of organizational size in the adoption of green supply chain management practices in China. *Corporate Social Responsibility and Environmental Management*, *15*(6), 322–337.

12 Integrating Sustainability into Humanitarian Procurement
A Prioritization of Barriers and Enablers

Mohammad Hossein Zarei,
Ruth Carrasco-Gallego, and Stefano Ronchi

CONTENTS

12.1 INTRODUCTION

Humanitarian supply chain and logistics have been gaining momentum considerably in academic literature within the past decade. Humanitarian supply chain and logistics can be defined as "the process of planning, implementing and controlling the efficient, cost-effective flow and storage of goods and materials, as well as related information, from the point of origin to the point of consumption for the purpose of alleviating the suffering of vulnerable people" (Thomas & Kopczak, 2005, p. 2). While having some similarities with commercial supply chains, they are fundamentally different in several perspectives (Kunz et al., 2017): First, the timing and the location of final customers, called beneficiaries in humanitarian supply chains, are unpredictable. Second, humanitarian supply chain management is often subject to high time pressure. Third, beneficiaries do not have a strong voice in stating their needs.

All these characteristics relate to purchasing and supply management function in humanitarian operations. The procurement department has to deal with the unpredictability of demand by estimating the location and storing prepositioned inventory to be used for emergency first response. The department should also estimate and supply necessary items, ranging from food to shelter, essential living aid, and medical items, needed by the impacted population. Therefore, purchasing and supply management function performs atypical activities different from its counterparts in commercial supply chains. In order to address these peculiarities, the academic literature has focused on different aspects of humanitarian procurement hitherto.

A stream of literature is devoted to improving the bidding and contracting in humanitarian procurement. Falasca and Zobel (2011) proposed a two-stage procurement decision model to address the uncertainty in disaster relief operations. Ertem et al. (2010) proposed a procurement auctions-based framework for humanitarian procurement that involved announcement, construction, and evaluation of bids. In their next study (Ertem et al., 2012), the authors focused on the announcement phase of the framework and introduced substitution and partial fulfillment options to diversify suppliers and include suppliers with fewer inventories. Bagchi et al. (2011) proposed an auction mechanism in supplying food for emergencies that decreased the possibility of gaming through a uniform price option. Their mechanism led to higher participation of actors and increased delivered food aid volumes. The study of Trestrail et al. (2009) outlined a mixed-integer program decision tool aimed at enhancing ocean carrier and food supplier bid pricing strategy. Finally, Balcik and Ak (2014) modeled the supplier selection problem for a humanitarian organization with framework agreements. The results show that the supplier selection and subsequent costs are sensitive to contract agreement terms in situations with high-impact disasters.

The second stream of work on humanitarian procurement has focused on the issue of inter-organizational cooperation for purchasing humanitarian items. Vaillancourt (2017) studied the procurement consolidation in humanitarian supply chains and concluded that inter-organizational collaboration for consolidated purchasing leads to reduced costs and higher quality of procured items. Herlin and Pazirandeh (2012) studied the impact of nonprofit and humanitarian organizations on shaping the market and the dominance over suppliers. Later (Herlin & Pazirandeh, 2015), they expanded their study to provide a framework for successful cooperative purchasing. Interestingly, they also addressed the pitfalls that might impede cooperative inter-organizational purchasing in humanitarian context. In another study (Pazirandeh & Herlin, 2014), they approached the cooperative purchasing from the buyer's perspective and studied how cooperative purchasing impacts on buyer's purchasing power. The study generated insights about the reasons that cooperative purchasing might fail due to inappropriate procurement strategy. More recently, Nikkhoo et al. (2018) addressed the issue of coordination of relief items procurement through a quantity flexibility contract in multi-echelon humanitarian supply chains. Their proposed model decreases losses of relief procurement and improves the satisfaction level of the affected area.

The third stream in humanitarian procurement literature focuses on sustainable procurement. This stream, although currently limited, is growing. Most of the works in this stream address the social pillar of sustainability, especially through ethical procurement. Schultz and Søreide (2008) delved into the problem of corruption in humanitarian emergency procurement. Their study revealed that internal agency control mechanisms, conflict-sensitive management, and the need for common systems among operators can be appropriate mechanisms to prevent corruption. The study of Wild and Zhou (2011) explicated ethical procurement strategies for humanitarian aid organizations. The results of their study expressed that concerns about ethical risks in humanitarian supply chains are different from commercial ones. The research about environmental sustainability in humanitarian procurement is even more limited as compared to social pillar. The work of Van Kempen et al. (2017) is among the few studies that explicitly considers all aspects of sustainability in humanitarian procurement. They conducted life cycle sustainability assessment to compare different humanitarian procurement strategies in terms of sustainability. Their results show that local sourcing is more environmentally and socially sustainable than international sourcing. Local sourcing contributes to social sustainability by supporting and empowering local societies while it also contributes to environmental sustainability by decreasing the carbon emissions emanating from international logistics. Looking at local procurement from that perspective, the study of Matopoulos et al. (2014) and more recently Piotrowicz (2018) investigate and propose deploying local sourcing in humanitarian supply chains. Other topics covered by previous research on humanitarian procurement are: vehicle procurement policy (Eftekhar et al., 2014), e-procurement (Walker & Brammer, 2012), joint procurement and inventory decision (Hu et al., 2017; Torabi et al., 2018), and supplier partner selection (Venkatesh et al., 2018).

From the review of literature on humanitarian procurement several gaps are identifiable. First, the literature around the topic of humanitarian procurement is still at its early stages. While there is a rich amount of literature about commercial procurement, the insights are often inapplicable for humanitarian organizations due to fundamental differences (Park et al., 2018; Venkatesh et al., 2018). Within the topic of humanitarian procurement, the issues of bidding and contracting are probably the leading ones with higher number of publications and citations while other issues such as sustainability have received considerably less attention. As for the methodology, the majority of reviewed papers adopted a quantitative approach to model and solve a specific procurement situation, while qualitative methods that study humanitarian specificities (barriers and enablers) in procurement and how to address them are missing. Finally, the literature of humanitarian procurement overlooks sustainability. However, previous research highlighted that sustainability in general, and specifically environmental sustainability, has to be considered for the future research in the area of humanitarian procurement (Sarkis et al., 2012; Abrahams, 2014).

This chapter tries to address the aforementioned gaps by integrating sustainability into humanitarian procurement. It aims at identifying the humanitarian procurement specificities in form of barriers and enablers and distinguish the most impacting ones. In doing so, it contends to respond to the following research question:

What are the most important enablers and barriers for integrating sustainability into humanitarian procurement?

The rest of this chapter is as follows. Section 12.2 reviews the concept of sustainable procurement and how it can be linked to humanitarian procurement. Section 12.3 introduces group analytic hierarchy process (AHP) as the methodology of the study. Section 12.4 describes the application of the methodology by synthesizing and ranking enablers and barriers of sustainable procurement in humanitarian supply chains. Section 12.5 discusses the findings and provides suggestions to tackle the barriers. Finally, Section 12.6 concludes the chapter, provides the limitations of the study, and suggestions for future research.

12.2 SUSTAINABLE PROCUREMENT IN HUMANITARIAN SUPPLY CHAINS

After the review of humanitarian procurement in previous section, this section explores the integration of sustainability in humanitarian procurement. It reviews the literature on the sustainable procurement both from academic and practitioners' perspectives and then discusses the importance and the need for sustainable procurement in humanitarian sector.

In this chapter, we take the notion of triple bottom line (Elkington, 1998) for sustainability including social, environmental, and economical sustainability, also known as the 3Ps: people, profit, and planet (Kleindorfer et al., 2005). Increasingly, organizations and academics realize that sustainable procurement is the key to addressing sustainability issues along the supply chains. Walker et al. (2012, p. 201) defined sustainable procurement as "the pursuit of sustainable development objectives through the purchasing and supply process. Sustainable procurement is consistent with the principles of sustainable development, such as ensuring a strong, healthy and just society, living within environmental limits, and promoting good governance." The literature review of Schneider and Wallenburg (2012) on the implementation of sustainable sourcing showed that the purchasing and supply management function needs to change its internal and external relationships in order to duly adapt and implement sustainable sourcing. In the same vein, the review study of Hoejmose and Adrien-Kirby (2012) identified socially and environmentally responsible procurement as salient managerial issues that managers need to address in the twenty-first century.

Probably the closest studies about sustainable procurement to this study are Walker and Brammer (2009), Brammer and Walker (2011), and Hasselbalch et al. (2015). The study of Walker and Brammer (2009) investigated sustainable procurement in the UK public sector and found that there is a significant variation across different public organizations in the implementation of sustainable procurement. Brammer and Walker (2011) investigated the same problem but in an international context of public bodies. Both studies found that cost and lack of management support can be the main barriers on the way of sustainable procurement. Similarly, Hasselbalch et al. (2015) studied the barriers of sustainable procurement in the United Nations and developed a framework which included different classes of barriers. These works offer valuable insights on the identification of barriers and enablers of sustainable procurement in public sector, which is different from humanitarian environment. Moreover, they have not prioritized the identified barriers according to their importance and potential for improvement.

From practical point of view, the recent release of the ISO 20400 on sustainable procurement shows the prominence of sustainable procurement implementation for organizations. The ISO 20400 defines sustainable procurement as "the procurement that has the most positive environmental, social and economic impacts on a whole life basis, which involves the sustainability aspects related to the goods or services and to the suppliers along the supply chains and contributes to the achievement of organizational sustainability objectives and goals and to sustainable development in general" (International Organization for Standardization, 2017). According to the standard, sustainable procurement should not be viewed as an abstract idealistic goal, but as a reasonable pragmatic aim. In order to integrate sustainability into the procurement process, it proposes five cyclical steps: (1) planning (preparing a sustainable sourcing strategy), (2) integrating sustainability requirements in the specification, (3) selecting suppliers (awarding contract), (4) managing the contract sustainably, and (5) reviewing and learning from the contract (evaluating and improving sustainability performance). The process is shown in Figure 12.1.

Humanitarian context is different from commercial supply chains in many ways. Looking to sustainability, the social aspect (people) is present in humanitarian supply chains by default. While commercial organizations focus on making profit, which sometimes lead to negligence of the social pillar, humanitarian supply chain's prime objective is alleviating the suffering of vulnerable people (Thomas & Kopczak, 2005). The bold social aspect of humanitarian context is at odds with the sustainability trend in commercial supply chain research, where environmental sustainability (planet) is more mature, partly because it contributes to waste reduction and profit generation. Therefore, it is imperative that research pays more attention to environmental sustainability in humanitarian supply chains (Sarkis et al., 2012). In this paper, we focus on both social and environmental bottom lines in humanitarian procurement.

There are different views towards humanitarian procurement. Some authors (Oloruntoba & Gray, 2006) argue that humanitarian supply chains should be lean on the upstream of the supply chain and agile on the downstream with a decoupling point in between in order to satisfy the beneficiaries' need in the most efficient way possible. They consider the leanness as the "value-adding processes unencumbered by waste." This view is in accordance with the previous stream of research in commercial supply chains, which argue "lean is green" (Colicchia et al., 2017) due to elimination of waste. On the other hand, the study of Matopoulos et al. (2014) considered procurement as no different from other humanitarian supply chain stages with similar characteristics and importance.

Taking one view or the other, integration of sustainability into humanitarian procurement seems imperative. Historic antecedents of humanitarian unsustainable operations suggest that if appropriate sustainable measures were taken into account during procurement, the environmental or social adverse impacts could have been avoided in subsequent stages of supply chain. For example, mosquito nets are common humanitarian items procured and provisioned to the areas impacted by a disaster, such as refugee camps. These mosquito nets are treated with insecticides to repel mosquitos. After the nets are torn, sometimes they are used as fishnets by local population; thus, releasing the hazardous chemicals into aquatic life

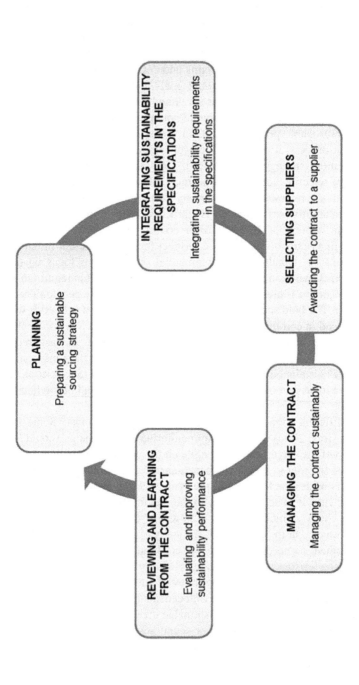

FIGURE 12.1 Integrating sustainability into the procurement process. (From International Organization for Standardization, ISO 20400:2017 Sustainable procurement: Guidance, 52, 2017.)

(Minakawa et al., 2008). Another example with regards to social sustainability is provisioning of cook stoves to beneficiaries by humanitarian organizations without proper consideration about the fuel. For the cook stoves using charcoal and wood fuel, if the fuel is not supplied properly and sufficiently, beneficiaries gravitate to the ecological environment around refugee camps and temporary houses to collect wood. It exposes them, especially women and children, to high risks of gender-based violence, wildlife attack, kidnapping, and human trafficking (Barbieri et al., 2017). If these issues had been considered through sustainable procurement, the drastic consequences could be avoided.

Moreover, since procurement of aid items is at the start of the humanitarian supply chain, there are more opportunities for integrating sustainability as compared to later supply chain stages. When an aid item with harmful material in the packaging has already been purchased, distributed, and disposed in the field by beneficiaries, there is little to be done to deal with the waste in a region impacted by disaster and limited recycling facilities. While proper consideration of such issues through procurement, for example by purchasing aid items with biodegradable packaging, could address the root of the problems.

However, with the volatility and unpredictability inherent in humanitarian supply chains, implementation of sustainable procurement is challenging. The specificities of humanitarian context impose that integrating sustainability into procurement requires different solutions from commercial procurement. Therefore, it is important to identify the contextual barriers on the way of sustainable procurement in humanitarian supply chain and prioritize the most important ones. This is the objective of this chapter: to elicit the ideas of humanitarian experts about barriers and enablers of sustainable procurement and rank them according to their potential opportunity for improvement, so that future research and practice efforts for integrating sustainability into humanitarian procurement focus on addressing the most pressing barriers.

12.3   METHODOLOGY

The methodology of this paper has two parts. First, in order to identify the main barriers and enablers of sustainable procurement in humanitarian supply chains, the ideas of a group of humanitarian experts were elicited through qualitative questionnaires. Second, when the ideas are gathered and complied, an analytic hierarchy process (AHP) was used to prioritize the identified enablers.

AHP is a suitable prioritization method that has widely been used in the literature of operations management and research for different aims. It has shown to be a viable tool in humanitarian logistics and has been applied for ranking critical success factors (CSFs) of humanitarian relief organizations (Celik & Taskin Gumus, 2018) and prioritizing emergency shelter areas (Trivedi & Singh, 2017a, 2017b), *inter alia*. Similarly, from the perspective of sustainable supply chain literature, AHP is strongly justified as a fitting prioritization method. The recent literature review of Maditati et al. (2018) critically reviews the most cited papers and provides a stepwise guideline for future research in green supply chain management. The guideline expresses that developing AHP-based ranking for drivers and barriers of sustainable supply chain should be a main future research agenda. Moreover, group decision-making through the application of group AHP is suggested for addressing multi-faceted and complex problems in new areas of research (Dong & Cooper, 2016). Since sustainable humanitarian procurement is a nascent area of research with little extant knowledge, combining the ideas of experts by group AHP prevents individual subjectivity to creep into the decision-making process and ensures obtaining a meaningful and reliable ranking. The following subsections describes AHP and group AHP in a nutshell. The application of the methodology to a real-world case is described in detail in the Section 12.4.

12.3.1 ANALYTIC HIERARCHY PROCESS (AHP)

AHP is a multiple-criteria decision-making (MCDM) method introduced by Saaty in 1972, which breaks down a problem into several constituent sub-problems, solves each sub-problem, and then aggregates them to solve the main problem (Saaty, 1990). The results offer a prioritized list of criteria from which the best one(s) can be selected. The main steps of conducting AHP are as follows: (1) *defining the goal and hierarchy,* (2) *making pairwise comparisons,* (3) *normalization and synthetization,* and (4) *consistency check.* In the following, a brief description of the procedure of AHP is provided from Zarei and Wong (2014):

1. *Defining the goal hierarchy:* First, the objective of conducting AHP should be defined. Then, the criteria that satisfy the objective and the alternatives should be defined.
2. *Making pairwise comparisons:* Pairwise comparison is the heart of AHP, making it distinct from a mere scoring technique. The decision maker makes pairwise comparisons between each possible pair of criteria and then between each pair of alternatives. The standard preference scale used for AHP is shown in Table 12.1. For example, if criterion a is strongly preferred to criterion b, then this comparison receives a value of 5. The comparisons are then recorded in a matrix (Equation 12.1), where a_{ij} is the result of the pairwise comparison between criterion i and j. It is noteworthy that after a is compared against b, the result of the reverse comparison (b against a) is simply the inverted value of the first comparison.

$$A = \begin{bmatrix} 1 & a_{12} & \cdots & a_{1n} \\ a_{21} & \cdots & a_{ij} & \cdots \\ \cdots & a_{ji} = 1/a_{ij} & \cdots & \cdots \\ a_{n1} & \cdots & \cdots & 1 \end{bmatrix} \tag{12.1}$$

3. *Synthetization:* After completing all the pairwise comparisons, the pairwise comparison matrices of criteria and alternatives are normalized by dividing the value of each column by its corresponding sum and then taking the average of the values in each row. The obtained matrices

TABLE 12.1

Preference Scale for Pairwise Comparisons

Preference Level	Numeric Value
Equally preferred	1
Equally to moderately preferred	2
Moderately preferred	3
Moderately to strongly preferred	4
Strongly preferred	5
Strongly to very strongly preferred	6
Very strongly preferred	7
Very strongly to extremely preferred	8
Extremely preferred	9

Source: Ishizaka, A., and Labib, A., *Expert Syst. Appl.*, doi:10.1016/j.eswa.2011.04.143, 2011.

(in form of one-column matrices), called preference matrices, show the priorities of criteria and alternatives. This ranking of decision alternatives is referred to as synthetization. Then, the preference matrices of criteria and alternatives are multiplied to determine the overall ranking of alternatives.

4. *Consistency check:* Since the decision maker might make inconsistent pairwise comparisons, it is imperative to check the consistency as follows:
   - Multiply the unnormalized pairwise comparison matrix by the matrix resulted from synthetization.
   - Divide each of the obtained values by its corresponding weight derived from the synthetization matrix and then sum up all the values.
   - Divide the sum by the number of criteria and call it λ.
   - Calculate the consistency index, CI, using Equation 12.2.

$$CI = (\lambda - n)/(n-1) \qquad (12.2)$$

   - If $CI = 0$, the decision-making process is perfectly consistent; otherwise, determine the level of consistency using Equation 12.3. In that equation, RI is the random index that is obtained from Table 12.2.

$$Consistency\,Level = CI\,/\,RI \qquad (12.3)$$

   - If the value of consistency level is 0.1 or less, it means that the pairwise comparisons are acceptably consistent. For greater values, the pairwise comparisons should be made again to achieve an acceptable degree of consistency.

TABLE 12.2
RI Values for *n* Items Being Compared

n	2	3	4	5	6	7	8	9	10
RI	0	0.58	0.90	1.12	1.24	1.32	1.41	1.45	1.51

Source: Ishizaka, A., and Labib, A., *Expert Syst. Appl.*, doi:10.1016/j.eswa.2011.04.143, 2011.

12.3.2 GROUP AHP

AHP has shown to be an effective tool for group decision-making by preventing personal bias during individual decision-making (Ishizaka & Labib, 2011). Group AHP is an extension of AHP that combines the preferences and pairwise comparisons elicited from several participants. There are different methods proposed for aggregation of individuals' ideas. Since in the case of this chapter, the respondents were making pairwise comparisons independent and separate from each other, aggregation of individual priorities using geometric mean is suitable (Forman & Peniwati, 1998).

If the ideas of individuals are believed to have different importance, group AHP makes it possible to weigh each individual's idea. Weighting ideas and aggregating them using geometric mean is called *weighted geometric mean method*, which is the most common method for aggregation of preferences in the literature of AHP (Xu, 2000). It is shown that if the individual judgments are fairly consistent (based on consistency evaluation described in Section 12.3.1), the resulting group AHP using weighted geometric mean is also acceptably consistent, dismissing the need to perform another consistency check for the group AHP (Xu, 2000). Next section introduces the application of group AHP for prioritizing the barriers on the way of sustainable humanitarian procurement.

12.4 IDENTIFYING AND PRIORITIZING ENABLERS AND BARRIERS

The data for this study was collected during a conference on humanitarian logistics in the United Nations City, Copenhagen and from an ongoing action research project with an international humanitarian organization. First, the eligibility criteria for the inclusion of respondents were defined by the researchers. Since the research objective was to identify real-world barriers and enablers that humanitarian organizations encounter, the researchers decided that the respondent community should be practitioners working in the procurement department of a humanitarian organization with at least five years of experience. Moreover, the pertaining organizations from which the respondents were selected had to have sustainability in their agenda, reflected either through annual reports, organizational website, or any other document that is publicly available.

Next, the researchers developed a project description explaining the objective and scope of the study. The objective was identifying barriers and enablers of sustainability in humanitarian procurement. The respondents, based on their experience, had to express the barriers and enablers that impact on integrating sustainability into humanitarian procurement. The researchers identified several respondents with the aforementioned eligibilities and approached them with the project objective and description. Before drawing their ideas, a description of sustainability was provided based on the triple bottom line to ensure that a unique understanding of sustainability exists among all the respondents. Out of several eligible people approached during the conference, three practitioners provided their insights and later participated in pairwise comparison for group AHP. The first respondent was the procurement assistant of a large international humanitarian organization with seven years of work experience as a contract manager and procurement assistant. The second respondent was the logistics and fleet manager at a European delegation of a large humanitarian organization with more than 12 years of work experience. Previous to his current position, he worked as the procurement officer in another international humanitarian organization. The third respondent was the procurement and logistics specialist at the country office of an international humanitarian organization in the Middle-East with around six years of work experience. The diversity of geographical regions and work descriptions of the respondents ensured collection of an inclusive list of barriers and enablers.

After gathering the ideas, the barriers and enablers were synthesized. The barriers and enablers stated by respondents were highly alike, which showed that the ideas of the respondents were convergent, even though they were from different humanitarian organizations. Similar barriers that were articulated differently were compiled. Finally, a total of 20 barriers and 8 enablers were identified. The researchers categorized them under 6 categories based on the similarity and relevance. The list of identified barriers and enablers are presented in Table 12.3. They will be discussed in more detail in Section 12.5.

TABLE 12.3
List of Identified Barriers and Enablers for Sustainable Humanitarian Procurement

Category	Barriers	Enablers
Local procurement	Higher frequency of issues such as child labor, unfair working times, and sweat shops in developing countries	Reduced transportation emissions through local procurement
	Damaged infrastructure in countries impacted by disasters and crises hampering local procurement	Empowering local communities and local capacity building through local procurement
	Lower quality of locally produced products compared to offshore buying	Better addressing of local communities' needs through local procurement
	Lower production capacity of local suppliers	
	Higher price of local suppliers	
Humanitarian context	Time-pressure and urgency overshadowing on sustainable procurement	
	Unpredictability of demand and location	
	Volatility of humanitarian context requires more packaging (for example for airdrops)	

(Continued)

TABLE 12.3 (*Continued*)
List of Identified Barriers and Enablers for Sustainable Humanitarian Procurement

Category	Barriers	Enablers
Intra-organizational	Limited managerial support and will for integrating sustainability into procurement	Recent introduction of ISO 20400 on sustainable procurement
	Limited knowledge and training among procurement staff for sustainable procurement	
	Higher cost of sustainable procurement	
	Autonomous purchase of local humanitarian offices focusing only on the lowest price	
Inter-organizational and supply chain	Different organizations follow different standards and requirements for their procured items	Cooperative purchasing and pooling procurement resources
		Similarity of many humanitarian items procured by different humanitarian organizations
		Humanitarian organizations benchmark and follow each other's successful practices
		Smaller community of humanitarian organizations compared to commercial counterparts
Funding	Low donors' awareness about sustainability	
	Earmarked funding	
	Funding fluctuation	
Supplier relationship	Short-term supplier relationships due to urgency and intermittence of humanitarian procurement	
	Low purchasing power of humanitarian organizations over suppliers of critical products (such as vaccines)	
	Extended humanitarian supply networks and difficulty in tracking indirect suppliers	
	Corruption in supplier selection and purchasing especially by local offices	

After the barriers and enablers were synthesized and categorized, the final list was sent to the three respondents for verification and conducting AHP. Since the number of identified enablers were relatively small, only prioritization of barriers was considered. The respondents were asked to make pairwise comparisons between the barriers under each category and then between the categories themselves. The basis for pairwise comparisons was *the possibility to overcome the barrier in practice*. After the results were received, the comparisons were checked for consistency. All the comparison from three respondents showed a satisfactory consistency level with a value under 0.1.

Finally, the preferences were aggregated using group AHP. Since the second respondent was a more senior procurement manager with longer experience and he also was in charge of a sustainable procurement project, higher weight was assigned to his preferences. Hence, the weight assigned to the senior respondent was 0.5 and the two other respondents' preferences were given the weight of 0.25. Then a weighted geometric mean for each pairwise comparison was calculated using Equation 12.4, where λ_i is the weight assigned to respondent i.

$$\bar{x} = \prod_{i=1}^{3} x_i^{\lambda_i} \tag{12.4}$$

For example, if the set of $X = (3, 4, 7)$ shows the preferences of the three respondents for a given pairwise comparison within the category of supplier relationship, while the weights are $\lambda_1 = \lambda_2 = 0.25$ and $\lambda_3 = 0.5$, then the aggregated importance for this set using weighted geometric mean is $(3^{0.25} \times 4^{0.25} \times 7^{0.5}) = 5.37$.

The rest of preferences were aggregated in the same way. Figure 12.2 shows the final priorities for each category of barriers (criteria) and the barriers within that category (sub-criteria). Next section discusses the identified barriers and enablers and the ranking in more detail.

12.5 DISCUSSION

This section discusses the identified barriers and enablers and the prioritization results of group AHP. According to Table 12.3, local procurement is the category with the highest number of barriers and enablers (5 barriers and 3 enablers identified). It means that while there are some challenges for local procurement, it is a promising area to start integrating sustainability. The enablers offer several advantages in terms of sustainability for local procurement over oversees procurement. First, local procurement of humanitarian items reduces the need for international freight transport by reducing the distances that items travel; thus, it contributes to environmental sustainability through lowering the emissions emanating from the logistics. Second, buying locally empowers the local community, local manufacturers, and workers by creating jobs and involving the society in value-adding activities. Third, locally produced items better satisfy the needs of the local population. For example, previous studies showed that cook stoves that are produced locally for displaced people and refugees are more widely accepted and put into use by beneficiaries due to consideration of the local cooking culture (Barbieri et al., 2017).

However, humanitarian organizations need to address several local procurement impediments, such as higher costs, social issues, lower quality, and lower production capacity, often associated with local suppliers due to the impact of a crisis or weak economic situation in developing countries. The importance of these impediments was so high that local procurement impediments was ranked as the most important category of barriers based on respondents' preferences (see Figure 12.2). Based on the prioritization, it seems that humanitarian organizations procuring locally have to ensure that social issues such as child labor, unfair working hours, and unhealthy working conditions are not present in local suppliers' operations. This can

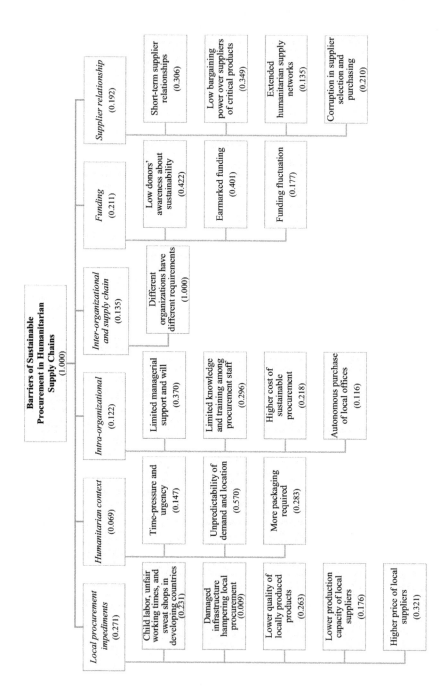

FIGURE 12.2 Prioritization of barriers for sustainable procurement in humanitarian supply chains.

be achieved through frequent supplier audits and evaluation. Moreover, the issues concerning cost, quality, and manufacturing capacity can be addressed through local capacity building and shifting towards longer-term constructive supplier development, rather than short-term intermittent procurement. For example, the relationship with a supplier of blankets or kitchen sets from which items were supplied for emergency situation can be extended through long frame agreements for post-disaster procurement.

The second important category of barriers concerns the funding environment. Funding and donations are the life blood of humanitarian organizations that allow them to survive and sustain their operations. Donor requirements are considered as important mandates in humanitarian context. When donors have limited awareness or require little about sustainable procurement, it is often difficult to take sustainability into account for the humanitarian organizations. In addition, donors usually tie up their funding to special spent purposes, which limits humanitarian organizations' flexibility to spend the donations elsewhere. This is known as earmarked funding and has been found to be negatively impacting humanitarian supply chain performance (Besiou et al., 2014). Sustainability typically comes with a cost and when donors do not allow funds to be spent for sustainability or have little knowledge about it, integrating sustainability into procurement is far to achieve.

The third important category is supplier relationship management. Similar to commercial procurement, if buyer has a higher purchasing power over its suppliers, it would be more feasible to ask for sustainability requirements (Meqdadi et al., 2018). The respondents believed that the most important barrier within this category is the low bargaining power of humanitarian organizations over suppliers of critical products. For some humanitarian products, such as vaccines, humanitarian organizations sometimes have few options for supplier selection, which reduces the leverage they have over their suppliers and consequently the bargaining power for incorporating sustainability into suppliers' processes. Moreover, humanitarian procurement, in contrast to commercial procurement, is often based on shorter supplier relationships due to the emergency and intermittence inherent in the humanitarian context. Integrating sustainability into procurement processes often requires long-term collaboration with suppliers. As mentioned before, developing long-term supplier frameworks and continuing procurement after the emergency subsides is a promising way for sustainable procurement in humanitarian supply chains.

The next important barriers within the category of supplier relationship according to the respondents' preferences is corruption in supplier selection and purchasing, especially in local offices purchases and extended humanitarian supply networks. The corruption in local offices' purchases can be tackled by increasing the visibility of headquarters over local offices' procurement. Some humanitarian organizations leave purchases below a financial threshold to local offices and do not bother to investigate such small purchases, increasing the risk of corruption for small purchases. Moreover, the extension of humanitarian supply chains to several tiers and echelons is an inherent humanitarian characteristic, which makes it difficult for humanitarian organizations to track down the sustainability of indirect suppliers in the second tier and beyond. The use of non-coercive power, such as rewards, can contribute to dissemination of sustainability in indirect suppliers (Meqdadi et al., 2018).

The fourth and fifth categories belong to intra- and inter-organizational barriers ranked almost equally. Within an organization, managerial support and awareness is the most important driver of sustainable procurement. This result is supported by the seminal study of Walker and Brammer (2009) in for-profit organizations. The next barriers in the intra-organization category include *limited knowledge and training of procurement staff* and *higher cost of sustainable procurement*. Both of these barriers are again associated with managerial commitment and interest in the sustainable procurement. By conducting employee training about sustainable procurement and willingness to spend more for sustainable products, managers can overcome these barriers on the way of sustainable procurement. Here, absence of pragmatic and clear guidelines to propel an organization towards sustainable procurement plays a major role in holding back organizations from putting sustainable procurement into practice. However, the introduction of ISO 20400 on sustainable procurement, listed as an enabler by respondents, can help mangers with a practical and implementable procedure. It should be noted that the standard is designed for for-profit organizations and its implementation in humanitarian procurement requires care in order to be aligned with humanitarian specificities and objective.

At the inter-organizational and supply chain level, the respondents believed that there were more enablers than barriers. Humanitarian organizations can pool their resources for joint procurement and achieve sustainable advantages. Joint procurement seems more feasible in humanitarian procurement due to the similarity and standardization of many aid items. Unlike commercial procurement, the aesthetics of a blanket or food packaging is not a major concern for beneficiaries. This reduces the diversity of aid products and leads to higher similarity, which facilitates joint procurement among different humanitarian organizations. It also increases the bargaining power over suppliers due to higher procured quantity and economies of scale. Another enabler at the inter-organizational and supply chain level is that humanitarian organizations often copy their best practices through benchmarking. A successful sustainable procurement implemented in one humanitarian organization can be benchmarked and adopted by another humanitarian organizations quickly with little modifications due to similarity of context. The smaller size of the humanitarian organization community, compared to their commercial counterparts, and higher transparency in reporting makes such benchmarking practices more feasible.

Finally, the respondents ranked humanitarian specificities as the least important category. The reason behind this is the basis of pairwise comparison: respondents were asked to prioritize the categories based on the possibility to overcome the barriers in practice. Obviously, barriers such as urgency, time-pressure, and unpredictability of demand, while distinguishing humanitarian supply chains from commercial ones, are inherent in humanitarian supply chains and cannot be changed or eliminated. Therefore, there are less opportunities for aligning them with sustainable procurement. With regards to packaging material and higher packaging consumption in humanitarian setting, the use of technology can be helpful. For example, when humanitarian organizations perform airdrops in an afflicted area, they cover the items with extra packaging to prevent it from being torn open when hitting the ground. New technologies on the use of parachutes for such items is developing, but they are not yet financially justifiable.

By and large, our identified barriers and enablers are supported by the literature of humanitarian supply chains as impacting factors on the way of institutionalizing sustainability. The importance of local procurement for sustainability (Van Kempen et al., 2017), the role of training procurement staff about sustainability (Abrahams, 2014), negative impact of earmarked funds on sustainability (Kunz & Gold, 2017), and the impeding role of corruption in humanitarian procurement (Schultz & Søreide, 2008) are some examples of advocacy by extant literature.

12.6   CONCLUSION

This paper unearthed and ranked the main barriers and enablers on the path of integrating sustainability to humanitarian procurement. The study offers several contributions to the literature of humanitarian logistics and supply chain management and to practitioners. First, sustainable humanitarian procurement is a nascent area of research and practice. In order to gravitate towards sustainability, it is imperative that the barriers and enablers in the humanitarian context are duly identified. The identification and categorization offered by this study deepens the understanding of sustainable humanitarian procurement for academics and sheds light on the specificities of humanitarian supply chains and how to deal with them to make humanitarian procurement more sustainable. Second, the application of group AHP and aggregation of preferences from several experts prevents from creeping subjectivity into the judgments made (Saaty, 1990) and ensures that the prioritization of barriers is sound, robust, and comprehensive enough to include all the procurement aspects.

Our work also offers valuable insights to practitioners. With sustainability moving to the center of attention for procurement practitioners, the main question for the integration of sustainability is simply "where to start from?" The findings of this study can help humanitarian procurement staff to initiate their efforts for sustainable procurement from tackling the most important barrier categories and the main barriers within each category, while benefiting the most from the identified enablers. Since the basis for the prioritization in this study was the possibility to overcome barriers in practice, it can help procurement managers to assign more resources to the barriers with higher importance, which subsequently leads to the biggest improvement while refraining from spending resources on barriers with lower opportunities for improvement.

The study had its own limitations. It does not claim that the identification and ranking of barriers and enablers is comprehensive. Eliciting the ideas of different experts might return different barriers and enablers from the ones identified by this study. Moreover, humanitarian operations include a wide range of operations from mitigation, to preparation, response, and reconstruction (Van Wassenhove & Pedraza Martinez, 2012). In this study, we have not considered all the phases in the disaster cycle and focused merely on disaster response.

Sustainable humanitarian procurement is an overlooked area of research and offers promising future research. Some suggestion for future research, specifically for barriers and enablers of sustainable humanitarian procurement, are: eliciting the ideas of more experts, using other MCDM methods for prioritization, including enablers in the ranking, and studying the interdependence among barriers and among enablers. In the wider scope of sustainable humanitarian procurement, future research can adopt and contextualize the already existing sustainable practices in commercial setting for humanitarian procurement. Moving from conceptual studies towards more empirical studies in collaboration with practitioners is a dire need of humanitarian supply chain research.

REFERENCES

Abrahams, D. (2014). The barriers to environmental sustainability in post-disaster settings: A case study of transitional shelter implementation in Haiti. *Disasters, 38*(s1), S25–S49. doi:10.1111/disa.12054.

Bagchi, A., Aliyas Paul, J., & Maloni, M. (2011). Improving bid efficiency for humanitarian food aid procurement. *International Journal of Production Economics, 134*(1), 238–245. doi:10.1016/j.ijpe.2011.07.004.

Balcik, B., & Ak, D. (2014). Supplier selection for framework agreements in humanitarian relief. *Production and Operations Management, 23*(6), 1028–1041. doi:10.1111/poms.12098.

Barbieri, J., Riva, F., & Colombo, E. (2017). Cooking in refugee camps and informal settlements: A review of available technologies and impacts on the socio-economic and environmental perspective. *Sustainable Energy Technologies and Assessments, 22*, 194–207. doi:10.1016/j.seta.2017.02.007.

Besiou, M., Pedraza-Martinez, A. J., & Van Wassenhove, L. N. (2014). Vehicle supply Chains in humanitarian operations: Decentralization, operational mix, and earmarked funding. *Production and Operations Management, 23*(11), 1950–1965. doi:10.1111/poms.12215.

Brammer, S., & Walker, H. (2011). Sustainable procurement in the public sector: An international comparative study. *International Journal of Operations & Production Management, 31*(4), 452–476. doi:10.1108/01443571111119551.

Celik, E., & Taskin Gumus, A. (2018). An assessment approach for non-governmental organizations in humanitarian relief logistics and an application in Turkey. *Technological and Economic Development of Economy, 24*(1), 1–26.

Colicchia, C., Creazza, A., & Dallari, F. (2017). Lean and green supply chain management through intermodal transport: Insights from the fast moving consumer goods industry. *Production Planning & Control, 28*(4), 321–334. doi:10.1080/09537287.2017.1282642.

Dong, Q., & Cooper, O. (2016). A peer-to-peer dynamic adaptive consensus reaching model for the group AHP decision making. *European Journal of Operational Research, 250*(2), 521–530. doi:10.1016/j.ejor.2015.09.016.

Eftekhar, M., Masini, A., Robotis, A., & Van Wassenhove, L. N. (2014). Vehicle procurement policy for humanitarian development programs. *Production and Operations Management, 23*(6), 951–964. doi:10.1111/poms.12108.

Elkington, J. (1998). *Cannibals with Forks: The Triple Bottom Line of 21st Century Business*: New Society Publishers.

Ertem, M. A., Buyurgan, N., & Pohl, E. A. (2012). Using announcement options in the bid construction phase for disaster relief procurement. *Socio-Economic Planning Sciences, 46*(4), 306–314. doi:10.1016/j.seps.2012.03.004.

Ertem, M. A., Buyurgan, N., & Rossetti, M. D. (2010). Multiple-buyer procurement auctions framework for humanitarian supply chain management. *International Journal of Physical Distribution & Logistics Management, 40*(3), 202–227. doi:10.1108/09600031011035092.

Falasca, M., & Zobel, C. W. (2011). A two-stage procurement model for humanitarian relief supply chains. *Journal of Humanitarian Logistics and Supply Chain Management, 1*(2), 151–169. doi:10.1108/20426741111188329.

Forman, E., & Peniwati, K. (1998). Aggregating individual judgments and priorities with the analytic hierarchy process. *European Journal of Operational Research, 108*(1), 165–169. doi:10.1016/S0377-2217(97)00244-0.

Hasselbalch, J., Costa, N., & Blecken, A. (2015). Investigating the barriers to sustainable procurement in the United Nations. In M. Klumpp, S. D. Leeuw, A. Regattieri, & R. D. Souza (Eds.), *Humanitarian Logistics and Sustainability* (pp. 67–86): Springer.

Herlin, H., & Pazirandeh, A. (2012). Nonprofit organizations shaping the market of supplies. *International Journal of Production Economics, 139*(2), 411–421. doi:10.1016/j.ijpe.2011.04.003.

Herlin, H., & Pazirandeh, A. (2015). Avoiding the pitfalls of cooperative purchasing through control and coordination: Insights from a humanitarian context. *International Journal of Procurement Management, 8*(3), 303–325.

Hoejmose, S. U., & Adrien-Kirby, A. J. (2012). Socially and environmentally responsible procurement: A literature review and future research agenda of a managerial issue in the 21st century. *Journal of Purchasing and Supply Management, 18*(4), 232–242. doi:10.1016/j.pursup.2012.06.002.

Hu, S.-L., Han, C.-F., & Meng, L.-P. (2017). Stochastic optimization for joint decision making of inventory and procurement in humanitarian relief. *Computers & Industrial Engineering, 111*, 39–49. doi:10.1016/j.cie.2017.06.029.

International Organization for Standardization. (2017). ISO 20400:2017 Sustainable procurement—Guidance. In (pp. 52).

Ishizaka, A., & Labib, A. (2011). Review of the main developments in the analytic hierarchy process. *Expert Systems with Applications.* doi:10.1016/j.eswa.2011.04.143.

Kleindorfer, P. R., Singhal, K., & Wassenhove, L. N. V. (2005). Sustainable operations management. *Production and Operations Management, 14*(4), 482–492.

Kunz, N., & Gold, S. (2017). Sustainable humanitarian supply chain management: Exploring new theory. *International Journal of Logistics Research and Applications, 20*(2), 85–104. doi:10.1080/13675567.2015.1103845.

Kunz, N., Van Wassenhove, L. N., Besiou, M., Hambye, C., & Kovács, G. (2017). Relevance of humanitarian logistics research: Best practices and way forward. *International Journal of Operations & Production Management, 37*(11), 1585–1599. doi:10.1108/IJOPM-04-2016-0202.

Maditati, D. R., Munim, Z. H., Schramm, H.-J., & Kummer, S. (2018). A review of green supply chain management: From bibliometric analysis to a conceptual framework and future research directions. *Resources, Conservation and Recycling, 139*, 150–162. doi:10.1016/j.resconrec.2018.08.004.

Matopoulos, A., Kovács, G., & Hayes, O. (2014). Local resources and procurement practices in humanitarian supply Chains: An empirical examination of large-scale house reconstruction projects. *Decision Sciences, 45*(4), 621–646. doi:10.1111/deci.12086.

Meqdadi, O. A., Johnsen, T. E., & Johnsen, R. E. (2018). Power and diffusion of sustainability in supply networks: Findings from four in-depth case studies. *Journal of Business Ethics.* doi:10.1007/s10551-018-3835-0.

Minakawa, N., Dida, G. O., Sonye, G. O., Futami, K., & Kaneko, S. (2008). Unforeseen misuses of bed nets in fishing villages along lake victoria. *Malaria Journal, 7*, 165. doi:10.1186/1475-2875-7-165.

Nikkhoo, F., Bozorgi-Amiri, A., & Heydari, J. (2018). Coordination of relief items procurement in humanitarian logistic based on quantity flexibility contract. *International Journal of Disaster Risk Reduction, 31*, 331–340. doi:10.1016/j.ijdrr.2018.05.024.

Oloruntoba, R., & Gray, R. (2006). Humanitarian aid: An agile supply chain? *Supply Chain Management: An International Journal, 11*(2), 115–120. doi:10.1108/13598 540610652492.

Park, J. H., Kazaz, B., & Webster, S. (2018). Surface versus air shipment of humanitarian goods under demand uncertainty. *Production and Operations Management, 27*(5), 928–948. doi:10.1111/poms.12849.

Pazirandeh, A., & Herlin, H. (2014). Unfruitful cooperative purchasing: A case of humanitarian purchasing power. *Journal of Humanitarian Logistics and Supply Chain Management, 4*(1), 24–42. doi:10.1108/JHLSCM-06-2013-0020.

Piotrowicz, W. D. (2018). In-kind donations, cash transfers and local procurement in the logistics of caring for internally displaced persons: The case of polish humanitarian NGOs and Ukrainian IDPs. *Journal of Humanitarian Logistics and Supply Chain Management.* doi:10.1108/JHLSCM-11-2017-0060.

Saaty, T. L. (1990). How to make a decision: The analytic hierarchy process. *European Journal of Operational Research, 48*(1), 9–26. doi:10.1016/0377-2217(90)90057-I.

Sarkis, J., Spens, K. M., & Kovács, G. (2012). A study of barriers to greening the relief supply chain. In K. M. Spens & G. Kovács (Eds.), *Relief Supply Chain Management for Disasters: Humanitarian Aid and Emergency Logistics* (pp. 196–207). Hershey PA, United States of America: IGI Global.

Schneider, L., & Wallenburg, C. M. (2012). Implementing sustainable sourcing: Does purchasing need to change? *Journal of Purchasing and Supply Management, 18*(4), 243–257. doi:10.1016/j.pursup.2012.03.002.

Schultz, J., & Søreide, T. (2008). Corruption in emergency procurement. *Disasters, 32*(4), 516–536. doi:10.1111/j.1467-7717.2008.01053.x.

Thomas, A. S., & Kopczak, L. R. (2005). From logistics to supply chain management: The path forward in the humanitarian sector. Retrieved from www.fritzinstitute.org/PDFs/WhitePaper/FromLogisticsto.pdf.

Torabi, A. S., Shokr, I., Tofighi, S., & Heydari, J. (2018). Integrated relief pre-positioning and procurement planning in humanitarian supply chains. *Transportation Research Part E: Logistics and Transportation Review, 113*, 123–146. doi:10.1016/j.tre.2018.03.012.

Trestrail, J., Paul, J., & Maloni, M. (2009). Improving bid pricing for humanitarian logistics. *International Journal of Physical Distribution & Logistics Management, 39*(5), 428–441. doi:10.1108/09600030910973751.

Trivedi, A., & Singh, A. (2017a). A hybrid multi-objective decision model for emergency shelter location-relocation projects using fuzzy analytic hierarchy process and goal programming approach. *International Journal of Project Management, 35*(5), 827–840. doi:10.1016/j.ijproman.2016.12.004.

Trivedi, A., & Singh, A. (2017b). Prioritizing emergency shelter areas using hybrid multi-criteria decision approach: A case study. *Journal of Multi-Criteria Decision Analysis, 24*(3–4), 133–145. doi:10.1002/mcda.1611.

Vaillancourt, A. (2017). Procurement consolidation in humanitarian supply chains: A case study. *International Journal of Procurement Management, 10*(2), 178–193.

Van Kempen, E. A., Spiliotopoulou, E., Stojanovski, G., & De Leeuw, S. (2017). Using life cycle sustainability assessment to trade off sourcing strategies for humanitarian relief items. *The International Journal of Life Cycle Assessment, 22*(11), 1718–1730. doi:10.1007/s11367-016-1245-z.

Van Wassenhove, L. N., & Pedraza Martinez, A. J. (2012). Using OR to adapt supply chain management best practices to humanitarian logistics. *International Transactions in Operational Research, 19*(1–2), 307–322. doi:10.1111/j.1475-3995.2010.00792.x.

Venkatesh, V. G., Zhang, A., Deakins, E., Luthra, S., & Mangla, S. (2018). A fuzzy AHP-TOPSIS approach to supply partner selection in continuous aid humanitarian supply chains. *Annals of Operations Research.* doi:10.1007/s10479-018-2981-1.

Walker, H., & Brammer, S. (2009). Sustainable procurement in the United Kingdom public sector. *Supply Chain Management: An International Journal, 14*(2), 128–137. doi:10.1108/13598540910941993.

Walker, H., & Brammer, S. (2012). The relationship between sustainable procurement and e-procurement in the public sector. *International Journal of Production Economics, 140*(1), 256–268. doi:10.1016/j.ijpe.2012.01.008.

Walker, H., Miemczyk, J., Johnsen, T., & Spencer, R. (2012). Sustainable procurement: Past, present and future. *Journal of Purchasing and Supply Management, 18*(4), 201–206. doi:10.1016/j.pursup.2012.11.003.

Wild, N., & Zhou, L. (2011). Ethical procurement strategies for international aid non-government organisations. *Supply Chain Management: An International Journal, 16*(2), 110–127. doi:10.1108/13598541111115365.

Xu, Z. (2000). On consistency of the weighted geometric mean complex judgement matrix in AHP1Research supported by NSF of China.1. *European Journal of Operational Research, 126*(3), 683–687. doi:10.1016/S0377-2217(99)00082-X.

Zarei, M. H., & Wong, K. Y. (2014). Making the recruitment decision for fresh university graduates: A study of employment in an industrial organisation. *International Journal of Management and Decision Making, 13*(4), 380–402.

Section IV

Research Methods in Sustainable Procurement Operations

13 Text Mining Applied to Literature on Sustainable Supply Chain (1996–2018)
An Analysis Based on Scopus

Jenny-Paola Lis-Gutiérrez, Mercedes Gaitán-Angulo, Carlos Bouza, Manuel Ignacio Balaguera, Melissa Lis-Gutiérrez, Cristian Beltrán, and Doris Aguilera

CONTENTS

13.1 INTRODUCTION

This chapter aims to establish the differences between the production and the visibility of research papers about sustainable supply chain that appear in Scopus indexed journals. The text mining process was made by taking the geographical origin of the authors as search and classification criterion. In our analysis, we try to establish the relevance of the information given by the articles descriptors. The purpose of this paper is to provide researchers in the field with elements and analysis that facilitate the understanding of knowledge production dynamics in this subject matter.

According to IBM (2018), text mining is: "[T]he process of analyzing collections of text materials in order to capture key themes and concepts and discover hidden relationships and existing trends without needing to know the exact words or terms that authors have used to express these concepts."

The process includes the following IBM steps (2018):

1. Identification of the text that will be used
2. Apply the text mining algorithms
3. Create categories and concepts
4. Analyze structured data, using techniques such as cluster, classification and predictive models

Below are some of the most recent works that make use of text mining. A study carried out with data from the Central Bank of the Republic of Turkey (CBRT) allowed, through text mining, for analyzing a large amount of information on monthly reports about price development and to explore issues and internal conglomerates. Text mining allowed analyzing the statistical consistency of the development of prices with the inflation figures of the annual consumer price index for Turkey (Feldman & Sanger, 2007).

This technique has also been strongly used in the biomedical sector since it allows browsing through a large number of texts of different origin. However, the development of text mining applications depends largely on the availability of textual content sources and the capabilities of the current text mining algorithms. In this sense, the performance of the algorithms depends on the tasks that are to be addressed. According to a study by Rodríguez-Esteban (2019), text mining in this area is useful for the semi-automated healing of biological databases, pharmaco-vigilance, discovery of biomarkers, construction of signaling pathways, and prediction of function protein and similarity, among others.

From another perspective, a relatively new development in text mining is the use of the Recurrent Neural Network (RNN), which is a branch of Deep Learning that serves to process a sequence of text data while state changes are maintained for a particular sequence. The RNN has been widely applied in the areas of non-segmented writing recognition and speech recognition (Su & Chen, 2018).

Likewise, Westergaard (2018) affirmed that text mining has become a popular strategy to keep up with the rapid growth of scientific literature. The extraction of texts from the scientific literature has been carried out mainly in collections of abstracts, due to their availability. They present an analysis of 15 million full-text articles in English, published between 1823 and 2016.

Text mining is also used as a search and support method in consulting services companies so that they can respond to the client's demand regardless of the experience of the consultants. This method allows to solve future inconveniences with the clients in the companies, since it predicts the behavior of the users from the data supplied by the company. This is done using correspondence, discriminant and data envelopment analysis (DEA). The results of this study indicated that the discrimination rate of the learning data converges to 100% as the number of words extracted increases. For their part, Cheng and Jin (2019) investigated the attributes that influence the experiences of Airbnb users by analyzing a set of online review comments through the process of text mining and sentiment analysis. Methodologically, this study contributed by illustrating how big data can be used and interpreted visually in tourism and hospitality studios. This study complements that of Cheng and Edwards (2017), which also includes the application of text mining to the collaborative economy, and that of Zhao et al. (2019), who used a sample of 127,629 reviews from tripadvisor.com to predict overall customer satisfaction using the technical attributes of online text reviews and customer participation in the review community.

Another application that text mining has is that when combined with other types of methods, such as forecasting, the different sources of data can be considered more accurately. As shown by the results of a study conducted by Watanabe et al. (2018), text mining facilitated the detection and examination of emerging issues and technologies by broadening the forecast knowledge base. Therefore, it was established that new forecasting applications could be designed. In particular, text mining provides a solid basis for reflecting on possible futures (Kayser & Blind, 2017). On the other hand, Santana Mansilla (2014) indicated that this technique is a viable alternative for analysis due to its ability to manage the vagueness, fuzziness, diversity of structures, and large number of words that characterize natural language (Hotho et al., 2005).

13.2 METHODOLOGY

13.2.1 DATA

Our sources were academic documents indexed in the Scopus database (e.g., articles, conference proceedings, chapters of compiled books, textbooks, etc.). The documents that make up the sample have been selected from the search equation (TITLE-ABS-KEY ("Sustainable Supply Chain") OR TITLE-ABS-KEY ("Green Supply Chain") OR TITLE-ABS-KEY ("durable supply Chain") OR TITLE-ABS-KEY ("Cadena de suministro verde") OR TITLE-ABS-KEY ("cadena de suministro sustentable") OR TITLE-ABS-KEY ("cadena de suministro durable") OR TITLE-ABS-KEY ("cadena de suministro sostenible") OR TITLE-ABS-KEY ("Cadena de aprovisionamiento verde") OR TITLE-ABS-KEY ("cadena de aprovisionamiento sustentable") OR TITLE-ABS-KEY ("cadena de aprovisionamiento durable") OR TITLE-ABS-KEY ("cadena de aprovisionamiento sostenible") OR TITLE-ABS-KEY ("Chaîne d'approvisionnement verte") OR TITLE-ABS-KEY ("Chaîne d'approvisionnement durable") OR TITLE-ABS-KEY ("Catena di fornitura verde") OR TITLE-ABS-KEY ("Catena di fornitura sostenibile") OR TITLE-ABS-KEY ("Catena di approvvigionamento sostenibile") OR TITLE-ABS-KEY ("Catena di approvvigionamento verde") OR TITLE-ABS-KEY ("Cadeia de Suprimentos verde") OR TITLE-ABS-KEY ("Cadeia de fornecimento verde") OR TITLE-ABS-KEY ("Cadeia de Suprimentos Sustentável") OR TITLE-ABS-KEY ("Cadeia de fornecimento Sustentável")).

The delimitation of the interrelations between the highlighted keywords will be carried out according to the following formula proposed from the work of Lulewicz-Sas (2017):

$$P(A \mid B) = \sum_{i=1}^{n} p_i\left(A|B\right) \tag{13.1}$$

$$p_i\left(A|B\right) = 1 \ if \ A \in art_i \ \wedge \ B \in art_i \tag{13.2}$$

$$0 \ if \ A \notin art_i \ \wedge \ B \notin art_i$$

where $p_i(A|B)$ is the probability of selecting article i from conditions A and B, i is the Number of scientific articles ($i = 1, 2,..., n$), A is the set of words in the search equation, B is the condition of the publication period (2000–2018), and art_i is the set of preselected items that meet at least one of conditions A or B.

The documents consulted and indexed by Scopus covered the period from January 2000 to June 2018. The bibliographic information of the documents was downloaded in csv format and processed in vosviewer (Perianes-Rodriguez et al., 2016, Aria & Cuccurullo, 2017). The sample of documents ($N = 2.893$) was divided into different parts, according to the continent of origin of the authors: belonging to Latin America or not.

Another of the elements used for the analysis was taken from the work of Westergaard (2018), Franceschini et al. (2012) and Mørk et al. (2014), and is related to co-occurrences, which correspond to a weighted count for each pair of entities using the following formula:

$$C_{(i,j)} = \sum_{k=1}^{n} w_d \delta_{dk}\left(i, j\right) + w_p \delta_{pk}\left(i, j\right) + w_s \delta_{sk}\left(i, j\right) \tag{13.3}$$

where δ is an indicator function that takes into account whether the terms i, j co-occur within the same document (d), paragraph (p), or phrase (s). W is the concurrency weight set here in 1.0, 2.0, and 0.2, respectively. Based on the weighted count, the score $S(i, j)$ was calculated using an alpha of 0.6 ($\alpha = 0.6$).

$$S_{(i,j)} = C_{ij}^{\alpha} \left(\frac{C_{ij}C_{..}}{C_i.C_j}\right)^{1-\alpha} \tag{13.4}$$

Finally, for the elaboration of the word clouds, it was considered following the Justicia de la Torre (2017), a collection of documents Ω on which the analysis is intended. In this case, $\Omega = \{D\_1,..., D\_n\}$ is a collection of documents, where each D_1 is a document. For its part, the text is divided into minimum unit's document, word, concept, phrase, and paragraph.

The words that form a text can provide a first level of the corpus. In this context, a word will be identified as we identify it with the symbol m_j, where $j = 1,..., k$, where k is the total number of words in a document. Each m_j is a sequence of characters within a sentence. Therefore, VD $= \{m\_1,..., m\_k\}$, will be the set of words extracted from the document.

Now, a bag of words may have a representation by real numbers when the presence and absence of a term in the document corresponds to a real value. Therefore, the relative frequency would be given by the number of times that a certain term appears in a document, calculated on the number of terms of said document, it will be calculated as:

$$D_i\left(t_j\right) = \frac{f_{ij}}{\sum_{i=1}^{k} f_{ij}} \tag{13.5}$$

13.3   RESULTS

This section is divided into two segments. In the first part, the descriptive results are shown, and in the second, the analytical ones, using text mining.

13.3.1   DESCRIPTIVE RESULTS

Between 1996 and May 30, 2018, 2,893 articles related to the search equation have been published (Figure 13.1). Specifically, for the Latin American case, the results are limited to 155; that is, only 5.36% of the production of the period (Figure 13.2). According to Figure 13.1, bibliographic production on the sustainable supply chain starts with greater vigor in 2008, when the number of published articles is multiplied by more than double the bibliographic production of the immediately

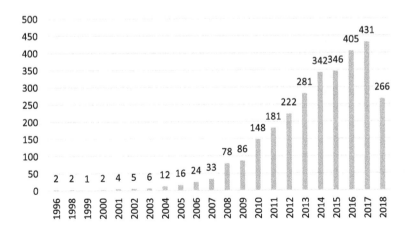

FIGURE 13.1   Academic production indexed in Elsevier on the subject (1996 June 2018). (From Own elaboration based on Elsevier, Scopus [Data set], https://www.scopus.com/search/form.uri?display=basic, 2018.)

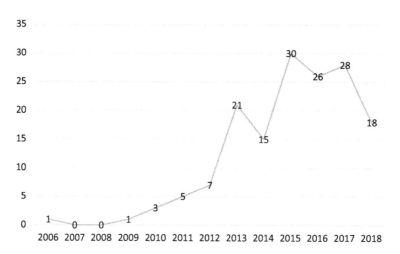

FIGURE 13.2   Academic production indexed in Scopus on the subject (1996 June 2018) and produced by authors of Latin American origin. (From Own elaboration based on Elsevier, Scopus [Data set], https://www.scopus.com/search/form.uri?display=basic, 2018.)

preceding year. The article that, according to Scopus, until June 2018 had received the highest number of citations was the Seuring and Müller (2008), who reviewed the literature on Sustainable Supply Chain (From a literature review to a conceptual framework for sustainable supply chain management), published in the *Journal of Cleaner Production*, and that counted until June 2018 with 1,591 citations in Scopus and 3,167 in Google Scholar. Secondly, there is Srivastava's (2007) work, called "Green supply-chain management: A state-of-the-art literature review," published in the *International Journal of Management Reviews*, and which had 1,412 citations in Scopus and 2,859 in Google academic.

At the aggregate level, the authors with the largest publications corresponded to the contents in Table 13.1 and Figure 13.3. All of them with more than 20 articles published on the subject. In the case of Latin American authors, only 2 have more than 20 articles: Jabbour, C.J.C., and De Sousa Jabbour, A.B.L., both are of Brazilian origin but are working at the Montpellier Business School. Professor Joseph Sarkis of the Worcester Polytechnic Institute in the United States is the author with greater number of papers.

In relation to the institutions with the largest number of works on the subject worldwide, the Dalian University of Technology (China) and Syddansk Universitet

TABLE 13.1

Authors with the Largest Number of Articles (1996 June 2018)

Author	Documents
Sarkis, J.	66
Govindan, K.	37
Jabbour, C.J.C.	34
Zhu, Q.	32
Gunasekaran, A.	23
Seuring, S.	22
Tseng, M.L.	22

Source: Own elaboration based on Elsevier, Scopus [Data set], https://www.scopus.com/search/form.uri?display=basic, 2018.

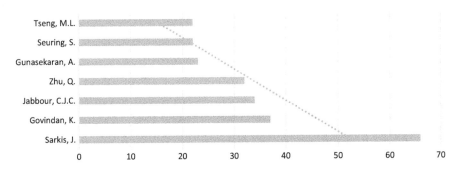

FIGURE 13.3 Authors with the largest number of articles (1996 June 2018). (From Own elaboration based on Elsevier, Scopus [Data set], https://www.scopus.com/search/form.uri?display=basic, 2018.)

(Denmark) stand out (Table 13.2 and Figure 13.4). In the case of the Latin American ones, the ones that stand out are the Universidade Estadual Paulista (with 37 works) and the Universidade de Sao Paulo (9 works).

With respect to the typology of the analyzed works, most of them are articles (60.63%) and conference papers (23.99%) (Table 13.3 and Figure 13.5). In the case of Latin American works, the number of articles corresponds to 102 (65.8%) and conference documents to 37 (23.9%) (Figure 13.6).

TABLE 13.2

Institutions with the Largest Number of Works (1996 June 2018)

Institution	Documents
Dalian University of Technology	49
Syddansk Universitet	45
Hong Kong Polytechnic University	41
Clark University	38
UNESP-Universidade Estadual Paulista	37
Worcester Polytechnic Institute	34
University of Tehran	34
Universitat Kassel	30

Source: Own elaboration based on Elsevier, Scopus [Data set], https://www.scopus.com/search/form. uri?display=basic, 2018.

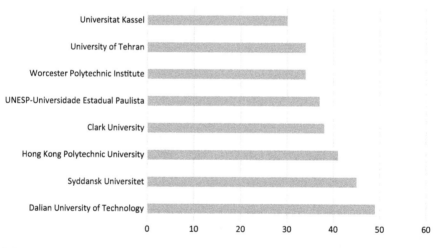

FIGURE 13.4 Institutions with the largest number of works (1996 June 2018). (From Own elaboration based on Elsevier, Scopus [Data set], https://www.scopus.com/search/form. uri?display=basic, 2018.)

TABLE 13.3
Typology of Published Works (1996 June 2018)

Typology	Documents	Percentage (%)
Article	1.754	60,63
Conference Paper	694	23,99
Book Chapter	149	5,15
Review	135	4,67
Article in Press	63	2,18
Conference Review	41	1,42
Editorial	23	0,80
Book	17	0,59
Erratum	7	0,24
Note	5	0,17
Short Survey	4	0,14
Letter	1	0,03

Source: Own elaboration based on Elsevier, Scopus [Data set], https://www.scopus.com/search/form.uri?display=basic, 2018.

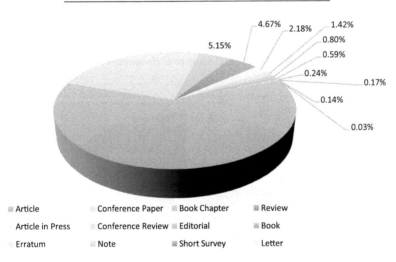

FIGURE 13.5   Typology of published works (1996 June 2018). (From Own elaboration based on Elsevier, Scopus [Data set], https://www.scopus.com/search/form.uri?display=basic, 2018.)

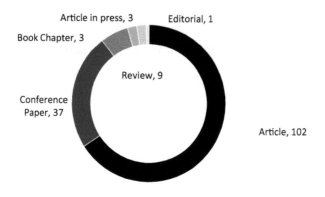

FIGURE 13.6 Typology of works published by authors from Latin America (1996 June 2018). (From Own elaboration based on Elsevier, Scopus [Data set], https://www.scopus.com/ search/form.uri?display=basic, 2018.)

With respect to the areas of knowledge, the majority corresponds to business (22%), engineering (20%), and decision sciences (10%) (Figure 13.7). For Latin America, the trend is reversed, since most of the documents correspond to engineering (78.25%), business (64.21%), and environmental sciences (15%) (Figure 13.8). The journals where most publications were from Latin American authors were: *Journal of Cleaner Production*, *Producao*, and *Gestao E Producao*.

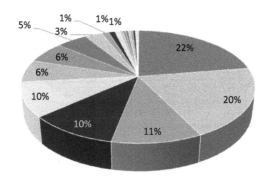

FIGURE 13.7 Areas of knowledge of the publications (1996 June 2018). (From Own elaboration based on Elsevier, Scopus [Data set], https://www.scopus.com/search/form. uri?display=basic, 2018.)

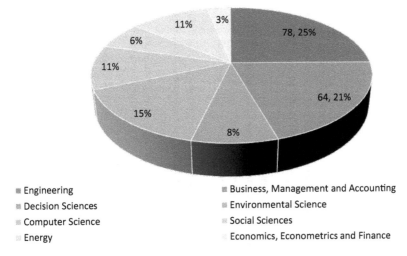

- Engineering
- Decision Sciences
- Computer Science
- Energy
- Business, Management and Accounting
- Environmental Science
- Social Sciences
- Economics, Econometrics and Finance

FIGURE 13.8 Areas of knowledge of Latin American publications (1996 June 2018). (From Own elaboration based on Elsevier, Scopus [Data set], https://www.scopus.com/search/form. uri?display=basic, 2018.)

13.3.2 CONTENT ANALYSIS

Of the 2,893 documents extracted from Scopus, the network of relationships of co-occurrences between keywords of titles and abstracts is presented in Figure 13.9. The formula used for the calculation is found in Section 1.2 (Van Eck & Waltman, 2011, 2014, 2018).

With respect to the content analysis of the titles of the complete database, it was indicated that the minimum frequency of appearance of the terms was 5. Initially, there were 5,277 terms and, under the imposed restriction, they were reduced to 284 terms. For the cluster analysis of Figure 13.9, 5 terms were selected as the minimum size and the most relevant 60% of all the terms; that is, 170 terms. In Figure 13.9 and Table 13.4, the 12 clusters can be seen.

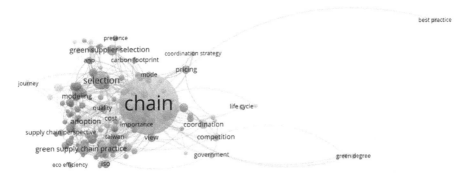

FIGURE 13.9 Cluster of words of the titles of the published works (1996 June 2018). (From Own elaboration using VOSviewer.)

TABLE 13.4

Cluster of Words of the Titles of the Published Works (1996 June 2018)

Cluster	Classification	Words
1	Industrial organization	Benefit, business performance, country, critical factor, eco efficiency, economic performance, electronics industry, empirical analysis, empirical evidence, environmental collaboration, evidence, firm performance, green innovation, green performance, green supply chain initiative, green supply chain practice, greening, institutional pressure, iso, Malaysia, medium enterprise, organizational performance, reprint, supply chain integration, Taiwan, view
2	Empirical research	analytic network process, anp, automobile industry, critical success factor, empirical investigation, enabler, fuzzy dematel, green supplier, green supply chain management adoption, green supply chain management implementation, green supply chain management performance, green supply chain performance, identification, indian automobile industry, indian perspective, interaction, interpretive structural modelling, ism, modeling, presence, pressure, prioritization, sustainable supplier evaluation, sustainable supply chain practice
3	Optimization and decisions	Carbon, carbon emission, decision support system, gas industry, genetic algorithm, green supply chain network, green supply chain network design, knowledge management, literature, multi objective optimization, oil, optimization model, risk management, social sustainability, solution, sscm, stochastic demand, sustainable supply chain management practice, taxonomy
4	Sustainability and carbon footprint	Activity, carbon management, CO_2 emission, competitive advantage, future, future direction, global supply chain, green logistic, green supply chain management strategy, part, quality, supply chain perspective, sustainable development, sustainable supply chain network design, transportation, trend
5	Analysis and measurement	Analytic hierarchy process, comparative analysis, control, determinant, environmental sustainability, exploratory study, green supply chain design, green supply chain performance measurement, importance, logistics service provider, manufacturing firm, planning, sustainable practice, systematic literature review, trust
6	Concepts	3rd international conference, challenges, conceptual framework, green, green manufacturing, international conference, logistics, multi criteria decision, proceeding, resilience, supply chain network, supply chain performance, sustainability assessment, technology

(Continued)

TABLE 13.4 (*Continued*)
Cluster of Words of the Titles of the Published Works (1996 June 2018)

Cluster	Classification	Words
7	Hierarchical analysis	carbon footprint, coordination strategy, dea, fuzzy ahp, fuzzy environment, green supplier selection, mode, models, pricing, sustainable supplier selection
8	Evaluation	Apparel industry, comparison, experience, green procurement, green supplier evaluation, knowledge, lesson, life cycle assessment, selection, supplier evaluation
9	Environmental impact	Biofuel, consideration, construction, environmental impact, green supply chain management system, life cycle, partner selection, sustainable supply chain design, system dynamic, textile industry
10	Sustainable supply chain network	Antecedent, art, designing, focus, integrated approach, journey, state, suprimento, sustainable supply chain network
11	Game theory and cooperation	Chain, competition, cooperation, coordination, game theoretic approach, government, green degree, green product design, pricing policy
12	Good practices	Adoption, brazil, cost, environmental management, medium sized enterprise, supply chain design, triple bottom line, best practice

Source: Own elaboration using VOSviewer.

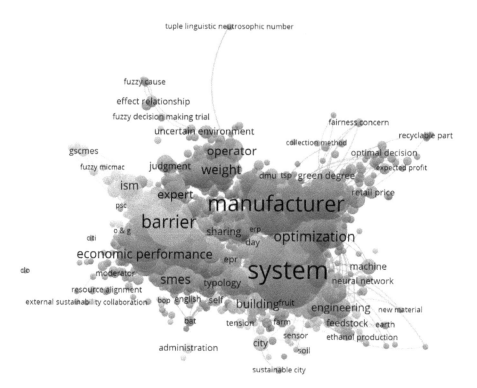

FIGURE 13.10   Cluster of words of the abstracts of the published works (1996 June 2018). (From Own elaboration using VOSviewer.)

With respect to the content analysis of the summaries of the complete database, it was indicated that the minimum frequency of appearance of the terms was 5. Initially, 38,122 terms appeared, and under the imposed restriction, they were reduced to 3,025 terms. For the cluster analysis of Figure 13.10, 5 terms were selected as the minimum size and the most relevant 60% of all the terms; that is, 1,815 terms. In Figure 13.10 and Table 13.5, 12 clusters of abstracts can be seen.

With respect to the previous table, it can be seen that cluster 1 relates the terms that relate to the strategy and management of the company. In this sense, it emphasizes terms such as consumer image, financial performance, industry, consumer, demand. Cluster 2 covers applications of mathematics and modeling the subject, while cluster 3 is related to hierarchical and classification analysis. In cluster 4, it is associated with case studies, while cluster 5 includes the use of qualitative analysis.

Cluster 6 is associated with the application of game theory to the topic of interest. Cluster 7 relates the terms that influence the performance, capacity, attributes,

TABLE 13.5

Cluster of Words of the Abstracts of the Published Works (1996 June 2018)

Cluster	Words
1	Adaptability, aee, anchor company, antecedent, asset specificity, benchmarking, best way, business ethic, business performance, buyer seller relationship, buyer supplier relationship, buying firm, chinese enterprise, chinese firm, chinese government, chinese manufacturer, citi, clo, cluster analysis, coercive, coercive pressure, collaborative paradigm, Colombia, company performance, comparative study, competitive environment, competitive pressure, competitive strategy, comprehensive literature review, computer parts industry, conceptualization, conceptualization, confirmatory factor analysis, consistency, construction sector, consumer demand, consumer perception, consumer pressure, contingency, contingency theory, contractor, contribute, corporate image, corporate performance, cpfr, crg, critical factor, current level, current trend, customer cooperation, customer integration, customer pressure, dearth, depth interview, descriptive, deviation, diffusion, direct effect, direct impact, dynamic capability, early stage, eco design, eco design practice, economic performance, effective supply chain management, electronic industry, electronics industry, empirical analysis, empirical evidence, empirical investigation, empirical result, ems, emss, enterprise performance, environmental capability, environmental collaboration, environmental improvement, environmental innovation, environmental innovation strategy, environmental management capability, environmental management maturity, environmental management practice, environmental monitoring, environmental sustainability practice, environmental training, environmental uncertainty, executive, exploratory factor analysis, external driver, external em, external green collaboration, external gscm practice, external practice, factor analysis, financial performance, firm competitiveness, firm performance, firm size, first phase, first study, fit, focal firm, food retailer, further study, ghrm, ghrm practice, green activity, green distribution, green human resource management, green implementation, green information system, green innovation, green innovation performance, green management, green operation, green performance, green process, green product development, green production, green purchase, green purchasing, green supply chain collaboration, green supply chain initiative, green supply chain integration, green supply chain management implementation, green supply chain management practice, green supply chain practice, green supply management, green transportation, greener product, gsc practice, gsci, gscm activity, gscm adoption, gscm approach, gscm dimension, gscm driver, gscm initiative, gscm practice, gscm practices implementation, gscm pressure, gscm strategy, gscm theory, gscmp, gscp, guanxi, hierarchical regression analysis, higher level, holistic view, hospital, hrm, hypothesis, important approach, important implication, important insight, independent variable, indian manufacturing organization, indirect relationship, individual firm, industry level, industry standard, industry type, information exchange, innovative strategy, institutional pressure, institutional theory, integral part, integrated framework, integrative framework, internal driver, internal environmental management, internal factor, internal green practice, internal gscm practice, internal integration, internal management, internal practice, internal process, *(Continued)*

TABLE 13.5 (Continued)

Cluster of Words of the Abstracts of the Published Works (1996 June 2018)

Cluster	Words
	international market, international organization, intersection, intra, investment recovery, italian firm, Italy, john wiley & sons, knowledge sharing, korea, large firm, large sample, leadership, lean manufacturing, low carbon management, lsp, lsps, mail survey, main aim, main gscm practice, major factor, major finding, management commitment, manufacturing enterprise, manufacturing firm, manufacturing manager, manufacturing organization, manufacturing sector, market performance, maturity, maturity level, mediating effect, mediating role, mediation effect, medium enterprise, medium sized enterprise, medium sized supplier, medwell journals, methodology approach, mimetic pressure, mismatch, moderating effect, moderating role, moderation effect, moderator, moral motive, motivator, multiple case study, multiple regression analysis, new insight, new opportunity, new product, nrbv, operational efficiency, operational performance, operationalization, organisational culture, organisational performance, organization involvement, organizational culture, organizational learning, organizational performance, orientation, outbound logistic, outsourcing, Pakistan, paper highlight, partnering, pathway, performance implication, performance outcome, pls sem, pollution prevention, positive effect, positive impact, positive influence, positive relationship, practical guidance, premise, previous literature, previous research, prior research, proactive approach, process development, processor, profile, quality standard, quantitative approach, quantitative data, quantitative method, questionnaire, questionnaire survey, rbv, rdt, reactive, regression analysis, regulatory pressure, relational bonding, relational capability, reputation, research framework, research limitation, research model, resource alignment, resource commitment, resource dependence theory, response rate, rest, results highlight, reverse logistics practice, reward, sc dynamic capability, sci, scp, second phase, sem, senior management, senior manager, sharing, significant contribution, significant driver, significant effect, significant impact, significant influence, significant positive effect, significant positive relationship, significant relationship, small, small firm, sme, smes, social capital, social performance, social science, social sustainability supply chain practice, sois, south korea, specific industry, spm, square, squares structural equation modeling, sscm adoption, sscm implementation, sscm initiative, sscm performance, sscm practice, sscm research, ssco, ssp, stakeholder theory, statistical analysis, statistical package, statistical result, statistical tool, structural equation, structural equation model, structural equation modeling, structural equation modelling, structured questionnaire, successful implementation, supplier assessment, supplier integration, supplier management, supplier relationship, supplier relationship management, supply base, supply chain complexity, supply chain integration, supply chain leadership, supply chain learning, supply chain level, supply chain management practice, supply chain partner, supply chain relationship, supply management, survey method, survey questionnaire, survey result, sustainable organization, sustainable packaging, sustainable performance, sustainable procurement, sustainable scm, sustainable supply chain initiative, tbl, technological innovation, testing, Thailand, theoretical contribution, theoretical perspective, theory building, tolerance evaluation index, top management, top management commitment, top management support, total quality environmental management, tqem, uae, sable response, useful insight, variance, Vietnam, void, volume uncertainty, waste reduction, willingness, wine industry, world scientific publishing company, wto

(Continued)

TABLE 13.5 (*Continued*)
Cluster of Words of the Abstracts of the Published Works (1996 June 2018)

Cluster	Words
2	Algorithm, american chemical society, announcement, assignment, Australia, aviation biofuel, best solution, better result, bio, biofuel, biomass, cap, carbon cap, carbon credit price, carbon dioxide emission, carbon emissions sensitive demand, carbon price, carbon tax, center, chance, closed loop, closed loop supply chain network, clsc network, CO_2, CO_2 emission, collaborative distribution, competitive market, complex problem, compromise solution, computational result, confidence level, considerable attention, constraint method, consumer product, cost effectiveness, cradle, current issue, cycle time, data envelopment analysis, dea, dea model, decision support system, demand rate, demand uncertainty, depot, design phase, disassembly, disaster, distribution center, distribution centre, dmu, dss, dual role factor, ease, eco indicator, economic activity, economic cost, economic factor, economic feasibility, economic objective, efficient frontier, efp, electricity generation, electronic product, emission cost, emissions reduction, environment factor, environmental attribute, environmental constraint, environmental friendly product, environmental influence, environmental investment, environmental legislation, environmental objective, environmental quality, equilibrium, erp, expense, experimental result, facility, facility location, factory, final product, finished product, first level, first stage, forward supply chain, fossil fuel, fuel consumption, functional unit, functionality, future supply chain, fuzzy comprehensive evaluation method, gams, generality, generalized model, genetic algorithm, gga, ghg, ghg emission, goal programming, goog, green component, green factor, green product design, green supply chain design, green supply chain issue, green supply chain network, greenhouse gas, greenhouse gas emission, greenhouse gases emission, gscm model, harm, heuristic, important decision, incentive mechanism, industrial case study, integrated supply chain, interplay, inventory, inventory model, inventory policy, job, key contribution, life product, logistic network, logistics cost, logistics system, main focus, main issue, major issue, major role, managerial insight, mathematical programming mathematical programming model, maximum, maximum profit, meta heuristic algorithm milp, minimization, minimum cost, mip, mixed integer linear programming, mixed integer linear programming model, modeling framework, multi objective, multi objective mathematical model, multi objective model, multi objective optimization model, multi objective problem, multi period, multi product, multiple objective, multiple retailer, multiple supplier, negotiation, net present value, network design, network structure, new challenge, new legislation, non, nsga ii, numerical example, numerical experiment, numerical result, numerical study, objective function, operational cost, operational strategy, optimal design, optimal flow, optimal price, optimal solution, optimal strategy, optimality, optimass, optimization, optimization model, optimization problem, order picking, overall cost, overconfidence, paper study, parameter, pareto front, pareto frontier, pareto optimal solution, past year,

(*Continued*)

TABLE 13.5 (Continued)

Cluster of Words of the Abstracts of the Published Works (1996 June 2018)

Cluster	Words
	penalty, petroleum supply chain, planning model, policy insight, pollutant, practical case, probability, process industry, product flow, product recovery, production capacity, production cost, production facility, production method, production planning, profit allocation, programming model, public sector, quantity, r & d, rapid growth, real case, real world, real world problem, relevant stakeholder, remanufacture, renewable energy source, repair, resource consumption, reverse chain, reverse logistics network, rework, risk attitude, route, routing, sc performance, scheduling, scnd, second layer, second level, second stage, semiconductor industry, service level, ship, shipment, significant reduction, single manufacturer, single product, site, social cost, solution approach, solution method, solution methodology, solution procedure, stage dea model, stakeholder group, stochastic demand, store, strategic customer behavior, strategic planning, structural property, superiority, supply chain configuration, supply chain cost, supply chain firm, supply chain model, supply chain network design, supply chain planning, supply chain system, sustainability consideration, sustainability target, sustainable network, sustainable supply chain network design problem, sustainable supply chain optimization, Sweden, tactical decision, target value, tariff, technology selection, time delivery, tire, total carbon emission, total cost, total logistics cost, total profit, tradeoff, transport mode, transportation activity, transportation cost, transportation mode, transportation sector, transportation system, truck, tsp, uncertain condition, uncertain demand, uncertain parameter, undesirable output, variability, variant, vehicle, vehicle routing, vertical integration, vmi, warehouse, whole process
3	Abstract, afdematel, alternative supplier, ambiguity, analytic network process, analytical network process, anp, appropriate supplier, basel, best green supplier, best supplier, calculation, carbon management, carbon policy, causal relationship, clustering, competence, competitive priority, complex relationship, criteria weight, current study, customer requirement, daily operation, dairy, dairy industry, dcs, decision framework, decision making model, decision making trial, dematel, dematel method, design requirement, developed approach, differential evolution, economic globalization, economic interest, effect relationship, effective approach, effective tool, efficacy, egypt, electronics company, empirical case study, evaluation criterium, evaluation index, evaluation index system, evaluation laboratory, evaluation laboratory method, evaluation method, evaluation process, evaluation result, expert, expert opinion, experts opinion, extensive literature review, fahp, final ranking, flexible systems management, fmea, following, fuzziness, fuzzy-ahp, fuzzy analytic hierarchy process, fuzzy analytical hierarchy process, fuzzy cause, fuzzy decision making trial, fuzzy delphi method, fuzzy dematel, fuzzy environment, fuzzy inference system, fuzzy logic, fuzzy number, fuzzy set, fuzzy set theory, fuzzy technique, fuzzy topsi, Ghana, global institute, gra, green criterium, green supplier, green supplier development, green supplier development program, green supplier evaluation, green supplier selection,

(Continued)

TABLE 13.5 (*Continued*)
Cluster of Words of the Abstracts of the Published Works (1996 June 2018)

Cluster	Words
	green supplier selection problem, GREY relational analysis, grsc, gsc capability, gsc management, gscm criterium, gscm performance, gss, guide, hierarchical structure, hoq, hsm, human being, hybrid method, hybrid model, ideal solution, illustrative case, illustrative example, important criterium, important driver, imprecision, incomplete information, index system, indian context, indian manufacturing industry, industrial application, industrial expertindustrial manager, influential factor, integrated approach, integrated methodology, integrated model, inter relationship, interdependency, interval, intuitionistic fuzzy set, ipa, irp, it2fss, judgment, key phase, key strategic consideration, linguistic preference, linguistic term, linguistic variable, literature survey, low carbon product, low carbon supply chain, managerial perspective, mas, mcdm, mcdm problem, mdpi, mining company, mining industry, modification, monte carlo simulation, multi criteria decision making, multi criteria decision making problem, multiple criteria decision making, multiple criterium, novel approach, novelty, order allocation, order performance, order preference, pair wise comparison, paper industry, performance analysis, performance level, phenomenon, plastic film, potential supplier, practicality, present work, primary data, prioritization, priority weight, process model, proposed approach, proposed method, protection, pulp, qfd, qualitative criterium, quality function deployment, rank, ranking, rationality, recent time, relational practice, relational view, relative importance, relative weight, research trend, research work, risk factor, risk mitigation strategy, selection criterium, selection process, sequence, similarity, social criterium, social problem, social risk, southern part, specific practice, strategic plan, sub criterium, subcriteria, subjectivity, suitable green supplier, suitable supplier, supplier evaluation, supplier performance, supplier selection criterium, supplier selection problem, supplier selection process, suppliers environmental performance, suppliers performance, supply chain risk, sustainability perspective, sustainable criterium, sustainable supplier, sustainable supplier selection, technical requirement, todim, topsi, topsis, traditional supply chain management, trapezoidal fuzzy number, trial, tscm, uncertain environment, uncertain influential factor, upstream supply chain, vagueness, vendor, vendor selection, vikor, vikor method, vital role, vlsekriterijumska optimizacija, weight
4	academic journal, agricultural product, animal farmer, automotive manufacturer, availability, banana, Bangladesh, bat, biodiversity, carrier, case analysis, case study approach, case study company, centrality, cfs, circular economy, clothing industry, collaborative relationship, compilation, correlation analysis, current situation, decrease, depth case study, description, destination, determination, developed country, differentiation, dilemma, directive, disturbance, document, downstream, drawing, e waste, eco management, economic issue, economic sector, ecosystem, eea, electric vehicle, electrical, elv, empirical example, energy use, environmental law, environmental point, environmental supply chain, epr, essential element, Europe, export, exposure, extended producer responsibility, fair trade, farm, fashion, fashion brand, fashion company, fashion industry, fashion supply chain, first tier supplier, freight transport, further development, further work, global value chain, governance mechanism, government intervention, green chemistry, green consumption,

(*Continued*)

TABLE 13.5 (*Continued*)

Cluster of Words of the Abstracts of the Published Works (1996 June 2018)

Cluster	Words
	green marketing activity, green supply chain performance measurement, gscm system, h & m, harvesting, hazard, heterogeneity, human health, import, important element, important subject, indirect impact, industrial chain, industrial development, industrial ecology, industrial symbiosis, inefficiency, information asymmetry, institutional environment, interchangeability, ire index, journey, key driver, key element, key theme, land, latin America, lead firm, lean practice, leather industry, lense, lesson, lettuce, life management, life vehicle, main reason, major theme, malaria, material flow, meaning, middle east, mineral industry, mining, module, motive, multi tier supply chain, n order supply chain, Netherlands, north America, oem, offer, origin, original equipment manufacturer, overall environmental impact, Patagonia, pesticide, post use, postponement, prerequisite, prevention, product remanufacturing, product stewardship, quest, framework, rdm, reference point, remanufactured product, remanufacturer, research proposition, retailing, return rate, road testing, rohs, scem, scm practice, sct, second tier supplier, service sector, significant attention, social development, social system, something, south east asia, spain, springer verlag, sscg, sscs, strategic advantage, sub supplier, supermarket, supply chain actor, supply chain governance, supply chain management literature, supply chain performance measurement, supply chain transparency, sustainability governance, sustainability performance measurement, sustainability requirement, sustainable building, sustainable fashion supply chain, sustainable manufacturing practice, sustainable society, sustainable supply chain governance, sustainable supply chain system, system structure, systematic review, systems theory, team, technological change, tendency, third party, tourism industry cluster, transport sector, triple bottom line dimension, usa, value chain aspect, vessel, waste disposal, waste generation, water use, weee, winter
5	academic researcher, action research, agile manufacturing, ahp method, apparel supply chain, banda aceh, bibliometric analysis, biotic resource, bop, bop paper, bop project, bottom, business operation, cas, chain sustainability, citation, claim, competitive edge, concentration, conceptual paper, corporate environment, critical analysis, critical aspect, current practice, current research, customer involvement, December, depletion, designer, doubt, e procurement technology, eco innovation, eco-design, energy conservation, English, enterprise network, environmental measure, external sustainability collaboration, finance, food mile, ford, frequency, further investigation, future city, future research opportunity, future trend, great importance, Greece, green corridor, green effort, green scm strategy, green supply chain activity, green supply chain management strategy, green supply chain strategy, green training, green transport corridor, gscm literature, gscns, humanitarian logistic, ijpr, important area, important strategy, impoverished community,

(Continued)

TABLE 13.5 *(Continued)*
Cluster of Words of the Abstracts of the Published Works (1996 June 2018)

Cluster	Words
	Indonesia, innovation leadership, intermediary, internal sustainability practice, international business, January, journal article, international business, key area, key aspect, key characteristic, key performance indicator, Keyword, kpi, kpis, labor right, lean production, length, lfp, lgscm, local community, local food, logistics, logistics company, low carbon, low level, lscm, main characteristic, major contributor, man, manufacturing supply chain, may, mcdm technique, measurement system, Medium, mnc, modelling, multiple stakeholder, neighborhood rough set, new product development, notion, palm oil, performance management system, platform, plm, pms, port, poultry industry, poverty, pre, product service system, production research, prototype, qualitative analysis, qualitative data, resc, research agenda, research direction, research effort, research gap, research project, research stream, research study, rough set, round, scf, second part, separation, sfsc, simulation result, social medium, special issue, specialist, spread, springer, springer verlag london limited, sscm construct, sscm theory, stakeholder perspective, structured approach, student, supplier participation, supply chain decision, supply chain research, sustainability measurement, sustainability research, sustainable consumption, sustainable initiative, sustainable supply chain management research, sustainable supply chain performance measurement, sustainable supply chain practice, systematic analysis, thorough review, timely topic, web
6	Acceptance, accuracy, american society, analytical model, analytical result, budget, cartelization, channel, channel member, channel profit, civil engineers, collection incentive, collection method, contract, coordination strategy, coordinator, decentralized supply chain, decision variable, dgsc,dynamic environment, dynamic nature, e commerce, echelon supply chain, economic efficiency, energy efficiency level, environmental awareness, equilibrium condition, expected profit, fairness concern, first mover advantage, follower, game, game model, game theory, government subsidy, governmental intervention, green degree, green image, green level, green manufacturer, green market, green supply, green supply chain inventory management, green supply chain member, gsc, high cost, higher profit, home appliance, ibwt green supply chain, igsc, important managerial insight, incentive, initial stage, intention, leader, low carbon promotion level, lower limit, main result, manufacturer, market demand, maximization, member, monetary penalty, non cooperative game, non green product, numerical analysis, operational decision, operational inefficiency effect, operations performance, opportunistic behavior, optimal decision, optimal order quantity, order quantity, own profit, paper analyze, paper construct, part tariff contract, power structure, price, pricing, pricing policy, pricing strategy, product demand, product green degree, product sustainability, profit, quality level, ratio, recyclable part, recycled part, recycler, refund amoun, retail price, retailer, revenue, revenue sharing contract, rfid, rival, sale, sales price, saving, sensitivity, share, social welfare, spss, stackelberg game, subsidy, supply chain coordination, supply chain efficiency, sustainable investment, swm, system performance, system profit, tamilnadu coastal area, targeted advertising, theoretical guidance, used product, valuation, welfare, hole green supply chain, whole system, wholesale price

(Continued)

TABLE 13.5 (*Continued*)
Cluster of Words of the Abstracts of the Published Works (1996 June 2018)

Cluster	Words
7	3pls, absence, average, balanced scorecard, bank, bsc, capita income, capture, change, gent, coffee, complement, comprehensive understanding, core enterprise, corporate interest, countermeasure, critical element, csc, current paper, divergence, ecological environment, education, effective implementation, electric, energy demand, engineering management, environmental supply chain management, environmental trade off Escm, evolutionary game model, evolutionary game theory, experiment, explanation, extensive review, external environment, facet, first, focal organization, formula, furniture industry, further research direction, game analysis, ge adoption, Germany, gis, global economy, green capability, green paradigm, green supply chain coordination surplus, green supply chain innovation, green supply chain model, green supply chain performance, green supply chain system, grscm, high degree, home appliance industry, hour, income, indication, information systems, infrastructure, intermodal transport, internal capability, international trade, inventory management, iot, isolation, logic, logistics function, logistics performance, logistics performance index, logistics provider, logistics sector, logistics service, multiple attribute decision, occurrence, operational practice, operator, organizational structure, packaging material, panel, pmm, practical example, premium, product recycling, pythagorean hesitant fuzzy hamacher, qualitative research, rail, rapid development, recent study, research challenge, research india, publications, restaurant, road, shipper, significant difference, simulation experiment, simulation study, single case study, sss, supply chain literature, sustainability agenda, sustainability integration, sustainability literature, sustainability principle, sustainable supply chain management approach, sustainable supply chain operation, top, transport, triz, tuple linguistic neutrosophic number, vbo, water consumption, wcsscm, wood waste, yale university, year period
8	3d printing, air, architecture, assembly, automotive company, bullwhip effect, characterization, class, cloud, compatibility, comprehensive analysis, computational analysis, conference, control, conversion, cycle, data mining, decomposition, distribution network, drive, earth, ecology, energy saving, engineering, equipment, estimation, evaluation system, exploration, extraction, fabrication, finite element analysis, forest, formation, fruit, fuzzy comprehensive evaluation, graph, green environment, green material, hotel, human factor, hydrogen, influencing factor, installation, internal pressure, internet, machine, maintenance, management system, mass, matlab, mechanical property, metal, microstructure, mine, mode, multi agent system, neural network, new material, node, noise, numerical simulation, prediction, preparation, proceeding, processing, property, pso, recognition, resistance, risk analysis, safety, scheme, security, simulation, simulation analysis, special focus, stability, steel, suitability, synthesis, system, system, dynamic, technology transfer, thickness, thing, tin, tls application, treatment, tree, udt, variation, water, wheel, workshop

(Continued)

TABLE 13.5 (*Continued*)
Cluster of Words of the Abstracts of the Published Works (1996 June 2018)

Cluster	Words
9	Twenty-first century, advance, agricultural waste, agrifood sector, biodiesel, biorefinery, brief overview, burden, carbon, carbon dioxide, cellulosic biomass, chemical, collection, corn, crop, day, deal, deterioration, disadvantage, dollar, ecological footprint, economic, electricity, emergy, energy crop, energy security, environmental benefit, ethanol, ethanol production, exchange, expansion, farmer, feedstock, fertilizer, food production, fuel, government policy, green energy, greenhouse gase, grid, heat, important difference, international standard, land use, limit, major source, manuscript, natural gas, new method, organic cotton, graph framework, pacific northwest, paradigm shift, pcf, percent, petroleum, plastic, potential advantage, promise, range, removal, renewable resource, specificity, standardization, storage, sugar, sustainability metric, sustainable production, sustainable way, syrup, textile, ton, transition, united states, yield
10	Academician, administration, anova, barrier, brand new product, consultation, contextual relationship, corporate picture, critical success factor, csf, csfs, customer relationship, management, dependence, dependence power, dependent variable, discovery, driving, enabler, everyone, fuzzy Micmac, fuzzy micmac analysis, gas industry, gscm concept, gscm implementation, gscmbs, gscme, gscmes, hierarchy, indian automobile industry, indian company, indian industry, indian mining industry, indirect effect, industries, industry expert, influential barrier, integrated logistic, interdependence, interpretive structural modeling, interpretive structural modelling, ism, ism model, key barrier, key enabler, key variable, major barrier, Micmac, micmac analysis, money, mutual influence, mutual relationship, psc, recycled material, regulator, sensor, sscp, strategic importance, structural model, sustainable growth, tism, total interpretive structural modeling, useful tool, warranty
11	Adaptation, air pollution, architect, assessment framework, big data, bridge, building, carbon emissions tax, certification scheme, city, collector, computer, conservation, construction project, council, end user, fsc, fsc wood, interpretation, Ireland, japan, key stakeholder, landscape, macro level, mexico, million, new paradigm, Nigeria, non product, power plant, private sector, prospect, public procurement, purchaser, resiliency, self, sense, smart city, soil, space, special emphasis, speed, stock, strategic management, sustainable city, tension, tourism, trip, turkey, typology, urban planning, urban, sustainability, wood
12	academic expert, adverse impact, bwm, dynamic characteristic, economic stability, energy transition, external factor, external force, gas, gsc initiative, long term survival, o & g, o & g industry, oil, political stability, product effectiveness, sscm strategy, stakeholder pressure, stewardship theory, sustainability goal, sustainable product development, sustainable supply chain strategy, take

Source: Own elaboration using VOSviewer.

and sustainability of the banking sector. In this sense, it emphasizes terms such as sustainability agenda, supply chain, supply chain performance, and corporate interest among others. It also links terms related to environmental development, waste management, green engineering, green capacity, and sustainable management approach. Cluster 8 is related to infrastructure, both physical and organizational, in that sense it contains terms such as distribution network, integral analysis, human factor, security, energy saving, property, exploration, extraction, noise, scheme, security, and analysis of risk among others.

Cluster 9 contains words related to the environment, agricultural production, and natural products such as agricultural waste, agro-food sector, biodiesel, carbon, carbon dioxide, corn, crop, ecological footprint, economic, energy cultivation, energy security, benefit environmental, ethanol, ethanol production, exchange, expansion, farmer, raw material, fertilizer, food production, fuel, green energy, greenhouse, natural gas, and organic cotton among others. Cluster 10 corresponds mostly to words related to research and the results of these processes with respect to the administrative area. In this sense, it is composed of terms such as academic, administration, anova, barrier, new product, consultation, contextual relationship, corporate image, critical success factor, relationship with the client, management, dependence, power of dependence, dependent variable, discovery, driving, facilitator, everyone. On the other hand, it relates terms concerning the industry as an Indian company, Indian industry, Indian mining industry, indirect effect, industries, expert barrier industry, influential, integrated logistics, interdependence, modeled interpretative structure, interpretive structural modeling, ism, ism model, key barrier, key enabler, variable key. Cluster 11 consists mainly of terms related to environmental sustainability and mitigation of impacts. Therefore, it relates terms such as tax on carbon emissions, certification scheme, collector, conservation, construction project, council, wood, power plant, resilience, self-knowledge, smart city, land, space, strategic management, and sustainable. Finally, cluster 12 relates terms associated with the results of research on environmental impacts; therefore, contains terms such as academic expert, adverse impact, dynamic characteristic, economic stability, energy transition, external factor, external force, gas, initiative, survival long-term, industry, oil, political stability, product effectiveness, CSSCC strategy, stakeholder pressure, custody theory, sustainability objective, sustainable product development, sustainability, supply chain strategy, among others.

It is also worth mentioning that the most frequent themes of the publications between January and June of 2018 corresponded to country case studies; decision, optimization or evaluation models; mathematical modeling; applications to industrial sectors or services.

With respect to international co-authorships, papers that had at least 5 publications were selected. In this way, from 4,917 authors to 211 (Figure 13.11) are passed and the results obtained in Table 13.1 and Figure 13.3 are reinforced. With respect

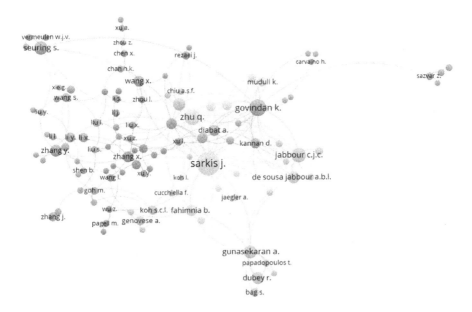

FIGURE 13.11   International co-authorship of the published works (1996 June 2018). (From Own elaboration using VOSviewer.)

FIGURE 13.12   Relations between institutions based on international co-authorship of published works (1996 June 2018). (From Own elaboration using VOSviewer.)

to collaboration between organizations, 4,676 institutions were identified, and those that had at least 2 articles were 458 (Figure 13.12). On the other hand, in Table 13.6 and Figures 13.13 through 13.15, the cluster of countries according to the international co-authorizes is shown. The countries with the largest number of jobs were China and the United States, the production of Latin America, is within the same ranges of Africa, except in the case of Brazil.

TABLE 13.6

Agglomeration of Countries, According to International Co-Authorship

Cluster	Countries
1	Argentina, Dominican Republic, Ecuador, Greece, Israel, Norway, Palestine, Puerto Rico, Spain
2	Botswana, Finland, Hungary, Netherlands, Panama, Poland, Republic of Korea, South Africa, United States
3	Belarus, China, Laos, Macau, Nicaragua, Philippines, Taiwan, Viet Nam
4	Australia, Bangladesh, Canada, Hong Kong, Lebanon, Qatar, Singapore, Thailand, United Arab Emirates
5	Indonesia, Iraq, Japan, Jordan, Malaysia, Oman, South Korea, Tanzania
6	Cyprus, Egypt, Ghana, Pakistan, Saudi Arabia, Sri Lanka, Tunisia, Turkey
7	Austria, Chile, Colombia, Cuba, Czech Republic, Mexico, Sweden
8	Algeria, Brazil, France, Luxembourg, Morocco
9	Ethiopia, Ireland, Italy, New Zealand, Portugal, United Kingdom
10	Denmark, India, Iran, Lithuania, Serbia
11	Belgium, Croatia, Estonia, Germany, Switzerland

Source: Own elaboration using VOSviewer.

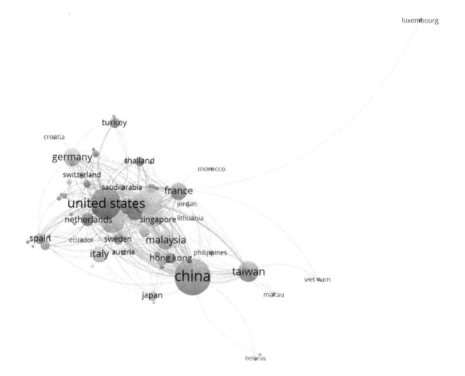

FIGURE 13.13 Agglomeration of countries, according to international co-authorship. (From Own elaboration using VOSviewer.)

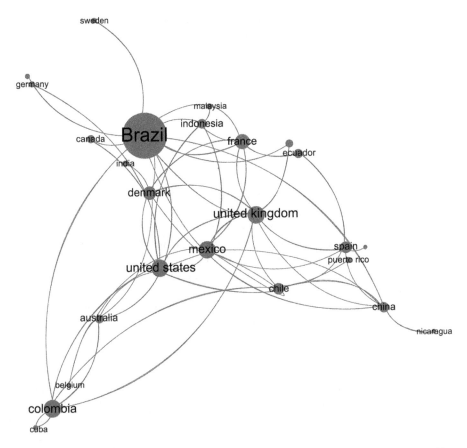

FIGURE 13.14 Agglomeration of countries, according to international co-authorship. (From Own elaboration using VOSviewer.)

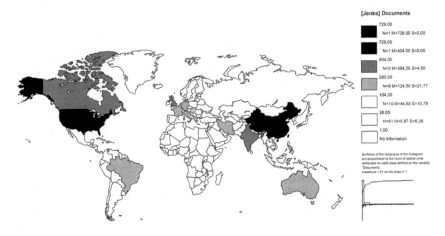

FIGURE 13.15 Production of documents by country. (From Own elaboration using Waniez, P., Philcarto [software], http://philcarto.free.fr, 2018.)

Cluster 1 and 7 (Figure 13.15) show more cooperation in terms of co-authorship in the Latin American region. For example, Colombia, Chile, Mexico, and Cuba have a high level of cooperation in the publication, while Argentina has more inter-action with the Dominican Republic, Ecuador, Puerto Rico, and Spain. The case of Brazil (Figure 13.14), which corresponds to the country in Latin America with the largest number of works, publishes more with authors of French-speaking institutions: Algeria, France, Luxembourg, and Morocco.

13.4 CONCLUSIONS

In this work, an exercise has been carried out for the mining of textual data, which extracted practical use information from a Big Data Source, such as Scopus. They tried to discover the relationships and the trends on sustainable supply chain.

In the area of supply chain, there are working persons coming from different areas: economists, engineers, and mathematicians. Some of them developing basic research but many of them dealing with partial problems. Evaluating the distortion of the best publications as well as fitting leadership's measures sources of information and their utility.

Our research sought to establish some regularities in the information structure, of papers covered by Scopus, which the similarities and differences between research institutions and universities. Analyzing scientific publications qualitatively and quantitatively and evaluating and comparing them with the research abroad (in this them) must provide of an important source of data for the researchers wishing to publish.

First, the annual variation of publications of authors of all countries were explored and estimated the visibility of the research carried out by Latin authors. Then, the co-authorship networks were analyzed by setting connections of at least two publications per country per year. Finally, text mining techniques were applied to semi-automatically identify the most recurrent topics or topics of the subject, based on the information available in the abstracts of the documents. Regarding this last point, we used the techniques as textual analysis: words counting (Qualitative analysis), texts Clustering (Text similarities), and texts classification (Text differences).

Among the main findings is that the issue of sustainable supply chain is not so recent in the academic literature and has been working extensively since 2008. The production of authors from Latin America, only representing 5.36%, which implies that it may not yet have fully permeated the interest of academics in the region. In this sense, it is the Brazilian researchers who write with researchers from international laboratories that lead the production of the region. In contrast, China is the country with the most academic production about the subject analysed; however, despite this, its citations are not high.

The analysis of text mining established that the topics of greatest interest and work during the period corresponded to application to the industrial organization; empirical research and good practices; forms of optimization and decision making; sustainability and carbon footprint; analysis and measurement; concepts; hierarchical analysis; evaluation; environmental impact; sustainable supply chain network; and Game theory and cooperation.

Colombia, Chile, Mexico, and Cuba present a high level of cooperation in the publication, while Argentina has more interaction with the Dominican Republic, Ecuador, Puerto Rico and Spain. The case of Brazil, which corresponds to the country in Latin America with the largest number of works, publishes more with authors of French-speaking institutions: Algeria, France, Luxembourg and Morocco.

Within the options of future work, it would be worthwhile to carry out this analysis by studying the interaction between the cluster of countries and the agglomerations of identified themes. It would be very interesting to apply hybrid analysis methods, which combine the cluster analysis presented here, and distance algorithms in order to attain better predictions about the behavior and evolution of the subject matter in the coming years.

Future work will include the use of other algorithms and comparing the effectiveness, both considering terms in abstracts and titles, in complete texts. Closely related to this idea is another way for future research corresponding to the mix between co-occurrence-based mining and the development of networks and graphs used to find functional associations based on criteria such as national origin of authors, publication journals, and language.

REFERENCES

Aria, M. & Cuccurullo, C. (2017). Bibliometrix: An R-tool for comprehensive science mapping analysis. *Journal of Informetrics, 11*(4), 959–975. doi:10.1016/j.joi.2017. 08.007.

Cheng, M. & Jin, X. (2019). What do Airbnb users care about? An analysis of online review comments. *International Journal of Hospitality Management, 76*(A), 58–70. Retrieved from https://www.sciencedirect.com/science/article/pii/S0278431917307491.

Cheng, M. & Edwards, D. (2017). A comparative automated content analysis approach on the review of the sharing economy discourse in tourism and hospitality. *Current Issues in Tourism*, 1–15. doi:10.1080/13683500.2017.1361908.

Elsevier. (2018). Scopus [Data set]. Retrieved from https://www.scopus.com/search/form. uri?display=basic.

Eskici, H. B. & Koçak, N. A. (2018). A text mining application on monthly price developments reports. *Central Bank Review, 18*(2), 51–60. doi:10.1016/j.cbrev.2018.05.001. Retrieved from https://www.sciencedirect.com/science/article/pii/S1303070118300271.

Feldman, R. & Sanger, J. (2007). *The Text Mining Handbook. Advanced Approaches in Analyzing Unstructured Data.* New York: Cambridge University Press.

Franceschini, A., Szklarczyk, D., Frankild, S., Kuhn, M., Simonovic, M., Roth, A., Lin, J., Minguez, P., Bork, P., von Mering, C., & Jensen, L. J. (2012). STRING v9. 1: Protein-protein interaction networks, with increased coverage and integration. *Nucleic Acids Research, 41*(D1), D808–D815.

Hotho, A. Nürnberger, A., & Paaß, G. (2005). Brief survey of text mining. LDV Forum GLDV. *Journal for Computational Linguistics and Language Technology, 20*(1), 19–62.

IBM. (2018). Acerca de la minería de textos. Retrieved from https://www.ibm.com/support/ knowledgecenter/en/SS3RA7_sub/ta_guide_ddita/textmining/shared_entities/tm_ intro_tm_defined.html.

Justicia de la Torre, M. D. C. (2017). *Nuevas técnicas de minería de textos: Aplicaciones.* Granada: Universidad de Granada. Retrieved from: http://digibug.ugr.es/handle/10481/46975.

Kannan, D., de Sousa Jabbour, A. B. L., & Jabbour, C. J. C. (2014). Selecting green suppliers based on GSCM practices: Using fuzzy TOPSIS applied to a Brazilian electronics company. *European Journal of Operational Research, 233*(2), 432–447.

Kayser, V. & Blind, K. (2017). Extending the knowledge base of foresight: The contribution of text mining. *Technological Forecasting and Social Change, 116*, 208–215. Retrieved from https://www.sciencedirect.com/science/article/pii/S0040162516304784.

Lulewicz-Sas, A. (2017). Corporate social responsibility in the light of management science—Bibliometric analysis. *Procedia Engineering, 182*, 412–417. doi:10.1016/j.proeng.2017.03.124.

Mørk, S., Pletscher-Frankild, S., Caro, A. P., Gorodkin, J., & Jensen, L. J. (2014). Protein-driven inference of miRNA–disease associations. *Bioinformatics, 30*(3), 392–397.

Perianes-Rodriguez, A., Waltman, L., & Van Eck, N. J. (2016). Constructing bibliometric networks: A comparison between full and fractional counting. *Journal of Informetrics, 10*(4), 1178–1195.

Rodriguez-Esteban, R. (2019). Text mining applications. *Encyclopedia of Bioinformatics and Computational Biology, 3*, 996–1000. doi:10.1016/B978-0-12-809633-8.12372-6.

Santana Mansilla, P., Costaguta, R., & Missio, D. (2014). Aplicación de Algoritmos de Clasificación de Minería de Textos para el Reconocimiento de Habilidades de E-tutores Colaborativos. Inteligencia Artificial. *Revista Iberoamericana de Inteligencia Artificial, 17*(53), 57–67.

Seuring, S. & Müller, M. (2008). From a literature review to a conceptual framework for sustainable supply chain management. *Journal of Cleaner Production, 16*(15), 1699–1710.

Srivastava, S. K. (2007). Green supply-chain management: A state-of-the-art literature review. *International Journal of Management Reviews, 9*(1), 53–80.

Su, C. J. & Chen, Y. A. (2018). Risk assessment for global supplier selection using text mining. *Computers & Electrical Engineering, 68*, 140–155. Retrieved from https://www.sciencedirect.com/science/article/pii/S0045790617328483.

Van Eck, N. J. & Waltman, L. (2011). Text mining and visualization using VOSviewer. *ISSI Newsletter, 7*(3), 50–54.

Van Eck, N. J. & Waltman, L. (2014). Visualizing bibliometric networks. In Y. Ding, R. Rousseau, & D. Wolfram (Eds.), *Measuring Scholarly Impact: Methods and Practice* (pp. 285–320). Springer.

Van Eck, N. J. & Waltman, L. (2018). *VOSviewer Manual.* Retrieved from http://www.vosviewer.com/getting-started#VOSviewer%20manual.

Waniez, P. (2018). Philcarto [software]. Retrieved from http://philcarto.free.fr.

Watanabe, R., Fujii, N., Kokuryo, D., Kaihara, T., Onishi, Y., Abe, Y., & Santo, R. (2018). A study on support method of consulting service using text mining. *Procedia CIRP, 67*, 569–573.

Westergaard, D., Stærfeldt, H. H., Tønsberg, C., Jensen, L. J., & Brunak, S. (2018). A comprehensive and quantitative comparison of text-mining in 15 million full-text articles versus their corresponding abstracts. *PLoS Computational Biology, 14*(2), e1005962. doi:10.1371/journal.pcbi.1005962.

Zhao, Y., Xu, X., & Wang, M. (2019). Predicting overall customer satisfaction: Big data evidence from hotel online textual reviews. *International Journal of Hospitality Management, 76*, 111–121. Retrieved from https://www.sciencedirect.com/science/article/pii/S0278431917310083.

14 Multi-criteria Decision-Making Techniques to Address Sustainable Procurement in Supply Chain Operations

Isabel M. João

CONTENTS

14.1 INTRODUCTION

A few years ago, procurement/purchasing was mainly concerned with quality versus price, but it has been evolving to integrate societal and environmental concerns into procurement activities mainly to reduce negative impacts upon social conditions, health, and environment (Giunipero et al., 2006). The procurement function is an influential agent of change in a company/organization due to the fact of being a process by which the companies acquire supplies, assets, and services taking into consideration several issues in addition to the value for money (e.g., the entire life cycle of products/services, the effect of the supplies/assets/services on the environment, the social aspects and the effects of factors such as labor conditions, human rights, poverty eradication, and use of recycled products) (Interagency Procurement Working Group, 2012).

The concept of sustainability in supply chain management (SCM) has gained relevance, at least in the last 20 years, and research on sustainable SCM has grown-up and gained more attention in the research community. Several authors defined sustainable SCM (Carter & Rogers, 2008; Spence & Bourlakis, 2009; Tate et al., 2010) and they all take into consideration the pillars of sustainable development in the complete coordination of crucial business processes related with core activities purchasing/procurement of products/services between organizations to satisfy the expectations of all stakeholders and increase the efficiency of the organizations and the whole supply chain (SC).

The literature review in multi-criteria decision-making methods (MCDM) applied to sustainable procurement/purchasing practices is challenging due to the commitment to evaluate the extent of MCDM usage in sustainable procurement/purchasing decisions. The complexity of green/sustainable procurement/purchasing in SC and sustainable development and the attainment of a sustainable future require efforts that must occur in a structured as well as transparent process. MCDM can contribute to achieve that goal. Despite the importance of MCDM to address sustainable procurement (SP) in SC operations much remains to be done in terms of the literature review to account for those issues. To facilitate further research on MCDM, this chapter presents a comprehensive literature review concerning MCDM approaches related to sustainability issues in SC with emphasis on suppliers and sourcing related decisions.

Sustainable procurement (SP) has been in the sights of researchers and practitioners, and the purchasing/procurement function nowadays tends to play a central role in considering environmental and social issues into all processes of an organization, thus having a huge impact on its performance (Green et al., 1998; Green et al., 2012). Topics such as green and sustainable procurement/purchasing are gaining prominence in the academic debate, in industry, and in private and public organizations. The Rio Earth Summit in 1992 strongly contributed the emergence of the SP concept during the 1990s. Green public policies started to appear at international levels, and some evolved into sustainable procurement policies.

The use of MCDM methods in supplier evaluation is crucial for developing and maintaining a long-term partnership with suppliers. There is some literature surveying the multi-criteria supplier evaluation approaches (Govindan, et al., 2015b; Huang et al., 2010). The modeling complexity of procurement decisions and the extensive nature of the process makes it heavily dependent on MCDM tasks. The multidimensionality of the sustainable development concept with environmental, social, and economic perspectives involves a high degree of conflict. This is the reason why MCDM approaches are suitable to deal with sustainability trade-offs at macro levels as well as micro levels of evaluation (Munda, 2005).

This chapter aims to collect and analyzing the existing contributions on MCDM techniques to address SP in SC operations. To this end questions to answer are:

1. What is the evolution of published articles on MCDM related to SP over the last 20 years?
2. What is the spread of the research studies based on publication source?
3. Which MCDM techniques and approaches have been used in the context of SP in SC operations?
4. Which are the main areas of application of MCDM techniques regarding SP in SC operations?

In addition to the literature review, this chapter will also analyze the decision-making methods used in the context of sustainable procurement while exploring the strengths and weaknesses of the MCDM methods. The chapter is organized as follows: Section 14.2 exhibits some definitions/terminology associated with SP including previous literature on green/sustainable procurement/purchasing and some of the most recent literature reviews on sustainable SCM—one of many responsibilities that can be faced by the procurement function. The procurement function has also a crucial role in the implementation of a sustainable strategy in the organization due to the case of dissemination of those values through the entire SC (Yamak et al., 2011).

Section 14.3 describes the main MCDM techniques. The MCDM techniques used in the literature review concerning SP in SC operations are those techniques illustrated in some well-known books devoted to multi-criteria decision-making. Previous literature reviews about MCDM, in the context of SP, are presented as well as a classification of hybrid MCDM techniques and approaches used in this literature review. Section 14.4 covers the research methodology and the literature review steps with the material collected and the analysis and synthesis of the selected articles. Section 14.5 presents the discussion and analysis of the results presented in Section 14.4 to answer the four research questions and explores the strengths and weaknesses of the main MCDM techniques used in SP. Lastly, Section 14.6 draws the final conclusions.

14.2 SUSTAINABLE PROCUREMENT

Sustainable procurement (SP) is a strategic issue that has been gaining acceptance through the years (Kumar et al., 2008) not solely in the public, but also in private sectors where firms have dealt with sustainable procurement/purchasing mostly on a voluntary basis (Beckmann et al., 2014; Quarshie et al., 2016). Guidelines on sustainability inspire the procurement professionals making decisions that cover the pillars of the Triple Bottom Line (TBL), but research shows that in most cases actions are restricted to the environmental dimensions of the TBL (Meehan & Bryde, 2011). However, the procurement can have a very important role to play in the sustainable development agenda due to its capability to engage external organizations in the SC (Green et al., 1996). In the present era of globalization, production and consumption take place in global markets outside the boundaries of a single nation, so a big challenge to procurement and SCM is the requirement to adopt sustainable practices (Johnsen et al., 2017). There are a growing number of internationally accepted standards such as ISO 14001 (Corbett & Kirsch, 2001), Eco-Management and Audit Scheme (EMAS) initiative, ISO 26000, or ISO 20400 that provides guidance to organizations (Sycamore & Stowers, 2015). These standards can be used by all types of organizations independently of their activity or size and guide them on integrating sustainability issues into procurement decision-making processes, as described in ISO 20400. Regarding the environmental dimension, the ISO 14001 stimulates organizations to review their purchasing decisions and inspect the environmental impacts of the goods and services that they buy enabling them to search for those purchases that have a good track record in terms of sustainability. Johnsen and co-workers have little doubt that sustainability is a very important current topic within procurement and SCM research (Johnsen et al., 2017). Some recent literature review concerning green/sustainable procurement shows that those issues have attracted attention from researchers and industry alike. The literature review of Appolloni et al. (2014) identified the main issues of green procurement (GP) excluding the public sector. Cheng et al. (2018) made a critical literature review related with green public procurement (GPP) to identify missing concepts and future trends concluding that the GPP debate is predominantly restricted to aspects related to implementation issues having still a long way to go. SP is currently in use in private and public sectors and in most fields of research, but a huge work is still going on engaging external organizations in the SC. Ghadimi et al. (2016) study the buyer-supplier relationship in SP framework carrying out a comprehensive review of articles between 2008 and 2014. Also, Johnsen et al. (2017) made a bibliographic review of sustainable purchasing and supply research evaluating and comparing the theory that strengthen them. Wetzstein et al. (2016) performed a systematic bibliographic review related to the assessment of suppliers, and their findings show that green and sustainable strategy to supplier selection and assessment represent an upcoming stream. There are challenges to overcome to ensure sustainability of the procurement/purchasing policy and more generally to ensure the sustainability of the value chain management. Appolloni et al. (2014) found out that research dedicated to GP is growing fast concerning the quantity and diversity of articles and journals addressing the theme of GP and exceeds the work developed in 2008 (Walker et al., 2008). The usage of GP in the public sector is covering new sectors, for example, the healthcare sector (Chiarini et al., 2017) and catering services (Neto & Gama Caldas, 2017).

Despite the work already developed, there is the need to better understand the procurement function to successfully undertake sustainable SC challenges. SCM is one of many responsibilities faced by the procurement function. Procurement decision-making opens a lot of opportunities to encompass the three pillars of sustainability into all processes and activities contributing to the increase of a social positive impact and a decrease of negative environmental impacts due to business activities (Giunipero et al., 2006; Lawson et al., 2009). Also, a lot of literature review work on Sustainable supply chain management (SSCM) has been produced with great emphasis in the recent years (Ahi & Searcy, 2013; Ansari & Kant, 2017; Ashby et al., 2012; Beske & Seuring, 2014; Brandenburg et al., 2014; Brandenburg & Rebs, 2015;

Bush et al., 2015; Carter & Liane Easton, 2011; Carter et al., 2008; Eskandarpour et al., 2015; Gold et al., 2010; Hassini et al., 2012; Khalid et al., 2015; Seuring & Gold, 2012; Seuring & Müller, 2008; Simon et al., 2015; Touboulic & Walker, 2015; Winter & Knemeyer, 2013). Companies need to continue their efforts to assimilate sustainability considerations, more methodically and meaningfully, into the procurement process. To accomplish this, the requirement to comply with SSCM practices of the organizations should be included in the regular contracts and tender documentation, and this framework should be effectively implemented.

14.2.1 DEFINITIONS OF SUSTAINABLE PROCUREMENT

The topic of SP is an extension of the GP topic, and several researchers defined and used different terminologies and concepts, for example, environmental procurement or environmental purchasing and green purchasing. Green et al. (1996) illustrated the environmental impacts of purchasing decisions by taking into consideration the many steps of the total procurement process. They noticed that environmental impacts are distributed through the chain, including the design/specification, the tenderer, the purchasing, and the distribution, rather than only in the production process. Since the mid-1990s the companies were increasingly pushed to re-define waste in the light of environmental legislation and regulation and encouraged to leave the old-style *end of pipe* solutions to embrace new and creative methods to source reduction and waste management. This was mainly due to the considerable interest in the emergence and impact of standards like BS 7750 and EMAS regulation that contributed to the reduction of environmental damages caused by every type of organization (Green & Irwin, 1996).

Carter and Carter (1998) defined environmental purchasing as:

> *the purchasing function's involvement in activities that include reduction, recycling, reuse, and substitution of materials.* (p. 659)

Zsidisin and Siferd (2001) proposed an improved definition for the environmental purchasing for a firm as:

> *the set of purchasing policies held, actions taken, and relationships formed in response to concerns associated with the natural environment. These concerns relate to the acquisition of raw materials, including supplier selection, evaluation and development; suppliers' operations; inbound distribution; packaging; recycling; reuse; resource reduction; and final disposal of the firm's products.* (p. 69)

The definitions used by different authors, countries, international and public organizations may be different, but the main idea is the same and gives emphasis to an oriented policy tool to reach the goals concerning optimal environmental performance promoting green products/services. The role of public and private procurement can be very important in changing the non-sustainable patterns of production/consumption (Bouwer et al., 2006).

The definitions provided by the different authors have been evolved over the years. The definition of Carter and Carter (1998) is restricted to technical issues such as reducing, recycling, reuse, and materials' substitution. Some more recent definitions have a holistic perspective of GP and are more open to the emerging new issues as accepted by Appolloni et al. (2014) that adopted the Large and Gimenez Thomsen (2011) definition of environmental purchasing or green purchasing or environmental purchasing as:

> *an integration of environmental considerations into purchasing policies, programs, and actions.* (p. 177)

SP can be defined in a few ways and some of the definitions are discussed to better understand the strategic role of the procurement function. The Department for Environment – DEFRA, UK (2006) defined SP as:

> *a process whereby organizations meet their needs for goods, services, works and utilities in a way that achieves value for money on a whole life basis in terms of generating benefits not only to the organization, but also to society and the economy, whilst minimizing damage to the environment.* (p. 10)

The formulation is a versatile way to define SP that embraces the dimensions of the TBL approach. The definition reflects the economic, environmental, and social consequences related to product/process design and development, use of non-renewable sources of raw materials, production, product delivery, and   product use and disposal. It also reflects the suppliers' abilities to take into consideration all the outcomes through the SC. Miemczyk, Johnsen and Macquet (2012) defined sustainable purchasing and supply management as:

> *the consideration of environmental, social, ethical and economic issues in the management of the organization's external resources in such a way that the supply of all goods, services, capabilities and knowledge that are necessary for running, maintaining and managing the organization's primary and support activities provide value not only to the organization but also to society and the economy.* (p. 489)

Oruezabala and Rico (2012) also defined SP as:

> *the efforts of an organization to achieve or simply improve performance of buying activities in three ways: environmentally, socially and economically.* (p. 574)

The public sector has a huge impact in sustainable development, and public procurement can be an important driver of SP. The integration of environmental criteria for public services and products was referred by some researchers (Evans et al., 2012) that corroborate the importance of green public procurement (GPP) tool to the achievement of environmental policy goals relating to climate change, use of resources, and sustainable patterns of production and consumption mainly due to the importance of the public sector spending on products/services. The European

Commission has begun activities to raise the level of Green Public Procurement (GPP) in all Member States in 2006 (Bouwer et al., 2006) and was defined in 2008 by the European Commission's communication as:

> *a process whereby public authorities seek to procure goods, services and works with a reduced environmental impact throughout their life cycle when compared to goods, services and works with the same primary function that would otherwise be procured.*
> (Commission of the European Communities, 2008)

Also, von Oelreich and Philp (2013) stated green public procurement (GPP) as the activity of procurement to use as a policy instrument for reaching environmental quality objectives. In the literature, sustainable public procurement (SPP) has a broader meaning when compared to GPP including the three pillars of sustainability.

According to ISO 20400:2017 (ISO, 2017) SP is defined as:

> *the procurement that has the most positive environmental, social and economic impacts possible over the entire life cycle.* (p. 6)

The ISO 20400:2017 is a guidance standard for SP that works well with ISO 14001:2015 and ISO 26000:2010 and can help organizations already working with either or both standards by encouraging them to meet their requirements to become able to achieve value for money raising benefits for the organization, society, while minimizing loss for the planet resource's.

Also, the EU Public Procurement reform introduced in 2016 provides more opportunities to choose socially responsible goods, services, and works (EPA Network, 2017). Ethical and responsible procurement will increase the transparency of the sourcing process, and SP should incorporate the benefits for the organization as well as for the society. While considering the impact of the TBL dimensions along with the price and quality, procurement professionals should also be aware of unacceptable practices (CIPS, n.d.). While in the past SP was largely focused on supplier selection and evaluation, compliance, and auditing, procurement professionals nowadays are a major force of change and can do much more with their SP efforts. Public and private organizations are gradually paying much more attention to the sustainable practices that will enhance social benefits, while minimizing environmental impacts and unacceptable practices to achieve long-term economic viability. Private and public organizations can significantly benefit from the development of effective decision support systems methodologies and approaches that successfully integrate all these criteria; therefore, improving the quality of the decision makers' decisions. Multi-criteria decision-making (MCDM) is as a suitable approach to address sustainability conflicts and trade-offs. Sustainability is by nature a multi/interdisciplinary issue, which is why a multi-criteria framework can help to create a common language facilitating the communication process that is always very difficult, mainly at the beginning when framing and structuring a decision problem (Figueira et al., 2005; Munda, 2005).

14.3   MULTI-CRITERIA DECISION-MAKING (MCDM) TECHNIQUES

Since the 1960s, MCDM has been an increasing vigorous research area within operations research (OR). The method of explicitly taking into consideration the multitude of different points of view of different individuals is the focus of MCDM. Regardless of the diversity of MCDM approaches and methods, the basic components of MCDM are very simple and consists of a set of alternatives (i.e., options, solutions, and so on), considers multiple criteria to assess (i.e., minimum two different criteria), and supports at minimum one decision maker (DM) or several decision makers (DMs) to evaluate different options (e.g., products, strategies, technologies, policies, processes, and so on) (Belton & Stewart, 2002). MCDM has been growing during the last half century and has as its purpose the development of adequate methodologies to aid DMs in solving situations in which multiple, and sometimes conflicting, factors (e.g., objectives, goals, criteria) need simultaneous assessment. MCDM helps DMs make decisions sometimes very difficult due to the complexity of the problem. MCDM is much more comprehensive than just a set of methods and theories. MCDM consists in a detailed framework to address the problems, and there is a huge amount of MCDM applications in real-world problems and very diverse fields (Figueira et al., 2005).

Mardani et al. (2015a) made a literature review concerning techniques and practical applications of MCDM from 2000 to 2014. They concluded that multiple criteria techniques have strong contributions in fields such as business, economy, production, energy and environment, among many others, and MCDM techniques can contribute to improve the quality of decisions. Diaz-Barriga-Fernandez et al. (2017) analyzed and critically assessed the available literature on multiple criteria methods explicitly to evaluate systems sustainability. The results showed that these techniques were used in very different situations, levels, and sectors, related to sustainability mainly due to the multidimensionality that is intrinsic to sustainability and sustainable development issues. The authors concluded that during the past few years (i.e., the study covers the period from 1999 to 2015) there has been a great propagation of studies adding sustainability criteria by the use of MCDM. This is a sign of the importance of these methods in problems devoted to the progress and application of systematic approaches to decision problems that involve the use of several criteria, goals, and objectives.

In the 1960s, basic MCDM concepts were explicitly considered, and it is worth to refer to the work on goal programming (GP) by Charnes and Cooper (1961) and the ELECTRE methods by Roy (1968). Since the 1970s, MCDM has been gaining a growing recognition with the conference on *Multiple Criteria Decision Making* by Cochrane and Zeleny (1973) as a starting point for the MCDM scientific field of research. In the last half century, MCDM has received huge attention from researchers and practitioners due to the diversity of MCDM methods that makes them very useful in different situations. The selection of the MCDM methods to use, or a combination of methods, will deeply depend on the specific problem under evaluation. Outranking methods applying the definition of outranking (Benayoun et al., 1966) are, for example, the ELECTRE (ELimination Et Choix Traduisant la REalité - ELimination and Choice Expressing REality) methods that are based on concordance analysis with ELECTRE I known as a reference for the first outranking method. A complete handling of ELECTRE techniques can be seen in detail in this references (Roy & Bouyssou, 1993; Vincke, 1989). Another class of very well-known outranking methods is PROMETHEE (Preference Ranking Organization Method for Enrichment Evaluations) techniques. PROMETHEE I and II refer to partial and complete ranking and emerged due to the work of Brans (1982). The ranking based on the interval and the continuous case related to PROMETHEE III and IV was developed later by J.P. Brans and B. Mareschal. The family of outranking methods, PROMETHEE, can be found in huge detail in Mareschal et al. (1984). In 1988, the authors developed the graphical representation called GAIA to support the PROMETHEE method. Behzadian et al. (2010) made an extensive literature review of methods and applications based on PROMETHEE, where they concluded the attractiveness of the method to academics and practitioners in areas including logistics and transportation, environmental management, energy management, manufacturing, and assembly, among others.

Multi-attribute utility theory (MAUT) is an approach that represents the preferences of a DM and needs the development of utility functions and scale constants for each attribute, which are then aggregated into a single synthesis criterion (Keeney & Raiffa, 1993). The MAUT is a valuable way of modeling preferences and it is appropriate to problems involving risk. To distinguish between theories for preference based on the concept of ordinal comparisons and strength of preference under conditions of certainty, the term value function is used instead of the utility function and multi-attribute value theory (MAVT) is the appropriate option. If the cardinal value function is additive, the weights can be calculated using the same procedure proposed for estimating the weights for the additive ordinal value function (Keeney & Raiffa, 1993). MACBETH (Measuring attractiveness by a categorical-based evaluation technique) allows to measure the differences of attractiveness between two options at a time against multiple criteria. The method only requires the use of qualitative judgments and is able to generate the options' scores in each criterion as well as weighing the criteria (Bana e Costa & Vansnick, 1994; Bana e Costa et al., 2011). For a literature review on MACBETH of the last 20 years see Ferreira and Santos (2018).

To build measurable multi-attribute value functions, the ratio judgments given by the AHP (analytic hierarchy process) method are a way to assess the functions. The AHP builds relative scales of numerical weights, known as priorities, from DM judgments expressed by reducing complex problems to a series of pairwise judgement comparisons and results aggregation (Saaty, 1987). The AHP structures the problem in a hierarchical structure, and the disaggregation of the problem allows the DM to evaluate the alternatives regarding to subsets of evaluation criteria. The method is based on three main steps: the disaggregation, pairwise judgements comparisons, and determination of priorities that requires the calculation of the maximum eigenvalue, consistency index, consistency ratios, as well as the normalized values for each criterion and option. The global priorities of options are obtained by additive aggregation of the local priorities with normalization of the sum to unity.

UTA (UTilitès Additives) techniques refer to the approach of evaluating a group of value/utility functions, presupposing the axiomatic basis of MAUT and assuming the preference disaggregation principle. The UTA method was developed by

Jacquet-Lagreze and Siskos (1982), and its goal is to infer additive value functions from a given ranking on a reference group of options.

The Technique for Order of Preference by Similarity to Ideal Solution (TOPSIS) was first proposed by Yoon and Hwang (1981), with some developments made by Hwang et al. (1993) and Yoon and Hwang (1995). Zavadskas et al. (2016a) present an overview on TOPSIS developments from 2000 to 2015 to solve complex decision making issues. In a simple form, the options are ranked having underlain their distances from positive/negative ideal solutions, and the best option is obtained by the smaller distance calculated from the positive ideal solution and by the largest distance in relation to the negative ideal solution. As pointed out by the bibliometric-based survey performed by Zyoud and Fuchs-Hanusch (2017), the technique is widely used in SCM and sustainability subjects and recognized as a powerful MCDM method to support strategic decisions. These authors made a literature review on AHP an TOPSIS and, globally, they verified the increasing recognition of the reviewed methods to support strategic decisions. Analytical network process (ANP) is a more universal approach to decision problems based on a generalization of hierarchies to networks with dependence and feedback.

There are other non-classical MCDM approaches like the decision rule (DR) approach. The mathematical foundation of the DR approach to MCDM is the Dominance-based rough set approach (DRSA) established by Greco et al. (2001, 2002). Contrary to the MAUT or MAVT and the outranking approach models that require specific preferential information, the DRSA uses some sort of preferential information closer to the natural way of thinking of the DM.

Another type of approach based on multi-objective optimization by ratio analysis (MOORA) was presented by Karel et al. (2006). Also, MOORA plus the full multiplicative form (MULTIMOORA) methods were subsequently developed by those authors (Brauers & Zavadskas, 2010).

The VIKOR method (Vise Kriterijumska Optimizacija I Kompromisno Resenje) allows for estimating the compromise solution of a decision problem when conflicting criteria exists, and it can be used to address complex systems. The method allows the determination of the compromise ranking/solution and the weight stability intervals for the compromise solution. (Opricovic & Tzeng, 2004). Gul et al. (2016) performed a bibliographic review on VIKOR method and fuzzy extensions with focus on the main research applications.

Another decision-making approach is DEMATEL, a technique that was first developed by Gabus and Fontela (1972, 1973) to envisage the structure of complex causal relations through matrices. DEMATEL is the acronym for decision-making trial and evaluation laboratory, and the method evaluates interrelations among factors through an intelligible structural model. Si et al. (2018) made a systematic review on the application of DEMATEL from 2006 to 2016. The literature review showed that a series of modified DEMATEL approaches have been developed, holding valuable insights for the use of the DEMATEL technique, and giving current research trends and directions for further research.

The work of Charnes et al. (1978) on data envelopment analysis (DEA) has been widely recognized as an effective technique for the measurement of the relative efficiency of a set of decision-making units (DMUs) that apply a multiplicity of inputs to produce a multiplicity of outputs, with many theoretical developments and practical applications being reported. The popularity of DEA is due to its capability to measure relative efficiencies of multiple-input and multiple-output DMUs without previous weights of the inputs and outputs. DEA makes use of a linear programming method to evaluate the relative efficiencies of options (Thanassoulis et al., 2012). The method evaluates the efficiencies between options, with the most efficient option having a value of 1.0, and the remaining options having a value that is a fraction of 1.0. The method has the ability to handle multiple inputs/outputs and calculate and analyze the efficiency. For a DEA conceptual exposition and literature review see (Lu & Wang, 2017). A wide list of DEA references was reported by Emrouznejad and Yang (2018). Also, Mardani et al. (2017a) performed a literature review of DEA related to energy efficiency, where DEA was evidenced to be a very promising technique to deal with the problems of energy efficiency, especially when the production function among inputs/outputs is nonexistent or extremely difficult to acquire.

Goal programming (GP) is a branch of optimization methods used when in the presence of multiple conflicting objectives. The method was properly introduced as a class of MCDM methods by Charnes et al. (1955), and the applications of goal programming as important tools in operations research and computer science increased a lot over the last years with applications in engineering, management, and social sciences (Colapinto et al., 2017).

There are several different approaches for dealing with uncertainty in MCDA. To apply fuzzy set theory for MCDA, the DMs must first identify ambiguous elements (e.g., specific criteria weights) to frame the decision problem. Then, it is necessary to define fuzzy sets and the membership functions to determine the ambiguity. The fuzzy set theory is defined a degree of membership to a set (Chen & Hwang, 1992; Zadeh, 1965). The degree of an element's membership to a fuzzy set is stated as a number ranging from zero (not belonging to the fuzzy set) and one (completely belonging to the fuzzy set). In their literature review, Broekhuizen et al. (2015) recognized five normally used approaches to measure and integrate uncertainty in MCDA: probabilistic and deterministic sensitivity analysis, Bayesian frameworks, fuzzy sets, and GREY theory. In GREY theory, the uncertainty is represented in terms of ranges named black, white, or GREY numbers representing the order of magnitude of uncertainty (Ju-Long, 1982).

Due to the vast number of available approaches, the selection of a detailed MCDM method, or combinations, must be made taking into attention the appropriate know-how of the fundamentals of the approach as well as the nature of the problem to study. This implies that the researcher needs to recognize that some aspects can be well covered by some methods and not by others, so the approach to adopt must be the object of a deep reflection by the researcher. Mardani et al. (2015a) in their systematic literature review based on MCDM approaches concluded that, over 15 years of research (i.e., 2000–2014), the MCDM methods were used in several applications and industrial sectors including SCM, quality management, manufacturing systems, safety and risk management, energy, environment, and sustainability, among other fields. Relating to MCDM approaches, the AHP technique was ranked in the first place, with hybrid MCDM techniques ranked in second place, but also TOPSIS. ELECTRE, ANP, DEMATEL, PROMETHEE among others were used in several applications.

14.3.1 PREVIOUS LITERATURE REVIEW ON MCDM
FOR PROCUREMENT/PURCHASING

There are several MCDM techniques and approaches with huge potential to address SP in SC operations, but despite the rapid growth of literature concerning green procurement/purchasing and the fast growing of MCDM techniques and their applications, there is still limited literature review studies addressing MCDM techniques applied specifically in green/sustainable procurement/purchasing.

Ho et al. (2010) made a bibliographic review of the MCDM approaches for supplier evaluation/selection from 2000 to 2008 and they concluded for the superiority of MCDM approaches to solve the supplier selection problems. Also, Wu and Barnes (2011) made a bibliographic review of MCDM for partner selection from the beginning of the twenty-first century until 2011, but without specific focus on green, environmental or a wider sustainability procurement emphasis, and they concluded that it is necessary to build up methods that can combine qualitative/ quantitative objectives that are usually addressed in partner selection problems. Chai et al. (2013) knowing the relevance of MCDM for supplier selection and being aware of the lack of systematic literature review on the subject conducted a review addressing the usage of decision-making techniques in supplier selection but not with strict sustainability focus. This was also the case of Guarnieri (2015) that made a summary of the main criteria, methods, and topics of multi-criteria supplier selection by a review of literature published in the period 2001–2012. Simić et al. (2017) completed the literature review of the last 50 years of fuzzy set theory, fuzzy decision-making, and hybrid solutions with a focus on fuzzy used in models for supplier assessment/selection. The research has shown that fuzzy hybrid approaches are useful to solve problems where the complexity is a key issue such as supplier evaluation. Banasik et al. (2016) focused their research on MCDM approaches for green supply chains, and the analysis showed that the use of MCDM methodologies for designing green supply chains is somehow a new and emerging research area. This is in accordance with the bibliographic review of MCDM and their applications done by Mardani et al. (2015a) with only 5.9% of papers from supply chain management published in the period of 2000–2014 in peer reviewed articles picked out from the Web of Science (WoS). The conclusions of Govindan et al. (2015b) are in the same line. Based on the MCDM approaches to green supplier selection/ evaluation in the period 1997–2011, they concluded that the employment of green topics within the supplier selection/evaluation problem is limited due to the relatively few papers identified. The authors found AHP among the most widely used MCDM approaches for green supplier evaluation/selection, and the most extensively used criteria was the environmental management system. According to those authors, many of the latest MCDM methods and tools integrate fuzzy logic and applications of Fuzzy MCDM (FMCDM). This is in accordance with the literature review performed in MCDM methods and applications by Mardani et al. (2015b) that highlighted the increasing importance of hybrid MCDM tools and approaches (HMCDM).

14.3.2 Classification of Hybrid MCDM Techniques

Hybrid multi-criteria decision-making (HMCDM) is a relatively new and increasingly adopted approach in decision-making developments and applications. HMCDM can support the decision maker, or decision makers, by jointing different methods and tools that allow the handling of miscellaneous material, interrelated criteria, or conflicting criteria, as well as different stakeholders' preferences and uncertain environments. Recently, Zavadskas et al. (2016b) made a review of HMCDM for sustainability issues and concluded that the HMCDM approaches have been progressively applied to take care of decisions in several sustainability areas, but the review was not with the focus in sustainable procurement. The authors emphasize the advantages of hybrid methods when compared to individual methods concluding that hybrid methods constitute an improvement to assist decision makers dealing with sustainability issues.

In the current research, the acronym HMCDM is used to identify the examined publications related to HMCDM involving MCDM and their combinations with other methodologies. Figure 14.1 illustrates the classification of HMCDM methods used in the current literature review.

The HMCDM is subdivided into three classes:

1. Based on the specificity of the problem at hand a mixture of MCDM approaches is used with the integration of results in the final decision-making (e.g., AHP and DEMATEL; DEMATEL and VIKOR; VIKOR and TOPSIS and SAW; PROMETHEE and AHP).
2. Real-life problems most of the time occur in a context of uncertainty with ambiguities, making difficult to reach the appropriate decision. Therefore, the integration of MCDM method(s) with fuzzy sets or GREY numbers can be the preferred approach of the researchers (e.g., fuzzy AHP (FAHP), FAHP and VIKOR, FAHP and Fuzzy VIKOR).
3. MCDM methods are also frequently combined with other approaches to solve a specific or more general problem. The combination of MCDM method(s) with other methods not classified as MCDM methods and different from fuzzy sets or GREY numbers.

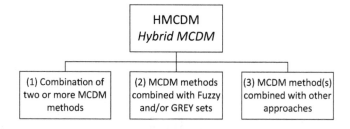

FIGURE 14.1 Classification of the HMCDM methods.

Resulting from the suggested structure of classification depicted in Figure 14.1, the hybrid MCDM publications of the current research are classified. Due to the existence of several distinct MCDM approaches and the suitability of different methods according to the type of problem at hand, it is also worth to explore the kinds of techniques and approaches used in HMCDM related to sustainable procurement.

14.4   RESEARCH METHODOLOGY

A review of existing scientific work on the subjects under investigation has the potential to guide scholars by providing examples, ideas, information, and insight to the field problems. The specific methodology to use in this literature review starts with the formulation of the research questions. The formulated questions guide the review by identifying the findings of the relevant individual studies addressing the research questions. This is achieved by specifically establishing to what extent the existing research concerning green/sustainable procurement/purchasing in the supply chain has progressed towards the use multi-criteria decision analysis methods over the last two decades. The literature review was based on five steps (Figure 14.2).

STEP 1 (i.e., question formulation) contains the preparation of the literature review with the establishment of the focus and the questions formulation, as previously defined (i.e., in Section 14.1). The question formulation is vital to clarify the criteria for the primary study inclusion in the review. STEP 2 (i.e., location of the studies) covers the location, selection, and appraisal of the research relevant to the review questions, the selection of bibliographic databases searches, the search strings selected, the grouping of keywords, the application of search conventions, and so on. The output is a comprehensive listing of articles that represent the core contribution, which helps to address the review questions. STEP 3 (i.e., study selection and evaluation) includes the selection of criteria and the reasons for the inclusion or exclusion from the selection criteria. STEP 4 (i.e., analysis and synthesis) refers to the analysis and synthesis of the information depicted in the articles developing knowledge concerning the magnitude and evolution of green/sustainable procurement/purchasing research issues using MCDM since 1997, the main journals where the research was published, and the evolution of the studies during the length of the study. STEP 5 is devoted to the reporting of the findings by reorganizing

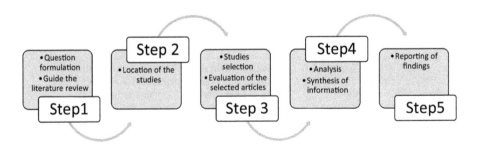

FIGURE 14.2   Literature review steps.

the material into a different arrangement through exploring the main applications concerning green/sustainable procurement/purchasing research using MCDA methods and tools, and by the identification of the main MCDM techniques used in the studies. An overview of the evolution and the main paths for further research in MCDM methods used for green/sustainable procurement/purchasing is depicted. Also described are the strengths and weaknesses of the MCDM methods identified in applications of SP in SC operations.

14.4.1 MATERIAL COLLECTION

The formulation of the questions defined in the introduction of the chapter was crucial to the preparation of the review, because the questions establish the focus of the research and help to clarify the criteria for primary study inclusion in the review. The location of the studies was made with search strings in electronic databases and finding research works relevant to the scope of the review. The bibliometric analysis was carried out on the widely popular Web of Science (WoS), which allows searching in several databases with cross-disciplinary research, and permitting an in-depth search of subjects including science, social science, arts, and humanities (i.e., supporting 256 disciplines). All subject areas were taken into consideration for the data search and selection. Google Scholar was also used to validate the previous searches to guarantee that all the appropriate studies, which were handled within the searching criteria, were included. The search criteria included peer reviewed articles and proceedings of conferences, excluding documents that did not comply with these criteria. The literature search was accomplished founded on various search strings selected by research topics. The search strings were selected to include terms related to procurement/purchasing (e.g., purchasing, procurement, sustainable procurement, sustainable purchasing, supplier, green procurement, green purchasing, green SC, SSC) and combined with terms related to multi-criteria decision-making techniques (e.g., MCDA, MCDM, outranking methods, ELECTRE methods, multi-attribute utility theory, UTA, AHP, ANP, fuzzy multiple criteria, PROMETHEE, DEMATEL, MAVT). This lead to the definition of a research focus, excluding studies when it was recognized that they did not belong to both areas or that they did not represent any relation. Even with this search and selection, it was still necessary to perform a manual check for all the articles that were selected according to the search strings to remove the articles that clearly did not address the topic under research. The manual check was performed based on the abstract of each journal article or conference article.

As previously mentioned, there is some evidence that green/sustainable procurement/purchasing have been discussed by practitioners and researchers mainly since the early 1990s, and one central issue concerning the way that organizations perceived the environmental dimension was the ISO 14001 released in 1996 (Darnall et al., 2000). Therefore, the time span for this research was 20 years, from 1997 to 2017, not excluding articles in press and searched until March 2018.

An initial sample of 533 articles was first selected, and after performing a manual check of the articles based on the abstract of each paper or conference paper, some articles were removed because they were not specifically related to the topic and were not considered to be within the context of the literature review. A huge number of articles (i.e., a total of 354 articles) corresponding to 66.4% were discarded because, despite being related to multi-criteria decision-making techniques applied to procurement/purchasing, they did not address the topic of green/environmental or sustainable supply issues. Also, a total of 48 articles corresponding to 9.0% of the total initial sample selected corresponds to the topic of sustainable supply management but not specifically related with procurement or purchasing decisions and evaluation; thus, they were also discarded. A final sample of 131 articles (i.e., corresponding to 24.6% of the initial selected papers) complied with the topic of the literature review and considered for further evaluation and analysis.

14.4.2 Analysis and Synthesis

The analysis and synthesis of the selected articles is presented with the main objective of developing knowledge concerning the size and progress of green/sustainable procurement/purchasing research issues using MCDM methods for the last 20 years. The distribution of the selected articles is presented based on the source of publication and the number of publications per year. Also, the publications on the topic of MCDM techniques and tools selected from the Web of Science database were sorted for MCDM techniques and for hybrid MCDM according to the classification described in Section 14.3.2. The articles were also organized by area of application and type of criteria related to environmental criteria or including the TBL framework with environmental, economic, and societal criteria.

14.4.2.1 Distribution Based on Source of Publication and Years

The articles selected were classified by source of publication and the number of publications/journal/conference and by year were identified. Figure 14.3 points out

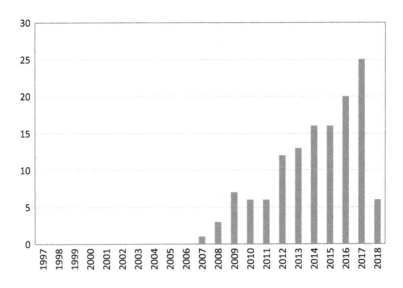

FIGURE 14.3 Number of publications per year.

the publications' number per year to see the evolution of the publications across the period and the trend of research over the years. It is possible to observe the evolution of articles in peer-reviewed journals/conferences within the last two decades.

The recent evolution concerning the number of publications is evidenced in Figure. 14.3, mainly after 2011. The bar corresponding to 2018 is shorter because it corresponds only to articles searched until March 2018.

Figure 14.4 presents the distribution of articles included in this literature review based on the name of the specific journal or aggregated by conference proceedings. The articles selected that are related to MCDM methods and tools for SP are distributed across journals (Figure 14.4) that cover a wide range of disciplines including operations research, intelligent systems, production management, cleaner production, and environmental and sustainability research areas.

The J Clean Prod is the top ranked journal, followed by the Int J Prod Res and by the Int J Prod Econ *ex aequo* in the second position of the top ranked journals. The next journal in the ranking is the Sustainability followed by the Comput Ind Eng *ex aequo* with Expert Syst Appl and immediately followed by the Eur J Oper Res. This top ranked research areas contain most of the articles corresponding to a total of 45.8% of the selected articles in this literature review. The *Ann Oper Res*, the *Int J Environ Sci Technol*, and the *Resour Conserv Recycl* with the same number of articles, come next in the ranking list contributing to this research field.

FIGURE 14.4   Distribution of publication by source.

14.4.3 CLASSIFICATION OF THE REFERENCES BY TYPE OF CRITERIA AND APPLICATION AREA

To model and support decision-making in green or sustainable procurement/purchasing several methods, different tools were used by researchers. From the total of 131 articles selected and related to MCDM studies on SP issues, the hybrid multi-criteria decision-making (HMCDM) approaches, with a total of 88 articles (67.2%), was the most used approach and showed to be the preferred approach used by the researchers. Among the HMCDM approaches, a total of 62 articles (70.4%) corresponds to the class (2), and 11.4% correspond to the class (1) with 18.2% of articles belonging to the class (3) according to the classification depicted in Figure 14.1. After a deep analysis of the articles to have a better understanding of the most frequently used methods in an uncertain environment and classified according to class (2), it was found that the main methods were namely AHP, TOPSIS, ANP, DEMATEL, and VIKOR.

The second top ranked was AHP alone with 10 articles corresponding to 7.6% of articles, followed by outranking methods with 3.1%, and followed by ANP, DEA, DEMATEL, and TOPSIS. If the AHP and ANP were grouped together, the number of articles rose to 15 corresponding to a total of 11.5%.

Table 14.1 presents the classification of the 131 references regarding the type of work and application area. The table presents the segmentation of articles by application area, type of criteria, and authors. The key effort is to identify the main applications for MCDM approaches in green/sustainable procurement. A central issue is to identify which industries and type of applications in the selected literature of MCDM on SP attracted more attention and show the most prevalent industrial applications.

Most of the research studies are related with environmental criteria corresponding to a total of 80.2% of the studies with only 19.8% of the applications encompassing the pillars of the TBL framework with economic, environmental, and societal dimensions. Relating to the industrial sectors, the automotive industry represents the largest group followed by the electrical and electronic industries. These two top ranked industries account for 32.8% of the studies.

14.5 DISCUSSION

This work attempted to perform a bibliographic review of the last 20 years concerning MCDM methods and tools used in the context of SP in international journals/conferences with pear revision through all database systems of the WoS. The main goal of the study was to make a literature review and examine the results based on the four research questions identified.

To answer the first question related to the spread of papers referring to the publication years, a graphical representation of frequencies was made in sub-section 14.4.2.1. The distribution undoubtedly indicates the increasing number of research articles, mainly the ones that were published in the last decade, with a huge proliferation of studies from 2012 until the present (March 2018) with 82.4% of the selected literature research. In the year of 2012, the number of published articles increased significantly in comparison to the preceding years. In general, the results indicate that the

TABLE 14.1

Classification of the References by Application Area and Type of Criteria

Activity sector	Type of Criteria[a]	References
Illustrative/ numerical example	Green/27	Ashlaghi (2014); Awasthi et al. (2010); Awasthi & Kannan (2016); Bai & Sarkis (2010a); Bakeshlou et al. (2017); Bali et al. (2013); Cao et al. (2015); Dobos & Voeroesmarty (2014); Haji Vahabzadeh et al. (2015); Hamdan & Jarndal (2017); Lee et al. (2009); Liu et al. (2017); Lu et al. (2007); Mafakheri et al. (2011); Özgen et al. (2008); Shaik & Abdul-Kader (2011); Shaw et al. (2013); Shen et al. (2013); Sinha & Anand (2017); Tang et al. (2018); Tsai & Hung (2009); Tuzkaya et al. (2009); Yan (2009); Yu & Wong (2014); Zhao & Guo (2014); Zhou (2016); Zhu et al. (2010)
	TBL/9	Amindoust et al. (2012); Awasthi et al. (2018); Bai & Sarkis (2010b); Dai & Blackhurst (2012); Govindan et al. (2013); Keshavarz Ghorabaee et al. (2016, 2017); Orji & Wei (2015a); Sarkis & Dhavale (2015)
Automotive industry	Green/22	Gulsen Akman (2015); Gülşen Akman & Pışkın (2013); Basu et al. (2017); Büyüközkan (2012); Büyüközkan & Çifçi (2012); Çifçi & Büyüközkan (2011); Gavareshki et al. (2017); Govindan et al. (2015a); Gupta & Barua (2017); Hashemi et al. (2015); Jain et al. (2016); Kannan et al. (2013); Kumar et al. (2014, 2016); Kumar Sahu et al. (2014); Liu et al. (2018); Parthiban et al. (2013); Qin et al. (2017); Wang et al. (2017); Yazdani (2014); Yu & Hou (2016); Zhou et al. (2016)
	TBL/2	Azadnia et al. (2012); Luthra et al. (2017)
Electric/electronic industry	Green/16	Chen et al. (2016); Freeman & Chen (2015); He et al. (2007); Hsu & Hu (2009); Hsu et al. (2013); Hsu et al. (2014); Kannan et al. (2014); Kuo et al. (2015); Lin et al. (2012); Liou et al. (2016); Tsui et al. (2015); Tuzkaya (2013); Watrbski & Salabun (2016); Wittstruck & Teuteberg (2012); Yazdani, Chatterjee et al. (2017); Yu & Tsai (2008)
	TBL/3	Arabsheybani et al. (2018); Chiouy et al. (2011); Tajik et al. (2014)
Construction industry	Green/4	Chen et al. (2008); Hsueh & Yan (2013); Kozik (2017); Mohammadi et al. (2017)
	TBL/2	Ahmadian et al. (2017); Xu et al. (2010)
Textile industry	Green/4	Boufateh et al. (2009); Fallahpour et al. (2016); Hashim et al. (2017); Shaw et al. (2012)
	TBL/2	Baskaran et al. (2012); Jia et al. (2015)
Food industry	Green/5	Banaeian et al. (2018); Govindan et al. (2017); Rezaei et al. (2016); Shi et al. (2018); Yazdani, Chatterjee et al. (2017)
	TBL/2	Azadnia et al. (2015); Tavana et al. (2017)

(Continued)

TABLE 14.1 (*Continued*)
Classification of the References by Application Area and Type of Criteria

Activity sector	Type of Criteria[a]	References
Petroleum/ petrochemical/ chemical	Green/4	Bai et al. (2017); Govindan & Sivakumar (2016); Kannan et al. (2015); Senthil et al. (2014)
	TBL/2	Azadnia et al. (2013); Figueiredo Barata et al. (2014)
Manufacturing industry	Green/7	Chung et al. (2016); Eydi & Bakhtiari (2017); Hashemzahi et al. (2017); Hou (2012); Kuo et al. (2010); Lee et al. (2009); Lima et al. (2013)
	TBL/3	Büyüközkan & Çifçi (2011); Mukherjee (2016); Song et al. (2017)
Healthcare industry	Green/2	Ahsan & Rahman (2017); Malik et al. (2016)
	TBL/1	Ghadimi & Heavey (2014)
High tech industry	Green/2	Kuo & Lin (2012); Lee et al. (2013)
Others	Green/12	Azarnivand & Azarnivand (2016); Dos Santos et al. (2017); Dou (2014); Etraj & Jayaprakash (2016); Hamdan & Cheaitou (2017); Kumar et al. (2014); Lee et al. (2011); Mahdiloo et al. (2015); Sivakumar et al. (2015); Theißen & Spinler (2014); Venkatesh et al. (2018); Yazdani et al. (2016)

[a] Green—Environmental criteria; TBL—Environmental, economic and social criteria/number of works.

use of MCDM techniques in SP increase popularity among the academic research community since the last decade, as most of the articles from the last 20 years have been published after 2007. The observed results may be explained by different reasons. One of the reasons is related to the evolution of MCDM techniques and tools in almost all areas of research studies. This is in accordance with the literature review on MCDM techniques and applications (Mardani et al., 2015a, 2017b; Zavadskas et al., 2016b), where the authors conclude that from 2006 the number of studies on MCDM substantially increased. Another reason is related to the evolution of the studies of sustainability in SCM. Looking at literature review studies concerning the integration of sustainability into SCM, the conclusion is that the distribution of articles per year was significantly augmented within the last few years (Brandenburg et al., 2015). The same authors refer that sustainable SCM is quite a recent area of research when in comparison to other disciplines of economics and management. Also, Walker et al. (2012) focus the attention on SP activities and mention that SP is increasingly growing interest in research concerning purchasing and supply chain issues and they noticed that the number of articles published since 1994 until 2012 suffered a strong increase from 2010 until 2012 suggesting a continuous increase trend. However, further analysis should be performed to identify if the growth trend will continue. Banasik et al. (2016) also made a literature review concerning use of MCDM approaches for green supply chains and they concluded that the utilization of MCDM approaches with this focus is a rather new but emerging field with a majority of articles related to production and distribution problems. The results in the current research regarding the number of articles on MCDM methods and tools used in SP studies were already expected and corroborate the above studies suggesting a rapid expansion of the research topic. This is mainly due to the usefulness of MCDM methods to support the multi-criteria natural topic of SP related with the TBL dimensions.

To answer the second question, the articles were distributed based on the research work by source of publication. The information depicted in Figure 14.4 allows a better idea of the distribution of articles based on the research area of the journal as presented on the WoS. The scientific area of the top ranked journal is related with cleaner production and technical issues, green/sustainable topics of engineering, and green/sustainable SC, as well as topics related to energy use and consumption and life cycle assessment of products and processes. Topics related to manufacturing, production, and operations management research, including complex decision problems coming up in the development control and management of production and logistics systems as well as research focus in the expert and intelligent systems technology applied in areas such as network management, engineering, project management, procurement, and strategic management are also among the research areas of the selected journals.

The publications were also classified based on MCDM methods and tools to answer the third question related to a better understanding in view of the multi-criteria decision-making (MCDM) techniques used in SP in SC operations. The conclusions of the study are in accordance with the Govindan et al. (2015b) study devoted to the use of MCDM methods and tools for green supplier evaluation, where the authors found that many of the latest approaches used in this research field integrate fuzzy logic.

Also, Mardani et al. (2015b) in their wide-ranging review on MCDM reported the increasing number of publications and applications of fuzzy MCDM to deal with uncertainties in problems. Zavadskas et al. (2016b) made a comprehensive review of hybrid multi-criteria decision-making methods for sustainability issues where they found the use of HMCDM methods increasingly growing to support decisions in different domains of sustainability. The main findings of the current literature review concerning MCDM for addressing SP in supply chain operations, which is a subtopic of the more embracing topic of sustainability, follows the same tendency with an increasing trend in the utilization of HMCDM methods to aid DMs in private and public SP decisions. The methods allow taking into consideration the TBL framework by the organization so that management takes into consideration the supply of products/services and the know-how required to run a sustainable business increasing the value for all stakeholders and to society.

From publications related to industrial real applications used, it is possible to reach the conclusion, based on data summarized in Table 14.1, that there are some leading industrial areas involved in the SP issues. The automotive industry is the leading industry with a total of 24 articles out of 131 where MCDM methods and tools were used to address SP. Also, areas such as electric and electronic industries, construction, textile, food and petroleum, and petrochemical and chemical industries are areas where those studies are representative. Studies are beginning to grow in healthcare industry, and this is accordance with the work of Chiarini et al. (2017), where the authors refer that the usage of GP in the public sector is covering new sectors, for example, the healthcare sector. Looking at sustainability dimensions covered in the articles, the focus is changing from the green dimension to address also the social dimension, but the data presented in Table 14.1 confirm that the number of studies following the TBL framework is still small when compared to green dimension.

14.5.1 STRENGTHS AND WEAKNESSES OF THE MAIN MCDM TECHNIQUES USED IN SP

Several MCDM approaches have been established and used over the last half century to deal with complex issues by dividing the whole problem into small sub-problems and to deal with discrete options that can be described and evaluated by a group of criteria, whose values can be assessed based on ordinal or cardinal data that can be evaluated exactly or can be fuzzy. Considering the real-world complexity of sustainable procurement, current research tends to integrate multiple decision-making methods and tools into a hybrid MCDM approach. Based on the current investigation, the most used MCDM methods used in sustainable procurement were identified and the combination of different and multiple methods is a trend that is expected to continue. The researchers need to address and evaluate the main strengths/weaknesses of the selected methods to be chosen for solving a specific problem and need to be aware of them in the identification and selection of the suitable methods and combinations (hybrid MCDM) to address a specific problem. Table 14.2 presents some strengths and weaknesses of the most used MCDM methods in sustainable procurement problems, as well as some representative articles of the current literature review where those methods were used.

TABLE 14.2
Strengths and Weaknesses of MCDM Methods

Method	Strengths	Weaknesses	Representative Articles[a]
AHP	Easy to use and broadly spread; the hierarchy structure allows to fit to different sized problems; simple route to take care of complex issues.	Problems can occur due to interdependence between criteria; rank reversal can be an issue. The method can give rise to inconsistencies in judgements as well as in the criteria ranking.	Ahsan & Rahman (2017) Chiouy et al. (2011)[b] Hou (2012) Azarnivand & Azarnivand (2016)[b] Basu et al. (2017) Buyukozkan et al. (2012)[b] Dai & Blackhurst (2012) Hsueh & Yan (2013) Lu et al. (2007) Sivakumar et al. (2015) Shaw et al. (2013) Çifçi & Büyüközkan (2011)[b] Lee et al. (2009)[b] Lee et al. (2011)[b] Mafakheri et al. (2011) Malik et al. (2016) Zhu et al. (2010)
DEA	Nonparametric technique able to deal with multiple input/output models and not requiring the need to have a functional way to relate inputs with outputs.	Assume that all input/output are exactly known. Big problems can be computationally intensive due to the need to have a linear program for each DMU.	Kumar et al. (2014) Dobos & Vocroesmarty (2014) Kumar et al. (2016) Jain et al. (2016)

(Continued)

TABLE 14.2 (*Continued*)
Strengths and Weaknesses of MCDM Methods

Method	Strengths	Weaknesses	Representative Articles[a]
DEMATEL	Ability to handle complex relationships between components of the system; can evidence the cross dependencies between factors; support the visual structuring of a model to reflect the factors' relations.	The relative weights of experts are not considered in aggregating personal judgments of experts into group assessments; cannot consider the aspiration level of alternatives.	Etraj & Jayaprakash (2016); Çifçi et al. (2011) Hsu et al. (2013) Govindan et al. (2015a)[b] Yazdani, Zarate et al. (2017)
TOPSIS	Cardinal ranking of the options and does not require the independence of attribute preferences. Simple process using the concept of distance measures. Measuring the distance from the ideal solution simplifies the choice of the options.	It requires all values of attributes to be numeric, and commensurability of all attributes is a requirement. Doesn't consider the correlation of attributes and the relative importance of the two distance measures (two reference points). Rank reversal may be an issue.	Yu & Wong (2014) Orji & Wei (2015a) Shen et al. (2013)[b] Watrbski et al. (2016)[b] Xu et al. (2010)[b] Eydi & Bakhtiari (2017) Zhao & Guo (2014)[b] Awasthi et al. (2010)[b] Ahmadian et al. (2017) Govindan et al. (2013) Yu & Wong (2014) Jia et al. (2015) Govindan et al. (2016)[b] Dos Santos et al. (2017)[b] Orji & Wei (2015b)[b]

(*Continued*)

TABLE 14.2 (*Continued*)
Strengths and Weaknesses of MCDM Methods

Method	Strengths	Weaknesses	Representative Articles[a]
VIKOR	Is a compromise ranking technique that makes use of a ranking list by measuring the proximity of the solution by linear normalization giving rise to a maximum *group utility* for the *majority* with a minimum individual regret for the *opponent*.	The input weights given by the decision makers sometimes may be an issue requiring the use of other MCDM methods.	Haji Vahabzadeh et al. (2015)[b] Zhou et al. (2016)[b] Awasthi et al. (2016)[b]
PROMETHEE	Outranking method easy to rank the options requiring a preference function and weights for each criterion.	Don't provide a clear procedure to allocate the weights. The outranking doesn't explain the strengths and weaknesses of the options.	Govindan et al. (2017)
ELECTRE	Easy to make comparisons among options for each of the criteria and establishing outranking relations between the options.	The result of the process may be difficult to explain; the outranking does not explain the strengths and weaknesses of the options.	Figueiredo Barata et al. (2014) Kozik (2017)
GP	Can handle multiple objectives and large-scale problems. Able to deal with conflicting objective measures.	The weight of the coefficients can be an issue usually solved in combination with other MCDM methods.	Tsai et al. (2009)
SMART	Use a simple procedure based on additive aggregation procedure. Capable of evaluate qualitative/quantitative criteria.	May not be convenient to deal with the complexities of some real problems.	

[a] Some representative articles of the current literature review concerning SP.
[b] Fuzzy + MCDM method.

The AHP is an MCDM method widely used in different types of problems by using a pairwise comparison at each node of the structure and making possible the evaluation of quantitative/qualitative criteria and options on the nine levels of verbal preference scale to convert verbal comparisons into numerical values, making the method very easy to use. Despite the weaknesses of the method, the simplicity of the method and the widespread dissemination places the method on the top ranking of MCDM approaches applied to sustainable procurement. To deal with uncertainties in real-world problems, some fuzzy extensions of the traditional AHP were developed based on fuzzy set theory. The fuzzy AHP was found by Mardani et al. (2015b) to be the preferred approach among the MCDM applications, mainly due to the ability to deal with uncertainty in real problems and in the human judgements. This is the reason why it is one of the most preferred approaches for decision problems where fuzziness is an issue (Wang & Chin, 2011).

In the current investigation, a total of 62 articles corresponding to 47.3% make use of the MCDM combination to include uncertainty theories. Among MCDM methods that utilize fuzzy decision-making approaches, fuzzy TOPSIS represented the most used approach with a total of 22 articles (i.e., 35.5%). Also, fuzzy AHP with a total of 20 articles (i.e., 32.3%) occupies the second position in the ranking. The most prevailing use of fuzzy TOPSIS makes the approach widely used in sustainable procurement in line with many other areas of research as per example the computer sciences and engineering (Kahraman et al., 2015). Recently, some other improvements of the TOPSIS method with extensions as well as variations of the former TOPSIS have been deployed, and some hybrid methods associate TOPSIS with different MCDM methods like AHP, DEA, PROMETHEE, and ELECTRE (Zavadskas et al., 2016b). In the current review of applications for sustainable procurement, TOPSIS was integrated with MCDM methods like VIKOR, GRA, DEMATEL, AHP, and ANP just to refer the most used combinations. In goal programming (GP), the goal (i.e., aspiration levels) is considered precise and concise, but there are some cases where the DM is unable to identify the goal values precisely, so fuzzy GP takes the vagueness into account by using the membership functions based on fuzzy set theory (Aouni et al., 2010). Many of the sustainable procurement studies employed applications of hybrid MCDM joining different methods to assist the decision makers in handling complex information and miscellaneous topics involving the critical issues related with DM preferences, the interdependence of criteria, the presence of conflicting criteria, or the vagueness or uncertain environment where decision takes place.

The selection of the adequate method is a challenge, and the specificity of the problem and the complexity of each condition requires the decision on the method or combinations to use.

14.6 CONCLUSION

This work has reviewed the literature relating to MCDM techniques in the context of SP. The main findings identify a fast growing body of MCDM techniques and approaches to support decisions in sustainable procurement and proved to be very useful to help organizations to meet their own needs for services, goods, and utilities achieving the value for money on the whole life cycle basis. A lot methods have been deployed and used almost over the last half century to account in an explicit way for multiple criteria to aid individuals/groups to rank/select/compare different alternatives/options or actions. Some of the strengths and weaknesses of the main MCDM techniques are summarized in Table 14.2, as well as some representative articles on the methods that can be used successfully to allow for the evaluation/ selection of processes and help the sustainable procurement decisions. However, a huge number of the reviewed articles involved more than one decision-making method and a combination of decision-making methods with other methods. These methods were classified as presented in Section 14.3.2 to simplify the analysis of the hybrid methods used in sustainable procurement to assist decision makers in handling complex information and vagueness and uncertain environments involving individual and group preferences and conflicting or interrelated criteria.

Despite the encouragement of TBL (i.e., framework encompassing social, economic and environmental criteria) guidelines on sustainability, the current literature review shows that, in most cases, only the green (i.e., environmental) dimension of the TBL has been taken into consideration in the selected articles concerning MCDM usage in SP. This is in line with previous literature studies related to SP as depicted in Section 14.2. Most studies are related to supplier qualification, tendering, and purchasing with studies related to supplier selection and evaluation. The qualification and tendering are very important to the existence of supplier qualification procedures reflecting the practices and policies of the organization. Most organizations already use ISO 14001 or other environmental standards, where this process is relatively straightforward, and justifies the higher number of studies related to GP, especially in light of the existence of environmental management systems in the organizations. The integration of the social dimension of the TBL is still an issue reflected by the reduced number of studies encompassing the TBL framework. More difficulties might be encountered by companies striving for TBL improvements in the absence of management systems that take into consideration the sustainable development pillars. Some organizations started already taking into account the environmental and social criteria into their procurement policies, but much work still needs to be performed concerning the design of supplier assessment management information taking into consideration the TBL supplier efficiency and changes over time. The procurement function is central to put forward the sustainable development agenda in light of its role in the engagement of all the external organizations to reach an SSC.

REFERENCES

Ahi, P., & Searcy, C. (2013). A comparative literature analysis of definitions for green and sustainable supply chain management. *Journal of Cleaner Production, 52*, 329–341.

Ahmadian, A. F. F., Rashidi, T. H., Akbarnezhad, A., & Waller, S. T. (2017). BIM-enabled sustainability assessment of material supply decisions. *Engineering Construction and Architectural Management, 24*(4), 668–695.

Ahsan, K., & Rahman, S. (2017). Green public procurement implementation challenges in Australian public healthcare sector. *Journal of Cleaner Production, 152*, 181–197.

Akman, G. (2015). Evaluating suppliers to include green supplier development programs via fuzzy c-means and VIKOR methods. *Computers & Industrial Engineering, 86*, 69–82.

Akman, G., & Pişkın, H. (2013). Evaluating green performance of suppliers via analytic network process and TOPSIS. *Journal of Industrial Engineering, 2013*, 1–13.

Amindoust, A., Ahmed, S., Saghafinia, A., & Bahreininejad, A. (2012). Sustainable supplier selection: A ranking model based on fuzzy inference system. *Applied Soft Computing, 12*(6), 1668–1677.

Ansari, Z. N., & Kant, R. (2017). A state-of-art literature review reflecting 15 years of focus on sustainable supply chain management. *Journal of Cleaner Production, 142*, 2524–2543.

Aouni, B., Martel, J.-M., & Hassaine, A. (2010). Fuzzy goal programming model: An overview of the current state-of-the art. *Journal of Multi-Criteria Decision Analysis, 16*(5–6), 149–161.

Appolloni, A., Sun, H., Jia, F., & Li, X. (2014). Green procurement in the private sector: A state of the art review between 1996 and 2013. *Journal of Cleaner Production, 85*, 122–133.

Arabsheybani, A., Paydar, M. M., & Safaei, A. S. (2018). An integrated fuzzy MOORA method and FMEA technique for sustainable supplier selection considering quantity discounts and supplier's risk. *Journal of Cleaner Production, 190*, 577–591.

Ashby, A., Leat, M., & Hudson-Smith, M. (2012). Making connections: A review of supply chain management and sustainability literature. *Supply Chain Management: An International Journal, 17*(5), 497–516.

Ashlaghi, M. J. (2014). A new approach to green supplier selection based on fuzzy multi-criteria decision making method and linear physical programming. *Tehnicki Vjesnik-Technical Gazette, 21*(3), 591–597.

Awasthi, A., & Kannan, G. (2016). Green supplier development program selection using NGT and VIKOR under fuzzy environment. *Computers & Industrial Engineering, 91*, 100–108.

Awasthi, A., Chauhan, S. S., & Goyal, S. K. K. (2010). A fuzzy multicriteria approach for evaluating environmental performance of suppliers. *International Journal of Production Economics, 126*(2), 370–378.

Awasthi, A., Govindan, K., & Gold, S. (2018). Multi-tier sustainable global supplier selection using a fuzzy AHP-VIKOR based approach. *International Journal of Production Economics, 195*, 106–117.

Azadnia, A. H., Ghadimi, P., Saman, M. Z. M., Wong, K. Y., & Heavey, C. (2013). An integrated approach for sustainable supplier selection using fuzzy logic and fuzzy AHP. *Applied Mechanics and Materials, 315*, 206–210.

Azadnia, A. H., Saman, M. Z. M., & Wong, K. Y. (2015). Sustainable supplier selection and order lot-sizing: An integrated multi-objective decision-making process. *International Journal of Production Research, 53*(2), 383–408.

Azadnia, A. H., Saman, M. Z. M., Wong, K. Y., Ghadimi, P., & Zakuan, N. (2012). Sustainable supplier selection based on self-organizing map neural network and multi criteria decision making approaches. *Procedia—Social and Behavioral Sciences, 65*, 879–884.

Azarnivand, A., & Azarnivand, A.-R. (2016). Discussion of "Green vendor evaluation and selection using AHP and Taguchi loss functions in production outsourcing in mining industry" by Sivakumar, R., Kannan, D., Murugesan, P. Resour. Policy 46 (2015) (1) 64–75. *Resources Policy, 49*, 51–52.

Bai, C., & Sarkis, J. (2010a). Green supplier development: Analytical evaluation using rough set theory. *Journal of Cleaner Production, 18*(12), 1200–1210.

Bai, C., & Sarkis, J. (2010b). Integrating sustainability into supplier selection with grey system and rough set methodologies. *International Journal of Production Economics, 124*(1), 252–264.

Bai, C., Rezaei, J., & Sarkis, J. (2017). Multicriteria green supplier segmentation. *IEEE Transactions on Engineering Management, 64*(4), 515–528.

Bakeshlou, E. A., Khamseh, A. A., Asl, M. A. G., Sadeghi, J., & Abbaszadeh, M. (2017). Evaluating a green supplier selection problem using a hybrid MODM algorithm. *Journal of Intelligent Manufacturing, 28*(4), 913–927.

Bali, O., Kose, E., & Gumus, S. (2013). Green supplier selection based on IFS and GRA. *Grey Systems: Theory and Application, 3*(2), 158–176.

Bana e Costa, C. A., Corte, J.-M., & Vansnick, J.-C. (2011). MACBETH (Measuring Attractiveness by a Categorical Based Evaluation Technique). In Cochrane, J.J. (Ed.), *Wiley Encyclopedia of Operations Research and Management Science*, Vol. 4 (pp. 2945–2950).

Bana e Costa, C. A., & Vansnick, J.-C. (1994). MACBETH: An interactive path towards the construction of cardinal value functions. *International Transactions in Operational Research, 1*(4), 489–500.

Banaeian, N., Mobli, H., Fahimnia, B., Nielsen, I. E., & Omid, M. (2018). Green supplier selection using fuzzy group decision making methods: A case study from the agri-food industry. *Computers & Operations Research, 89*, 337–347.

Banasik, A., Bloemhof-Ruwaard, J. M., Kanellopoulos, A., Claassen, G. D. H., & van der Vorst, J. G. A. J. (2016). Multi-criteria decision making approaches for green supply chains: A review. *Flexible Services and Manufacturing Journal*, 1–31.

Baskaran, V., Nachiappan, S., & Rahman, S. (2012). Indian textile suppliers' sustainability evaluation using the grey approach. *International Journal of Production Economics, 135*(2), 647–658.

Basu, R. J., Subramanian, N., Gunasekaran, A., & Palaniappan, P. L. K. (2017). Influence of non-price and environmental sustainability factors on truckload procurement process. *Annals of Operations Research, 250*(2), 363–388.

Beckmann, M., Hielscher, S., & Pies, I. (2014). Commitment strategies for sustainability: How business firms can transform trade-offs into win-win outcomes. *Business Strategy and the Environment, 23*(1), 18–37.

Behzadian, M., Kazemzadeh, R. B., Albadvi, A., & Aghdasi, M. (2010). PROMETHEE: A comprehensive literature review on methodologies and applications. *European Journal of Operational Research, 200*(1), 198–215.

Belton, V., & Stewart, T. J. (2002). *Multiple Criteria Decision Analysis*. Berlin, Germany: Springer US.

Benayoun, R., Roy, B., & Sussmann, B. (1966). *Manuel de référence du programme Electre*. Paris: Note de Synthèse et Formation n° 25 de la Direction Scientifique de la SEMA.

Beske, P., & Seuring, S. (2014). Putting sustainability into supply chain management. *Supply Chain Management: An International Journal, 19*(3), 322–331.

Boufateh, I., Perwuelz, A., Rabenasolo, B., & Jolly-Desodt, A.-M. (2009). Multiple Criteria Decision Making for environmental impacts optimization. In *2009 International Conference on Computers & Industrial Engineering* (pp. 606–611). IEEE.

Bouwer, M., Jonk, M., Berman, T., Bersani, R., Lusser, H., Nappa, V., Nissinen, A., Parikka, K., Szuppinger, P., & Viganò, C. (2006). Green public procurement in Europe 2006— Conclusions and recommendations. Haarlem: Virage Milieu & Management (http://ec. europa. eu/environment/gpp/pdf/take_5. pdf.

Brandenburg, M., & Rebs, T. (2015). Sustainable supply chain management: A modeling perspective. *Annals of Operations Research, 229*(1), 213–252.

Brandenburg, M., Govindan, K., Sarkis, J., & Seuring, S. (2014). Quantitative models for sustainable supply chain management: Developments and directions. *European Journal of Operational Research, 233*(2), 299–312.

Brans, J. P. (1982). L'ingénierie de la décision. Elaboration d'instruments d'aide à la décision. Méthode PROMETHEE. In R. Nadeau & M. Landry (Eds.), *L'aide à la décision: Nature, instruments et perspectives d'avenir*. Québec, Canada: Les Presses de l'Université Laval.

Brauers, W. K. M., & Zavadskas, E. K. (2010). Project management by multimoora as an instrument for transition economies. *Technological and Economic Development of Economy, 16*(1), 5–24.

Broekhuizen, H., Groothuis-Oudshoorn, C. G. M., van Til, J. A., Hummel, J. M., & IJzerman, M. J. (2015). A review and classification of approaches for dealing with uncertainty in multi-criteria decision analysis for healthcare decisions. *Pharmaco Economics, 33*(5), 445–455.

Bush, S. R., Oosterveer, P., Bailey, M., & Mol, A. P. J. (2015). Sustainability governance of chains and networks: A review and future outlook. *Journal of Cleaner Production, 107*, 8–19.

Büyüközkan, G. (2012). An integrated fuzzy multi-criteria group decision-making approach for green supplier evaluation. *International Journal of Production Research, 50*(11), 2892–2909.

Büyüközkan, G., & Çifçi, G. (2011). A novel fuzzy multi-criteria decision framework for sustainable supplier selection with incomplete information. *Computers in Industry, 62*(2), 164–174.

Büyüközkan, G., & Çifçi, G. (2012). A novel hybrid MCDM approach based on fuzzy DEMATEL, fuzzy ANP and fuzzy TOPSIS to evaluate green suppliers. *Expert Systems with Applications, 39*(3), 3000–3011.

Buyukozkan, G., Arsenyan, J., & Ruan, D. (2012). Logistics tool selection with two-phase fuzzy multi criteria decision making: A case study for personal digital assistant selection. *Expert Systems with Applications, 39*(1), 142–153.

Cao, Q., Wu, J., & Liangb, C. (2015). An intuitionsitic fuzzy judgement matrix and TOPSIS integrated multi-criteria decision making method for green supplier selection. *Journal of Intelligent and Fuzzy Systems, 28*(1), 117–126.

Carter, C. R., & Carter, J. R. (1998). Interorganizational determinants of environmental purchasing: Initial evidence from the consumer products industries. *Decision Sciences, 29*(3), 659–684.

Carter, C. R., & Liane Easton, P. (2011). Sustainable supply chain management: Evolution and future directions. *International Journal of Physical Distribution & Logistics Management, 41*(1), 46–62.

Carter, C. R., & Rogers, D. S. (2008). A framework of sustainable supply chain management: Moving toward new theory. *International Journal of Physical Distribution & Logistics Management, 38*(5), 360–387.

Chai, J., Liu, J. N. K., & Ngai, E. W. T. (2013). Application of decision-making techniques in supplier selection: A systematic review of literature. *Expert Systems with Applications, 40*(10), 3872–3885.

Charnes, A., & Cooper, W. W. (1961). *Management models and industrial applications of linear programming.* John Wiley & Sons.

Charnes, A., Cooper, W. W., & Ferguson, R. O. (1955). Optimal estimation of executive compensation by linear programming. *Management Science, 1*(2), 138–151.

Charnes, A., Cooper, W. W., & Rhodes, E. (1978). Measuring the efficiency of decision making units. *European Journal of Operational Research, 2*(6), 429–444.

Chen, H. M., Chou, S.-Y., Quoc Dat, L., & Yu, T. H.-K. (2016). A fuzzy MCDM approach for green supplier selection from the economic and environmental aspects. *Mathematical Problems in Engineering, 2016*, 10. doi:10.1155/2016/8097386.

Chen, S.-J., & Hwang, C.-L. (1992). Fuzzy sets and their operations. In *Fuzzy Multiple Attribute Decision Making. Lecture Notes in Economics and Mathematical Systems*, Vol. 375 (pp. 42–100). Berlin, Germany: Springer.

Chen, Z., Li, H., Ross, A., Khalfan, M. M. A., & Kong, S. C. W. (2008). Knowledge-driven ANP approach to vendors evaluation for sustainable construction. *Journal of Construction Engineering and Management-Asce, 134*(12), 928–941.

Cheng, W., Appolloni, A., D'Amato, A., & Zhu, Q. (2018). Green public procurement, missing concepts and future trends—A critical review. *Journal of Cleaner Production, 176*, 770–784.

Chiarini, A., Opoku, A., & Vagnoni, E. (2017). Public healthcare practices and criteria for a sustainable procurement: A comparative study between UK and Italy. *Journal of Cleaner Production, 162*, 391–399.

Chiouy, C.-Y., Chou, S.-H., & Yeh, C.-Y. (2011). Using fuzzy AHP in selecting and prioritizing sustainable supplier on CSR for Taiwan's electronics industry. *Journal of Information and Optimization Sciences, 32*(5), 1135–1153.

Chung, C.-C., Chao, L.-C., & Lou, S.-J. (2016). The establishment of a green supplier selection and guidance mechanism with the ANP and IPA. *Sustainability, 8*(3), 259.

Çifçi, G., & Büyüközkan, G. (2011). A fuzzy MCDM approach to evaluate green suppliers. *International Journal of Computational Intelligence Systems, 4*(5), 894–909.

CIPS. (n.d.). Sustainable and Ethical Procurement—The Chartered Institute of Procurement and Supply. Retrieved 6 July 2018, from https://www.cips.org/en/knowledge/procurement-topics-and-skills/sustainability/sustainable-and-ethical-procurement/.

Cochrane, J. L., & Zeleny, M. (1973). *Multiple Criteria Decision Making*. University of South Carolina Press.

Colapinto, C., Jayaraman, R., & Marsiglio, S. (2017). Multi-criteria decision analysis with goal programming in engineering, management and social sciences: A state-of-the art review. *Annals of Operations Research, 251*(1–2), 7–40.

Commission of the European Communities. (2008). Public procurement for a better environment. Brussels, Belgium.

Corbett, C. J., & Kirsch, D. A. (2001). International diffusion of ISO 14000 certification. *Production and Operations Management, 10*(3), 327–342.

Dai, J., & Blackhurst, J. (2012). A four-phase AHP-QFD approach for supplier assessment: A sustainability perspective. *International Journal of Production Research, 50*(19), 5474–5490.

Darnall, N., Rigling Gallagher, D., Andrews, R. N. L., & Amaral, D. (2000). Environmental management systems: Opportunities for improved environmental and business strategy? *Environmental Quality Management, 9*(3), 1–9.

DEFRA, Department for Environment, F. and R. A. (2006). Procuring the future sustainable procurement national action plan: Recommendations from the sustainable procurement task force.

Diaz-Barriga-Fernandez, A. D., Santibanez-Aguilar, J. E., Radwan, N., Napoles-Rivera, F., El-Halwagi, M. M., & Ponce-Ortega, J. M. (2017). Strategic planning for managing municipal solid wastes with consideration of multiple stakeholders. *ACS Sustainable Chemistry & Engineering, 5*(11), 10744–10762.

Dobos, I., & Voeroesmarty, G. (2014). Green supplier selection and evaluation using DEA-type composite indicators. *International Journal of Production Economics, 157*, 273–278.

Dos Santos, B. M., Prato Neto, C. R., Ferreira, A. R., Bueno, W. P., Soares, M., Blesz Junior, A. E., Borchardt, M., & Godoy, L. P. (2017). Performance of green suppliers in supply chain management. *Interciencia, 42*(12), 805–811.

Dou, Y. (2014). Evaluating green supplier development programs with a grey-analytical network process-based methodology. *European Journal of Operational Research, 233*(2), 420–431.

Emrouznejad, A., & Yang, G. (2018). A survey and analysis of the first 40 years of scholarly literature in DEA: 1978–2016. *Socio-Economic Planning Sciences, 61*, 4–8.

EPA Network. (2017). Recommendations towards the EU plastics strategy.

Eskandarpour, M., Dejax, P., Miemczyk, J., & Péton, O. (2015). Sustainable supply chain network design: An optimization-oriented review. *Omega, 54*, 11–32.

Etraj, P., & Jayaprakash, J. (2016). Prioritizations of GSCM criteria by DEMATEL method for Government Public Procurement in Indian Perspective. In *Intelligent Systems and Control (ISCO), 2016 10th International Conference*. (pp. 1–6). IEEE.

Evans, S., Hills, S., & Orme, J. (2012). Doing more for less? Developing sustainable systems of social care in the context of climate change and public spending cuts. *British Journal of Social Work, 42*(4), 744–764.

Eydi, A., & Bakhtiari, M. (2017). A multi-product model for evaluating and selecting two layers of suppliers considering environmental factors. *Rairo-Operations Research, 51*(4), 875–902.

Fallahpour, A., Olugu, E. U., Musa, S. N., Khezrimotlagh, D., & Wong, K. Y. (2016). An integrated model for green supplier selection under fuzzy environment: Application of data envelopment analysis and genetic programming approach. *Neural Computing and Applications, 27*(3), 707–725.

Ferreira, F. A. F., & Santos, S. P. (2018). Two decades on the MACBETH approach: A bibliometric analysis. *Annals of Operations Research*, 1–25.

Figueira, J., Greco, S., & Ehrogott, M. (2005). *Multiple Criteria Decision Analysis: State of the Art Surveys* (Vol. 78). New York: Springer.

Figueiredo Barata, J. F., Goncalves Quelhas, O. L., Costa, H. G., Gutierrez, R. H., Lameira, V. de J., & Meirino, M. J. (2014). Multi-criteria indicator for sustainability rating in suppliers of the oil and gas industries in Brazil. *Sustainability, 6*(3), 1107–1128.

Freeman, J., & Chen, T. (2015). Green supplier selection using an AHP-Entropy-TOPSIS framework. *Supply Chain Management: An International Journal, 20*(3), 327–340.

Gabus, A., & Fontela, E. (1972). *World problems, an Invitation to Further Thought within the Framework of DEMATEL*. Geneva, Switzerland: Battelle Geneva Research Centre.

Gabus, A., & Fontela, E. (1973). *Perceptions of the World Problem Atique: Communication Procedure, Communicating with Those Bearing Collective Responsibility. DEMATEL Report No.1*. Geneva, Switzerland.

Gavareshki, M. H. K., Hosseini, S. J., & Khajezadeh, M. (2017). A case study of green supplier selection method using an integrated ISM-Fuzzy MICMAC analysis and multi-criteria decision making. *Industrial Engineering and Management Systems, 16*(4), 562–573.

Ghadimi, P., & Heavey, C. (2014). Sustainable supplier selection in medical device industry: Toward sustainable manufacturing. *Procedia CIRP, 15*, 165–170.

Ghadimi, P., Azadnia, A. H., Heavey, C., Dolgui, A., & Can, B. (2016). A review on the buyer—Supplier dyad relationships in sustainable procurement context: Past, present and future. *International Journal of Production Research, 54*(5), 1443–1462.

Giunipero, L., Handfield, R. B., & Eltantawy, R. (2006). Supply management's evolution: Key skill sets for the supply manager of the future. *International Journal of Operations & Production Management, 26*(7), 822–844.

Gold, S., Seuring, S., & Beske, P. (2010). Sustainable supply chain management and inter-organizational resources: A literature review. *Corporate Social Responsibility and Environmental Management, 17*(4), 230–245.

Govindan, K., & Sivakumar, R. (2016). Green supplier selection and order allocation in a low-carbon paper industry: Integrated multi-criteria heterogeneous decision-making and multi-objective linear programming approaches. *Annals of Operations Research, 238*(1–2), 243–276.

Govindan, K., Kadzinski, M., & Sivakumar, R. (2017). Application of a novel PROMETHEE-based method for construction of a group compromise ranking to prioritization of green suppliers in food supply chain. *Omega-International Journal of Management Science, 71*, 129–145.

Govindan, K., Khodaverdi, R., & Jafarian, A. (2013). A fuzzy multi criteria approach for measuring sustainability performance of a supplier based on triple bottom line approach. *Journal of Cleaner Production, 47*, 345–354.

Govindan, K., Khodaverdi, R., & Vafadarnikjoo, A. (2015a). Intuitionistic fuzzy based DEMATEL method for developing green practices and performances in a green supply chain. *Expert Systems with Applications, 42*(20), 7207–7220.

Govindan, K., Rajendran, S., Sarkis, J., & Murugesan, P. (2015b). Multi criteria decision making approaches for green supplier evaluation and selection: A literature review. *Journal of Cleaner Production, 98*, 66–83.

Greco, S., Matarazzo, B., & Slowinski, R. (2001). Rough sets theory for multicriteria decision analysis. *European Journal of Operational Research, 129*(1), 1–47.

Greco, S., Matarazzo, B., & Slowinski, R. (2002). Data mining tasks and methods: Classification: Multicriteria classification. In W. Klösgen & J. M. Zytkow (Eds.), *Handbook of Data Mining and Knowledge Discovery* (pp. 318–328). New York: Oxford University Press, Inc.

Green, K., & Irwin, A. (1996). Clean technologies. In P. Groenewgen, K. Fischer, E. G. Jenkins, & J. Schot (Eds.), *The Greening of Industry Resource Guide and Bibliography* (pp. 169–194). Washington, DC: Island Press.

Green, K., Morton, B., & New, S. (1996). Purchasing and environmental management: interactions, policies and opportunities. *Business Strategy and the Environment, 5*(3), 188–197.

Green, K., Morton, B., & New, S. (1998). Green purchasing and supply policies: Do they improve companies' environmental performance? *Supply Chain Management: An International Journal, 3*(2), 89–95.

Green, K. W., Zelbst, P. J., Meacham, J., & Bhadauria, V. S. (2012). Green supply chain management practices: Impact on performance. *Supply Chain Management: An International Journal, 17*(3), 290–305.

Guarnieri, P. (2015). Synthesis of Main Criteria, Methods and Issues of Multicriteria Supplier Síntese dos Principais Critérios, Métodos e Subproblemas da Seleção de Fornecedores Multicritério. *Revista de Administração Contemporânea, 19*(1), 1–25.

Gul, M., Celik, E., Aydin, N., Taskin Gumus, A., & Guneri, A. F. (2016). A state of the art literature review of VIKOR and its fuzzy extensions on applications. *Applied Soft Computing, 46*, 60–89.

Gupta, H., & Barua, M. K. (2017). Supplier selection among SMEs on the basis of their green innovation ability using BWM and fuzzy TOPSIS. *Journal of Cleaner Production, 152*, 242–258.

Haji Vahabzadeh, A., Asiaei, A., & Zailani, S. (2015). Reprint of "Green decision-making model in reverse logistics using FUZZY-VIKOR method". *Resources, Conservation and Recycling, 104*, 334–347.

Hamdan, S., & Cheaitou, A. (2017). Supplier selection and order allocation with green criteria: An MCDM and multi-objective optimization approach. *Computers & Operations Research, 81*, 282–304.

Hamdan, S., & Jarndal, A. (2017). A two stage green supplier selection and order allocation using AHP and multi-objective genetic algorithm optimization. In *2017 7th International Conference on Modeling, Simulation, and Applied Optimization.* IEEE.

Hashemi, S. H., Karimi, A., & Tavana, M. (2015). An integrated green supplier selection approach with analytic network process and improved Grey relational analysis. *International Journal of Production Economics, 159*, 178–191.

Hashemzahi, P., Musa, S. N., & Yusofi, F. (2017). A hybrid fuzzy multi-criteria decision making model for green supplier selection. *Journal of Fundamental and Applied Sciences, 9*, 417–429.

Hashim, M., Nazam, M., Yao, L., Baig, S. A., Abrar, M., & Zia-ur-Rehman, M. (2017). Application of multi-objective optimization based on genetic algorithm for sustainable strategic supplier selection under fuzzy Environment. *Journal of Industrial Engineering and Management-Jiem, 10*(2), 188–212.

Hassini, E., Surti, C., & Searcy, C. (2012). A literature review and a case study of sustainable supply chains with a focus on metrics. *International Journal of Production Economics, 140*(1), 69–82.

He, J., Zhang, Y., & Shi, Y. (2007). A multi-criteria decision support system of water resource allocation scenarios. *Knowledge Science, Engineering and Management, 4798,* 593–598.

Ho, W., Xu, X., & Dey, P. K. (2010). Multi-criteria decision making approaches for supplier evaluation and selection: A literature review. *European Journal of Operational Research, 202*(1), 16–24.

Hou, F. (2012). Supplier selection and evaluation in environmental purchase. In W. Z. Chen, P. Dai, Y. L. Chen, Q. T. Wang, & Z. Jiang (Eds.), *Advanced Mechanical Design, Pts 1–3* (Vol. 479–481, pp. 352–356). Trans Tech Publications.

Hsu, C. W., Kuo, R. J., & Chiou, C. Y. (2014). A multi-criteria decision-making approach for evaluating carbon performance of suppliers in the electronics industry. *International Journal of Environmental Science and Technology, 11*(3), 775–784.

Hsu, C.-W., & Hu, A. H. (2009). Applying hazardous substance management to supplier selection using analytic network process. *Journal of Cleaner Production, 17*(2), 255–264.

Hsu, C.-W., Kuo, T.-C., Chen, S.-H., & Hu, A. H. (2013). Using DEMATEL to develop a carbon management model of supplier selection in green supply chain management. *Journal of Cleaner Production, 56,* 164–172.

Hsueh, S.-L., & Yan, M.-R. (2013). A Multimethodology Contractor Assessment Model for Facilitating Green Innovation: The View of Energy and Environmental Protection. *Scientific World Journal.*

Huang, C.-Y., Tzeng, G.-H., Lin, Y.-F., & Ho, S. (2010). Configuring the next generation handset by using an emotional design based MCDM framework. *In Technology Management for Global Economic Growth (PICMET), 2010 Proceedings of PICMET'10.* (pp. 1–10). IEEE.

Hwang, C. L., & Yoon, K. (1981). *Multiple Attribute Decision Making: Methods and Applications A State-of-the-Art Survey.* Berlin Heidelberg: Springer.

Hwang, C.-L., Lai, Y.-J., & Liu, T.-Y. (1993). A new approach for multiple objective decision making. *Computers & Operations Research, 20*(8), 889–899.

Interagency Procurement Working Group. (2012). *UN Procurement Practitioner's Handbook.*

ISO. (2017). *ISO 20400:2017—Sustainable procurement—Guidance. ISO (the International Organization for Standardization).*

Jacquet-Lagreze, E., & Siskos, J. (1982). Assessing a set of additive utility functions for multicriteria decision-making, the UTA method. *European Journal of Operational Research, 10*(2), 151–164.

Jain, V., Kumar, S., Kumar, A., & Chandra, C. (2016). An integrated buyer initiated decision-making process for green supplier selection. *Journal of Manufacturing Systems, 41,* 256–265.

Jia, P., Govindan, K., Choi, T.-M., & Rajendran, S. (2015). Supplier selection problems in fashion business operations with sustainability considerations. *Sustainability, 7*(2), 1603–1619.

Johnsen, T. E., Miemczyk, J., & Howard, M. (2017). A systematic literature review of sustainable purchasing and supply research: Theoretical perspectives and opportunities for IMP-based research. *Industrial Marketing Management, 61,* 130–143.

Ju-Long, D. (1982). Control problems of grey systems. *Systems & Control Letters, 1*(5), 288–294.

Kahraman, C., Onar, S. C., & Oztaysi, B. (2015). Fuzzy multicriteria decision-making: A literature review. *International Journal of Computational Intelligence Systems, 8*(4), 637–666.

Kannan, D., Govindan, K., & Rajendran, S. (2015). Fuzzy axiomatic design approach based green supplier selection: A case study from Singapore. *Journal of Cleaner Production, 96,* 194–208.

Kannan, D., Jabbour, A. B. L. de S., & Jabbour, C. J. C. (2014). Selecting green suppliers based on GSCM practices: Using fuzzy TOPSIS applied to a Brazilian electronics company. *European Journal of Operational Research, 233*(2), 432–447.

Kannan, D., Khodaverdi, R., Olfat, L., Jafarian, A., & Diabat, A. (2013). Integrated fuzzy multi criteria decision making method and multi-objective programming approach for supplier selection and order allocation in a green supply chain. *Journal of Cleaner Production, 47*, 355–367.

Karel, W., Brauers, M., & Zavadskas, E. K. (2006). The MOORA method and its application to privatization in a transition economy. *Control and Cybernetics, 35*(2).

Keeney, R. L., & Raiffa, H. (1993). *Decisions with Multiple Objectives: Preferences and Value Trade-Offs.* Cambridge University Press.

Keshavarz Ghorabaee, M., Amiri, M., Zavadskas, E. K., Turskis, Z., & Antucheviciene, J. (2017). A new multi-criteria model based on interval type-2 fuzzy sets and EDAS method for supplier evaluation and order allocation with environmental considerations. *Computers & Industrial Engineering, 112*, 156–174.

Keshavarz Ghorabaee, M., Zavadskas, E. K., Amiri, M., & Esmaeili, A. (2016). Multi-criteria evaluation of green suppliers using an extended WASPAS method with interval type-2 fuzzy sets. *Journal of Cleaner Production, 137*, 213–229.

Khalid, R. U., Seuring, S., Beske, P., Land, A., Yawar, S. A., & Wagner, R. (2015). Putting sustainable supply chain management into base of the pyramid research. *Supply Chain Management: An International Journal, 20*(6), 681–696.

Kozik, R. (2017). Application of multi-criteria decision analysis in selecting of sustainable investments. In *International Conference on Numerical Analysis and Applied Mathematics* (*ICNAAM*) (Vol. 1863).

Kumar Sahu, N., Datta, S., & Sankar Mahapatra, S. (2014). Green supplier appraisement in fuzzy environment. *Benchmarking: An International Journal, 21*(3), 412–429.

Kumar, A., Jain, V., & Kumar, S. (2014). A comprehensive environment friendly approach for supplier selection. *Omega, 42*(1), 109–123.

Kumar, A., Jain, V., Kumar, S., & Chandra, C. (2016). Green supplier selection: A new genetic/immune strategy with industrial application. *Enterprise Information Systems, 10*(8), 911–943.

Kumar, A., Ozdamar, L., & Ning Zhang, C. (2008). Supply chain redesign in the healthcare industry of Singapore. *Supply Chain Management: An International Journal, 13*(2), 95–103.

Kumar, D. T., Palaniappan, M., Kannan, D., & Shankar, K. M. (2014). Analyzing the CSR issues behind the supplier selection process using ISM approach. *Resources Conservation and Recycling, 92*, 268–278.

Kuo, R. J. J., Wang, Y. C. C., & Tien, F. C. C. (2010). Integration of artificial neural network and MADA methods for green supplier selection. *Journal of Cleaner Production, 18*(12), 1161–1170.

Kuo, R. J., & Lin, Y. J. (2012). Supplier selection using analytic network process and data envelopment analysis. *International Journal of Production Research, 50*(11), 2852–2863.

Kuo, T. C., Hsu, C.-W., & Li, J.-Y. (2015). Developing a green supplier selection model by using the DANP with VIKOR. *Sustainability, 7*(2), 1661–1689.

Large, R. O., & Gimenez Thomsen, C. (2011). Drivers of green supply management performance: Evidence from Germany. *Journal of Purchasing and Supply Management, 17*(3), 176–184.

Lawson, B., Cousins, P. D., Handfield, R. B., & Petersen, K. J. (2009). Strategic purchasing, supply management practices and buyer performance improvement: An empirical study of UK manufacturing organizations. *International Journal of Production Research, 47*(10), 2649–2667.

Lee, A. H. I. I., Kang, H. Y., Lin, C. Y., & Wu, H. W. (2013). Selecting candidate suppliers using a multiple criteria decision making model. *Advanced Materials Research, 694–697*, 3472–3475.

Lee, A. H. I. I., Kang, H.-Y., Hsu, C.-F., & Hung, H.-C. (2009). A green supplier selection model for high-tech industry. *Expert Systems with Applications, 36*(4), 7917–7927.

Lee, C. K. M. K. M., Lau, H. C. W. C. W., Ho, G. T. S. T. S., & Ho, W. (2009). Design and development of agent-based procurement system to enhance business intelligence. *Expert Systems with Applications, 36*(1), 877–884.

Lee, T. (Jiun-Shen), Phuong Nha Le, T., Genovese, A., & Koh, L. S. C. (2011). Using FAHP to determine the criteria for partner's selection within a green supply chain. *Journal of Manufacturing Technology Management, 23*(1), 25–55.

Lima, F. R., Osiro, L., & Carpinetti, L. C. R. (2013). A fuzzy inference and categorization approach for supplier selection using compensatory and non-compensatory decision rules. *Applied Soft Computing, 13*(10), 4133–4147.

Lin, C.-T., Chen, C.-B., & Ting, Y.-C. (2012). A green purchasing model by using ANP and LP methods. *Journal of Testing and Evaluation, 40*(2), 203–210.

Liou, J. J. H., Tamosaitiene, J., Zavadskas, E. K., & Tzeng, G.-H. (2016). New hybrid COPRAS-G MADM Model for improving and selecting suppliers in green supply chain management. *International Journal of Production Research, 54*(1), 114–134.

Liu, A., Liu, H., Xiao, Y., Tsai, S.-B., & Lu, H. (2018). An empirical study on design partner selection in green product collaboration design. *Sustainability, 10*(1).

Liu, J., Wu, X., Zeng, S., & Pan, T. (2017). Intuitionistic linguistic multiple attribute decision-making with induced aggregation operator and its application to low carbon supplier selection. *International Journal of Environmental Research and Public Health, 14*(12).

Lu, B., & Wang, S. (2017). DEA conceptual exposition and literature review. In *Container Port Production and Management* (pp. 7–17). Singapore: Springer Singapore.

Lu, L. Y. Y., Wu, C. H., & Kuo, T. C. (2007). Environmental principles applicable to green supplier evaluation by using multi-objective decision analysis. *International Journal of Production Research, 45*(18–19), 4317–4331.

Luthra, S., Govindan, K., Kannan, D., Mangla, S. K., & Garg, C. P. (2017). An integrated framework for sustainable supplier selection and evaluation in supply chains. *Journal of Cleaner Production, 140*, 1686–1698.

Mafakheri, F., Breton, M., & Ghoniem, A. (2011). Supplier selection-order allocation: A two-stage multiple criteria dynamic programming approach. *International Journal of Production Economics, 132*(1), 52–57.

Mahdiloo, M., Saen, R. F., & Lee, K.-H. (2015). Technical, environmental and eco-efficiency measurement for supplier selection: An extension and application of data envelopment analysis. *International Journal of Production Economics, 168*, 279–289.

Malik, M. M., Abdallah, S., & Hussain, M. (2016). Assessing supplier environmental performance: Applying analytical hierarchical process in the United Arab Emirates healthcare chain. *Renewable & Sustainable Energy Reviews, 55*, 1313–1321.

Mardani, A., Jusoh, A., Nor, K. M., Khalifah, Z., Zakuan, N., & Valipour, A. (2015a). Multiple criteria decision-making techniques and their applications – A review of the literature from 2000 to 2014. *Economic Research, 28*(1), 516–571.

Mardani, A., Jusoh, A., & Zavadskas, E. K. (2015b). Fuzzy multiple criteria decision-making techniques and applications – Two decades review from 1994 to 2014. *Expert Systems with Applications, 42*(8), 4126–4148.

Mardani, A., Zavadskas, E. K., Khalifah, Z., Zakuan, N., Jusoh, A., Nor, K. M., & Khoshnoudi, M. (2017b). A review of multi-criteria decision-making applications to solve energy management problems: Two decades from 1995 to 2015. *Renewable & Sustainable Energy Reviews, 71*, 216–256.

Mardani, A., Zavadskas, E. K., Streimikiene, D., Jusoh, A., & Khoshnoudi, M. (2017a). A comprehensive review of data envelopment analysis (DEA) approach in energy efficiency. *Renewable and Sustainable Energy Reviews, 70*, 1298–1322.

Mareschal, B., Brans, J. P., & Vincke, P. (1984). *Prométhée: A new family of outranking methods in multicriteria analysis* (ULB Institutional Repository). ULB—Universite Libre de Bruxelles.

Meehan, J., & Bryde, D. (2011). Sustainable procurement practice. *Business Strategy and the Environment, 20*(2), 94–106.

Miemczyk, J., Johnsen, T. E., & Macquet, M. (2012). Sustainable purchasing and supply management: A structured literature review of definitions and measures at the dyad, chain and network levels. *Supply Chain Management: An International Journal, 17*(5), 478–496.

Mohammadi, H., Farahani, F. V., Noroozi, M., & Lashgari, A. (2017). Green supplier selection by developing a new group decision-making method under type 2 fuzzy uncertainty. *International Journal of Advanced Manufacturing Technology, 93*(1–4), 1443–1462.

Mukherjee, K. (2016). An integrated approach of sustainable procurement and procurement postponement for the multi-product, assemble-to-order (ATO) production system. *Production, 26*(2), 249–260.

Munda, G. (2005). Multiple criteria decision analysis and sustainable development. In *Multiple Criteria Decision Analysis: State of the Art Surveys* (pp. 953–986). New York: Springer-Verlag.

Neto, B., & Gama Caldas, M. (2017). The use of green criteria in the public procurement of food products and catering services: a review of EU schemes. *Environment, Development and Sustainability*, 1–29.

Opricovic, S., & Tzeng, G.-H. (2004). Compromise solution by MCDM methods: A comparative analysis of VIKOR and TOPSIS. *European Journal of Operational Research, 156*(2), 445–455.

Orji, I. J., & Wei, S. (2015a). An innovative integration of fuzzy-logic and systems dynamics in sustainable supplier selection: A case on manufacturing industry. *Computers & Industrial Engineering, 88*, 1–12.

Orji, I. J., & Wei, S. (2015b). Dynamic modeling of sustainable operation in green manufacturing environment. *Journal of Manufacturing Technology Management, 26*(8), 1201–1217.

Oruezabala, G., & Rico, J.-C. (2012). The impact of sustainable public procurement on supplier management—The case of french public hospitals. *Industrial Marketing Management, 41*(4), 573–580.

Özgen, D., Önüt, S., Gülsün, B., Tuzkaya, U. R., & Tuzkaya, G. (2008). A two-phase possibilistic linear programming methodology for multi-objective supplier evaluation and order allocation problems. *Information Sciences, 178*(2), 485–500.

Parthiban, P., Zubar, H. A., & Katakar, P. (2013). Vendor selection problem: A multi-criteria approach based on strategic decisions. *International Journal of Production Research, 51*(5), 1535–1548.

Qin, J., Liu, X., & Pedrycz, W. (2017). An extended TODIM multi-criteria group decision making method for green supplier selection in interval type-2 fuzzy environment. *European Journal of Operational Research, 258*(2), 626–638.

Quarshie, A. M., Salmi, A., & Leuschner, R. (2016). Sustainability and corporate social responsibility in supply chains: The state of research in supply chain management and business ethics journals. *Journal of Purchasing and Supply Management, 22*(2), 82–97.

Rezaei, J., Nispeling, T., Sarkis, J., & Tavasszy, L. (2016). A supplier selection life cycle approach integrating traditional and environmental criteria using the best worst method. *Journal of Cleaner Production, 135*, 577–588.

Roy, B. (1968). Classement et choix en presence de points de vue multiples. *Revue Française d'informatique et de Recherche Opérationnelle, 2*(6), 57–75.

Roy, B., & Bouyssou, D. (1993). *Aide multicritère à la décision: méthodes et cas.* Paris, France: Economica.

Saaty, R. W. (1987). The analytic hierarchy process—What it is and how it is used. *Mathematical Modelling, 9*(3–5), 161–176.

Sarkis, J., & Dhavale, D. G. (2015). Supplier selection for sustainable operations: A triple-bottom-line approach using a Bayesian framework. *International Journal of Production Economics, 166,* 177–191.

Senthil, S., Srirangacharyulu, B., & Ramesh, A. (2014). A robust hybrid multi-criteria decision making methodology for contractor evaluation and selection in third-party reverse logistics. *Expert Systems with Applications, 41*(1), 50–58.

Seuring, S., & Gold, S. (2012). Conducting content-analysis based literature reviews in supply chain management. *Supply Chain Management: An International Journal, 17*(5), 544–555.

Seuring, S., & Müller, M. (2008). From a literature review to a conceptual framework for sustainable supply chain management. *Journal of Cleaner Production, 16*(15), 1699–1710.

Shaik, M., & Abdul-Kader, W. (2011). Green supplier selection generic framework: A multi-attribute utility theory approach. *International Journal of Sustainable Engineering, 4*(1), 37–56.

Shaw, K., Shankar, R., Yadav, S. S., & Thakur, L. S. (2012). Supplier selection using fuzzy AHP and fuzzy multi-objective linear programming for developing low carbon supply chain. *Expert Systems with Applications, 39*(9), 8182–8192.

Shaw, K., Shankar, R., Yadav, S. S., & Thakur, L. S. (2013). Global supplier selection considering sustainability and carbon footprint issue: AHP multi-objective fuzzy linear programming approach. *International Journal of Operational Research, 17*(2), 215.

Shen, L., Olfat, L., Govindan, K., Khodaverdi, R., & Diabat, A. (2013). A fuzzy multi criteria approach for evaluating green supplier's performance in green supply chain with linguistic preferences. *Resources Conservation and Recycling, 74,* 170–179.

Shi, H., Quan, M.-Y., Liu, H.-C., & Duan, C.-Y. (2018). A novel integrated approach for green supplier selection with interval-valued intuitionistic uncertain linguistic information: A case study in the agri-food industry. *Sustainability, 10*(3), 733.

Si, S.-L., You, X.-Y., Liu, H.-C., & Zhang, P. (2018). DEMATEL technique: A systematic review of the state-of-the-art literature on methodologies and applications. *Mathematical Problems in Engineering.*

Simić, D., Kovačević, I., Svirčević, V., & Simić, S. (2017). 50 years of fuzzy set theory and models for supplier assessment and selection: A literature review. *Journal of Applied Logic, 24,* 85–96.

Simon, A. T., Serio, L. C. Di, Pires, S. R. I., & Martins, G. S. (2015). Evaluating supply chain management: A methodology based on a theoretical model. *Revista de Administração ContemporâNea, 19*(1), 26–44.

Sinha, A. K., & Anand, A. (2017). Towards fuzzy preference relationship based on decision making approach to access the performance of suppliers in environmental conscious manufacturing domain. *Computers & Industrial Engineering, 105,* 39–54.

Sivakumar, R., Kannan, D., & Murugesan, P. (2015). Green vendor evaluation and selection using AHP and Taguchi loss functions in production outsourcing in mining industry. *Resources Policy, 46,* 64–75.

Song, W., Xu, Z., & Liu, H.-C. (2017). Developing sustainable supplier selection criteria for solar air-conditioner manufacturer: An integrated approach. *Renewable and Sustainable Energy Reviews, 79,* 1461–1471.

Spence, L., & Bourlakis, M. (2009). The evolution from corporate social responsibility to supply chain responsibility: The case of Waitrose. *Supply Chain Management: An International Journal, 14*(4), 291–302.

Sycamore, I., & Stowers, K. (2015). Safe-Ice Literature Review On Sustainable Procurement Grant Agreement number: Project acronym: SAFE-ICE Project title: Research, Innovation and Business Support for a Low Carbon Economy Funding Scheme: Interreg Iva 2 Seas.

Tajik, G., Azadnia, A. H., Ma'aram, A., & Hassan, S. A. H. S. (2014). A hybrid fuzzy MCDM approach for sustainable third-party reverse logistics provider selection. In D. Kurniawan (Ed.), *Materials, Industrial, and Manufacturing Engineering Research Advances 1.1, 845*, 521–526.

Tang, X., Huang, Y., & Wei, G. (2018). Approaches to multiple-attribute decision-making based on pythagorean 2-tuple linguistic bonferroni mean operators. *Algorithms, 11*(1).

Tate, W. L., Ellram, L. M., & Kirchoff, J. F. (2010). Corporate social responsibility reports: A thematic analysis related to supply chain management. *Journal of Supply Chain Management, 46*(1), 19–44.

Tavana, M., Yazdani, M., & Di Caprio, D. (2017). An application of an integrated ANP–QFD framework for sustainable supplier selection. *International Journal of Logistics Research and Applications, 20*(3), 254–275.

Thanassoulis, E., Kortelainen, M., & Allen, R. (2012). Improving envelopment in data envelopment analysis under variable returns to scale. *European Journal of Operational Research, 218*(1), 175–185.

Theißen, S., & Spinler, S. (2014). Strategic analysis of manufacturer-supplier partnerships: An ANP model for collaborative CO2 reduction management. *European Journal of Operational Research, 233*(2), 383–397.

Touboulic, A., & Walker, H. (2015). Theories in sustainable supply chain management: A structured literature review. *International Journal of Physical Distribution & Logistics Management, 45*(1/2), 16–42.

Tsai, W.-H., & Hung, S.-J. (2009). Treatment and recycling system optimisation with activity-based costing in WEEE reverse logistics management: An environmental supply chain perspective. *International Journal of Production Research, 47*(19), 5391–5420.

Tsui, C.-W., Tzeng, G.-H., & Wen, U.-P. (2015). A hybrid MCDM approach for improving the performance of green suppliers in the TFT-LCD industry. *International Journal of Production Research, 53*(21), 6436–6454.

Tuzkaya, G. (2013). An intuitionistic fuzzy Choquet integral operator based methodology for environmental criteria integrated supplier evaluation process. *International Journal of Environmental Science and Technology, 10*(3), 423–432.

Tuzkaya, G., Ozgen, A., Ozgen, D., & Tuzkaya, U. R. (2009). Environmental performance evaluation of suppliers: A hybrid fuzzy multi-criteria decision approach. *International Journal of Environmental Science & Technology, 6*(3), 477–490.

Venkatesh, V. G., Zhang, A., Deakins, E., Luthra, S., & Mangla, S. (2018). A fuzzy AHP-TOPSIS approach to supply partner selection in continuous aid humanitarian supply chains. *Annals of Operations Research*, 1–34.

Vincke, P. (1989). *L'aide multicritère à la décision.* Editions de l'Université de Bruxelles.

von Oelreich, K., & Philp, M. (2013). *Green Public Procurement A tool for Achieving National Environmental Quality Objectives Report 6600 • November 2013.* Bromma, Sweden.

Walker, H., Di Sisto, L., & McBain, D. (2008). Drivers and barriers to environmental supply chain management practices: Lessons from the public and private sectors. *Journal of Purchasing and Supply Management, 14*(1), 69–85.

Walker, H., Miemczyk, J., Johnsen, T., & Spencer, R. (2012). Sustainable procurement: Past, present and future. *Journal of Purchasing and Supply Management, 18*(4), 201–206.

Wang, K.-Q., Liu, H.-C., Liu, L., & Huang, J. (2017). Green supplier evaluation and selection using cloud model theory and the qualiflex Method. *Sustainability, 9*(5).

Wang, Y.-M., & Chin, K.-S. (2011). Fuzzy analytic hierarchy process: A logarithmic fuzzy preference programming methodology. *International Journal of Approximate Reasoning, 52*(4), 541–553.

Watrbski, J., & Salabun, W. (2016). Green supplier selection framework based on multi-criteria decision-analysis approach. *Sustainable Design and Manufacturing 2016, 52*, 361–371.

Wetzstein, A., Hartmann, E., Benton Jr., W. C., Hohenstein, N.-O., Benton W. C., J., & Hohenstein, N.-O. (2016). A systematic assessment of supplier selection literature— State-of-the-art and future scope. *International Journal of Production Economics, 182*, 304–323.

Winter, M., & Knemeyer, A. M. (2013). Exploring the integration of sustainability and supply chain management. *International Journal of Physical Distribution & Logistics Management, 43*(1), 18–38.

Wittstruck, D., & Teuteberg, F. (2012). Integrating the concept of sustainability into the partner selection process: A fuzzy-AHP-TOPSIS approach. *International Journal of Logistics Systems and Management, 12*(2), 195.

Wu, C., & Barnes, D. (2011). A literature review of decision-making models and approaches for partner selection in agile supply chains. *Journal of Purchasing and Supply Management, 17*(4), 256–274.

Xu, M., Chen, Q., & Cui, L. (2010). An Improved Approach for Supplier Selection in Project Material Bidding Procurement. In Modeling Risk Management in Sustainable Construction. (pp. 3–10). Berlin, Heidelberg: Springer.

Yamak, A., Souchon, T., & Fröhlich, E. (2011). An empirical analysis of the benefits and significance of strategy enablers in the implementation of purchasing strategies. In R. Bogaschewsky, M. Eßig, R. Lasch, & W. Stölzle (Eds.), *Supply Management Research: Aktuelle Forschungsergebnisse 2011* (pp. 269–289). Wiesbaden, Germany: Gabler.

Yan, G. (2009). Research on green suppliers' evaluation based on AHP & Genetic Algorithm. In *2009 International Conference on Signal Processing Systems* (pp. 615–619). IEEE.

Yazdani, M. (2014). An integrated MCDM approach to green supplier selection. *International Journal of Industrial Engineering Computations, 5*(3), 443–458.

Yazdani, M., Chatterjee, P., Zavadskas, E. K., & Hashemkhani Zolfani, S. (2017). Integrated QFD-MCDM framework for green supplier selection. *Journal of Cleaner Production, 142*, 3728–3740.

Yazdani, M., Hashemkhani Zolfani, S., & Zavadskas, E. K. (2016). New integration of MCDM methods and QFD in the selection of green suppliers. *Journal of Business Economics and Management, 17*(6), 1097–1113.

Yazdani, M., Zarate, P., Coulibaly, A., & Zavadskas, E. K. (2017). A group decision making support system in logistics and supply chain management. *Expert Systems with Applications, 88*, 376–392.

Yoon, K., & Hwang, C. L. (1995). *Multiple attribute decision making: An introduction.* Sage Publications.

Yu, C., & Wong, T. N. (2014). A supplier pre-selection model for multiple products with synergy effect. *International Journal of Production Research, 52*(17), 5206–5222.

Yu, J.-R., & Tsai, C.-C. (2008). A decision framework for supplier rating and purchase allocation: A case in the semiconductor industry. *Computers & Industrial Engineering, 55*(3), 634–646.

Yu, Q., & Hou, F. (2016). An approach for green supplier selection in the automobile manufacturing industry. *Kybernetes, 45*(4), 571–588.

Zadeh, L. A. (1965). Fuzzy sets. *Information and Control, 8*(3), 338–353.

Zavadskas, E. K., Mardani, A., Turskis, Z., Jusoh, A., & Nor, K. M. (2016a). Development of TOPSIS method to solve complicated decision-making problems—An overview on developments from 2000 to 2015. *International Journal of Information Technology & Decision Making, 15*(3), 645–682.

Zavadskas, E. K., Govindan, K., Antucheviciene, J., & Turskis, Z. (2016b). Hybrid multiple criteria decision-making methods: A review of applications for sustainability issues. *Economic Research, 29*(1), 857–887.

Zhao, H., & Guo, S. (2014). Selecting green supplier of thermal power equipment by using a hybrid MCDM method for sustainability. *Sustainability, 6*(1), 217–235.

Zhou, F., Lin, Y., Wang, X., Zhou, L., & He, Y. (2016). ELV recycling service provider selection using the hybrid MCDM method: A case application in China. *Sustainability, 8*(5).

Zhou, G. (2016). Research on supplier performance evaluation system based on data mining with triangular fuzzy information. *Journal of Intelligent & Fuzzy Systems, 31*(3), 2035–2042.

Zhu, Q., Dou, Y., & Sarkis, J. (2010). A portfolio-based analysis for green supplier management using the analytical network process. *Supply Chain Management: An International Journal, 15*(4), 306–319.

Zsidisin, G. A., & Siferd, S. P. (2001). Environmental purchasing: A framework for theory development. *European Journal of Purchasing & Supply Management, 7*(1), 61–73.

Zyoud, S. H., & Fuchs-Hanusch, D. (2017). A bibliometric-based survey on AHP and TOPSIS techniques. *Expert Systems with Applications, 78*, 158–181.

15 The Investigation of Sustainable Procurement System Satisfying Turkey's Energy Demand by Means of Multi-period Interval Parameter Integer Linear Programming (MP-IPILP)

Miraç Eren, Erkan Oktay, and Ümran Şengül

CONTENTS

15.1   INTRODUCTION

Energy, the ability to do work, is the source of life for human life. In order to be able to handle human life, it needs energy for many vital activities such as heating, cooling, cooking, transportation, lighting...etc. In nature, energy can be obtained directly from primary sources (coal, natural gas, hydro, wind, solar, nuclear, etc.). Secondary energy sources can also be obtained through the conversion of primary energy sources. Electricity, which is one of the greatest discoveries of the last two centuries, is the most important secondary energy source in this way (Kumbaroğlu et al. 2008).

The twentieth century witnessed the transition from coal to petroleum-based resources, predominantly in terms of meeting energy needs. At the same time, along with industrialization and globalization, there has been a rapid increase in energy demand. Many countries and regions are currently using fossil-fuel (coal, oil, and natural gas) resources to meet rising energy demands. Therefore, fossil fuels continue to be the dominant sources for energy supply (Cofala et al. 2004). For example, the use of coal for electricity generation increased by approximately 235% between 2000 and 2010 (Lean and Smyth 2010). According to the latest International Energy Agency (IEA) report, world energy demand is increasing by about 1.6% per annum and still, more than 80% of the world's main energy production is provided from the burning of fossil fuels. By the year 2030, at the same time, this demand is expected to reach about 700×1018 j/year (Pękala et al. 2010). Clearly, the energy sector will play an important role in climate change as the consumption of fossil fuels causes large-scale greenhouse gas emissions, because the increasing intensity of the greenhouse gas in the atmosphere is expected to accelerate the rate of global warming (Chen et al. 2010). For example, the carbon dioxide (CO_2) concentration in the atmosphere was approximately 280 ppm in 1750 but increased to 367 ppm in 1999. Global CO_2 emissions are expected to exceed 30×109 tons per year in the near future (Griggs and Noguer 2002). Accordingly, it is expected that the average global temperature would increase by 1.1°C and become 6.4°C over the next 100 years (Lean and Smyth 2010). In such a scenario, protection of the environment is very important for sustainable life (Suganthi and Samuel 2012). How to reduce these bad conditions and effectively manage energy systems, accordingly, and the issue of how regional and economic development will be affected are all of great importance (Li et al. 2011b). For this reason, many countries try to reduce carbon intensity by choosing renewable energy sources (hydro, wind, solar, and biomass) instead of fossil fuels and/or increase energy conversion technology efficiency (Doukas et al. 2007). Along with that, the use of energy resources is limited by environmental and greenhouse gas concerns as well as the high costs and resource availability of new technologies. Although coal, compared to other sources, is relatively dirtier, it creates a significant portion of the system due to its low processing cost. Although electricity generated by natural gas fuels is more expensive, it has lower greenhouse gas and air pollution emissions than coal. Hydroelectric, wind, and solar are the primary sources of renewable energy without greenhouse gas emissions. However, they are also associated with high costs due to resource availability and stability concerns. Nuclear energy is also produced at high costs due to its safety reasons (Li et al. 2011b).

Therefore, in order to meet the changing demand levels of end consumers, political decision makers are responsible for the optimal allocation of the primary energy sources, which can be converted to secondary energy sources through various conversion and processing technologies from many power plants. Thus, energy strategy and policy are strongly influenced by political decision makers in the direction of sustainability and supply security objectives. In this context, the application of eco-friendly energy technologies, such as renewable energy systems in terms of supply and energy efficiency technologies and demand, is one of the key ways to achieve these goals (van Beeck et al. 2009).

Energy demands come from many end users (e.g., residential, commercial, industrial, transportation, and agricultural users). Electricity demand for each sector is represented by fuel and electricity input. Many impact factors, such as population growth rate, economic development, end-user habits, supply/service policy, and their interactions, lead to uncertain energy demand levels. In addition, uncertainties regarding various economic and technical parameters as well as facility expansion dynamics including timing, sizing, and location selection that can affect decision-making in the energy optimization analysis are complex. Further, there may also be uncertainties in many system components, such as energy supply/demand, conversion/processing technology, pollution emission intensity, and utility/cost parameters. In such a complex system, when local resources cannot meet their demands, energy imports. If the targeted energy production level is exceeded, it causes excessive production and/or export. In addition, the desired energy production and flow pattern can vary according to different demand levels and greenhouse gas reduction strategies. Moreover, energy saving and processing costs, investment costs for capacity development and expansion, and income from related energy products may not be deterministic. Finally, looking at a long-term planning period, if the energy supply does not adequately meet end-users' growing demands, the decision makers remain in a dilemma at the point of either transferring more resources on capacity expansion of existing plants or returning to other energy production options with higher costs (Li et al. 2011b). It is desirable for decision makers to develop effective methods to support the planning decisions of energy systems in which all these issues are addressed together.

Energy models containing very complex planning decisions have been developed to meet different requirements for energy planning and have been used by developing countries for many years (Sadeghi and Mirshojaeian Hosseini 2006). So far, as mentioned before, there are two important problems related to energy planning and, naturally, energy models. The first is political dimensions such as economic, political, and environmental, and the second is the uncertainty for the parameters. The first one can be solved with the help of a multi-objective model, and the other by the concept of fuzzy numbers.

In the literature review, under constraints satisfying the current demand through limited resources, it is observed that problems of planning and distribution of energy resources have been solved by linear programming models constructed through approaches either the single-objective (such as cost minimization/profit maximization) or the multi-objective optimizing many criteria simultaneously (such as minimization of production cost and greenhouse gas emissions/maximization of social acceptance).

Linear programming problems can be shaped according to the state of the parameters. The parameters may be deterministic or they may have a special distribution with a certain mean and standard deviation (probabilistic state) or a random distribution (possibilistic-fuzzy condition). Apart from these, they can possibly have a completely uncertain (that is, a certain average and non-distributed) interval. Mathematical models can be constructed in various ways according to the structure of the parameters that are formed according to the present state of the data.

The studies in which the parameters related to single-objective linear programming models established for energy planning are deterministic (Madlener et al. 2005; Sirikitputtisak et al. 2009; Lin et al. 2010; Mirzaesmaeeli et al. 2010; SULUKAN et al. 2010; Arnette and Zobel 2012; Flores et al. 2014) and outstanding at first glance. In addition, there are mathematical modeling studies constructed with the fuzzy parameters due to the lack of available data (Sadeghi and Mirshojaeian Hosseini 2006) and with the interval parameters in which the data is within a certain range but whose average and distribution is uncertain (Cai, Huang, Lin et al. 2009; Cai, Huang, Yang et al. 2009; Lin and Huang 2009a, 2009b; Dong et al. 2011; Wang et al. 2013). Recently, the single-objective hybrid studies involving stochasticity, fuzziness, and uncertainty related to the planning of energy resources (Lin et al. 2009; Chen et al. 2010; Li et al. 2010, 2011a, 2011b; Lin and Huang 2010; Xie et al. 2010; Jin et al. 2014) have also steadily increased.

There are also multi-objective linear programming models with deterministic parameters (Mezher et al. 1998; Pokharel and Chandrashekar 1998; Chedid et al. 1999; Mavrotas et al. 1999; Agrawal and Singh 2001; Antunes et al. 2004; Jana and Chattopadhyay 2004; Oliveira and Antunes 2004; Deshmukh and Deshmukh 2009; Daim et al. 2010; Jinturkar and Deshmukh 2011a, 2011b; Oliveira and Antunes 2011; Suresh et al. 2011; San Cristóbal 2012; Wu and Xu 2013; Namdari et al. 2014) and fuzzy parameters (Borges and Antunes 2003; Mavrotas et al. 2003; Nakata et al. 2011; Kazemi et al. 2012) related to the planning of energy resources. In addition, the models constructed as multi-objective, two or more aims can be solved simultaneously by utility approach, goal programming, fuzzy approach, and interactive approaches, and a compromise solution can be obtained (Yaralıoğlu and Umarusman 2010). When examined from the perspective of solution approaches, it is anticipated that studies will reach a compromised optimal solution using goal programming (Mezher et al. 1998; Pokharel and Chandrashekar 1998; Mavrotas et al. 1999; Agrawal and Singh 2001; Antunes et al. 2004; Oliveira and Antunes 2004; Deshmukh and Deshmukh 2009; Jinturkar and Deshmukh 2011b; Oliveira and Antunes 2011; Suresh et al. 2011; San Cristóbal 2012) and fuzzy approach (Chedid et al. 1999; Borges and Antunes 2003; Mavrotas et al. 2003; Jana and Chattopadhyay 2004; Daim et al. 2010; Jinturkar and Deshmukh 2011a; Kazemi et al. 2012; Wu and Xu 2013; Namdari et al. 2014).

It is possible to incorporate the Interval Parameter Programming (IPP) structure into the linear programming framework to reflect such multiple uncertainties when supply and/or demands of existing energy sources, economic data (i.e., utility and cost), conversion efficiency, and production capacities are not deterministic values. For this purpose, interval parameter linear programming (IPLP) has been developed as the basic algorithm of parametric mathematical programming, which is valid for optimization under incomplete uncertainty (i.e., known fluctuation interval, but unknown possible or probable distributed information) (Huang et al. 1992). The IPLP is useful for handling interval uncertainties on both the left and right sides of the constraints as well as the coefficients in the objective function (Huang 1996).

In the direction of the reduction of greenhouse gas emissions, there is a need for energy production and distribution planning, which takes into account the dynamic characteristics expressed by temporal and spatial changes of the relations between supply and demand. It is desirable for states to develop effective methods to support the planning decisions of energy systems in which all these issues are addressed together. This study also aims at the planning of sustainable production and procurement system satisfying Turkey's energy demand, under profit maximization scenario taking into account the increased export, in the period between the years 2000 and 2011. Thus, the findings obtained by energy system resource planning are compared with the resource distribution data made because of public policies in the current period, and some proposals are made. Therefore, firstly, under the heading "Theoretical Method," Multi-Period Interval Parameter Mixed Integer Linear Programming MP-IPMILP methodology is mentioned. Afterward, in line with the related aim, the restricted and linear equation system is proposed according to the simplified energy reference system of Turkey being determined by considering the data from the previous periods. The results obtained based on the mathematical model generated under the heading "Findings and Discussions" are compared with the results of state policy allocations. In the "Conclusions" section, some suggestions are presented.

15.2 THEORETICAL METHOD

15.2.1 INTERVAL PARAMETER LINEAR PROGRAMMING (IPLP) METHODOLOGY

An interval parameter $x^{\pm}$ value for x value defined as a member of a closed and restricted set of real numbers is defined as a range where only the upper and lower bounds of x are known but the distribution information is unknown.

$$x^{\pm} = \left[x^-, x^+ \right] = \left\{ t \in x \mid x^- \le t \le x^+ \right\} \tag{15.1}$$

where x^- and x^+ are the upper and lower bounds of $x^{\pm}$, respectively. When $x^- = x^+$ $x^{\pm}$ is the deterministic number.

The sign function $sign(x^{\pm})$ for $x^{\pm}$ is defined as in Equation 15.2.

$$sign(x^{\pm}) = \begin{cases} 1 & if \quad x^{\pm} \ge 0 \\ -1 & if \quad x^{\pm} < 0 \end{cases} \tag{15.2}$$

The $|x|^{\pm}$ absolute value is defined as in Equation 15.3.

$$|x|^{\pm} = \begin{cases} x^{\pm} & if \quad x^{\pm} \ge 0 \\ -x^{\pm} & if \quad x^{\pm} < 0 \end{cases} \tag{15.3}$$

Thus,

$$|x|^- = \begin{cases} x^- & \text{if} & x^\pm \geq 0 \\ -x^+ & \text{if} & x^\pm < 0 \end{cases} \tag{15.4}$$

$$|x|^+ = \begin{cases} x^+ & \text{if} & x^\pm \geq 0 \\ -x^- & \text{if} & x^\pm < 0 \end{cases} \tag{15.5}$$

According to these definitions, the constrained linear programming model consisting of interval parameters is as the equation system in Equation 15.6.

$$min / max \quad f^\pm = C^\pm.X^\pm \text{ subject to}: A^\pm.X^\pm \leq B^\pm \ X^\pm \geq 0 \tag{15.6}$$

where $R^\pm$ corresponds to a set of interval parameter numbers, $A \in \{R^\pm\}^{mxn}$, $B \in \{R^\pm\}^{mx1}$, $C \in \{R^\pm\}^{1xn}$ and $X \in \{R^\pm\}^{nx1}$.

15.2.2 ENERGY SYSTEM PLANNING VIA THE MULTI-PERIOD, INTERVAL PARAMETER INTEGER LINEAR PROGRAMMING (MP-IPMILP)

This study aims the planning of sustainable production and procurement system satisfying Turkey's energy demand, under profit maximization scenario taking into account the increased export, in the period between the years 2000 and 2011. For this, firstly, the simplified energy reference system of Turkey has been determined by considering the data from the previous periods. For this, it has benefited from the energy balance sheets compiled by the International Energy Agency. Moreover, the needed parameters have been obtained by manipulating statistical data, compiled from the databases Turkey Statistical Institute, Turkey's Energy and Natural Resources Ministry's databases. According to this, Turkey's simplified reference energy system is as shown in Figure 15.1.

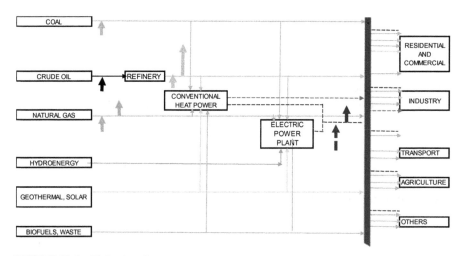

FIGURE 15.1 Turkey's reference energy system.

The simplified reference energy system is represented by a directed network of initial energy in the form of primary energy that gradually flows into another useful energy source to meet a specific external demand. In the reference energy system, seven sources have been identified as coal, crude oil, natural gas, hydro, geothermal-solar-wind, biofuel-waste, and electricity. Usage areas have been divided into groups as residential and commercial, industrial, transportation, agriculture and other (non-energy use-petrochemical raw material stocks) sectors.

According to the determined reference energy system, the desired objective is to provide an optimal attitude within a policy framework that is either the transfer of more resources to increase the capacity of existing plants or the orientation to other energy production options with higher costs, or whether energy from other countries should be imported or not. For this, the mathematical model in a form of the multi-periodic, interval parameterized, integer and linear (MP-IPILP) is proposed to solve the desired target. The proposed mathematical model is

- Multi-period due to investigate the period between the years 200 and 2011 in Turkey
- Interval parameters because the parameters such as economic data (i.e., benefit and cost), conversion efficiency and production capacities have incomplete uncertainty (i.e., known fluctuation interval but unknown possibility or probability distribution information)
- Includes integer constraints because it refers to the fact that existing plants have more resources to increase capacity or to return to other energy production options with higher costs or to decide whether to import energy from other regions

The constrained linear equations system determined in the study was established in line with the goal of maximizing the exports (i.e., the difference between the total export incomes of the sources and the total import expenditures in the 2000–2011 period). According to this,

15.2.2.1 Indices
Primary and secondary energy sources:

$i = 1$: Coal
$i =\ 2$: Oil
$i = 3$: Natural Gas
$i = 4$: Hydro
$i = 5$: Geothermal, Solar, Wind
$i = 6$: Biofuel, Waste
$i = 7$: Oil Products
$i = 8$: Electric
$i = 9$: Heat

Sectors:
  $s = 1$: Residential and Commercial
  $s = 2$: Industry
  $s = 3$: Transportation
  $s = 4$: Agriculture
  $s = 5$: Other

Years
  $t = 2000$–2011 (12 years)

15.2.2.2 Parameters

$ET_i^{\pm}$: The utilization efficiencies for electricity generation of primary energy resources ($i = 1, 3, 4, 5, 6, 7$), (kTOE/GWh)

$IET_i^{\pm}$: The usage efficiencies for heat generation of the primary energy sources ($i = 1, 3, 5, 6$), (kTOE/TJ)

BK_i : The initial (i.e., the year 2000) capacity of the power plant in which the source i is used (GW)

$MAKK_i$: The maximum power generation capacity of the power plant in which the source i is used (GW)

$SURE_i^{\pm}$: The daily working hours of the power plant in which the source i is used (hour/day)

$MAKPK_i$: Maximum energy potential for electric energy that can be generated by using energy source i (GW)

$SGAM_i$: The annual greenhouse gas reduction (here CO_2) amount of the power generating plant using the energy source i (tonnes/GWh)

$YTKSM_i$: The new plant installation cost of the power plant generating electricity using the energy source i ($\$$)

$IBYM_i$: operational maintenance and investment costs for increased capacity and new plant installation capacity of the electricity generating power plants using the energy source i ($\$$/GW)

$URETIM_{it}$: The energy production of energy source i in period t ($i = 1,..., 6$), (kTOE)

$AARZ_{it}$: The expandable energy supply of energy source i in period t ($i = 1,..., 6$), (kTOE)

$TALEP_{its}$: The demand by sectors s of energy source i in period t (kTOE for $i = 1,..., 6$, GWh for $i = 8$ and TJ for $i = 9$)

ITF_{it} : The import price of source i in period t ($\$$/binTEP for $i = 1,2,3,7$ and $\$$/GWh for $i = 8$)

IHF_{it} : The export price of source i in period t ($\$$/binTEP for $i = 3,7$ and $\$$/GWh for $i = 8$)

$RE^{\pm}$: Refinery efficiency (refinery input/refinery output)

DK : Unit conversion coefficient (kTOE/GWh)

$GEL\dot{I}R$: Revenue from greenhouse gas (here CO_2) reduction ($\$$/tonne)

15.2.2.3 Variables

$Z_{it}^{\pm}$: Energy produced from source i in period t ($i = 1, 2, 3, 4, 5, 6$), (kTOE)

$IT_{it}^{\pm}$: The energy imported from source i in period t (GWh for $i = 1,2,3,7$ and 8)

$IH_{it}^{\pm}$: The energy exported from source i in period t (GWh for $i = 3,7$ and 8)

$X_{it}^{\pm}$: Energy value of oil products obtained from the refinery (kTOE for $i = 7$)

$Xs_{ist}^{\pm}$: The energy value of the quantity sent in sector s from energy source i in period t (kTOE for $i = 1,..., 6$, GWh for $i = 8$ and TJ for $i = 9$)

$Xelek_{it}^{\pm}$: The electricity obtained from source i in period t (GWh)

$Xisi_{it}^{\pm}$: The heat obtained from source i in period t (TJ)

$XKelek_{it}^{\pm}$: The expandable capacity of the power conversion plants generating electricity using the energy source i in period t (GW)

$XTKelek_{it}^{\pm}$: The new plant installation capacity of the power conversion plants generating electricity using the energy source i in period t (GW)

$Y_{it}^{\pm}$: Binary decision variable to build a new plant or not for the power plant where the energy source i is used

15.2.2.4 Aims

$$max \quad f_1^{\pm} = (Aim1) + (Aim2) - (Aim3) - (Aim4) \tag{15.7}$$

$$(Aim1) = \sum_{t=1}^{12} \left(\sum_{i}^{3,7} DK.IHF_{it}^{\pm}.IH_{it}^{\pm} + IHF_{8t}^{\pm}.IH_{8t}^{\pm} \right) \tag{15.8}$$

$$(Aim2) = \sum_{t=1}^{12} \left(\sum_{\substack{i=1 \\ (i\neq 2)}}^{7} GELIR.SGAM_i.Xelek_{it}^{\pm} \right) \tag{15.9}$$

$$(Aim3) = \sum_{t=1}^{12} \left(\sum_{i}^{1,2,3,7} DK.ITF_{it}^{\pm}.IT_{it}^{\pm} + ITF_{8t}^{\pm}.IT_{8t}^{\pm} \right) \tag{15.10}$$

$$(Aim4) = \sum_{t=1}^{12} \left(\sum_{\substack{i=1 \\ (i\neq 2)}}^{7} YTKSM_{it}^{\pm}.Y_{it}^{\pm} + IBYM_i.(XKelek_{it}^{\pm} + XTKelek_{it}^{\pm}) \right) \tag{15.11}$$

Subjects to:

Supply Constraints
- Coal supply

$$-Z_{1t}^{\pm} - DK.IT_{1t}^{\pm} + ET_1^{\pm} \cdot Xelek_{1t}^{\pm} + IET_1^{\pm} \cdot Xisi_{1t}^{\pm} + \sum_{s=1}^{5} Xs_{1st}^{\pm} \le URETIM_{1t}, \forall t \tag{15.12}$$

- Crude oil supply

$$URETIM_{2t} + Z_{2t}^{\pm} + DK.IT_{2t}^{\pm} = RE^{\pm}.X_{7t}^{\pm} \tag{15.13}$$

$$X_{7t}^{\pm} + DK.IT_{7t}^{\pm} - DK.IH_{7t}^{\pm} - ET_{7}^{\pm}.Xelek_{7t}^{\pm} - IET_{7t}^{\pm}.Xls1_{7t}^{\pm} - \sum_{s=1}^{5} Xs_{7st}^{\pm} \geq 0, \quad \forall t \tag{15.14}$$

- Natural gas supply

$$-Z_{3t}^{\pm} - DK.IT_{3t}^{\pm} + DK.IH_{3t}^{\pm} + ET_{3}^{\pm} \cdot Xelek_{3t}^{\pm} + IET_{3}^{\pm} \cdot Xls_{3t}^{\pm}$$

$$+ \sum_{s=1}^{5} Xs_{3st}^{\pm} \leq URETIM_{3t}, \quad \forall t \tag{15.15}$$

- Hydro, geothermal, solar, wind, biofuels, and waste supply

$$-Z_{it}^{\pm} + ET_{i}^{\pm} \cdot Xelek_{it}^{\pm} + IET_{i}^{\pm} \cdot Xls_{it}^{\pm} + \sum_{s=1}^{5} Xs_{ist}^{\pm} \leq URETIM_{it}, \quad \forall t; i = 4,5,6 \tag{15.16}$$

- Restrictions on enhanced supply sources

$$Z_{it}^{\pm} \leq AARZ_{it} \quad \forall t; i = 1,2,3,4,5,6 \tag{15.17}$$

Capacity expansion constraints

$$Xelek_{it}^{\pm} - SURE_{it}^{\pm}.\left\{ \sum_{t'=1}^{t} \left(XKelek_{it'}^{\pm} + XTKelek_{it'}^{\pm} \right) \right\}.365$$

$$\leq BK_{i}.SURE_{it}^{\pm}.365 \quad \forall t; i = 1,3,4,5,6,7 \tag{15.18}$$

$$\sum_{t=1}^{12} XKelek_{it}^{\pm} \leq MAKK_{i} - BK_{i} \quad i = 1,3,4,5,6,7 \tag{15.19}$$

$$\sum_{t'=1}^{t} XKelek_{it}^{\pm} \geq \left(MAKK_{i} - BK_{i} \right).Y_{it}^{\pm} \quad \forall t; i = 1,3,4,5,6,7 \tag{15.20}$$

$$XTKelek_{it}^{\pm} \leq \left(MAKP_{i} - MAKK_{i} \right).Y_{it}^{\pm} \quad \forall t; i = 1,3,4,5,6,7 \tag{15.21}$$

$$\sum_{t=1}^{12} XTKelek_{it}^{\pm} \leq MAKP_{i} - MAKK_{i} \quad i = 1,3,4,5,6,7 \tag{15.22}$$

Demand constraints
- Total electric energy demand

$$\sum_{i}^{1,3,4,5,6,7} Xelek_{it}^{\pm} + IT_{8t}^{\pm} - IH_{8t}^{\pm} \geq \sum_{s=1}^{5} TALEP_{8st}, \quad \forall t \qquad (15.23)$$

- Total heat energy demand

$$\sum_{i}^{1,3,6,7} Xtst_{it}^{\pm} - Xs_{92t}^{\pm} \geq TALEP_{9t2}, \quad \forall t \qquad (15.24)$$

- Meeting all sectors' demands with all resources

$$Xs_{ist}^{\pm} \geq TALEP_{its} \quad \forall i (i \neq 2), s, t \qquad (15.25)$$

- Technical constraints

$$Z_{it}^{\pm}, IT_{it}^{\pm}, IH_{it}^{\pm}, X_{it}^{\pm}, Xs_{ist}^{\pm}, Xelek_{it}^{\pm}, Xtst_{it}^{\pm}, XKelek_{it}^{\pm}, XTKelek_{it}^{\pm} \geq 0 \quad ve \quad Y_{it}^{\pm} = 0,1 \quad (15.26)$$

15.3 FINDINGS AND DISCUSSIONS

Linear programming model with interval parameter constructed under profit maximization goal was resolved by the branch boundary algorithm-BDMLP solver in the GAMSide 23.5 optimization program and obtained the lower limit and upper limit values for each variable. Thus, the results obtained for each resource are summarized and compared with resource allocation made through government policies. Accordingly, the results are as follows.

15.3.1 COAL

Solid fossil fuels such as anthracite, other bituminous coals, sub-bituminous coals, lignite, patent fuel, coke, oven coke, gas coke, coal tar, peat, oil shale, and oil sands are generally named as coal. According to energy system planning that demonstrates the continuing adventure relating in which coal is the production, import, export, transfer to conversion plants, and use of sectors, the optimal resource procurement recommendation and the performed resource allocations are shown in Figures 15.2 through 15.5.

The optimal energy procurement model suggests that the production that should be between 2000 and 2011 should be at the maximum production limit provided in 2011. In other words, the recommended way to reduce both external dependence and increase profitability is to keep production at the highest level in each period. Whereas, despite the increase made in the coal production during the specified period (referred to here as URETIMit), it has only caught the capacity that should have been in recent years.

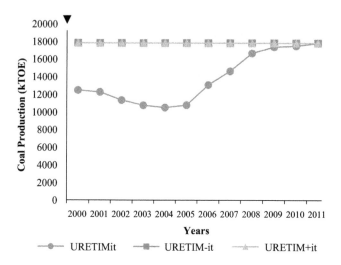

FIGURE 15.2 The coal productions performed and recommended by the energy procurement system model.

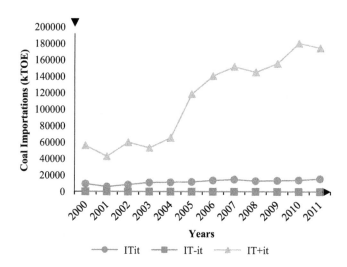

FIGURE 15.3 The coal importations performed and recommended by the energy procurement system.

Coal is a product that can be exported, but between 2000 and 2011 as a result of government policies, it was used only for domestic consumption, and exports were not realized. The optimal energy procurement model suggests that the coal should either never be imported or should be imported in high quantities, depending on its calorific value (i.e., its quality), in terms of increasing profitability. The import made according to government policies (referred to here as ITit) is between the specified lower and upper import band. The recommended model suggests that imports

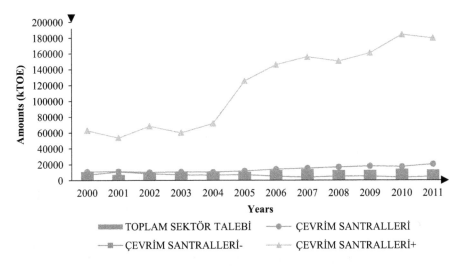

FIGURE 15.4 The total coal demand as well as the amounts transferred to power conversion plants according to the energy procurement system and in the past.

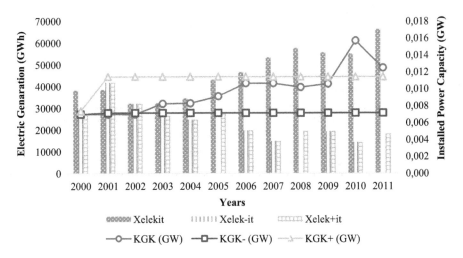

FIGURE 15.5 Electricity generated from as well as installed power capacities of coal-fired power plants both according to energy procurement system model and in the past.

should be made within a certain interval depending on the maximum level of production and the calorific value of coal, and accordingly that the total sector demand is met and the remaining amount is transferred to the conversion plants to increase the profitability. The amount to be sent to the conversion plants is in the specific lower and upper limit band due to the difference of its calorific value. The amount transferred to the conversion plants because of the policies implemented (referred to herein as "ÇEVRİM SANTRALLERİit") is also within the specified upper and lower limit band.

According to the recommended model, it is suggested that the total capacities of these plants have been between 7,149 and 11,439 GW since 2001, due to the variation according to the calorific value of coal for electricity produced in coal-fired conversion plants. Contrary to what is recommended, electricity quantity produced by coal-fired power plants because of past term policies has tended to increase during the period and accordingly, there has been an increase in the power plant capacities. Overall, however, total power capacity is observed at the optimal scale level.

15.3.2 CRUDE OIL AND PETROLEUM PRODUCTS

The optimal resource procurement recommendation and the performed resource allocations because of past term policies are shown in Figures 15.6 through 15.12.

The optimal energy procurement model suggests that the required production between 2000 and 2011 should be at the maximum production limit in 2000. In other words, as a way to both reduce external dependence and increase profitability, it is necessary to keep production at the highest level for each period. Crude oil production during the specified period (indicated here by URETIMit) showed a trend below the amount that had to be produced in the following years when it was at its maximum level in 2000. Although crude oil is a product that can be exported, it had only been used for domestic consumption between 2000 and 2011, and no exports have been realized. In addition, it is suggested that crude oil is required for imports at certain intervals in each period. The current crude oil import (referred to here as ITit) and the recommended import demonstrate a close trend to each other during the periods. However, due to the 2008 global economic crisis, the current crude oil imports declined slightly in 2009.

Optimal resource procurement system prefers to minimize imports to keep costs as low as possible in the direction of increasing profitability. Further, wherein, it is

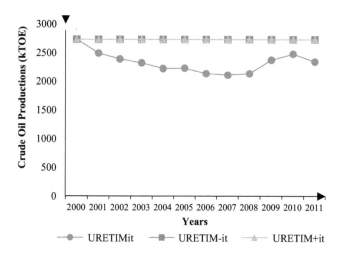

FIGURE 15.6 The crude oil productions performed and recommended by the energy procurement system model.

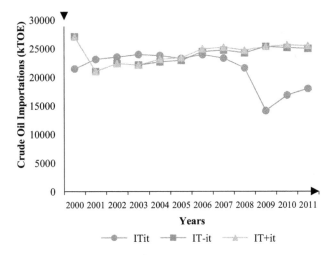

FIGURE 15.7 The crude oil importations performed and recommended by the energy procurement system model.

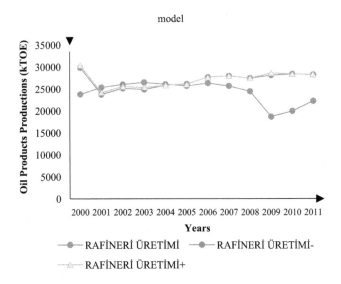

FIGURE 15.8 The oil products productions performed and recommended by the energy procurement system model.

observed that the import values are within a certain interval because of the solution of the mathematical model in which interval numbers expresses the parameter values. Performed imports have been more than recommended and have increased gradually with a certain trend. Especially in 2009, more petroleum products were imported. However, the optimal model suggests that importing crude oil from oil product imports is optimal in that year. In addition, petroleum products

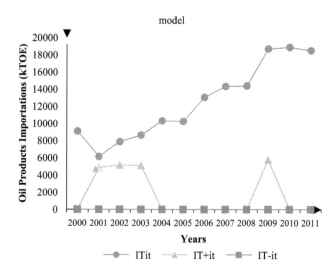

FIGURE 15.9   The oil products importations performed and recommended by the energy procurement system model.

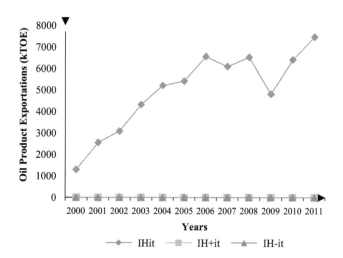

FIGURE 15.10   The oil products exportations performed and recommended by the energy procurement system model.

were imported and exported for many years. According to the optimal allocation, the export of petroleum products would not be a suitable behavior. However, it is observed that the performed exports are in an increasing tendency as in imports. Again, energy planning suggests that the production and imports of petroleum products should be carried out at certain intervals to meet the total sector demand, and the remaining amount should be transferred to the conversion plants. Since parameters are the interval, the amounts to be sent to the conversion plants are at certain lower and upper limit intervals but, in the real. The amounts transferred to the

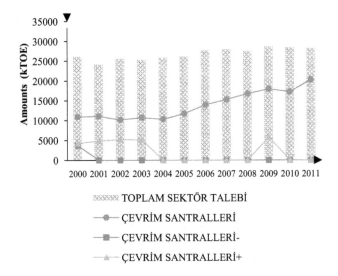

FIGURE 15.11 The total demand of oil-products as well as the amounts transferred to power conversion plants according to the energy procurement system and in the past.

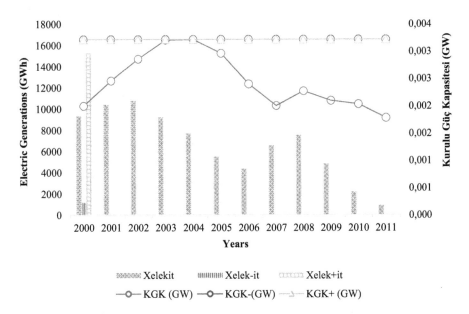

FIGURE 15.12 Electricity generated from as well as installed power capacities of oil products-fired power plants both according to energy procurement system model and in the past.

conversion plants (referred to herein as "ÇEVRİM SANTRALLERİit") have been above the upper and lower limit band determined by the optimal allocation planning and exhibited an upward trend over time.

It would be reasonable to produce electricity between 1,173.5 and 15,255 GWh from power plants based on petroleum products in 2000 in order to maximize total profitability according to the energy resources distribution plan. For this, the initial total power capacity (2000 capacity) needs to be increased from 1,996 GW to 3,216 GW, the maximum power capacity between 2000 and 2011. However, in reality, the amount of electricity generated from power plants based on petroleum products tended to fall over time. As a result, the installed power returned to its initial capacity in 2011, despite the increase during the period.

15.3.3 Natural Gas

According to the recommended energy model besides the performed resource allocations made for which natural gas is production, importation, exportation, transferring to conversion power plants and total use of sectors, optimal resource procurements are shown from Figures 15.13 to 15.17.

The recommended energy model suggests that the required production between 2000 and 2011 should be at the maximum production limit in 2008. In other words, as a way to both reduce external dependence and increase profitability, it is necessary to keep production at the highest level for each period. However, with the increase in natural gas production (indicated here by URETIMit) in the current period, it reached the capacity that should have been in 2008, but it started to decrease again. In addition, modeling suggests that importing natural gas by a certain amount and with increasing tendency increases profitability. The imports (expressed here as ITit) are generally higher than recommended. Exports were also made during periods when natural gas was imported. According to the model established with the goal of profit maximization, it is concluded that natural gas exportation would not be a suitable behavior. However, export has been on the rise since 2007 and seems to

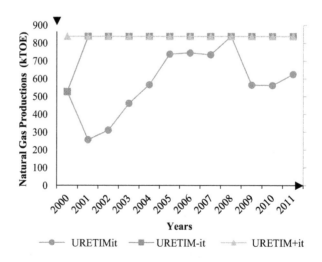

FIGURE 15.13 The natural gas productions performed and recommended by the energy procurement system model.

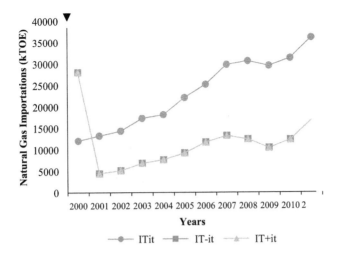

FIGURE 15.14 The natural gas importations performed and recommended by the energy procurement system model.

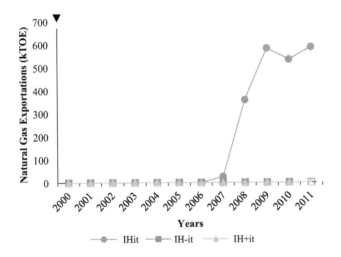

FIGURE 15.15 The natural gas exportations performed and recommended by the energy procurement system model.

be in an upward trend. In addition, the optimal model recommends increasing the profitability by realizing certain values of natural gas production and importation and, accordingly, only meeting the demand of the total sector and transferring some natural gas to the conversion power plants only in 2000. According to the modeling, the amount transferred to the conversion plants (referred to here as "ÇEVRİM SANTRALLERİit") is above the generally stated values and exhibits an upward tendency over time. Already the reason for the high imports is due to electricity production from natural gas. However, according to the model, it is observed that the

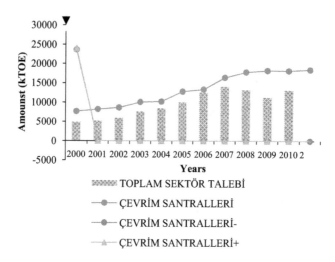

FIGURE 15.16 The total demand for natural gas as well as the amounts transferred to power conversion plants according to the energy procurement system and in the past.

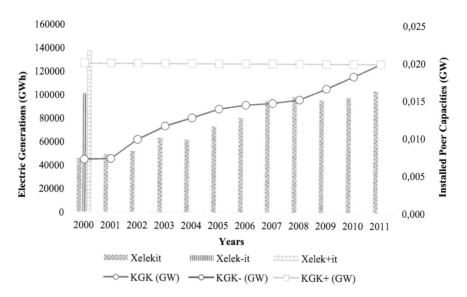

FIGURE 15.17 Electricity generated from as well as installed power capacities of natural gas-fired power plants both according to energy procurement system model and in the past.

total sector demand can be met with the current production, and it is not reasonable to make electricity production by importing natural gas.

Again, according to the modeling, it is concluded that it would be reasonable to produce electricity between 101,377.5 and 137,583 GWh from natural gas-based power plants in 2000. For this, the starting total power capacity (2000 capacity) needs to be increased from 7,044 GW to 19.83 GW, the maximum power capacity between 2000 and 2011. However, electricity generated from natural gas-fired power plants has also tended to increase over time. As a result, an increase in initial power capacity has been observed since 2001.

15.3.4 HYDROPOWER

Some of the products like hydro and solar energy need to be converted to electricity to get obtained. As a result, energy production from these products is naturally restricted to electricity generation only. Therefore, renewable electricity sources and technologies such as hydro, solar photovoltage, tide, wave, ocean, and wind can only be measured in terms of electricity output (usually kilo-, megawatt, gigawatt hours). Naturally, a trade related to this production is not the trade of renewable and wastes, is only the electricity and heat trade. However, it is still very difficult (or impossible) to determine the source of the electricity being traded. Accordingly, their current values, as well as the power plant capacity and electric energy to be required according to the optimal energy procurement model aiming to maximize the total profitability by utilizing the hydropower, are shown in Figure 15.18.

According to the optimal model, it would be reasonable to generate electricity between 37,517 and 52,337 GWh from the hydroelectric power plants in total, for which the initial total power capacity (2000 capacity) of 11,175 GW should be 14,335–17,135 GW. Electricity generated from hydroelectric power plants has also tended to increase over time. Depending on this, an increase in initial power capacity has been observed since 2001.

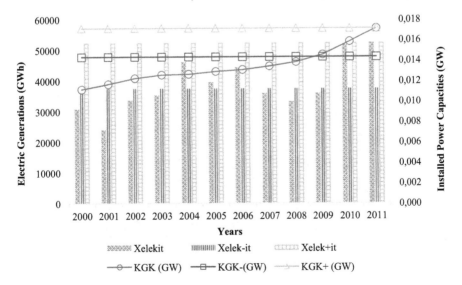

FIGURE 15.18 Electricity generated from as well as installed power capacities of hydro-electric plants both according to energy procurement system model and in the past.

15.3.5 Geothermal, Solar, and Wind Energy

There are products such as geothermal and solar heat both produced and used in the conversion and end-user consumption sectors, as well as wind and solar photovoltaic products that need to be converted to electricity to be obtained. According to the recommended energy model besides the performed resource allocations made for which these products are production, importation, exportation, transferring to conversion power plants and total use of sectors, Optimal resource procurements are shown from Figures 15.19 to 15.21.

The recommended energy model suggests that the required production between 2000 and 2011 should be within the maximum production limit in 2011. In other words, as a way to both reduce external dependence and increase profitability, it is necessary to keep production at the highest level for each period. However, geothermal, solar and wind energy production (here indicated by URETIMit) has increased, but only recently caught up with the capacity that should have been. Especially after 2008, the increasing tendency became even more. According to the proposed model, it is necessary to meet the increasing sector demand by via the maximum production to be made between 2000 and 2011 and be transferred the remaining amount to the conversion plants. Therefore, it is observed that in the first period when there is little sector demand, giving priority to conversion plants by making maximum production is reasonable for electricity generation. However, it is necessary to reduce the amounts transferred to the conversion plants in response to the increasing sector demand in time. In the current period, the sector request was met depending on the production made over time and the share of conversion plants has also increased with increasing production. Especially after 2008, the amount transferred to the conversion plants, showing a high trend, has begun to approach the recommended optimal value.

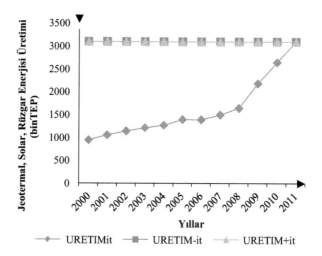

FIGURE 15.19 The geothermal, solar, and wind energy productions performed and recommended by the energy procurement system model.

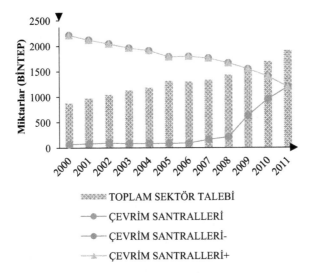

FIGURE 15.20   The total demand for geothermal, solar, wind energy as well as the amounts transferred to power conversion plants according to the energy procurement system and in the past.

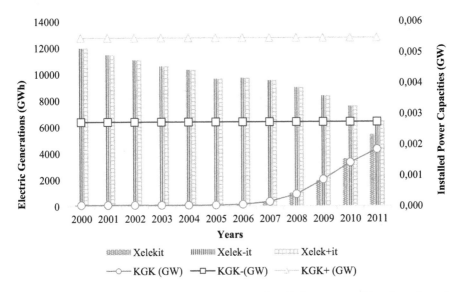

FIGURE 15.21   Electricity generated from as well as installed power capacities of geothermal, solar, wind energy plants both according to energy procurement system model and in the past.

According to the recommended model, it would be reasonable to generate 11,973 GWh of electricity from geothermal, solar and wind energy plants. Therefore, it is concluded that the initial total power capacity of 0.036 GW (capacity of 2000) should be in the range of 2,736–5,466 GW due to the interval condition of the related parameters.

However, electricity generated from geothermal, solar and wind energy plants also tended to increase over time. As a result, an increase in initial power capacity has been observed since 2001. The electricity production from these plants reached 5417 GWh in 2011 from 108.9 GWh in 2000. Thus, the total power capacity increased from 0.036 GW in 2000 to 1.843 GW in 2011. However, according to the recommended model, more wind, geothermal and solar energy need to be exploited to produce more electricity. In particular, it is recommended to use the idle capacity, which is called the unused power plant capacity, to increase the profitability.

15.3.6 BIOFUELS AND WASTES

There are energy resources containing products such as waste, firewood, biogas, and liquid biofuels produced and used multipurpose in the conversion and end-user consumption sectors. According to the recommended energy model besides the performed resource allocations made for which these energy resources are production, importation, exportation, transferring to conversion power plants and total use of sectors, optimal resource procurements are shown from Figures 15.22 to 15.24.

The recommended energy model suggests that the required production between 2000 and 2011 should be within the maximum production limit in 2000. In other words, as a way to both reduce external dependence and increase profitability, it is necessary to keep production at the highest level for each period. However, the biofuels and waste production (indicated here by URETIMit) has declined, and especially after 2010, the declining trend has been greater. Also according to the recommended model, by via the maximum production to be made between 2000 and 2011, it is concluded that the sector demand should be met and the remaining amount should be transferred to the conversion plants. Therefore, it is observed that in the first period when there is high sector demand, giving priority to conversion plants by making

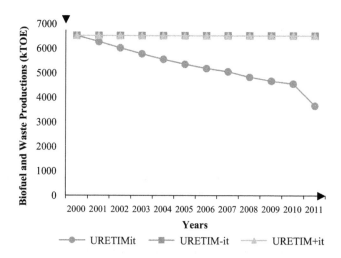

FIGURE 15.22   The biofuel and waste energy productions performed and recommended by the energy procurement system model.

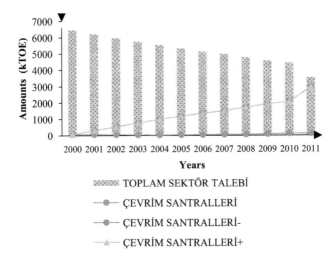

FIGURE 15.23   The total demand for biofuel and waste as well as the amounts transferred to power conversion plants according to the energy procurement system and in the past.

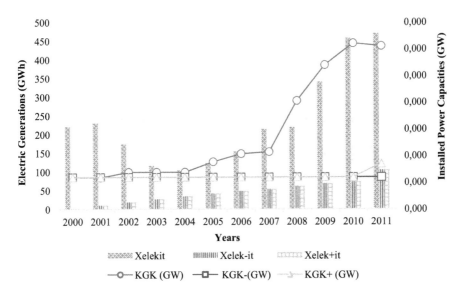

FIGURE 15.24   Electricity generated from as well as installed power capacities of biofuel and waste energy plants both according to energy procurement system model and in the past.

maximum production is reasonable for electricity generation. However, it is necessary to increase the amounts transferred to the conversion plants in response to the decreasing sector demand in time. However, depending on the production made, over time, the demand for the current sector was met and the share of the conversion plants increased in spite of declining production.

According to the recommended model, it would be reasonable to generate electricity with an increasing trend over time from biofuel and waste-based power plants. Therefore, it is concluded that the initial total power capacity of 0.024 GW (capacity of 2000) should be in the range of 0,024–0,034 GW due to the interval condition of the related parameters. Electricity generated from biofuels and waste-based power plants tended to increase over time. As a result, since 2000, the initial power plant capacity has increased in the relevant period. Electricity generation based on biofuels and waste energy reached 469 GWh in 2011 from 220 GWh in 2000. Thus, the total power capacity increased from 0,024 GW in 2000 to 0,122 GW in 2011. Optimal modeling suggests that biofuels and waste products should only meet the total sector demand and be exported instead of generating electricity with the remaining portion due to their high production costs. However, exports of these products have not been realized before, and because of the low thermal values of most of these products, transportation is not economical over long distances.

15.3.7 ELECTRICITY

Electricity can be produced as secondary energy as well as primary energy. Primary electricity is derived from the mechanical energy of natural resources such as hydro, wind, tide, and wave power. Secondary energy is produced from the heat generated by the nuclear division of nuclear fuels, geothermal, and solar heat, and by the combustion of primary fuels such as coal, natural gas, petroleum, renewable, and waste. It is also capable of being distributed to end users through national and international transmission and distribution lines after electricity is produced. The recommended optimal resource procurement, as well as the performed resource allocations according to the energy distribution plan from the production, import, and export of the electricity to the final use, are shown in Figures 15.25 through 15.28.

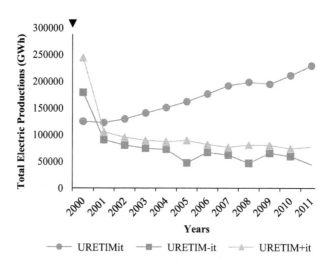

FIGURE 15.25 The electric energy generations performed and recommended by the energy procurement system model.

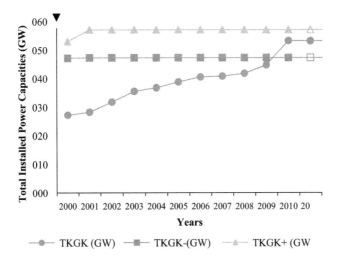

FIGURE 15.26 Total installed power capacities of electric energy plants both according to energy procurement system model and in the past.

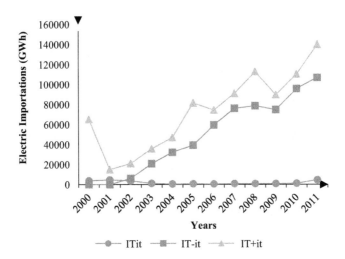

FIGURE 15.27 The electric energy importations performed and recommended by the energy procurement system model.

It is concluded that it would be more appropriate to follow a decreasing trend over time in electricity generation according to the recommended model. However, electricity produced has followed a trend of continuous increase. In addition, according to the model, it turns out that the total power capacities should be at a certain average level depending on the amount of electricity to be generated. However, due to the electricity generated, the total power capacity tended to increase over time.

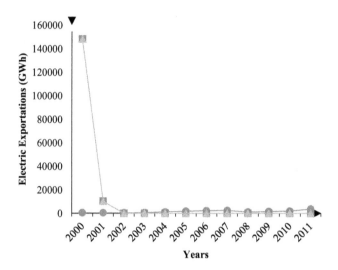

FIGURE 15.28 The electric energy exportations performed and recommended by the energy procurement system model.

Further, to increase profitability, the recommended model suggested that electric should be imported within a certain amount and within increasing tendency. So that it is more convenient to import direct electricity instead of importing other primary energy sources in order to generate electricity at the point of increasing profitability. In fact, the import amount is so large that it is recommended to import 20–30 times more than the amount imported, especially in 2011. Here, it is observed that the amounts to be imported due to the interval data of cost parameters are within a certain lower and upper limit. However, the current imports (referred to here as ITit) are generally low. Moreover, as well as electricity is imported, its exportation has also been carried out in periods. At the point of profit maximization, it is concluded that the export of electricity for the recommended model would not be a suitable behavior (but it is the case that exports in 2000 increased profitability). The current exportation, however, exhibited an increasing tendency, in general, despite the decline in 2008.

15.4 CONCLUSIONS

Turkey remains insufficient in terms of meeting the total energy demand of the end users because of its fast-growing population and economy. Therefore, its external dependence on fossil fuels such as oil and natural gas is increasing day by day. In such a case, it is inevitable to import some sources. But what is important here is how much and when to import from which source, together with maximum production, to meet the end-user demand in line with the goal of increasing profitability and reducing external dependence. Therefore, this study has recommended the planning of a sustainable production and procurement system satisfying Turkey's energy demand, under a profit maximization scenario taking into account the increased export. As a process, the period between the years 2000 and 2011 were examined. Thus, the findings obtained by energy system resource planning were compared with the resource distribution data made because of public policies in the current period, and some recommendations were made. Here, the desired aim is to provide an optimal manner to a policy framework that is either the transfer of more resources to increase the capacity of existing plants, the orientation to other energy production options with higher costs, or whether energy from other countries should be imported or not. Therefore, the mathematical model proposed in a form of the Multi-Period Interval Parameter Integer Linear Programming (MP-IPILP) has been established to reach the desired aim.

When the findings are examined, according to the energy distribution model created to increase the profitability, it is concluded that the coal importation must be within certain ranges according to its calorific value in order to meet the direct need of conversion plants and total sector demand. However, in the current period, electricity generation in coal-fired power plants has increased, and this increase in the generation has been achieved through the import of coal. It is concluded that the import of crude oil must remain stable; therefore, it would be appropriate in terms of optimality that petroleum products should be either not imported or imported within a certain range according to their calorific value. In contrast, electricity generation in power plants based on petroleum products has declined over time and approached the recommended optimal level. Especially in 2008, there was a serious reduction in oil imports. The recommended model suggests that natural gas imports should have increased in line with increasing demand. Accordingly, it is observed that natural gas-based electricity generation has not been very appropriate in terms of optimality. However, it has been observed that the import of natural gas in the current period has increased in correlation with the recommended optimal amount, and this increased amount is slightly higher than the optimal solution. The reason for this is that as much of the difference has been used in the generation of electricity. As a result, the established mathematical model does not suggest electricity generation from natural gas. To meet the total electricity demand of the sectors, the recommended model suggests that electricity generation based on petroleum products and natural facilities should not be done while coal-based electricity production should be reduced over time and that the requirement is eliminated by importing electricity as high as possible. The findings of the analysis revealed that it is important to utilize hydroelectric power plants, especially for electricity generation. In order to achieve this, the hydroelectric power plant capacity should be as high as possible. Because of state policies, it has been observed that the power plant capacity has reached an optimal level in recent years, and accordingly, the electricity production is in the optimal level. It has also been found that generating electricity from geothermal, solar, and wind energy has previously been reasonable, but it would be beneficial to reduce this generating towards recent years. Because of state policies, it is seen that the electricity generation from these sources approached the optimum. In general, the optimized model demonstrates that optimality would be achieved by reducing dependence on fossil resources and focusing on renewable resources (especially hydropower). Turkey has already Europe's largest hydro, wind, and geothermal energy potential. The goal of the MENR 2023 vision is to use the whole of the 36,000 MW of hydroelectric and 600 MW of geothermal potential technically and economically considered, as well as to increase the wind power to 20,000 MW and to reach 600 MW installed capacity from solar power plants.

REFERENCES

Agrawal, R. K., and S. P. Singh. 2001. Energy allocations for cooking in UP households (India): A fuzzy multi-objective analysis. *Energy Conversion and Management* 42 (18):2139–54.

Antunes, C. H., A. G. Martins, and I. S. Brito. 2004. A multiple objective mixed integer linear programming model for power generation expansion planning. *Energy* 29 (4):613–27. doi:10.1016/j.energy.2003.10.012.

Arnette, A., and C. W. Zobel. 2012. An optimization model for regional renewable energy development. *Renewable and Sustainable Energy Reviews* 16 (7):4606–15. doi:10.1016/j.rser.2012.04.014.

Borges, A. R., and C. H. Antunes. 2003. A fuzzy multiple objective decision support model for energy-economy planning. *European Journal of Operational Research* 145 (2):304–16.

Cai, Y. P., G. H. Huang, Q. G. Lin, X. H. Nie, and Q. Tan. 2009. An optimization-model-based interactive decision support system for regional energy management systems planning under uncertainty. *Expert Systems with Applications* 36 (2):3470–82. doi:10.1016/j.eswa.2008.02.036.

Cai, Y. P., G. H. Huang, Z. F. Yang, and Q. Tan. 2009. Identification of optimal strategies for energy management systems planning under multiple uncertainties. *Applied Energy* 86 (4):480–95. doi:10.1016/j.apenergy.2008.09.025.

Chedid, R., T. Mezher, and C. Jarrouche. 1999. A fuzzy programming approach to energy resource allocation. *International Journal of Energy Research* 23 (4):303–17.

Chen, W. T., Y. P. Li, G. H. Huang, X. Chen, and Y. F. Li. 2010. A two-stage inexact-stochastic programming model for planning carbon dioxide emission trading under uncertainty. *Applied Energy* 87 (3):1033–47. doi:10.1016/j.apenergy.2009.09.016.

Cofala, J., M. Amann, F. Gyarfas, W. Schoepp, J. C. Boudri, L. Hordijk, C. Kroeze, L. Junfeng, D. Lin, and T. S. Panwar. 2004. Cost-effective control of SO_2 emissions in Asia. *Journal of Environmental Management* 72 (3):149–61.

Daim, T. U., G. Kayakutlu, and K. Cowan. 2010. Developing Oregon's renewable energy portfolio using fuzzy goal programming model. *Computers & Industrial Engineering* 59 (4):786–93. doi:10.1016/j.cie.2010.08.004.

Deshmukh, S. S., and M. K. Deshmukh. 2009. A new approach to micro-level energy planning—A case of northern parts of Rajasthan, India. *Renewable and Sustainable Energy Reviews* 13 (3):634–42. doi:10.1016/j.rser.2007.11.015.

Dong, C., G. H. Huang, Y. P. Cai, and Y. Xu. 2011. An interval-parameter minimax regret programming approach for power management systems planning under uncertainty. *Applied Energy* 88 (8):2835–45. doi:10.1016/j.apenergy.2011.01.056.

Doukas, H., W. Mannsbart, K. D. Patlitzianas, J. Psarras, M. Ragwitz, and B. Schlomann. 2007. A methodology for validating the renewable energy data in EU. *Renewable Energy* 32 (12):1981–98.

Flores, J. R., J. M. Montagna, and A. Vecchietti. 2014. An optimization approach for long term investments planning in energy. *Applied Energy* 122:162–78. doi:10.1016/j.apenergy.2014.02.002.

Griggs, D. J., and M. Noguer. 2002. Climate change 2001: The scientific basis. Contribution of working group I to the third assessment report of the intergovernmental panel on climate change. *Weather* 57 (8):267–9.

Huang, G., B. W. Baetz, and G. G. Patry. 1992. A grey linear programming approach for municipal solid waste management planning under uncertainty. *Civil Engineering Systems* 9 (4):319–35.

Huang, G. H. 1996. IPWM: An interval parameter water quality management model. *Engineering Optimization+ A35* 26 (2):79–103.

Jana, C., and R. Chattopadhyay. 2004. Block level energy planning for domestic lighting—A multi-objective fuzzy linear programming approach. *Energy* 29 (11):1819–29. doi:10.1016/j.energy.2004.03.095

Jin, L., G. H. Huang, D. Cong, and Y. R. Fan. 2014. A robust inexact joint-optimal α cut interval type-2 fuzzy boundary linear programming (RIJ-IT2FBLP) for energy systems planning under uncertainty. *International Journal of Electrical Power & Energy Systems* 56:19–32. doi:10.1016/j.ijepes.2013.10.029.

Jinturkar, A. M., and S. S. Deshmukh. 2011a. A fuzzy mixed integer goal programming approach for cooking and heating energy planning in rural India. *Expert Systems with Applications* 38 (9):11377–81. doi:10.1016/j.eswa.2011.03.006.

Jinturkar, A. M., and S. S. Deshmukh. 2011b. Optimization of energy resources for cooking and heating—A fuzzy goal programming approach. *Journal of Renewable and Sustainable Energy* 3 (4):043117. doi:10.1063/1.3625251.

Kazemi, A., M. R. Mehregan, H. Shakouri G., and M. Hosseinzadeh. 2012. Energy resource allocation in Iran: A fuzzy multi-objective analysis. *Procedia—Social and Behavioral Sciences* 41:334–41. doi:10.1016/j.sbspro.2012.04.038.

Kumbaroğlu, G., R. Madlener, and M. Demirel. 2008. A real options evaluation model for the diffusion prospects of new renewable power generation technologies. *Energy Economics* 30 (4):1882–908.

Lean, H. H., and R. Smyth. 2010. CO_2 emissions, electricity consumption and output in ASEAN. *Applied Energy* 87 (6):1858–64.

Li, G. C., G. H. Huang, Q. G. Lin, X. D. Zhang, Q. Tan, and Y. M. Chen. 2011. Development of a GHG-mitigation oriented inexact dynamic model for regional energy system management. *Energy* 36 (5):3388–98. doi:10.1016/j.energy.2011.03.037.

Li, M. W., Y. P. Li, and G. H. Huang. 2011a. An interval-fuzzy two-stage stochastic programming model for planning carbon dioxide trading under uncertainty. *Energy* 36 (9):5677–89. doi:10.1016/j.energy.2011.06.058.

Li, Y. F., Y. P. Li, G. H. Huang, and X. Chen. 2010. Energy and environmental systems planning under uncertainty—An inexact fuzzy-stochastic programming approach. *Applied Energy* 87 (10):3189–211. doi:10.1016/j.apenergy.2010.02.030.

Li, Y. P., G. H. Huang, and X. Chen. 2011b. Planning regional energy system in association with greenhouse gas mitigation under uncertainty. *Applied Energy* 88 (3):599–611. doi:10.1016/j.apenergy.2010.07.037.

Lin, Q. G., and G. H. Huang. 2009a. A dynamic inexact energy systems planning model for supporting greenhouse-gas emission management and sustainable renewable energy development under uncertainty—A case study for the City of Waterloo, Canada. *Renewable and Sustainable Energy Reviews* 13 (8):1836–53. doi:10.1016/j.rser.2009.01.021.

Lin, Q. G., and G. H. Huang. 2009b. Planning of energy system management and GHG-emission control in the Municipality of Beijing—An inexact-dynamic stochastic programming model. *Energy Policy* 37 (11):4463–73. doi:10.1016/j.enpol.2009.05.066.

Lin, Q. G., and G. H. Huang. 2010. An inexact two-stage stochastic energy systems planning model for managing greenhouse gas emission at a municipal level. *Energy* 35 (5):2270–80. doi:10.1016/j.energy.2010.01.042.

Lin, Q. G., G. H. Huang, B. Bass, and X. S. Qin. 2009. IFTEM: An interval-fuzzy two-stage stochastic optimization model for regional energy systems planning under uncertainty. *Energy Policy* 37 (3):868–78. doi:10.1016/j.enpol.2008.10.038.

Lin, Q. G., G. H. Huang, B. Bass, X. H. Nie, X. D. Zhang, and X. S. Qin. 2010. EMDSS: An optimization-based decision support system for energy systems management under changing climate conditions—An application to the Toronto-Niagara Region, Canada. *Expert Systems with Applications* 37 (7):5040–51. doi:10.1016/j.eswa.2009.12.007.

Madlener, R., G. Kumbaroğlu, and V. Ş Ediger. 2005. Modeling technology adoption as an irreversible investment under uncertainty: The case of the Turkish electricity supply industry. *Energy Economics* 27 (1):139–63. doi:10.1016/j.eneco.2004.10.007.

Mavrotas, G., D. Diakoulaki, and L. Papayannakis. 1999. An energy planning approach based on mixed 0–1 multiple objective linear programming. *International Transactions in Operational Research* 6 (2):231–44.

Mavrotas, G., H. Demertzis, A. Meintani, and D. Diakoulaki. 2003. Energy planning in buildings under uncertainty in fuel costs: The case of a hotel unit in Greece. *Energy Conversion and Management* 44 (8):1303–21.

Mezher, T., R. Chedid, and W. Zahabi. 1998. Energy resource allocation using multi-objective goal programming: The case of Lebanon. *Applied Energy* 61 (4):175–92.

Mirzaesmaeeli, H., A. Elkamel, P. L. Douglas, E. Croiset, and M. Gupta. 2010. A multi-period optimization model for energy planning with CO(2) emission consideration. *Journal of Environmental Management* 91 (5):1063–70. doi:10.1016/j.jenvman.2009.11.009.

Nakata, T., D. Silva, and M. Rodionov. 2011. Application of energy system models for designing a low-carbon society. *Progress in Energy and Combustion Science* 37 (4):462–502. doi:10.1016/j.pecs.2010.08.001.

Namdari, M., J. H. Yoon, A. Abadi, S. M. Taheri, and S. H. Choi. 2014. Fuzzy logistic regression with least absolute deviations estimators. *Soft Computing* 19 (4):909–17. doi:10.1007/s00500-014-1418-2.

Oliveira, C., and C. H. Antunes. 2004. A multiple objective model to deal with economy–energy–environment interactions. *European Journal of Operational Research* 153 (2):370–85. doi:10.1016/s0377-2217(03)00159-0.

Oliveira, C., and C. H. Antunes. 2011. A multi-objective multi-sectoral economy–energy–environment model: Application to Portugal. *Energy* 36 (5):2856–66. doi:10.1016/j.energy.2011.02.028.

Pękala, Ł. M., R. R. Tan, D. C. Y. Foo, and J. M. Jeżowski. 2010. Optimal energy planning models with carbon footprint constraints. *Applied Energy* 87 (6):1903–10.

Pokharel, S., and M. Chandrashekar. 1998. A multiobjective approach to rural energy policy analysis. *Energy* 23 (4):325–36.

Sadeghi, M., and H. M. Hosseini. 2006. Energy supply planning in Iran by using fuzzy linear programming approach (regarding uncertainties of investment costs). *Energy Policy* 34 (9):993–1003.

San Cristóbal, J. R. 2012. A goal programming model for the optimal mix and location of renewable energy plants in the north of Spain. *Renewable and Sustainable Energy Reviews* 16 (7):4461–4. doi:10.1016/j.rser.2012.04.039.

Sirikitputtisak, T., H. Mirzaesmaeeli, P. L. Douglas, E. Croiset, A. Elkamel, and M. Gupta. 2009. A multi-period optimization model for energy planning with CO_2 emission considerations. *Energy Procedia* 1 (1):4339–46. doi:10.1016/j.egypro.2009.02.247.

Suganthi, L., and A. A. Samuel. 2012. Energy models for demand forecasting—A review. *Renewable and Sustainable Energy Reviews* 16 (2):1223–40.

SULUKAN, E., M. SAĞLAM, T. S. UYAR, and M. KIRLIDOĞ. 2010. Determining optimum energy strategies for Turkey by MARKAL model. *Journal of Naval Science and Engineering* 6 (1):27–38.

Suresh, H. B., L. K. Sreepathi, and H. M. Ravikumar. 2011. Optimal allocation of rural energy resources using goal programming—A case study. *International Journal on Electrical & Power Engineering* 2 (3):19.

van Beeck, N., H. Doukas, M. Gioria, C. Karakosta, and J. Psarras. 2009. Energy RTD expenditures in the European Union: Data gathering procedures and results towards a scientific reference system. *Applied Energy* 86 (4):452–9.

Wang, X., Y. Cai, J. Chen, and C. Dai. 2013. A grey-forecasting interval-parameter mixed-integer programming approach for integrated electric-environmental management—A case study of Beijing. *Energy* 63:334–44. doi:10.1016/j.energy.2013.10.054.

Wu, Z., and J. Xu. 2013. Predicting and optimization of energy consumption using system dynamics-fuzzy multiple objective programming in world heritage areas. *Energy* 49:19–31. doi:10.1016/j.energy.2012.10.030.

Xie, Y. L., Y. P. Li, G. H. Huang, and Y. F. Li. 2010. An interval fixed-mix stochastic programming method for greenhouse gas mitigation in energy systems under uncertainty. *Energy* 35 (12):4627–44. doi:10.1016/j.energy.2010.09.045.

Yaralıoğlu, K., and N. Umarusman. 2010. Çok Amaçli Doğrusal Programlamadan Sistem Tasarimina: De Novo. *Dokuz Eylül Üniversitesi Sosyal Bilimler Enstitüsü Dergisi* 12 (4):61–74.

16 A Mixed-Integer Sustainable Public Transportation Model of Procurement Purposes Surrounding a Construction Site

Payam Khazaelpour and Mahdi Chini

CONTENTS

16.1   INTRODUCTION

In recent years, there has been a growing interest in sustainability issues that are particularly intertwined with procurement of companies and organizations of a community (Walker & Phillips, 2008). There also needs to be a controlling force, namely a few restrictions, as to special preventing factors of negative impacts when it comes to sustainability, which results in a twisting tie with both supplying and transporting procurements to a construction site as stated by Walker and Phillips (2008). Similarly, Brammer and Walker (2011) confirmed that the negative environmental impacts should be prohibited by taking economic and social limitations into a sustainable community system. In fact, economic and social limitations should play the role of protecting indictors of an environmentally sustainable constructing site as a network, implementing the sustainable main notions economically, environmentally, and socially all together, which is referred to as the triple aspects of sustainability (Brammer & Walker, 2011).

Cost reduction of public transportation in the public sector can positively recognize the influence on procurement transportation of a construction site under certain conditions. According to an indirect notion, which starts up mathematical integer programming, it is assumed that the reduction of public transportation costs surrounding a construction network indirectly helps procurement transportation to be optimized in a myriad of ways. Reduction of public transportation costs cause the surrounding area of the construction site to economically, environmentally, and socially reach the sustainability set goals. The different notion did not directly optimize procurement transportation toward and outward the construction site, but indirectly considered optimization of procurement transportation minimizing public transportation costs of the surrounding area. Furthermore, the indirect notion of cost reduction on public transportation multilaterally implements sustainable aspects and minimizes the public transportation costs, which results in a far more real procurement transportation.

There are many applications of integer programming where integer solutions are required. This qualitative tool is used modeling several situations to control and optimize the objectives that are desired (Bixby, 2012). Considering the wide variety of integer programming applications that originates from Linear Programming (LP) relaxations in a general concept, the bottlenecks that threatens the future of LP relaxations in finding a solution approach to real-world problems is overcome as claimed by Bixby (2012). Afterwards, integer programming stepped forward to basing Mixed-integer programing (MIP) fundamentals through which pure integer programming was far easier when dealing with solution-finding. But regardless, modeling some rare situation of state-of-the-art, heuristic algorithms of MIP, including disjunctive programming established primarily by Balas, and cutting-plan techniques by Van Roy and Wolsey were developed and made use of to approximately give rise to optimum solution or mostly cause to meet near to optimum solution (Bixby, 2012). As discussed in this research study, the MIP used to model sustainability fundamentals appears logically derived in accordance with globally known sustainable aspects, is considered easy to understand, and meets the sustainable desired objective, namely cost reduction of public transportation to indirectly optimize procurement transportation of surrounding area of the construction site.

Mixed-integer programming as a specific and useful accurate tool to implement many thoughts mathematically is used in this research study to approach assumptions of sustainability as a mixed-integer model. The purpose behind the use of mixed-integer modeling is the fact that it is readily solvable using different solver developers including CPLEX, GROUBI, and LPSOLVER as a number of solution approaches to integer and mixed-inter programming (Schouwenaars, De Moor, Feron, & How, 2001). However, the CPLEX solution approach has appeared the most accurate algorithm among other solution approaches (Leyton-Brown, Nudelman, Andrew, McFadden, & Shoham, 2003). Moreover, there are many more approaches to integer and mixed-integer solutions in which the LP relaxation of the first CPLEX solution finding does not take an integer solution. For this reason, the solution approach has already taken steps into the next stage where the obtained solution is performed through further calculation using branch-and-bound/cut to final integer solution (Tomlin, 1971).

This research serves to emphasize that MIP appears to be an appropriate modeling tool to deal with sustainability in regard to a construction site. The modeling process is then aimed to approach via simplex, to achieve the LP relaxation, and further improve the obtained solution if it is to be a non-integer solution. Clarifying the solution process, Excel CPLEX solver is used to obtain the LP relaxation to reduce public transportation costs given the fact that sustainability aspects are modeled mathematically. In the next step, advantages of the mixed-integer model is discussed to show the model of mixed-integer programming outweighing other existing models in a myriad of aspects in this research study.

16.2 LITERATURE REVIEW

16.2.1 SUSTAINABLE PROCUREMENT TRANSPORTATION

Firms in today's world are inextricably intertwined with the procurement market, which includes goods, services, utilities, and works, known as logistics services (Large, Kramer, & Hartmann, 2013). Furthermore, procurement supply of the firms is counted as a logistic service and has become a significant issue allowing them a constant life in the recent years. Bilateral relations between suppliers and buyers, which introduce supplier-buyer relations, are managed to fit a new approach, calling for integration of sustainable affairs in both sides simultaneously (Large et al., 2013). Requesting sustainable affairs along with integration in a myriad of fields, including the procurement process at a construction site over the second half of twentieth century, have made scholars and experts give a broad range of definitions so that the firms can gain a better understanding of sustainable development and to transparently implement the rules and regulations of sustainability issues according to what was supposed to be defined beforehand (Beynaghi et al., 2014). Finally in 1987, the phrase *sustainable development* became popularized as to its global importance, citing a report printed through the World Commission on Environment and Development (WCED), which defines Sustainable Development (SD) through a development that satisfies the present needs without reaching a compromise on the capabilities of future generations to fulfill their own needs (Beynaghi et al., 2014). Similarly, Loosemore (2016) deduced that recent sustainable developments achieved in the procurement fields not only care for current needs without any interference with the abilities of future generations in solution-finding, but confidently contribute to the society in which firms are expected to create positive outcome.

The procurement process towards a construction site in recent years has met a change according to the new definition published for SD. Accordingly, traditional procurement process product delivery was merely aimed and considered other additional objectives (Loosemore, 2016). Containing new objectives, the SD approach along with the procurement process in a new, different viewpoint of this research not only states the product delivery needs to be well done, but also the products involved in the procurement process might provide job opportunities for which the disabled, unemployed, and ordinary people could take advantage (Large et al., 2013). This multi-perspective response to SD-based procurement supply of a network appears to be sufficiently strong in terms of the sustainability approach so that it assists with suppliers' sustainable activities as to their recently set goals. Lately, sustainable procurement supply, and its side effects on how positively it can affect the society, establishes social, environmental, and economic inner-connections, which have become of interest. For example, in the United Kingdom, 66% of real estate offices, which are responsible of providing for people's desired accommodations, appears willing to fetch social values to their society. However, just 23% of them took this notion as a priority and stepped towards the change (Loosemore, 2016).

This literature serves to emphasize the triple bottom lines of SD along with considering the procurement process as way to deal with a mixed-integer procurement transportation model towards a construction site, and thereby link the three most important criteria, namely environmental, economic, and social dimensions. However, they partially manage a conflict with one another (Aktin & Gergin, 2016). Through the Green Supply Chain Management (GSCM), the sustainably environmental dimension succeeds by controlling the amount of greenhouse gases emission of CO_2, consumption of renewable energies, and reducing waste. Though, it causes a number of costs by establishing a contradiction between the environmental dimension origination and the costs produced (Aktin & Gergin, 2016). In fact, Aktin and Gergin (2016) assumed the contradiction mentioned is manageable through Sustainable supply chain management (SSCM) and an involved company's collaboration, integrating all three dimensions of SD. Also, they point to a joint standard of environmental and social dimensions and collaborate with its providers as to a policy that mostly concentrates on social values, while minimizing the environmentally negative side effects. Environmental aspects in sustainable development constitutes a principle-based commitment in procurement judgements that even other aspects of SD as traditional criteria, namely flexibility, service improvement, and cost reduction, could not influence on (Large et al., 2013). Among the studies of environmental aspects in SD, there seems to be a collaborative enhancement of suppliers that confirms purchasers' environmentally sustainable abilities, called GSCM reinforced abilities. Similarly, Large et al. (2013) stated and consequently examined that environmental improvements act through collaboratively joint goal setting, taking advantage of SSCM, and collaboratively sustainable activities throughout the chain in the surrounding area of study. Given the economic and social dimensions of SD involving procurement process, both are fairly defined in a common sense as to giving positive outcomes to the community/network, flourishing economically, and improving social prosperity respectively (Aktin & Gergin, 2016).

16.2.2 ENVIRONMENTAL DIMENSION OF SUSTAINABILITY ABOUT A CONSTRUCTION SITE

The environmental dimension causes costs and services to implement a new vision, which supports conservation of Payment for Environmental Services (PES) (Jindal, Kerr, Ferraro, & Swallow, 2013). As an example, Green Public Procurement (GPP) as a PES project has been recently given attention so that it demands environmentally friendly services and products to simulate green procurement about a construction site (Testa, Annunziata, Iraldo, & Frey, 2016). Jindal et al. (2013) point out many PES projects not only enforce environmental objectives, but assist with elimination of poverty as to a main objective of socially propounded goal in SD. Moreover, its goal is to reduce the global emissions of carbon dioxide, known as a greenhouse gas, and graphically features particularly in United States that 28.1% CO_2 results from CO_2 the transportation system, while 33.5% of emission originates from fossil fuel intake (Hsu et al., 2015). Hsu et al. (2015) also impart reduction of emission depends on fuel CO_2 consumption and how effectively greenhouse gases growth is held back to either reduce the number of unnecessary journeys or traverse in fuel-efficient green vehicles approach. However, in this study a new approach is taken, particularly for procurement transportation, in which a mixed-integer programming CO_2 model using a different notion to deal with procurement transportation is developed. Despite many models created to reduce the amount of carbon dioxide or other greenhouse gases through various approaches, the model presented here focuses on traffic congestion, where an optimal control of road-traffic is of interest. Turning to details, the optimal control happens when procurement transporting vehicles have more choice to take different roads, and this notion occurs when public transportation vehicles increases. Though, it brings a considerable amount of costs to the society, which necessarily makes eligibility of several constraints, including a PES constraint, to the model of procurement transportation to control it in different aspects as will be discussed later.

There are many approaches that consider reduction of CO_2 emission in procurement transportation models. Technically speaking, they mostly focus on the sources of CO_2 emission and try to control them with different ideas. Considering the sources of CO_2 emission, Hsu et al. (2015) states that vehicles mostly reflect these sources and are environmentally-friendly improved when their mechanisms consume less CO_2. Similarly, Maheshwari, Kachroo, Paz, and Khaddar (2015) confirm that safety, efficiency, and reduction of environmentally destructive side effects happen if vehicles adopt new environmentally friendly transportation systems. For instance, advances recently show that attentions are drawn towards communication technology as to a proper tool used in vehicles to improve their performances in terms of positive effectiveness in a myriad of aspects (Maheshwari et al., 2015). Meanwhile, the notion presented here in this study looks into the situation where procurement transportation becomes more applicable, being less time-consuming through minimizing public transportation costs, which also include a reduction of greenhouse gases emission cost. A few models have taken a different approach, other than minimizing the sources emission, and could adopt a methodology that considers multivariate state space in traffic flow modeling to reduce either traffic congestion or CO_2 emission (Maheshwari et al., 2015), while dealing with the presented idea in this study covering public transportation accessibility to the people living near a construction site to reduce the number of people's non-public vehicles for easier procurement transportation purposes towards construction site. Consequently, reduction of fuel consumption, or CO_2 emission, to more environmentally-friendly procurement transportation has bound a few repetitively known approaches together for optimal transportation purposes in developing vehicles (Hsu et al., 2015). However, the presented chapter takes environmental dimension of SD as one of the main sources to make optimization occur.

Environmentally speaking, there have been many studies, which search for suitable ways of reducing CO_2 emission, although there are a few of them that interpret those of green initiatives and environmental performance alignment with suitable standards, adopting environmental attitudes (Centobelli, Cerchione, & Esposito, 2017). One of them undertaken by Vincent and Jerram (2006) clearly declares that Bus Rapid Transit (BRT) appeared far more efficient than a Light Rail Transit (LRT) system in terms of CO_2 emission. The study in (2006) concludes that a BRT system of public transportation using either 40-ft or 60-ft buses emits less amount of CO_2 than any usual bus public transportation. In addition to all efforts made to reduce CO_2 emissions, road space for pedestrians and prevention of road traffic accidents happen to be other objectives that can be taken to maintain the environmentally sustainable dimension of SD, either reducing the number of oversize vehicles or adding up a number of green public vehicles for transportation purposes (Browne, Allen, Nemoto, Patier, & Visser, 2012). Likewise, it is environmentally-friendly understood that the integration of land, sea, and air transport network will help decrease the amount of CO_2 emitted; referred to as push-and-pull measures (Nocera & Cavallaro, 2011). Basically, this notion is focused on by EU committee as noticed by Nocera and Cavallaro (2011) and meant to influence individual decisions, dividing into two groups. The former contains financial instruments, including taxes, charges, and tolls. While the latter forms pull measures, which encourage people to mostly use public transportation green vehicles, and negatively argue for detrimentally massive sources of CO_2 emission, known as private transporting vehicles. This research method, named CO_2 efficiency, carried out in (2011) also introduced an approach through which the impacts, noted as economic parameters, of polluting gases such as CO_2 is evaluated using the classical techniques. However, this method shows some inconsistencies when local gases impact the environment globally. Regardless of the approaches described, there are still many more notions and ideas capable of either reducing CO_2 emission or keeping a certain level of emission not presented yet, such as reducing public transportation costs (Bailey, Mokhtarian, & Little, 2008).

16.2.3 Economic Dimension about a Construction Site

In addition to the environmental dimension of the sustainability approach, the presented model in this research study also takes advantage of sustainable economic and social dimensions, where linear and non-linear programming is able to define the concept of sustainable systems (Maheshwari, Khaddar, Kachroo, & Paz, 2016). Economically speaking, the presented model considers an approach in which cost reduction of public transportation was considered as the objective function. Cost reduction in the public transportation segment will provide an opportunity of wide public accessibility by which people living near the construction site can easily travel as fast as possible. However, easier access to public transportation as a course of action to reduce transportation costs led to a greater challenge, which transportation organization needs to tackle. Wide accessibility of public transportation and the reduction of costs in this area of study mean the government must provide more subsidy to the transportation organization so that they are financially protected to overcome their existing and upcoming costs. While, more subsidy on behalf of the government accounts for a governmental cost increase, which is noticed against sustainability dimensions. Therefore, there needs to be a savings course of action through which the government prevents extra costs imposed. An economically efficient approach, which also reduces more than the amount government subsidized transportation organization before the wide accessibility approach is introduced, describes a financial incentive given to low-income people. In fact, the subsidy given to the transportation organization is replaced with a plan in which the subsidy cares for low-income people as their transportation subsidy to support their intercity movements as Sohail, Maunder, and Cavill (2006) claim different types of public transportation vehicles, including motorized and non-motorized, can tackle low-income people's intercity trips. Correspondingly, this approach allows the government to save more money and spend the savings to achieve positive externalities of sustainable dimensions. Moreover, a breakdown in transportation subsidies can lead the government towards assisting a struggling sector of industry with a financial encouragement to be capable of remaining in the international competitions, as these areas are not effectively supported through general economy and may be less likely to boost as much value as they need in the rival economy. Actually, when the transit system allows the low-income people to find a job and be economically self-sustainable, the government can save a lot of money by reducing the unemployment care programs (Jindal et al., 2013). Consequently, the presented model is not only backed up via a sustainable environmental dimension, in this research study, but also holds economical representativeness of positivity in a myriad of aspects.

There are three major economical determinants to measure the impacts of a transport project on selections of poverty, particularly on low-income peoples' transportation. These three side effect indicators are noted as efficiency criterion, transport market along with institutional failures, and government regulations (Gannon & Liu, 1997). Firstly, efficiency criterion is widely accepted as a measure of performance by which a reduction in poverty is measured. In fact, cost-benefit analysis of the transport sector operation is performed to rationally guide the transporting project selection and design. However, poverty reduction becomes complicated as we move forward, presenting inaccurate cost-benefit analysis because sometimes poverty reduction is directly opposed to efficiency measures. Secondly, transport demand elastically follows income, indicating travel expenditures is larger than total communities' transportation income, which means that the government must spend a round sum of money as the transportation subsidy in regular periods. Given the poor are lacking in their daily expenses, transport and institutional failures put economic pressure on them as well. Thirdly, government regulations are usually place more limitation when it comes to providing transport opportunities for the poor, which also ignores them economically. Furthermore, the main three components of the poor's financial transport situation appear to be less supportive of the poor globally in the public transportation province.

Increasing economic opportunities are indelibly linked with transportation investment as part of a greater network in which there will exist no outpacing movements, if sufficient investments and renewal traversing methods are not to be injected into the transportation system (Kaewunruen, Sussman, & Matsumoto, 2016). Current investments are earned from different projects by which the widening of public transportation travel is reinforced containing interstates, freeways, busways, and heavy railways (Murray, Davis, Stimson, & Ferreira, 1998). One of these projects that takes creativity to implementation is through incorporation of personalized public transit options. In other words, Murray et al. (1998) mentions through this option people, low-incomes in particular, become capable of choosing the public mood they are economically willing to travel with as taxis, public buses, and street trains, which comprise different moods of intercity travel. Accordingly, a new program is set, locating jobs and workers near one another to increase employment access and improve low-income people's financial status (Wachs & Taylor, 1998). Wachs and Taylor (1998) also confirm that distance to workplace is closely related to the financial earnings of people. Evidences show low-income people, who work within four miles of home, can earn $634 at most in averages. While other people working within four to ten miles of home earn about $620, and those of over ten miles are the lowest wage recipients of this society with the average amount of $433 per month (Wachs & Taylor, 1998). Furthermore, economic opportunities are part of the cost reduction of public transport incentives as upcoming mixed-integer model represents.

16.2.4 REALIZED SOCIAL DIMENSION ABOUT A CONSTRUCTION SITE

Besides economically positive perspectives, the model presented also accompanies social representativeness of positivity in a number of dimensions as sustainability development rules have been legislated and defined beforehand (Beynaghi et al., 2014). Socially speaking, wide accessibility itself provides people with the opportunity of quick travel, which saves a lot of time and reduces traffic congestion as public transportation green vehicles are being mainly used to deliver passengers to their final destination. Moreover, the amount of money saved from the new approach of sustainable procurement transportation elaborated on as a transportation subsidy for low-income people can play a role of a financial support to broaden education in society. The new approach that takes low-income people as the heart of transportation subsidy inclusion, particularly with a falling employment rate and rising fuel prices are increasingly fluctuating, appears to be challenging. However, it is still justified to be a sufficiently money-saving approach, compared to previous multilateral transportation subsidy plans. Concerning this critical period, four general areas have been examined through a number of interviewees, including transportation spending patterns, travel behavior, advantages and disadvantages of different means of travel, and cost management strategies. As a result, these four areas remained underdeveloped due to inadequate data. However, they were socially proven to have negative effects on low-income people's lifestyle. Despite all financial pressures that are carried on by poor people, they might find financial support to overwhelm transportation expenditures and be capable of keeping up with other financial-social problems. A solution that might financially cover their transportation costs is to provide them with public transportation vehicles for free so that they physiologically become capable of finding a job, attend cultural activities, and financially caring for themselves as economically ordinary people. Access to affordable transportation for low-income people, as wide as economically moderate people, accounts for a few underlying benefits, including self-sustainability, independency, and other household essentials. Similarly, Jindal et al. (2013) supports that other local customer-based facilities such as medical care, rural transportation, local education, and recreational activities can reinforce economic growth as well. Socioeconomically, wide access to public transportation also brings about positive externalities as much as it reduces traffic congestion and economically saving other sustainable dimensions, such as fuel consumption for the environmental dimension and time of traveling as fast as possible (Richardson, 2005).

Discussed as economically as sustainable measures appear comparing, there happens to be a few social measures of performance with which social benefits and costs compare to each other (Layard, 1994). Socially measured, cost effectiveness is capable of tracking transportation services over time in the financially poor provinces in megacities. However, it is a useful performance in other city segments and inextricably accompanied with some other measures of social benefits, which fails to present an accurate performance solely when transport pricing and investment decisions are considered. Accurate cost-benefit analysis is a constructive tool to evaluate new transporting infrastructure projects. Accurately, the cost of widening roads or railways and constructing them are calculated separately. The general costs of operating, capital, and external are incrementally important to the social cost of each project option. The costs of the previously existing system will only be important to the extent that they are merely closely reliant on the project outcomes. There happens to be a significant gap between private and social costs, if the level of activity performed by the economy differs from that of optimum. This gap needs to be narrowed down by establishing the influence of external costs, including taxes and subsidies, into the effectively paid cost. This way, a stably parallel pair of line representing private and social costs are kept up with each other, giving an account for the need that minimization of procurement transportation cost maintains the parallel lines as close as possible.

16.3 SUSTAINABLE LINEAR AND MIXED-INTEGER PROGRAMMING IN PROCUREMENT TRANSPORTATION OF A CONSTRUCTION SITE

Sustainable development in procurement transportation of linear programming has recently emerged with the integration of SD dimensions to propose a multi-dimensional approach of procurement transportation, in particular. Considering a SD multi-dimensional aspect, Maheshwari et al. (2016) confirmed that mathematical linear or non-linear modeling with the notion of green transportation is applied effectively covering the environmentally-friendly dimension of transportation in public transportation. A recent study that is pointed out considers environmental impacts and public transportation linear modeling together as a sustainable linear programming model in the surrounding area of study (Maheshwari et al., 2016). However, Maheshwari et al. (2016) relatively proved in a study that dynamic programming in relation with transportation system can be a suitable package of sustainable non-linear programming to take the environmental dimension of sustainable development as major criteria. In this research, a linear programming model is more of interest as the linearly constructed model, which covers both the transportation system and triple dimensions of SD when surrounding a construction site matters.

Recently, there are a few studies that focus on a combination of mixed-integer programming and SD dimensions. Mostly, they have dealt with the environmental aspect of sustainable development in different ways of reducing energy consumption (d'Amore & Bezzo, 2016). Moreover, d'Amore and Bezzo (2016) similarly suggested that studies mostly are carried out, focusing on environmental aspects of sustainable development in particular, but taking SD multi-dimensional aspects to draw a constructive conclusion in terms of sustainability as environmental and economic aspects of SD, both together, are simultaneously in favor of scholars' certain goal, currently absorbs many attentions. Accordingly, in this study as procurement transportation along with sustainable dimensions are both taken into consideration, a research study is formed, taking credit for mixed-integer programming as a trustworthy optimization tool through which a multi-dimensional problem appears solvable. Carried out by DiJoseph and Chien (2013), a similar public transport problem considering the notion of combined sustainable dimensions is modeled and maximizes the number of public transport customers in a local surrounding area by using mixed-integer programming optimization technique. Broadly, mixed-integer programming has been used in different problems ranging from the simplest ones, a simple transportation network, to more complex as those of fuzzy mixed-integers programming problems where triangular fuzzy numbers are involved to include uncertainty to a sustainable model (Pourrousta, Dehbari, Tavakkoli-Moghaddam, Imenpour, & Naderi-Beni, 2012). Paying special attention to a wide range of mixed-integer programming application in science, a combination of different sustainable desires is applied according to sustainability rules to minimize total public transportation costs around the construction site in this research study.

Mixed-integer programming as a successful tool of optimization technique and has been used in many applications including business and engineering for over 50 years (Vielma, 2015). Identically, Vielma (2015) also suggested that the two main reasons mixed-integer programming has become successful among other optimization techniques are, firstly, take linearity factor and, secondly, claim flexibility of MIP. However, linearity of MIP, which takes a straightforward formulation into account, sometimes negatively affects efficiency of LP formulation solvers (Vielma, 2015). As mentioned, LP relaxation appears to have already been reasoned a significant factor to frank and easy MIP modeling, so that the MIP modeling happens, benefiting from the strength of LP modeling relaxation. Moreover, there are many types of constraints to MIPs, which all originate from polyhedral analysis of MIP basis, relying on the strength of their LP formulations (Atamtürk, 2005). Correspondingly, the analysis of MIP fundamental structures leads to strong cutting plans as discussed by Atamtürk (2005), and the more we apply stronger cuts the better MIP formulation provide you with near-to-optimum solutions. However, regardless of the MIP strength, research has been rarely carried out in terms of integer and mixed-integer programming since seventies. This is due to the complexities that integer variables might introduce to the model presenting untrustworthy results as an optimum solution (Atamtürk, 2005). MIP is also a solution after modeling that takes a few solvers to find the optimum or near-to-optimum result including, CPLEX, GROUBI, LPSOLVER, and a few more, which have been still under development (Hutter, Hoos, & Leyton-Brown, 2010).

In this research study, a mixed-integer programming model is built, accompanying seven constraints through which seven different limitations of a mixed-integer model are added to make it seem more realistic in comparison with a real public transportation problem. This model represents all public transportation costs, objective function, the need to be minimized so that procurement transportation is more likely to travel faster towards construction site because public transportation vehicles with lower costs encourage individuals to take them on more and more, and most importantly, the government is contracted to provide people with public transportation at preferential prices (Ševrović, Brčić, & Kos, 2015). For this main reason, Ševrović et al. (2015) claim that the government imposes policies on equal rights to public transportation accessibility for all citizen of both the local and main areas. Moreover, subsidies on behalf of the government for public transportation is restricted and cannot cover all costs estimated. Furthermore, related municipalities merely subsidize public transportation that is being provided for people of the certain areas, but if there were not to be any subsidies from municipalities, carriers are obliged to charge full costs to the passengers, and this would lead to a significant reduction of citizen numbers benefitting from using public transportation (Ševrović et al., 2015). Dealing with constraints, seven different limitations are introduced step by step and established for a specific purpose in the sustainable procurement transportation model. Truly, the intensions of constraints take different purposes based on different needs, and several methods of constraint construction by which different needs are satisfied, including linear, Parametric curve, Hyperplanar, Parametric surface, side, and so on. However, realistically speaking, the linear type satisfies the purpose for this research study (Bowyer, Davies, & Baena, 2014). Additionally, linear or non-linear programming including mixed-integer programming has always come with an objective or multi-objective function, which determines either a minimization or maximization problem to be optimized. Technically, objective functions in both extremely large and quite small scales link different components of a model to achieve the optimum or near-to-optimum solution (Holguín-Veras, Pérez, Jaller, Van Wassenhove, & Aros-Vera, 2013).

Despite the fact that mostly the linearity of mixed-integer programming makes it easier for us to overcome math modeling, other approaches are also applied in this certain area of mathematics, specifically when it comes to large scale logistics problems (Vieira & Luna, 2016). Vieira and Luna (2016) also named some approaches including, multi-criteria decision-making (MCDM), topology modeling, costs and services integration, dynamic programming, stochasticity, and reliability, which make the original model face new challenges in terms of modeling process. Nevertheless, Campbell and O'Kelly (2012) stated that although modeling processes have met many successes and developments throughout the recent years, they are still far away from realistic modeling to stimulate real logistics problems as drawn within real network activities. As recently drawing much attention, multi-objective functions have substituted for single objective models and make for a better way to find solutions for real problems and optimal solutions (Campbell & O'Kelly, 2012). Considering MCDM as a logical approach to a logistics problem, the criteria including cost or coverage of all logistics network is considered. Meanwhile, a conflicting one, such as environmental dimension of sustainability, is also on the side with those of considered ones, applying MCDM as a result of a multi-objective problem (Campbell & O'Kelly, 2012). But regardless of all known approaches to modeling, the mixed-integer programming approach, which also originates from linear programming, is still popularly used among all other approaches due to its oversimplification (Balconi, Brusoni, & Orsenigo, 2010).

16.4 MATHEMATICAL MODEL CLARIFICATION OF PROCUREMENT TRANSPORTATION ABOUT A CONSTRUCTION SITE

$$\text{Min} \quad \sum_{i=1}^{n}\sum_{j=1}^{m}\sum_{t=1}^{T}\sum_{k=1}^{K} C_{ijt}^{k} * x_{ijt}^{k} + \sum_{i=1}^{n}\sum_{j=1}^{m}\sum_{t=1}^{T}\sum_{k=1}^{K} V_{ijt}^{k} * y_{ijt}^{k} \tag{16.1}$$

S.t

$$l_{ijt}^{k} * x_{ijt}^{k} \le b_{ijt}^{k} * y_{ijt}^{k} \qquad \forall i,j,k,t \tag{16.2}$$

$$x_{ijt}^{k} \ge y_{ijt}^{k} \qquad \forall i,j,t,k \tag{16.3}$$

$$\sum_{i=1}^{n}\sum_{k=1}^{K} y_{ijt}^{k} \ge 1 \qquad \forall j,t \tag{16.4}$$

$$\sum_{t=1}^{T} y_{ijt}^{k} = \sum_{t=1}^{T} y_{jit}^{k} \qquad \forall i,j,k \tag{16.5}$$

$$\sum_{t=1}^{T} y_{ijt}^{k} \le \sum_{t=1}^{T} f_{jit}^{k} \qquad \forall i,j,k \tag{16.6}$$

$$(1 - z_{ij}^{k}) + Mz_{t} \ge \frac{\sum_{t=1}^{T} x_{ijt}^{k}}{\sum_{t=1}^{T} l_{ijt}^{k}} \ge \frac{\left(\sum_{t=1}^{T} y_{ijt}^{k}\right)}{\left(\sum_{t=1}^{T} f_{ijt}^{k}\right)} * z_{ij}^{k} + (\varepsilon) * (1 - z_{ij}^{k}) \qquad \forall i,j,k \tag{16.7}$$

$$\frac{l_{ijt}^{k}}{p_{ijt}^{k}} * y_{ijt}^{k} \le 1 \qquad \forall i,j,k,t \tag{16.8}$$

$$x_{ijt}^{k} \ge 0, f_{ijt}^{k}, y_{ijt}^{k} \in \{0,1\}, z_{ij}^{k} \in \{0,1\} \qquad \forall i,j,k,t \tag{16.9}$$

16.4.1 MODEL INTRODUCTION AND METHODOLOGY

16.4.1.1 Objective Function

The model presented in this section is firstly conceptually described and mathematically drawn afterwards on parameters, decision variables, constraints, and objective function.

The sustainable model has taken a different notion, which indirectly leads to comfortable procurement transportation around the construction site in this research study. Technically speaking, models presented recently either take economic and social dimensions of sustainability and sacrifice environmental aspect or put effort into keeping the environmental dimension first and let other dimensions follow after (Naganathan & Chong, 2017). However, this model attempts to eliminate negative impacts and spread positivity over all three sustainable dimensions of development at the same time. A mixed-integer model was drawn from the triple bottom lines of sustainability concept as it appears easy to solve in the first place using a wide variety of software choices from the simplest, Microsoft Excel, to one of the most sophisticated, called AIMMS. Secondly, as integer programming happens to be a strong tool of the solution finding process, either optimum solution or near-to-optimum solution is mostly obtained benefiting from CPLEX solver, which is highly trustworthy when it comes to an optimum or a near-to-optimum solution. Moreover, an integer programming tool is capable of development when it's linked with other optimization techniques, such as Tabu search, and enjoys induced decomposition, learning strategies, and controlled randomization in details (Glover, 1986). Similarly, Beyer, Dujardin, Watts, and Possingham (2016) affirmed through their constructed model that not only is integer linear programming as strong as meta-heuristics, but somehow outperforms in some modeling cases so that a near-to-optimum solution is obtainable within 0.5% of optimality through integer/mixed-integer modeling, while meta-heuristics such as Simulated Annealing (SA) underperformed within 19% of optimality rate. Social, economic, and environmental impacts as to the promising outcomes of sustainable development to mathematical implementation of our model are verbally elaborated on as follows.

As Naganathan and Chong (2017) truly claimed, social and economic impacts are mostly bound and fulfilled together towards positivity. Accordingly, a certain thought is involved in the model to prove this concept that costs of public transportation are supposed to be reduced in the surrounding construction site as an objective function. It economically-socially claims that procurement transportation towards a construction site happens faster as the traffic congestion decreases. Furthermore, availability of public transport increases, while fossil fuel consumption economically-environmentally decreases as traffic reduces. The objective function consists of two parts, containing known and unknown cost all together. Known costs relate to those of which depend on traversing distances from (i) to (j) as noted by C_{ijt}^k, meanwhile unknown costs noted by V_{ijt}^k are not governed by distance traversing and are fixed during any travel from (i) to (j). The summation of known and unknown costs over all indices noted as $(i),(j),(t)$, and (k) in a row indicates the objective function as to a minimization problem coming as follows.

$$\text{Min} \quad \sum_{i=1}^{n}\sum_{j=1}^{m}\sum_{t=1}^{T}\sum_{k=1}^{K} C_{ijt}^k * x_{ijt}^k + \sum_{i=1}^{n}\sum_{j=1}^{m}\sum_{t=1}^{T}\sum_{k=1}^{K} V_{ijt}^k * y_{ijt}^k \qquad (16.10)$$

16.4.1.2   Decision Variables

The presented model is composed of three decision variables, which specifies the conditions under which the optimal solution appears differently. These three variables are described as follows and shown within the model performing. Moreover, each condition to which the model presents a different solution is elaborated on mathematically.

16.4.1.2.1   Traversing Distances

This decision variable noted as x_{ijt}^k identifies the amount of distance traversed from (i) to (j) for each different type of transporter (k) in each period (t) over the considered transporting network.

x_{ijt}^k : traveling distances from node starting (i) to destination node (j) for each public transporter (k) in each period (t).

16.4.1.2.2   Binary Variables (0–1)

There are two different binary variables in this model, which presents two different conditions. Firstly, y_{ijt}^k is the binary variable, which states if transporter (k) arrives at node (j), it should be 1 otherwise it takes on 0.

$y_{ijt}^k = \begin{cases} 1 \\ 0 \end{cases}$ if the transporter type (k) arrives at certain node (j), the binary variable y_{ijt}^k chooses (1) otherwise (0).

z_t is another binary variable, which accounts for whether or not an environmentally-friendly transporter has been used to perform in the system?

If the answer is yes, this binary variable needs to take on (1), otherwise it should be (0).

$z_t = \begin{cases} 1 \\ 0 \end{cases}$ if vehicles are green the value of this binary variable should be (1), otherwise (0) is required to be taken on.

f_{ijt}^k is a binary decision variable through which the existence of transporter type (k) is investigated and equivalent to the binary variable y_{ijt}^k to complete fuel consumption constraint.

$f_{ijt}^k = \begin{cases} 1 \\ 0 \end{cases}$ if the transporter type (k) is available to travel from (i) to (j), (1) is accepted, otherwise 0 needs to be substituted for f_{ijt}^k.

In fact, the constraint $\sum_{t=1}^{T} y_{ijt}^k \leq \sum_{t=1}^{T} f_{ijt}^k$ appears completing for the fractional fossil fuel constraint in all periods of transporter type (k) availability.

16.4.1.3   Parameters

Parameters of the presented model are assumed hypothetically. Moreover, the parameters considered are based on the assumptions that all are predetermined and discussed over the supposed network in this segment.

16.4.1.3.1   Price of the Variable Costs

Variable costs price in the model is that of known price and dependent on the distances traversed from location (i) to (j). As used in the objective function, it is noted C_{ijt}^k and multiplied by x_{ijt}^k to show the variable cost increases when the distance x_{ijt}^k increases and, accordingly, it decreases when the distance x_{ijt}^k from (i) to (j) reduces.

16.4.1.3.2   Price of the Fixed Costs

On the contrary, fixed costs price in this model is that of the price that is unknown and independent on the distances traversed from location (i) to (j). As it is also used in the objective function, it is noted V_{ij}^k and multiplied by the binary variable y_{ijt}^k to show firstly it is independent of the distance traversed, and if the route between (i) to (j) exists, the fixed price is accounted for, otherwise it takes on zero.

16.4.1.3.3   Capacity for Transporters Type (k) per Liter

Transporters type (k) have to be bound by a certain fuel capacity b_{ijt}^k, identified for each type (k) from location (i) to location (j) per period (t) in the model as to limit their fuel shares within the system.

16.4.1.3.4   Consumption of Fuel per Liter

There is a parameter noted l_{ijt}^k, which shows how many liters are needed from location (i) to location (j) per transporters type (k) per period (t).

16.4.1.3.5   Standard Amount of CO_2 Emission

The standard amount of emission for CO_2 is accounted for as noted p_{ijt}^k, defined for transporters type (k), per period (t) for each liter of fuel consumption.

16.4.1.3.6   An Insignificant Value

An insignificant value ε is elaborated between zero and one to provide a bilateral condition for fuel limitation constraint.

16.4.1.4   Constraints

The presented model contains seven constraints, each for different reasons, and each constraint imposes a limitation to the model through which minimization of public transportation costs turns out to be processing. All constraints are presented and mathematically shown as follows.

16.4.1.4.1   Capacity Constraint of Fossil Fuel Consumption

This constraint states that each transporter type (k) has a capacity of fuel consumption by which a fuel share is given to transporters type (k) per each period (t). Mathematically driven, the first constraint is tied up with amount of distance traversed from location (i) to location (j) and mathematically shown as follows.

$$l_{ijt}^k * x_{ijt}^k \leq b_{ijt}^k * y_{ijt}^k \qquad \forall i,j,k,t \qquad (16.11)$$

$$x_{ijt}^k \geq y_{ijt}^k \qquad \forall i,j,t,k \qquad (16.12)$$

These two constraints come together, completing each other conceptually. Technically, the second constraint is assumed first, which supports the first constraint. It claims if a route from location (i) to (j) exists, the amount of distance that is required to be traveled needs to be at least a kilometer, and per each kilometer a certain amount of liter is assigned throughout the route from (i) to (j). If there happens not to be any chosen route from location (i) to location (j), the distance that needs to be traversed should be zero as mathematically shown above in the constraints.

16.4.1.4.2   Accessibility Constrains

As claimed in the sustainable dimensions that sociability is one of the key factors in sustainable development, accessibility of public transporters type (k) to people who live in the surrounding of construction site is taken into account. In fact, accessibility to public transportation has several advantages in terms of society, environment, and economy. Firstly, a broad accessibility to public transportation causes people not to use their own private cars, which in turn holds back traffic congestion throughout the surrounding area of the construction site. Secondly, by eliminating private cars, CO_2 emission will decrease throughout the surrounding area, and this reduction on CO_2 emission purposefully takes an environmentally-friendly approach to sustainable development (SD). Thirdly, private cars elimination appears economic as many underlying costs including fuel consumption, amortization, and many taxes included in the buying of a car are prohibited. Therefore, the accessibility constraint not only aims at a socially wide range of public transportation, but accounts for other sustainable dimensions simultaneously and is mathematically described as follows.

$$\sum_{i=1}^{n}\sum_{k=1}^{K} y_{ijt}^{k} \geq 1 \qquad \forall j,t \tag{16.13}$$

16.4.1.4.3   Safety Constraint

This constraint is applied when transporters type (k) need to avoid traversing through the routes, which are not chosen by the binary variable from location (i) to location (j). Moreover, if a route is chosen and eligible for transporters type (k) to travel through, the distance between location (i) and location (j) needs to be at least a kilometer as logically assumed in the modeling process. Therefore, the third constraint is mathematically drawn as mentioned below.

$$x_{ijt}^{k} \geq y_{ijt}^{k} \qquad \forall i,j,t,k \tag{16.14}$$

16.4.1.4.4   Flow Constraint of Each Transporter Type (k)

Basically, the flow constraint in the model makes sure that each transporter type (k) enjoys a flow at each arriving node (j), which means the inflow and outflow at each different node should be equal for all periods (t). This notion is mathematically deduced as follows.

$$\sum_{t=1}^{T} y_{ijt}^{k} = \sum_{t=1}^{T} y_{jit}^{k} \qquad \forall i,j,k \tag{16.15}$$

Considering the concept of constraint, inflow and outflow for certain (i),(j), and (k) are put equally in either side of the constraint in all periods (t).

16.4.1.4.5 Fuel Consumption Constraint

This constraint is in turn composed of two constraints, which complete each other. Dealing with the details, the first constraint states that transporter type (k) arrived at a chosen node (j), starting from a certain node (i), over all periods (t) provided the route from (i) to (j) is chosen. Basically, if $y_{ijt}^{k} = 1$, it is assumed that the route from (i) to (j) has been chosen and mathematically shows the formulation below as follows.

$$\sum_{t=1}^{T} y_{ijt}^{k} \leq \sum_{t=1}^{T} f_{ijt}^{k} \quad \forall i, j, k \tag{16.16}$$

The second constraint, which completes the first constraint above, claims that fuel consumption for each transporter type (k) in all periods (t) is limited either between 1 and a quite a large number noted (M) or by a fraction between 0 and 1. It means the amount of distance traversed form (i) to (j) per kilometer should be more than the amount of fuel consumed per liter, which falls in two binding limits as described beforehand.

$$(1 - z_{ij}^{k}) + M z_{ij}^{k} \geq \frac{\sum_{t=1}^{T} x_{ijt}^{k}}{\sum_{t=1}^{T} l_{ijt}^{k}} \geq \left(\frac{\sum_{t=1}^{T} y_{ijt}^{k}}{\sum_{t=1}^{T} f_{ijt}^{k}} \right) * z_{ij}^{k} + (\varepsilon) * (1 - z_{ij}^{k}) \quad \forall i, j, k \tag{16.17}$$

The concept behind this constraint is the fact that the fuel consumption for each transporter (k) should be less than the amount of distance traversed if green vehicles are used according to $(z_t = 1)$. Another assumption is considered that the number of available vehicles to travel from (i) to (j) should be equal and greater than 1 for the summation of entire periods (t). This assumption reflects that if there are vehicles available, the numbers of routs from (i) to (j) should be less than and equal to the number of available vehicles in all periods. In other words, we ought to have a vehicle available in the least for all periods (t).

16.4.1.4.6 Limitation of CO_2 Emission Constraint

This constraint supports a fraction, which introduces l_{ijt}^{k} as the liter of consumption in the nominator of this fraction and p_{ijt}^{k} as the standard amount of emission from (i) to (j) per each period (t) for transporter type (k) as the denominator of this fraction. The whole concept of this constrain states that the fraction described should be at most one if the route from (i) to (j) is chosen, which means $y_{ijt}^{k} = 1$ to let the describing constraint be active in the model. The constraint is mathematically drawn as follows.

$$\frac{l_{ijt}^{k}}{p_{ijt}^{k}} * y_{ijt}^{k} \leq 1 \quad \forall i, j, k, t \tag{16.18}$$

16.4.1.4.7 Continuous and Binary Constraints

This segment describes that the decision variables should be either continuous or binary as shown below mathematically.

$$x_{ijt}^{k} \geq 0, f_{ijt}^{k}, y_{ijt}^{k} \in \{0,1\}, z_{ij}^{k} \in \{0,1\} \qquad \forall i, j, k, t \qquad (16.19)$$

16.5   INTENTION OF SOLUTION FINDING

Solution finding to any premodeled mixed-integer program takes various approaches and algorithms, giving the solution of their kinds as aimed at by the experts of this area of study in construction purposes, in particular. For example, branch-and-bound/cut is a usefully solving tool that has served as either optimum seeker or near-to-optimum seeker approach over many decades (Linderoth & Savelsbergh, 1999). Previous integer or mixed-integer models have not taken the notion that this research study has taken, and accordingly the effects that reduction of procurement transportation cost surrounding a construction site would shed on lower-income people still remain undiscussed in terms of sustainable transportation purposes in a society. Solution finding of the presented model gives rise to a mutual comparison of existing models and the presented one in this research study. Considering the mutual comparison solution to the presented model clarifies that the objective function including seven constraints is optimized in terms of transportation cost and is compared with other methodologies of solution finding. Furthermore, any introduced improvement of thought after the model implementation can be easily taken into account to any better targeted solution as calculated by objective function. The branch-and-bound/cut methodology to solution finding appears assured and trustworthy, if solution of the present model is to clear the optimization plate. But beforehand, the primary solution should be achieved using an optimization solver, namely CPLEX as an assured solver among all others (Hutter et al., 2010). Actually, CPLEX gives rise to the linear relaxation solution through which further improvements according to the presented objective function towards the most optimized solution may occur. Consequently, the supplementary approach which provide us with an optimum solution is of interest in the next section. Firstly, through CPLEX the linear relaxation is achieved, and then branch-and-bound/cut, the prior solution is improved as will be discussed later in other sections. The obtained result is then compared to the existing mixed-integer models solutions to reach an overall consequence based on optimum resulted numbers. Moreover, this optimum number can be interpreted socially, economically, and environmentally as follows.

Having achieved the optimum number from the presented model, it is socially concluded when the transportation cost is reduced in terms of time or money, and the accessibility increases as the propensity for public transportation between any two places of movement increases. Sequentially, if the accessibility increases between places in public transportation structures, the decentralization rate will be easily reachable. For this particular reason, the privilege of traffic congestion control is obtained giving an overview to procurement transportation green vehicles, which are planning to either move towards or outwards from the construction site. Moreover, timing concepts are precisely scheduled for transportation purposes.

Economically speaking, implementation of the reduced cost of procurement transportation model allows the government to be the main supportive care giver of low-income people in their transporting life. Moreover, the amount of money saved from subsidy plans is to be constructively injected into the society, providing citizens with cultural, recreational, and educational programs where the government can promote the values of pre-saved money and improve job safety requirements, particularly in transportation area (Weisbrod & Reno, 2009). As elaborated before, saved money from subsidy plans, achieved from cost reduction of public transportation, will positively affect economic propensity of the society. For this reason, there are measure of performances, which were also discussed before, giving clarification on how the four economic measures performed, known as access, travel, spending, and non-monetary measures, all noted before. In fact, these four measures of performance economically allow the government to financially support a particular category of people when it comes to their public transportation needs. Actually, the combination of two approaches as described in the first paragraph gives rise to the multilaterally pointed consequences if the mixed-integer model reaches an optimum solution.

Environmentally speaking, the carbon dioxide emission is bound to the amount of distance each transporter travels and should always hold a rate between 0 and 1 if a transportation path is chosen. The optimum cost achieved in the presented model assures the lowest CO_2 emission rate as assumed between 0 and 1. Also, there is a binary variable that is responsible for green vehicle transportation use, stating whether or not public transportation uses green vehicles. Lastly, the intension of finding the optimum solution has covered three dimensions of sustainability and guaranteed that the aforementioned consequences of sustainability aspects are all to be implemented by mixed-integer programming cost reduction of public transportation in surrounding area of construction site.

16.6   A NUMERICAL EXAMPLE (CASE STUDY)

16.6.1   Excel Implementation

16.6.1.1   Cost Parameters

In this section, a numerical example is provided to validate the presented model using Microsoft Excel software to reach the optimum cost of public transportation outside of the networking area of construction. Below is the table of variable costs form location (i) to location (j) if the path between (i) and (j) is chosen. All numbers in the table show the costs in thousands of dollars that the transporter type (k) has to carry if used for public transportation purposes in period (t) (Table 16.1).

In addition to variable costs of public transportation, which is based on the distances traversed from origin (i) to destination (j), there are fixed costs of traveling based on the distances traversed. Numbers in the table below account for fixed costs of public transportation from location (i) to (j), if transporter type (k) travels through it in period (t) (Table 16.2).

TABLE 16.1
Variable Costs of Public Transportation Surrounding a Construction Site from Location (i) to Location (j) (Distance Dependent)

C_{ij} $k = 1,\dots,4$	1	2	3	4	5	6
1	0	15	20	12	7	17
2	15	0	18	34	27	25
3	20	18	0	16	10	19
4	12	34	16	0	37	21
5	7	27	10	37	0	22
6	17	25	19	21	22	0

TABLE 16.2
Fixed Costs of Public Transportation Surrounding a Construction Site in 6 Periods with 4 Kinds of Transporters (Distance Independent)

V_{tk} $k = 1,\dots,4$	1	2	3	4
1	1	3	4	6
2	3	2	7	8
3	4	7	5	2
4	6	8	2	4
5	2	2	5	6
6	1	3	7	7

16.6.1.2 Capacity Parameters

Fuel capacity for each transporter (k) is given via the table below, which shows each transporter (k) traveling from (i) to (j) based on the amount of fuel dedicated to the certain routes of $(i), (j)$, and type (k) transporter in period (t) as follows (Table 16.3).

16.6.1.3 Fuel Consumption Parameters

Fuel consumption for each transporter type (k) is also given via table below, which shows how much fuel per kilometer each transporter type (k) consumes while traveling from location (i) to location (j) in period (t) (Table 16.4).

TABLE 16.3

Capacity Parameters of Transporters, Moving from Location (i) to Location (j), Including 4 Types of Transporters

b_{ij} $k = 1,...,4$	1	2	3	4	5	6
1	0	5	7	2	1	6
2	5	0	8	2	6	2
3	7	8	0	4	7	9
4	2	2	4	0	3	4
5	1	6	7	3	0	5
6	6	2	9	4	5	0

TABLE 16.4

Fuel Consumption Parameters Allocated from Location (i) to Location (j), Including 4 Types of Transporters

l_{ij} $k = 1,...,4$	1	2	3	4	5	6
1	0	5	7	2	1	6
2	5	0	8	2	6	2
3	7	8	0	4	7	9
4	2	2	4	0	3	4
5	1	6	7	3	0	5
6	6	2	9	4	5	0

16.6.1.4 Standard Emission Parameters

These parameters are standard numbers given per kilometer in the this table in period (t), measured from location (i) to location (j) as follows (Table 16.5).

16.6.1.5 The Small Number of ε

This small number is indicated assuming that it should hold a fractional value between 0 and 1, as explained beforehand in the relevant constraint section.

16.6.2 EXCEL REPORT

Firstly, it is assumed that the model presented enjoys just two dimensions as noted (i) and (j), and the other dimensions that happen to be (k), and (t) are numbered 1, meaning that there is just a single period of (t) and a single type of public transporter typed as (k) for optimization purposes.

All parameters, variables, constraints, and the objective function are introduced into Excel as they are described, and illustrated below.

TABLE 16.5

Standard Emission Parameters Assigned from Location (i) to Location (j), Including 4 Types of Transporter

p_{ij} $k = 1,...,4$	1	2	3	4	5	6
1	0	2	5	1	4	1
2	2	0	4	9	3	4
3	5	4	0	6	8	1
4	1	9	6	0	5	4
5	4	3	8	5	0	1
6	1	4	1	4	1	0

Parameters are as numbered, as they were in the above tables in the Excel. They are all illustrated here as they come as follows.

16.6.2.1 Variable Costs $\left(C_{ijt}^k\right)$ (Table 16.6)

TABLE 16.6

Variable Costs from Location (i) to location (j) Reported on Excel, Considering $k = 1$ and $t = 1$

$C_{ij}/k = 1/t = 1$	1	2	3	4	5	6
1	0	15	20	12	7	17
2	15	0	18	34	27	25
3	20	18	0	16	10	19
4	12	34	16	0	37	21
5	7	27	10	37	0	22
6	17	25	19	21	22	0

16.6.2.2 Fixed Costs $\left(V_{ijt}^k\right)$ (Table 16.7)

TABLE 16.7

Fixed Costs Reported on Excel, Considering $k = 1$ and $t = 1$

V_{kt}	1	2	3	4	5	6
1	1	3	4	6	2	1
2	3	2	7	8	2	3
3	4	7	5	2	5	7
4	6	8	2	4	6	7

16.6.2.3 Capacity Parameters $\left(b_{ijt}^{k}\right)$ (Table 16.8)

TABLE 16.8

Capacity Parameters from Location (i) to Location (j) Reported on Excel, Considering $k = 1$ and $t = 1$

$b_{ij}/k = 1/t = 1$	1	2	3	4	5	6
1	0	5	7	2	1	6
2	5	0	8	2	6	2
3	7	8	0	4	7	9
4	2	2	4	0	3	4
5	1	6	7	3	0	5
6	6	2	9	4	5	0

16.6.2.4 Fuel Consumption Parameters $\left(l_{ijt}^{k}\right)$ (Table 16.9)

TABLE 16.9

Fuel Consumption Parameters Allocated from Location (i) to Location (j), Reported on Excel Considering $k = 1$ and $t = 1$

$l_{ij}/k = 1/t = 1$	1	2	3	4	5	6
1	0	5	7	2	1	6
2	5	0	8	2	6	2
3	7	8	0	4	7	9
4	2	2	4	0	3	4
5	1	6	7	3	0	5
6	6	2	9	4	5	0

16.6.2.5 Standard Emission Parameters $\left(p_{ijt}^{k}\right)$

Now the variables are introduced to the Excel as follows (Table 16.10).

16.6.2.6 Binary Variables $\left(y_{ijt}^{k}\right)$

It is to notify that the other binary variables noted as f_{ijt}^{k} and z_{ij}^{k} are already included and do not need to be contained in a different constraint as the model presents (Table 16.11).

TABLE 16.10

Standard Emission Parameters Allocated from Location (i) to Location (j), Reported on Excel Considering $k = 1$ and $t = 1$

$p_{ij}/k = 1/t = 1$	1	2	3	4	5	6
1	0	5	5	2	4	6
2	2	0	4	9	3	4
3	5	4	0	6	8	1
4	1	9	6	0	5	4
5	4	3	8	5	0	1
6	1	4	1	4	1	0

TABLE 16.11

Binary Variables Reported on Excel, Considering $k = 1$ and $t = 1$

$y_{ij}/k = 1/t =$ 1	1	2	3	4	5	6	
1	0	1	0	0	0	0	1
2	0	0	0	0	0	1	1
3	0	0	0	0	1	0	1
4	0	0	1	0	0	0	1
5	1	0	0	0	0	0	1
6	0	0	0	1	0	0	1
	1	1	1	1	1	1	

16.6.2.7 Continuous Variables $\left(x_{ijt}^{k} \right)$

It is also to mention that the values at the bottom of the variable tables satisfy the constraints in which there need to be a single path from location (i) to location (j) as binary variable y_{ijt}^{k} chooses. Moreover, the condition that states the minimum distance between each stop should be a kilometer is also satisfied as illustrated by x_{ijt}^{k} in the table above (Table 16.12).

16.6.2.8 Capacity Constraint

It is to confirm that all constraints hold an identifiably specific fuel capacity, in regard to the distances traversed, are satisfied (Tables 16.13 and 16.14).

TABLE 16.12

Continuous Variables Reported on Excel, Considering $k = 1$ and $t = 1$

$x_{ij}/k = 1/t = 1$	1	2	3	4	5	6
1	0	1	0	0	0	0
2	0	0	0	0	0	1
3	0	0	0	0	1	0
4	0	0	1	0	0	0
5	1	0	0	0	0	0
6	0	0	0	1	0	0
	1	1	1	1	1	1

TABLE 16.13

Capacity Constraint Reported on Excel, Considering $k = 1$ and $t = 1$

$l_{ij}*x_{ij}$	1	2	3	4	5	6
1	0	5	0	0	0	0
2	0	0	0	0	0	2
3	0	0	0	0	7	0
4	0	0	4	0	0	0
5	1	0	0	0	0	0
6	0	0	0	4	0	0

TABLE 16.14

Capacity Constraint Reported on Excel, Considering $k = 1$ and $t = 1$

$b_{ij}*y_{ij}$	1	2	3	4	5	6
1	0	5	0	0	0	0
2	0	0	0	0	0	2
3	0	0	0	0	7	0
4	0	0	4	0	0	0
5	1	0	0	0	0	0
6	0	0	0	4	0	0

16.6.2.9 Standard Emission Constraint (Table 16.15)

TABLE 16.15

Standard Emission Constraint Given to the Binary Variables Reported on Excel, Considering $k = 1$ and $t = 1$

$(l_{ij}/p_{ij})*y_{ij}$	1	2	3	4	5	6
1	0	1	0	0	0	0
2	0	0	0	0	0	0.5
3	0	0	0	0	0.875	0
4	0	0	0.666667	0	0	0
5	0.25	0	0	0	0	0
6	0	0	0	1	0	0

16.6.2.10 Fuel Consumption Constraint

Confirming all constraints toward fuel consumption, a two-sided constraint reaches two different intervals in which one side confirms an interval between ε and 1. While, the other side shows a limitation $\left(\sum_{t=1}^{T} y_{ijt}^{k} \Big/ \sum_{t=1}^{T} f_{ijt}^{k} \right)$ in which fuel consumption is bound up to (M) and can be properly relevant to the distances traversed by green public transporters (Table 16.16).

16.6.2.11 Objective Function

The objective function is formed containing fixed costs and variable costs together and reach to an optimum solution based on the constraints considered. The objective function appears as follows in the presented model (Table 16.17).

TABLE 16.16

Fuel Consumption Constraint Reported on Excel, Considering $k = 1$ and $t = 1$

$(x_{ij}/l_{ij})*y_{ij}$	1	2	3	4	5	6
1	0	0.2	0	0	0	0
2	0	0	0	0	0	0.5
3	0	0	0	0	0.142857	0
4	0	0	0.25	0	0	0
5	1	0	0	0	0	0
6	0	0	0	0.25	0	0

TABLE 16.17
Optimized Objective Function

Variable and fixed Costs   100

16.7 ADVANTAGES OF THE PRESENTED MODEL

In this research study, which couples the knowledge of sustainability with mixed-integer programming, a few objectives are satisfied presenting a mixed-integer model when it comes to the costs of public transportation over the surrounding area. First and for most, the mixed-integer programming appears to be quite realistic and easy-to-solve programming, which even Microsoft Excel can find an optimum solution to. However, other software developers, including AIMMS, which is specifically used for optimization purposes, can overperform. Moreover, the binary variables that have been considered completing the model takes a bilateral condition, which also helps the model be two-sided while more constraints are introduced. In fact, lots of notions are situated within a few constraints with binary variables expressing a two-sided constraint conceptually. The solution presented in this research study has been achieved by integer parameters, which is also an integer solution. Although mixed-integer programming can also deal with non-integer parameters and reach a continuous objective function as realistically as the real-world modeling mostly fitting in non-integer solution as an optimum solution, it is targeted at an integer optimum solution and continues on restricting the feasible region to achieve the first optimal integer solution as objective function moves towards either minimization or maximization.

Mixed-integer programming is derived from linear programming in which LP relaxation plays the main role of further improvement in terms of either maximization or minimization. In fact, the linearity aspect of the modeling process is highly of interest due to its oversimplification, easy comprehension, and wide application in different fields of study. Coupled with linearity, the four main operations of mathematic including division, summation, multiplication, and subtraction are all applicable in mixed-integer programming as to which path to take towards the most optimal integer objective function.

Further improvements will be applied if the LP relaxation does not appear integer. Achieving a non-integer LP relaxation at first, the branch-and-cut approach can lead the already obtained solution to an integer optimum solution, where adding up several cuts including Gomory and Chavatal-Gomory cuts can restrict the feasible region to feasible integer solutions. Though, there is no need to add up cuts when the LP relaxation is in turn an integer. Improvement approaches are far wider than it seems when it comes to achieving an integer solution. Preprocessing is also another approach that comes before solution finding approaches through which the feasible region becomes smaller, and an optimum integer solution in terms of either maximization or minimization is easily reachable. In this research study, the optimum solution is already integer, and any improvements might lead to increasing the final cost as an objective function will be likely to jump up to the next integer solution. Furthermore, a previous integer objective function as an acceptable optimal solution is considered.

16.8   DISCUSSION

While investigating the consequences of the presented model in this research study, it reveals that the reduction of public transportation costs can indirectly help the presumed transport network save more money as part of their subsidy plans. Obviously, the amount of money saved in the total public transportation costs entitles authorities to being wiser towards other aspects of a community including cultural, recreational, educational, and many more. The obtained results of the presented model, which shows the inevitable amount of money that public transportation needs to be dealing with, is far from the amount of money paid as a total subsidy to the public sector of transportation over the surrounding area. Accordingly, the amount of money saved is employed to benefit societies economically, environmentally, and socially based on the triple aspects of sustainability.

As economically mentioned, there are indicators that economically assess the community performance, surrounding the considered transporting network in this research study, and plan the need for any economic improvement. They are performed using certain approaches to achieving final results through which economic interpretations are established, and emergency requirements on economic parts of a society are scheduled. Economically achieved results express that the financial help to lower-income people not only reduces the amount of public transportation costs that leads to a new governmental subsidy plan, but also enables the government to be developing some other programs in other sustainable aspects including environmental and social ones. Consequently, an economic aspect of a society financially supports other sustainable aspects of it, partly covering environmental and social sustainable dimensions.

Other aspects of sustainability noted as environmental and social are also intertwined with the cost reduction of the presented public transportation model in the surrounding area of study. Cost reduction of public transportation concerning CO_2 emission constraint and fuel consumption constraint include the environmental aspect of sustainability as to the spread in using green transporting vehicles for public transportation purposes. The reduced costs also help the environmental dimension to be more obvious over all transporting routes, as green vehicles take over public transportation responsibilities towards and outwards the construction site.

The social aspect of sustainability is satisfied considering the accessibility constraint of the public transporter over all the transporting area of the construction site. The economic aspect also helps the social dimension to implement wide accessibility of public transport and bound to reducing the cost of public transportation. In fact, the amount of money saved based on the presented model can be a good excuse to financially support accessibility and safety constraints in this model. Furthermore, the economic aspect is by far the most important dimension of sustainability in a community, as it partly provides other dimensions with a wide gateway for implementation and other programs that empower other sub-classified sustainable aspects to be developed, including recreational, cultural, and educational features of a nation.

In this research study, the model presented accounts for $100,000 as the achieved integer minimum public transportation cost including the seven constraints is described as follows.

The values for binary and continuous variables were determined using Microsoft Excel as it is shown in the tables above. Moreover, the network that public transporters structure delivering people at the nodes, meant as the bus stops. These binary values construct a network of six vertices and six edges in which the distance relevant costs and non-distance relevant costs all together come to $100,000 as objective function assures in this model. Based on the routes chosen by binary variables, continuous variables estimate distances of green public transporters traversing, introduced in a number of different types as noted (k). Given to the determined values of variables, the other constraints are given values as tables of standard emission, and fuel consumption are derived showing each route chosen is dedicated a fractional number as to satisfy these two constraints of sustainable main concepts. Besides, the fuel consumption constraint takes a two-sided notion in which a binary variable on using green vehicles for transportation purposes controls which side needs to be active and the counter side to be inactive.

Other regular constraints as many models are usually derived describe continuous variables as positive, and binary ones that consist a set of two members including 0 and 1 are constraint indicators of assumed real world conditions.

16.9   CONCLUSION

As the presented model is a multilateral one studying sustainability aspects of public transportation in the surrounding construction area, the objective function is satisfied containing seven constraints, each accounts for sustainable dimensions of a unity/network. Accessibility and safety together relate to social dimension, which is provided over the surrounding area of construction site for transportation purposes. Meanwhile, it also reduces the cost of public transportation as the objective function assures. The constraints including fuel consumption and standard emission of CO_2 are part of the presented model accounting for the environmental aspect of sustainability issue, which the latter holds CO_2 emission in a preferred interval of $[0,1]$ as environmentally looked into, and the former constructs an interval in which two different notions of fuel consumption are covered. This constraint, when respect to the value 1 that is substituted for the binary variable consists of an interval $\left(\sum_{t=1}^{T} y_{ijt}^{k} \Big/ \sum_{t=1}^{T} f_{ijt}^{k} \right)$ up to (M) as a quite great value of assumed model, and for 0 constraints an interval of (ε) to 1 as a very close limitation for non-green vehicles fuel consumption assumption. All other constraints are derived according to flow balance of the network that public transporters structure transporting people nearby and delivering them at each node.

The optimum solution calculated considers all different aspects of sustainability via the constraints introduced to the model of surrounding area of the construction site. The notion of mixed-integer programming, as it appears quite easy to solve and understandable, provides a very useful tool in mathematics to mix sustainability and integer programming skills to just emphasize that linear programming is still a reasonably strong skill to be partnering with other significant concepts, sustainability in particular. However, integer programming resulted from linear programming has in turn been developed, and brought about the creation of many approaches, including branch-and-bound/cut, to achieve the integer optimum solution in different integer models. Hence, several improvement layers can apply better solutions if the LP relaxation does not happen to be integer in general.

REFERENCES

Aktin, T., & Gergin, Z. (2016). Mathematical modeling of sustainable procurement strategies: Three case studies. *Journal of Cleaner Production, 113*, 767–780.

Atamtürk, A. (2005). Cover and pack inequalities for (mixed) integer programming. *Annals of Operations Research, 139*(1), 21–38.

Bailey, L., Mokhtarian, P. L., & Little, A. (2008). *The Broader Connection between Public Transportation, Energy Conservation and Greenhouse Gas Reduction*. Retrieved from https://www.apta.com/resources/reportsandpublications/Documents/land _use.pdf.

Balconi, M., Brusoni, S., & Orsenigo, L. (2010). In defense of the linear model: An essay. *Research Policy, 39*(1), 1–13.

Beyer, H. L., Dujardin, Y., Watts, M. E., & Possingham, H. P. (2016). Solving conservation planning problems with integer linear programming. *Ecological Modelling, 328*, 14–22.

Beynaghi, A., Moztarzadeh, F., Maknoon, R., Waas, T., Mozafari, M., Hugé, J., & Leal Filho, W. (2014). Towards an orientation of higher education in the post Rio + 20 process: How is the game changing? *Futures, 63*, 49–67.

Bixby, R. E. (2012). A brief history of linear and mixed-integer programming computation. *Documenta Mathematica*, 107–121.

Bowyer, S. A., Davies, B. L., & Baena, F. R. (2014). Active constraints/virtual fixtures: A survey. *IEEE Transactions on Robotics, 30*(1), 138–157.

Brammer, S., & Walker, H. (2011). Sustainable procurement in the public sector: An international comparative study. *International Journal of Operations & Production Management, 31*(4), 452–476.

Browne, M., Allen, J., Nemoto, T., Patier, D., & Visser, J. (2012). Reducing social and environmental impacts of urban freight transport: A review of some major cities. *Procedia-Social and Behavioral Sciences, 39*, 19–33.

Campbell, J. F., & O'Kelly, M. E. (2012). Twenty-five years of hub location research. *Transportation Science, 46*(2), 153–169.

Centobelli, P., Cerchione, R., & Esposito, E. (2017). Environmental sustainability in the service industry of transportation and logistics service providers: Systematic literature review and research directions. *Transportation Research Part D: Transport and Environment, 53*, 454–470.

d'Amore, F., & Bezzo, F. (2016). Strategic optimization of biomass-based energy supply chains for sustainable mobility. *Computers & Chemical Engineering, 87*, 68–81.

DiJoseph, P., & Chien, S. I. J. (2013). Optimizing sustainable feeder bus operation considering realistic networks and heterogeneous demand. *Journal of Advanced Transportation, 47*(5), 483–497.

Gannon, C. A., & Liu, Z. (1997). Poverty and transport.

Glover, F. (1986). Future paths for integer programming and links to artificial intelligence. *Computers & Operations Research, 13*(5), 533–549.

Holguín-Veras, J., Pérez, N., Jaller, M., Van Wassenhove, L. N., & Aros-Vera, F. (2013). On the appropriate objective function for post-disaster humanitarian logistics models. *Journal of Operations Management, 31*(5), 262–280.

Hsu, C.-Y., Yang, C.-S., Yu, L.-C., Lin, C.-F., Yao, H.-H., Chen, D.-Y., ... Chang, P.-C. (2015). Development of a cloud-based service framework for energy conservation in a sustainable intelligent transportation system. *International Journal of Production Economics, 164*, 454–461.

Hutter, F., Hoos, H. H., & Leyton-Brown, K. (2010). Automated configuration of mixed integer programming solvers. Paper presented at the International Conference on Integration of Artificial Intelligence (AI) and Operations Research (OR) Techniques in Constraint Programming.

Jindal, R., Kerr, J. M., Ferraro, P. J., & Swallow, B. M. (2013). Social dimensions of procurement auctions for environmental service contracts: Evaluating tradeoffs between cost-effectiveness and participation by the poor in rural Tanzania. *Land Use Policy, 31*, 71–80.

Kaewunruen, S., Sussman, J. M., & Matsumoto, A. (2016). Grand challenges in transportation and transit systems. *Frontiers in Built Environment, 2*, 4.

Large, R. O., Kramer, N., & Hartmann, R. K. (2013). Procurement of logistics services and sustainable development in Europe: Fields of activity and empirical results. *Journal of Purchasing and Supply Management, 19*(3), 122–133.

Layard, P. R. G. (1994). *Cost-Benefit Analysis*. Cambridge, UK: Cambridge University Press.

Leyton-Brown, K., Nudelman, E., Andrew, G., McFadden, J., & Shoham, Y. (2003). Boosting as a metaphor for algorithm design. Paper presented at the International Conference on Principles and Practice of Constraint Programming.

Linderoth, J. T., & Savelsbergh, M. W. (1999). A computational study of search strategies for mixed integer programming. *INFORMS Journal on Computing, 11*(2), 173–187.

Loosemore, M. (2016). Social procurement in UK construction projects. *International Journal of Project Management, 34*(2), 133–144.

Maheshwari, P., Kachroo, P., Paz, A., & Khaddar, R. (2015). Development of control models for the planning of sustainable transportation systems. *Transportation Research Part C: Emerging Technologies, 55*, 474–485.

Maheshwari, P., Khaddar, R., Kachroo, P., & Paz, A. (2016). Dynamic modeling of performance indices for planning of sustainable transportation systems. *Networks and Spatial Economics, 16*(1), 371–393. doi:10.1007/s11067-014-9238-6.

Murray, A. T., Davis, R., Stimson, R. J., & Ferreira, L. (1998). Public transportation access. *Transportation Research Part D: Transport and Environment, 3*(5), 319–328.

Naganathan, H., & Chong, W. K. (2017). Evaluation of state sustainable transportation performances (SSTP) using sustainable indicators. *Sustainable Cities and Society, 35*, 799–815.

Nocera, S., & Cavallaro, F. (2011). Policy effectiveness for containing CO_2 emissions in transportation. *Procedia-Social and Behavioral Sciences, 20*, 703–713.

Pourrousta, A., Dehbari, S., Tavakkoli-Moghaddam, R., Imenpour, A., & Naderi-Beni, M. (2012). A fuzzy mixed integer linear programming model for integrating procurement-production-distribution planning in supply chain. *International Journal of Industrial Engineering Computations, 3*(3), 403–412.

Richardson, B. C. (2005). Sustainable transport: Analysis frameworks. *Journal of Transport Geography, 13*(1), 29–39.

Schouwenaars, T., De Moor, B., Feron, E., & How, J. (2001). Mixed integer programming for multi-vehicle path planning. *Paper presented at the Control Conference (ECC)*, 2001 European.

Ševrović, M., Brčić, D., & Kos, G. (2015). Transportation costs and subsidy distribution model for urban and suburban public passenger transport. *PROMET-Traffic&Transportation, 27*(1), 23–33.

Sohail, M., Maunder, D., & Cavill, S. (2006). Effective regulation for sustainable public transport in developing countries. *Transport Policy, 13*(3), 177–190.

Testa, F., Annunziata, E., Iraldo, F., & Frey, M. (2016). Drawbacks and opportunities of green public procurement: An effective tool for sustainable production. *Journal of Cleaner Production, 112*, 1893–1900.

Tomlin, J. A. (1971). An improved branch-and-bound method for integer programming. *Operations Research, 19*(4), 1070–1075.

Vieira, C. L. D. S., & Luna, M. M. M. (2016). Models and methods for logistics hub location: A review towards transportation networks design. *Pesquisa Operacional, 36*(2), 375–397.

Vielma, J. P. (2015). Mixed integer linear programming formulation techniques. *SIAM Review, 57*(1), 3–57.

Vincent, W., & Jerram, L. C. (2006). The potential for bus rapid transit to reduce transportation-related CO_2 emissions. *Journal of Public Transportation, 9*(3), 12.

Wachs, M., & Taylor, B. D. (1998). Can transportation strategies help meet the welfare challenge? *Journal of the American Planning Association, 64*(1), 15–19.

Walker, H., & Phillips, W. (2008). Sustainable procurement: Emerging issues. *International Journal of Procurement Management, 2*(1), 41–61.

Weisbrod, G., & Reno, A. (2009). *Economic Impact of Public Transportation Investment.* Washington, DC: American Public Transportation Association.

17 A Sustainable Inventory Optimization Problem under Process Flexibility

Sedat Belbağ, Mustafa Çimen, Mehmet Soysal, and Ali Kemal Çelik

CONTENTS

17.1 INTRODUCTION

Though many academics and practitioners pay growing attention to the concept of sustainable supply chain management in recent years, one of the main challenges in sustainable supply chain management is considered as managing the perishability of products. Perishability can be simply defined as deterioration of products over time due to various reasons such as quality decay, spoilage, or evaporation. Perishability related constraints affect several supply chain processes (e.g., procurement, production, and inventory). The perishability of products prevents from keeping larger amounts of inventories as a buffer against uncertainty in demand (Ahumada and Villalobos, 2009). Specific tools such as rapid delivery or flexibility may help decision makers to deal with demand uncertainty in sustainable supply chains.

Flexibility is an ability to adopt to changes in internal and/or external factors. Among various flexibility types such as machine flexibility, volume flexibility, or product flexibility; the process flexibility in manufacturing systems aims to produce different products at the same production facility without bearing major setup costs (Sethi and Sethi, 1990). Accordingly, process flexibility also provides the ability to change the volume of products (Jordan and Graves, 1995). Early studies in the literature (e.g., Jaikumar, 1984; Andreou, 1990; Fine and Freund, 1990) often focus on two types of systems: (i) the dedicated system that produces a single type of product in each facility and (ii) the full flexible system that produces all types of products in each facility. Jordan and Graves (1995) introduce the concept of chaining design where a chain refers "a group of products and facilities, which are connected, directly or indirectly, by product assignment decisions." Within the proposed design, a facility can produce a subset of the products and a product can be produced in a subset of the facilities. The results of the study illustrate that the proposed so-called limited flexibility design may provide almost all benefits of full flexibility design with a relatively smaller amount of investment.

Since the seminal work of Jordan and Graves (1995), there is a growing interest in industrial process flexibility applications, particularly in the automotive industry as well as textile, electronic, and service industries (Chou et al., 2010). Process flexibility provides a competitive advantage but comes with a cost that production and inventory control become more complicated to manage. When a facility is able to produce multiple products and a product can be produced in multiple facilities, the production ability relations between facilities and products form a complex non-decomposable system. Therefore, production and inventory control become a multi-dimensional decision problem for production systems with process flexibility.

This study addresses a multi-product, multi-period, capacitated inventory optimization problem for perishable products with sequence-dependent setup costs and process flexibility. Demand is assumed to be deterministic and known in advance for the entire planning horizon. Inventory for each product is reviewed periodically. Capacity restrictions exist on production and inventory amounts. Whenever the product is changed in a facility, a sequence-dependent setup cost is incurred. In addition, a setup carryover is assumed when a product is the last to be produced in a period and the first in the following period. Unit production costs may differ among facilities due to several factors such as locations and technology. A product is valued according to its age; therefore, the unit selling price of products, which is known in advance, decreases over time. Backordering is not allowed, and any unsatisfied demand is lost with a penalty cost. Three different process flexibility settings will be addressed—dedicated, limited-flexibility, and full-flexibility. The objective is to determine when, how much, and in which sequence to produce for each product in each facility at the beginning of each period in order to maximize total profit. To the best of our knowledge, this is the first attempt to incorporate sequence-dependent setup costs to inventory optimization problems for perishable products under process flexibility assumption.

The problem has been formulated and solved using a Mixed Integer Linear Programming approach. Several numerical analyses have been carried out on three aforementioned process flexibility settings by means of the proposed decision support model. The results illustrate the potential benefits of respecting sequence-dependent setup costs and perishability of the products for the addressed problem in sustainable supply chains. The proposed enhanced inventory optimization model could aid decision makers in sustainable inventory management.

The remainder of this chapter is as follows. The following section gives information about inventory optimization with perishability, process flexibility, and sequence-dependent setup cost. The third section defines the existing problem to be solved mathematically in this study. The fourth section implements the underlying problem within a numerical study. The present chapter concludes with the discussion of results and recommendations for future studies.

17.2   INVENTORY OPTIMIZATION WITH PERISHABILITY, PROCESS FLEXIBILITY, AND SEQUENCE-DEPENDENT SETUP COST

Supply chain management concerns the design, management, and control of a network of organizations involved in various activities (e.g., procurement, production, inventory, and distribution) to add value in the form of goods and services for end or industrial customers (Aprile et al., 2005). The recent increased competition among companies has resulted in a growing interest in supply chain management. As product life cycles shorten and product varieties increase, supply chains contest with each other in order to achieve a competitive advantage.

Inventory has a strategic role in supply chain management, as the key performance indicator *supply chain profitability* is directly affected from inventory levels (Stadtler and Kilger, 2002). The amount of inventory kept also has an impact on supply chain responsiveness, i.e., high inventory levels allow companies to increase their responsiveness in return for increased inventory holding costs (Mendoza and Ventura, 2010). Additionally, a high inventory level enables a reduction in production and transportation costs due to economies of scale achievable through savings in setup cost or reductions in the number of *less than truck load* transportation. A low level of inventory contributes to reduced inventory cost but may result in lost sales (Chopra and Meindl, 2015). The trade-offs between the benefits and riskiness of keeping inventory have been widely discussed in the related literature.

Production and inventory management is generally studied within the scope of supply chain management. Sustainability and flexibility are the recent prominent topics addressed in the production and inventory management literature. Perishability, which is a key issue in terms of social sustainability, refers to the deterioration process of products over time due to several factors such as physical decay, spoilage, or evaporation. In practice, decision makers have to respect product perishability, while dealing with supply chain management of perishable products (Abad, 2000). A number of studies in the existing literature have already incorporated the perishability nature of products into the production and inventory control problems; an interested reader on the topic is referred to the reviews conducted by Nahmias (1982), Goyal and Giri (2001), Bakker et al. (2012), and Janssen et al. (2016). The proposed decision support models for perishable products mainly follow two approaches while addressing perishability: products with fixed lifetime and products with continuous quality decay (Goyal and Giri, 2001). The fixed lifetime assumption (e.g., Hwang and Hahn, 2000; Olsson and Tydesjö, 2010; Cheong, 2013) may be valid for several industries where quality and the utility of a product remain the same until the end of shelf-life, and afterwards goes down to zero, immediately. This is not the case for a number of product types in practice, such as electronics, fruits, fashionable clothes, where the quality of the product decreases over time even before the shelf-life. In both assumptions, fixed lifetime and continuous quality decay, product perishability consideration adds to the complexity of supply chain decisionmaking process.

Flexibility is the ability to adapt any internal and/or external changes without considerable effort. Flexibility types are classified according to their focus: product, volume, machine, process, etc. (Sethi and Sethi, 1990). Among these flexibility types, process flexibility aims to produce various product types in a facility with minor setup time and costs. Process flexibility enables the reduction of production batch sizes and inventory costs (Browne et al., 1984). Several studies address process flexibility issue within production and inventory management (e.g., Chou et al., 2008; Iravani et al., 2014; Çimen et al., 2016; Soysal et al., 2018). The ability to reduce inventory levels helps to diminish product waste; therefore, it provides additional benefit for sustainable supply chain management.

Another issue is that sequence-dependent setup costs could occur during product changeovers in flexible production systems, i.e., production setup costs might be dependent on the product production sequence. Similar to the aforementioned additional complexity derived from perishability consideration, both process flexibility and sequence-dependent setup cost complicate the control of production system.

In conclusion, this study addresses a multi-dimensional production and inventory control problem with perishability, process flexibility, and sequence-dependent setup cost. To the best of our knowledge, this study is the first attempt in the literature in terms of addressing the mentioned dimensions, simultaneously.

17.3 PROBLEM DEFINITION

Here, this study addresses a multi-period inventory optimization problem under process flexibility, perishability, and sequence-dependent setup cost assumptions. The considered production system consists of multiple facilities, where each facility may be able to produce more than one product. Three production systems are considered under different process flexibility settings; dedicated, limited-flexibility, and full-flexibility. In a dedicated production setting, a facility is devoted to producing one type of product, in particular, whereas a facility has the ability of producing each product type in a full-flexibility setting. In the limited-flexibility setting, which is employed, each facility can produce a fraction of products by forming a chaining design where production abilities are shared among the production system (Çimen and Kirkbride, 2017). Figure 17.1 illustrates the aforementioned process flexibility settings for five facilities and five products.

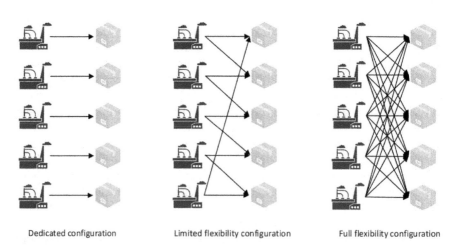

Dedicated configuration   Limited flexibility configuration   Full flexibility configuration

FIGURE 17.1   Flexibility settings for the considered production system.

Another concern of this chapter is the perishability nature of the products. The products physically deteriorate in each period, and the value of products decreases with the increase in the age of products, accordingly. The unit selling price of a product, therefore, depends on its age. New products have a higher selling price, which decreases over time. Products have a fixed shelf-life.

Facilities are ready to produce a product at the beginning of each period. When a facility has the ability to produce multiple products, there may be changeovers among products in a period. Whenever a changeover occurs (i.e., a change in production sequence) from one product to another in a period, a sequence-dependent setup cost is incurred. The setup settings for the last produced product is carried over to the beginning of the next period.

The inventory costs comprise production, holding, shortage, and waste costs. Due to various factors such as process technology and location, unit production costs vary among facilities. Inventory is periodically reviewed, and any products left at the end of each period yields a holding cost. For the sake of simplicity, it is assumed that the unit holding cost does not change with regard to the age of the product. A shortage cost is incurred for any unsatisfied demand. When the age of a product exceeds its shelf-life, it will be spoiled with a waste cost.

Demand is deterministic and known in advance. Demand is satisfied from production and/or inventory, backlogging is not allowed. The facilities have finite production capacities in each period.

The aim of the described problem is to determine (i) when, how much, and in which sequence to produce each product in each facility at the beginning of each period, and (ii) which age of products are used to satisfy demand in each period, in order to maximize total profit. Table 17.1 presents the notation for the considered model.

Now, the mathematical formulation of the addressed problem is formally described. The problem is formulated as a Mixed Integer Linear Programming model.

$$Max \sum_{t} \sum_{p} \left[\sum_{a} pr_{p,a} \Theta_{p,t,a} - \sum_{f} u_{f,p} Q_{f,p,t} - \right.$$

$$\left. \sum_{f} \sum_{p';p \neq p'} \phi_{f,p,p'} S_{f,p,p',t} - h_p \eta_{p,t} - \alpha_p \pi_{p,t} - \sigma_p \omega_{p,t} \right] \tag{17.1}$$

subject to

$$I_{p,0,a} = k_{p,a} \quad \forall p \in P, a \in A; a \neq 0 \tag{17.2}$$

$$I_{p,t,0} = 0 \quad \forall p \in P, t \in T \tag{17.3}$$

$$I_{p,t+1,1} = \sum_{f} Q_{f,p,t} - \Theta_{p,t,0} \quad \forall p \in P, t \in T \tag{17.4}$$

$$I_{p,t+1,a+1} = I_{p,t,a} - \Theta_{p,t,a} \quad \forall p \in P, a \in A \backslash \{0,m\}, t \in T \tag{17.5}$$

TABLE 17.1
Parameters and Decision Variables

F	Set of facilities, $F = \{1,2,\ldots,\lvert F \rvert\}$
P	Set of products, $P = \{1,2,\ldots,\lvert P \rvert\}$
T	Set of periods, $T = \{1,2,\ldots,\lvert T \rvert\}$
A	Set of product ages, $A = \{0,1,2,\ldots,\lvert A \rvert\}$
$pr_{p,t,a}$	Unit revenue of product p of age a in period t
$u_{p,t}$	Unit production cost of product p in period t
$\phi_{f,p,p'}$	Setup cost to change production from product p to p' in facility f
h_p	Unit holding cost for product p
α_p	Unit shortage cost for product p
σ_p	Unit waste cost for product p
c_f	Capacity of facility f
$l_{f,p}$	Production link between facility f and product p, $l_{f,p} \in \{0,1\}$
m	Fixed maximum shelf life, $m \geq 1$ and $m = \lvert A \rvert$, in periods
$k_{p,a}$	Beginning inventory of product p of age a
$d_{p,t}$	Demand for product p in period t
$\Theta_{p,t,a}$	Quantity of sold product of age a used to satisfy the demand of product p in period t
$Q_{f,p,t}$	Amount of product p produced by facility f in period t
$I_{p,t,a}$	Inventory of product p of age a at the beginning of period t
$\pi_{p,t}$	Derived decision variable to calculate negative inventory levels at the end of period t
$\eta_{p,t}$	Derived decision variable to calculate positive inventory levels at the end of period t
$\omega_{p,t}$	Amount of waste product p in period t
$S_{f,p,p',t}$	Binary variable equal to 1 if the production changes from product p to p' in facility f in period t and 0 otherwise
$Y_{f,p,t}$	Binary variable equal to 1 if facility f set up for product p at the beginning of period t, and 0 otherwise
$v_{f,p,t}$	Auxiliary variable

$$\eta_{p,t} = \sum_{a=0}^{m-1} I_{p,t+1,a+1} \qquad \forall p \in P, t \in T \tag{17.6}$$

$$\pi_{p,t} = d_{p,t} - \sum_{a} \Theta_{p,t,a} \qquad \forall p \in P, t \in T \tag{17.7}$$

$$\omega_{p,t} = I_{p,t,m} - \Theta_{p,t,m} \qquad \forall p \in P, t \in T, m \in A \tag{17.8}$$

$$\sum_{p} Q_{f,p,t} \leq c_f \qquad \forall f \in F, t \in T \tag{17.9}$$

$$Q_{f,p,t} \leq c_f l_{f,p} \left(\sum_{p';p \neq p'} S_{f,p',p,t} + Y_{f,p,t} \right) \qquad \forall f \in F, p \in P, t \in T \qquad (17.10)$$

$$\sum_{p} Y_{f,p,t} = 1 \qquad \forall f \in F, t \in T \qquad (17.11)$$

$$Y_{f,p,t} + \sum_{p';p \neq p'} S_{f,p',p,t} = Y_{f,p,t+1} + \sum_{p';p \neq p'} S_{f,p,p',t} \qquad \forall f \in F, p \in P, t \in T \quad (17.12)$$

$$\sum_{p';p \neq p'} S_{f,p,p',t} \leq 1 \qquad \forall f \in F, p \in P, t \in T \qquad (17.13)$$

$$\sum_{p';p \neq p'} S_{f,p',p,t} \leq 1 \qquad \forall f \in F, p \in P, t \in T \qquad (17.14)$$

$$v_{f,p,t} + (S_{f,p,p',t} \mid P \mid) - \mid P \mid + 1 \leq v_{f,p',t} \qquad \forall f \in F, p, p' \in P; p \neq p', t \in T \quad (17.15)$$

$$1 \leq v_{f,p,t} \leq \mid P \mid \qquad \forall f \in F, p \in P, t \in T \qquad (17.16)$$

$$Q_{f,p,t} \geq 0 \qquad \forall f \in F, p \in P, t \in T \qquad (17.17)$$

$$\Theta_{p,t,a} \geq 0 \qquad \forall p \in P, t \in T, a \in A, \qquad (17.18)$$

$$I_{p,t,a} \geq 0 \qquad \forall p \in P, t \in T, a \in A, \qquad (17.19)$$

$$\pi_{p,t} \geq 0 \qquad \forall p \in P, t \in T \qquad (17.20)$$

$$\eta_{p,t} \geq 0 \qquad \forall p \in P, t \in T \qquad (17.21)$$

$$v_{f,p,t} \geq 0 \qquad \forall f \in F, p \in P, t \in T \qquad (17.22)$$

$$S_{f,p,p',t} \in \{0,1\} \qquad \forall f \in F, p, p' \in P; p \neq p', t \in T \qquad (17.23)$$

$$Y_{f,p,t} \in \{0,1\} \qquad \forall f \in F, p \in P, t \in T \qquad (17.24)$$

The objective function (17.1) maximizes the total sales revenue, minus production, setup, holding, shortage, and waste costs. Constraints (17.2)–(17.8) relate to the inventory decisions. In particular, constraints (17.2) define the number of products at age a for each product at the beginning of the planning horizon. Constraints (17.3) guarantee that the inventory includes only product at an age greater than zero at the beginning of each period. Constraints (17.4) calculate the amount of inventory for each product at the end of the period, which gives the beginning inventory level of the following period. Constraints (17.5) are used for tracking the age of the products held. Constraints (17.6) and (17.7) calculate the amount of inventory on hand and the amount of shortage, respectively, for each product. Constraints (17.8) enable to track product waste amounts.

Constraints (17.9)–(17.10) relate to capacity restrictions. Constraints (17.9) ensure that the total amount of production in a facility cannot exceed the capacity of the facility. Constraints (17.10) allow to produce product p only if (i) facility f has the ability to produce the product, (ii) either the facility is already set up to produce the product, or a changeover to produce the product is tracked.

Constraints (17.11)–(17.16) determine the sequence of production in each period and keep track of setup carryover information for each period. Constraints (17.11) ensure that a facility is set up for only one product at the beginning of each period. Constraints (17.12) guarantee that if a facility is set up for a product at the beginning of a period or a changeover occurs in a period, the facility may either continue with the same setup configuration in the next period or a changeover occurs to another product. Constraints (17.13)–(17.16) eliminate the possible sub-tours of production sequence, which may prevent the model from determining the optimal solution. The study employs the same approach as in Nemhauser and Wolsey (1988), where a similar modeling trick is presented in the context of the traveling salesman problem. Constraints (17.17)–(17.24) represent the restrictions imposed on the decision variables.

17.4   NUMERICAL STUDY

This section presents an implementation of the proposed model using hypothetical data. The production systems consisting of five facilities and five products are considered for the aforementioned three flexibility settings (see Figure 17.1). In the limited flexibility setting, each factory can produce two products. Note that, each factory is able to produce a single product in a dedicated setting, whereas in a full flexibility setting, any products can be produced in each facility.

Each product has a different deterioration rate; therefore, prices vary based on the product age (see Table 17.2). The shelf-life is set to 4 periods for all products.

When a product changeover occurs in a facility, a fixed sequence-dependent setup cost is incurred. Different setup costs are assigned to each facility: Facility 1—100, Facility 2—120, Facility 3—80, Facility 4—140, Facility 5—110. Unit production costs of each facility for each product is given in Table 17.3. The unit holding and shortage costs of each product are equal to 0.5 and 5 per period, respectively.

TABLE 17.2
Prices with Regard to Product Age

Product	Age				
	0	1	2	3	4
1	25	22	19	16	13
2	21	15	9	0	0
3	24	23	22	21	20
4	18	16	14	7	3
5	25	20	15	14	13

TABLE 17.3
The Unit Production Costs for Each Product in Each Facility

Facilities	Products									
	1	2	3	4	5	6	7	8	9	10
1	1	1.1	1.21	1.33	1.46	1.61	1.77	1.95	2.14	2.36
2	2.36	1	1.1	1.21	1.33	1.46	1.61	1.77	1.95	2.14
3	2.14	2.36	1	1.1	1.21	1.33	1.46	1.61	1.77	1.95
4	1.95	2.14	2.36	1	1.1	1.21	1.33	1.46	1.61	1.77
5	1.77	1.95	2.14	2.36	1	1.1	1.21	1.33	1.46	1.61
6	1.61	1.77	1.95	2.14	2.36	1	1.1	1.21	1.33	1.46
7	1.46	1.61	1.77	1.95	2.14	2.36	1	1.1	1.21	1.33
8	1.33	1.46	1.61	1.77	1.95	2.14	2.36	1	1.1	1.21
9	1.21	1.33	1.46	1.61	1.77	1.95	2.14	2.36	1	1.1
10	1.1	1.21	1.33	1.46	1.61	1.77	1.95	2.14	2.36	1

The planning horizon comprises 12 periods. Following demand patterns are employed for the numerical analysis: Product 1—constant, Product 2—seasonal, Product 3—increasing, Product 4—decreasing, Product 5—hectic. All demand means are taken from Soysal & Çimen (2017) and presented in Table 17.4. At the beginning of the planning horizon, the inventory level for each product is set to zero. The capacity of each facility is given as: Facility 1—60, Facility 2—60, Facility 3—50, Facility 4—50, Facility 5—40.

The ILOG-OPL development studio and CPLEX 12.6 optimization package have both been used to develop and solve the presented formulation for the proposed model. Optimal solutions were obtained on a computer of Pentium(R) i7 2.4 GHz CPU with 8 GB memory.

TABLE 17.4

Demand Patterns

Demand	Periods											
	1	2	3	4	5	6	7	8	9	10	11	12
Fixed	50	50	50	50	50	50	50	50	50	50	50	50
Seasonal	68	83	88	83	69	50	31	17	12	17	32	50
Increasing	3	9	16	28	34	37	43	59	70	91	99	111
Decreasing	111	99	91	70	59	43	37	34	28	16	9	3
Hectic	27	22	94	27	17	74	120	12	50	28	19	110

The optimal values of cost components for the three flexibility settings are provided in Table 17.5. These results clearly indicate that it is beneficial to incorporate flexibility to the production system for reducing holding and shortage costs. Although product changeovers increase total sequence-dependent setup costs in limited flexibility and full flexibility designs, holding and shortage costs drastically decrease compared to the ones observed in dedicated configuration.[*] Another noteworthy result is that the optimal total profit of limited flexibility design is almost equal to that of full flexibility design (the profit gap between two configurations is approximately 1%). This result shows that limited flexibility design is nearly as beneficial as full flexibility design.

Table 17.6 presents the optimal production plan for limited flexibility design.[†] As seen in Table 17.6, despite the sequence-dependent setup costs and varying unit production costs, product changeovers occur to satisfy the demands and to reduce costs. Table 17.7 illustrates that enhanced flexibility provides the advantage of satisfying demands and reducing inventories by producing in various facilities.

TABLE 17.5

Optimal Values of Cost Components for Three Process Flexibility Designs

	Dedicated	Limited Flexibility	Full Flexibility
Revenue	57284	65046	65496
Production Cost	2609	3005.2	3320.2
Setup Cost	0	2120	1560
Holding Cost	814.5	206	131.5
Shortage Cost	1955	415	415
Waste Cost	0	0	0
Profit	51905.5	59299.8	60069.3

---

[*] Respectively, 74%–83% improvement in holding costs and 78% improvement in shorting costs for both settings

[†] The results of dedicated and full flexibility designs are given in Tables 17.7 through 17.13.

TABLE 17.6
Optimal Production Plan under Limited Flexibility

Products	1	2	3	4	5	6	7	8	9	10	11	12
1	F1(58)	F1(60)	F1(32)	F1(37) F5(13)	F1(51) F5(7)	F1(42)	F1(50)	F1(59)	F1(48)	F1(43)	F1(50)	F1(50)
2	F2(60)	F2(60)	F1(28) F2(60)	F1(23) F2(60)	F1(9) F2(60)	F1(18) F2(32)	F2(31)	F2(17)	F1(12)	F1(17)	F2(32)	F2(50)
3	—	F3(9)	F3(16)	F3(30)	F3(41)	F2(28)	F2(29) F3(14)	F2(32) F3(50)	F2(60) F3(50)	F2(60) F3(50)	F2(28) F3(50)	F3(50)
4	F3(50) F4(50)	F3(41) F4(50)	F3(34) F4(50)	F3(20) F4(50)	F3(9) F4(50)	F3(44)	F3(36)	F4(34)	F4(28)	F4(16)	F4(12)	—
5	F5(40)	F5(40)	F5(40)	F5(27)	F5(31)	F4(50) F5(40)	F4(50) F5(40)	F5(22)	F5(40)	F5(28)	F5(39)	F4(50) F5(40)

TABLE 17.7
Optimal Production Plan under Full Flexibility Design

Products						Periods						
	1	2	3	4	5	6	7	8	9	10	11	12
1	F1(58)	F1(60)	F1(32) F1(28)	F1(50) F1(10)	F1(55)	F1(46)	F1(49)	F1(50)	F1(50)	F1(50)	F1(50)	F1(50)
2	F2(60)	F2(60)	F2(60)	F2(60) F5(13)	F2(43) F5(36)	F5(40)	F5(31)	F5(17)	F5(12)	F5(19)	F5(40)	F5(40)
3	F5(13)	—	F3(15)	F3(30)	F3(41)	F3(50)	F2(1) F3(20)	F2(60)	F2(60) F3(9)	F2(60) F3(50)	F2(60) F3(50)	F2(60) F3(21)
4	F3(50) F4(50)	F3(49) F4(50)	F4(50)	F3(20) F4(50)	F3(9) F4(50)	F4(50)	F3(30)	F3(49)	F3(41)	—	—	—
5	F5(27)	F3(1) F5(40)	F3(35) F5(40)	F5(27)	F2(17)	F1(14) F2(60)	F1(11) F2(59) F4(50)	F4(12)	F4(50)	F4(28)	F4(50)	F3(29) F4(50)

TABLE 17.8
Product Changeovers during Planning Horizon under Limited Flexibility

Facilities	1	2	3	4	5	6	7	8	9	10	11	12
1	—	—	1–2	2–1	1–2	2–1	—	—	1–2	2–1	—	—
2	—	—	—	—	—	2–3	3–2	2–3	—	—	3–2	—
3	—	4–3	3–4	4–3	3–4	—	4–3	—	—	—	—	—
4	—	—	—	—	—	4–5	5–4	—	—	—	—	4–5
5	—	—	—	5–1	1–5	—	—	—	—	—	—	—

Table 17.8 illustrates product changeovers in each facility for each period (e.g., a changeover occurs from product 4 to product 3 in facility 3 in period 2). Note that there is a considerable number of changeovers for the entire planning horizon, although production system is associated with sequence-dependent setup costs. The number of changeovers decreases by 20% in full flexibility design as shown in Table 17.9 due to the enhanced flexibility.

Table 17.10 presents the amounts of initial and final inventories, the amount of quantity sold and the amount of demand for each product in each period under limited flexibility design. According to the Table 17.10, aged products are rarely used to satisfy demands in the entire planning horizon. In both limited flexibility and full flexibility designs, product age does not reach to the maximum shelf-life for each product; however, a considerable number of products are sold at the maximum shelf-life in the dedicated design (see Table 17.11).

Additionally, we would like to investigate the potential effects of taking sequence-dependent setup costs and the product deterioration into account. For this purpose, different parameter settings have been analyzed under limited flexibility design. Here, the initial parameter settings used for the previous analyses are regarded as the base setting.

First, sequence-dependent setup costs are decreased to 20% and increased to 500% of the initial costs, respectively. A comparison among the base, low setup

TABLE 17.9
Product Changeovers during Planning Horizon under Full Flexibility Design

Facilities	1	2	3	4	5	6	7	8	9	10	11	12
1	—	—	1–2	2–1	—	1–5	5–1	—	—	—	—	—
2	—	—	—	—	2–5	—	5–3	—	—	—	—	—
3	—	4–5	5–3	3–4	4–3	—	3–4	—	4–3	—	—	3–5
4	—	—	—	—	—	—	4–5	—	—	—	—	—
5	3–5	—	—	5–2	—	—	—	—	—	—	—	—

TABLE 17.10

Detailed Optimal Solution under Limited Flexibility

Products		Periods											
		1	2	3	4	5	6	7	8	9	10	11	12
1	B.I.	58	60	32	50	58	42	50	59	48	43	50	50
	Q.S.	50	42	32	50	50	42	50	50	41	43	50	50
				8ᵃ	18ᵃ		8ᵃ			9ᵃ	7ᵃ		
	D.	50	50	50	50	50	50	50	50	50	50	50	50
	F.I.	8	18	0	0	8	0	0	9	7	0	0	0
2	B.I.	60	60	88	83	69	50	31	17	12	17	32	50
	Q.S.	60	60	88	83	69	50	31	17	12	17	32	50
	D.	68	83	88	83	69	50	31	17	12	17	32	50
	F.I.	-8	-23	0	0	0	0	0	0	0	0	0	0
3	B.I.	0	9	16	30	41	28	43	82	110	110	78	50
	Q.S.	0	9	16	28	32	28	43	59	47	28	17	50
						2ᵃ	9ᵃ			23ᵃ	63ᵃ	82ᵃ	61ᵃ
	D.	3	9	16	28	34	37	43	59	70	91	99	111
	F.I.	-3	0	0	2	9	0	0	23	63	82	61	0
4	B.I.	100	100	84	70	59	44	36	34	28	16	12	0
	Q.S.	100	91	84	70	59	43	36	34	28	16	9	3ᵃ
								1ᵃ					
	D.	111	99	91	70	59	43	37	34	28	16	9	3
	F.I.	-11	-8	-7	0	0	1	0	0	0	0	3	0
5	B.I.	40	40	40	27	31	90	90	22	40	28	39	90
	Q.S.	27	9	40	27	17	60	90	12	40	28	19	90
			13ᵃ	31ᵃ			14ᵃ	30ᵃ		10ᵃ			20ᵃ
	D.	27	22	94	27	17	74	120	12	50	28	19	110
	F.I.	13	31	-23	0	14	30	0	10	0	0	20	0

Abbreviations: B.I.: Beginning inventory, Q.S.: Quantity sold, D.: Demand, F.I.: Final inventory.
ᵃ Represents the age of product which is equal to one period.

cost and high setup cost settings has been presented to illustrate the importance of considering sequent-dependent setup cost in the addressed problem.

Second, the case is analyzed where the model ignores the effect of product deterioration on price. In this setting, the model assumes that products do not lose any value before the maximum shelf-life but are regarded as waste afterwards. A comparison is presented to illustrate the effect of considering product deterioration on total revenue. We would like to note that the shelf-life is still assumed as four days.

Table 17.12 presents the optimal values of cost components for the base, low setup cost and high setup cost settings. Compared to the results of the base setting, the

TABLE 17.11

Detailed Optimal Solution under Dedicated System

Products		1	2	3	4	5	6	7	8	9	10	11	12
								Periods					
1	B.I.	50	50	50	50	50	50	50	50	50	50	50	50
	Q.S.	50	50	50	50	50	50	50	50	50	50	50	50
	D.	50	50	50	50	50	50	50	50	50	50	50	50
	F.I.	0	0	0	0	0	0	0	0	0	0	0	0
2	B.I.	60	60	60	60	60	50	31	17	12	17	32	50
	Q.S.	60	60	60	60	60	50	31	17	12	17	32	50
	D.	68	83	88	83	69	50	31	17	12	17	32	50
	F.I.	−8	−23	−28	−23	−9	0	0	0	0	0	0	0
3	B.I.	50	97	138	172	194	210	223	50	221	201	160	111
	Q.S.	3	9	16^b	27^a	13^c	37^d	20^b	9^c	49^a	41^c	39	50
					1^c	21^d		23^d	50^d	21^d	50^d	50^b	11^a
												1^c	50^b
													9^d
	D.	3	9	16	28	34	37	43	59	70	91	99	111
	F.I.	47	88	122	144	160	173	180	171	151	110	61	0
4	B.I.	50	50	50	50	50	43	37	34	28	16	9	3
	Q.S.	50	50	50	50	50	43	37	34	28	16	9	3
	D.	111	99	91	70	59	43	37	34	28	16	9	3
	F.I.	−61	−49	−41	−20	−9	0	0	0	0	0	0	0
5	B.I.	40	53	40	40	53	76	42	40	68	58	70	91
	Q.S.	27	9	40	27	17	40	40	12	40	28	7	40
				13^a	31^a		23^a			10^a		12^a	33^a
							11^b						18^d
	D.	27	22	94	27	17	74	120	12	50	28	19	100
	F.I.	13	31	−23	13	36	2	−78	28	18	30	51	−19

Abbreviations: B.I.: Beginning inventory, Q.S.: Quantity sold, D.: Demand, F.I.: Final inventory.
[a,b,c,d] Represents the age of product which are equal to one, two, three and four periods, respectively.

TABLE 17.12

Optimal Values of Cost Components for Different Setup Cost Settings

	Base Setting	Low Setup Cost Setting	High Setup Cost Setting
Revenue	65046	65821	63628
Production Cost	3005.2	3080.8	3151.4
Setup Cost	2120	880	3640
Holding Cost	206	82	232.5
Shortage Cost	415	405	565
Waste Cost	0	0	0
Profit	59299.8	61373.2	56039.1

increase of setup cost leads to 5% decrease of total profit, whereas its decrease results in 3% increase of total profit.

Tables 17.13 through 17.15 present the amounts of initial and final inventories, the amount of quantity sold and the amount of demand for each product in each period under low setup cost and high setup cost settings. According to the results,

TABLE 17.13
Detailed Optimal Solution under Full Flexibility Design

Products								Periods					
		1	2	3	4	5	6	7	8	9	10	11	12
1	B.I.	58	68	50	50	55	46	49	50	50	50	50	50
	Q.S.	50	42	32	50	50	45	49	50	50	50	50	50
			8^a	18^a			5^a	1^a					
	D.	50	50	50	50	50	50	50	50	50	50	50	50
	F.I.	8	18	0	0	5	1	0	0	0	0	0	0
2	B.I.	60	60	88	83	79	40	31	17	12	19	40	40
	Q.S.	60	60	88	83	69	40	31	17	12	17	30	40
							10^a					2^a	10^a
	D.	68	83	88	83	69	50	31	17	12	17	32	50
	F.I.	−8	−23	0	0	10	0	0	0	0	2	10	0
3	B.I.	13	10	15	30	41	50	21	60	69	110	129	81
	Q.S.	3	9^a	15	28	32	28	21	59	69	91	99	81
				1^b		2^a	9^a	22^a		1^a			11^a
													19^b
	D.	3	9	16	28	34	37	43	59	70	91	99	111
	F.I.	10	1	0	2	9	22	0	1	0	19	30	0
4	B.I.	100	99	50	70	59	50	30	49	41	28	12	3
	Q.S.	100	99	50	70	59	43	30	34	13	16^a	9^b	3^c
								7^a		15^a			
	D.	111	99	91	70	59	43	37	34	28	16	9	3
	F.I.	−11	0	−41	0	0	7	0	15	28	12	3	0
5	B.I.	27	41	75	27	17	74	120	12	50	28	50	79
	Q.S.	27	22	75	27	17	74	120	12	50	28	19	79
				19^a									31^b
	D.	27	22	94	27	17	74	120	12	50	28	19	110
	F.I.	0	19	0	0	0	0	0	0	0	0	31	0

Abbreviations: B.I.: Beginning inventory, Q.S.: Quantity sold, D.: Demand, F.I.: Final inventory.
[a,b,c] Represents the age of product which are equal to one, two and three periods, respectively.

TABLE 17.14

Detailed Optimal Solution under Limited Flexibility Design with Low Setup Costs

Products		Periods											
		1	2	3	4	5	6	7	8	9	10	11	12
1	B.I.	50	52	48	50	52	50	50	52	50	50	61	39
	Q.S.	50	50	48	50	50	48	50	50	48	50	50	39
				2^a			2^a			2^a			11^a
	D.	50	50	50	50	50	50	50	50	50	50	50	50
	F.I.	0	2	0	0	2	0	0	2	0	0	11	0
2	B.I.	68	83	72	83	69	50	31	17	12	17	32	50
	Q.S.	68	83	72	83	69	50	31	17	12	17	32	50
	D.	68	83	88	83	69	50	31	17	12	17	32	50
	F.I.	0	0	−16	0	0	0	0	0	0	0	0	0
3	B.I.	15	12	16	30	45	38	43	59	70	110	129	111
	Q.S.	3	9^a	13	28	34	28	42	59	70	91	80	81
				3^b			7^a	1^a				19^a	30^a
							2^b						
	D.	3	9	16	28	34	37	43	59	70	91	99	111
	F.I.	12	3	0	2	9	1	0	0	0	19	30	0
4	B.I.	100	99	37	70	59	43	37	34	28	16	12	3
	Q.S.	100	99	37	70	59	43	37	34	28	16	9	3^a
	D.	111	99	91	70	59	43	37	34	28	16	9	3
	F.I.	−11	0	−54	0	0	0	0	0	0	0	3	0
5	B.I.	27	26	90	27	31	90	90	12	50	28	39	90
	Q.S.	27	22	90	27	17	60	90	12	50	28	19	90
				4^a			14^a	30^a					20^a
	D.	27	22	94	27	17	74	120	12	50	28	19	110
	F.I.	0	4	0	0	14	30	0	0	0	0	20	0

Abbreviations: B.I.: Beginning inventory, Q.S.: Quantity sold, D.: Demand, F.I.: Final inventory.

[a,b]   Represents the age of product which are equal to one and two periods, respectively.

the value of setup cost directly affects the number of aged products used, i.e., the increase of setup costs leads to use more aged products than the ones used in the base setting, whereas reduction of setup costs leads to a decrease on the usage of aged products. Additionally, the demand is more frequently satisfied from inventory rather than by using new products in high setup cost setting compared to low setup cost setting.

TABLE 17.15

Detailed Optimal Solution under Limited Flexibility Design with High Setup Costs

Products		Periods											
		1	2	3	4	5	6	7	8	9	10	11	12
1	B.I.	60	70	52	50	50	50	57	90	80	70	60	50
	Q.S.	50	40	30	48	50	50	50	43	10	20	30	40
			10[a]	20[a]	2[a]				7[a]	40[a]	30[a]	20[a]	10[a]
	D.	50	50	50	50	50	50	50	50	50	50	50	50
	F.I.	10	20	2	0	0	0	7	40	30	20	10	0
2	B.I.	60	60	88	72	60	60	31	17	12	17	32	50
	Q.S.	60	60	88	72	60	50	21	17	12	17	32	50
								10[a]					
	D.	68	83	88	83	69	50	31	17	12	17	32	50
	F.I.	-8	-23	0	-11	-9	10	0	0	0	0	0	0
3	B.I.	0	0	0	30	52	41	43	60	70	91	100	111
	Q.S.	0	0	0	28	32	23	39	59	69	91	99	110
						2[a]	18[a]	4[a]		1[a]			1[a]
	D.	3	9	16	28	34	37	43	59	70	91	99	111
	F.I.	-3	-9	-16	2	18	4	0	1	0	0	1	0
4	B.I.	100	99	100	70	50	43	37	49	56	28	12	3
	Q.S.	100	99	91	61	50	43	37	34	13	16[a]	9[b]	3[c]
					9[a]	9[a]				15[a]			
	D.	111	99	91	70	59	43	37	34	28	16	9	3
	F.I.	-11	0	9	9	0	0	0	15	28	12	3	0
5	B.I.	40	53	71	34	47	104	120	19	57	57	79	110
	Q.S.	27	9	40	27	17	51	90	12	50	28	19	50
			13[a]	31[a]			23[a]	23[a]					31[a]
								7[c]					22[a]
													7[d]
	D.	27	22	94	27	17	74	120	12	50	28	19	110
	F.I.	13	31	-23	7	30	30	0	7	7	29	60	0

Abbreviations: B.I.: Beginning inventory, Q.S.: Quantity sold, D.: Demand, F.I.: Final inventory.

[a,b,c,d] Represents the age of product which are equal to one, two, three and four periods, respectively.

Table 17.16 presents the optimal values of cost components for the base and ignored product deterioration settings. The results reveal that taking product deterioration into account enables to increase total profit by nearly 1%.

Table 17.17 illustrates the amounts of initial and final inventories, the amount of quantity sold, and the amount of demand for each product in each period when the product deterioration is ignored. The table shows that ignoring product deterioration results in a higher usage of aged products and, correspondingly, a higher number of deteriorated products and lower sustainability performance and profits.

TABLE 17.16

Optimal Values of Cost Components for the Base Case and Ignored Deterioration Settings

	Base Setting	Ignored Deterioration Setting
Revenue	65046	63958 (65749[a])
Production Cost	3005.2	3167.6
Setup Cost	2120	1270
Holding Cost	206	226
Shortage Cost	415	470
Waste Cost	0	0
Profit	59299.8	58824.4

[a] The value in parenthesis is how much the revenue would be if the deterioration did not exist.

TABLE 17.17

Detailed Optimal Solution under Limited Flexibility Design with No Product Deterioration

Products		1	2	3	4	5	6	7	8	9	10	11	12
							Periods						
1	B.I.	60	70	50	50	50	50	60	90	80	70	60	50
	Q.S.	50	50	30	50	50	50	50	40	10	20	30	40
				10[a]	20[a]				10[a]	40[a]	30[a]	20[a]	10[a]
	D.	50	50	50	50	50	50	50	50	50	50	50	50
	F.I.	10	20	0	0	0	0	10	40	30	20	10	0
2	B.I.	60	60	90	72	69	64	31	17	12	17	32	50
	Q.S.	60	60	88	70	69	50	17	17	12	17	32	50
					2[a]			14[a]					
	D.	68	83	88	83	69	50	31	17	12	17	32	50
	F.I.	−8	−23	2	−11	0	14	0	0	0	0	0	0
3	B.I.	0	0	16	29	42	37	43	59	70	91	100	111
	Q.S.	0	0	16	28	33	29	43	59	70	91	100	110
						1[a]	8[a]						1[a]
	D.	3	9	16	28	34	37	43	59	70	91	99	111
	F.I.	−3	−9	0	1	8	0	0	0	0	0	1	0
4	B.I.	100	100	85	70	59	43	37	50	56	28	12	3
	Q.S.	100	99	84	70	59	43	37	34	12	16[a]	9[b]	3[c]
				1[a]						16[a]			
	D.	111	99	91	70	59	43	37	34	28	16	9	3
	F.I.	−11	1	−6	0	0	0	0	16	28	12	3	0

(*Continued*)

TABLE 17.17 (*Continued*)

Detailed Optimal Solution under Limited Flexibility Design with No Product Deterioration

Products		1	2	3	4	5	6	7	8	9	10	11	12
							Periods						
5	B.I.	40	54	72	40	53	104	120	19	57	57	79	110
	Q.S.	27	22	40	27	4	68	90	12	43	21	19[a]	50
				18[a]		13[a]	6[a]	30[b]		7[a]	7[a]		50[a]
				13[b]									10[b]
	D.	27	22	94	27	17	74	120	12	50	28	19	110
	F.I.	13	32	−23	13	36	30	0	7	7	29	60	0

Abbreviations: B.I.: Beginning inventory, Q.S.: Quantity sold, D.: Demand, F.I.: Final inventory.

[a,b,c] Represents the age of product which are equal to one, two and three periods, respectively.

17.5 CONCLUSION

This study has modeled and analyzed the production and inventory control problem under perishability, process flexibility, and sequence-dependent setup assumptions. To the best of our knowledge, the model is unique in (i) incorporating perishability in process flexibility problems and (ii) taking the sequence of production into account, which directly affects total setup costs. The proposed model can be used to aid decision-making processes in both tactical and operational level planning of production systems with perishable inventory (e.g., food production, electronics, etc.).

The added value of the proposed model based on hypothetical data has been illustrated in the numerical study. According to the results, investing in process flexibility for a production system may considerably reduce both holding and shortage costs. The numerical study also indicates that (i) product changeovers occur to satisfy the demands and reduce costs in limited and full flexibility designs despite the sequence-dependent setup costs and varying unit production costs, and (ii) process flexibility has a positive effect on reducing the amount of inventory.

One possible extension is to incorporate stochasticity (e.g., stochastic demand) in the considered problem. Note, when the demand for perishable products is uncertain, product wastes will be inevitable. Another possible extension is to apply traditional inventory policies such as FIFO and LIFO into the considered problem.

REFERENCES

Abad, P. L. (2000). Optimal lot size for a perishable good under conditions of finite production and partial backordering and lost sale. *Computers & Industrial Engineering, 38*(4), 457–465. doi:10.1016/S0360-8352(00)00057-7.

Ahumada, O., & Villalobos, J. R. (2009). Application of planning models in the agri-food supply chain: A review. *European Journal of Operational Research, 196*(1), 1–20. doi:10.1016/j.ejor.2008.02.014.

Andreou, S. A. (1990). A capital budgeting model for product-mix flexibility. *Journal of Manufacturing and Operations Management, 3*(1), 5–23.

Aprile, D., Garavelli, A. C., & Giannoccaro, I. (2005). Operations planning and flexibility in a supply chain. *Production Planning & Control, 16*(1), 21–31. doi:10.1080/095372804 12331313348.

Bakker, M., Riezebos, J., & Teunter, R. H. (2012). Review of inventory systems with deterioration since 2001. *European Journal of Operational Research, 221*(2), 275–284. doi:10.1016/j.ejor.2012.03.004.

Browne, J., Dubois, D., Rathmill, K., Sethi, S. P., & Stecke, K. E. (1984). Classification of flexible manufacturing systems. *The FMS Magazine, 2*(2), 114–117.

Cheong, T. (2013). Joint inventory and transshipment control for perishable products of a two-period lifetime. *The International Journal of Advanced Manufacturing Technology, 66*(9–12), 1327–1341. doi:10.1007/s00170-012-4411-x.

Chopra, S., & Meindl, P. (2015). *Supply Chain Management: Strategy, Planning, and Operation.* Harlow, UK: Pearson.

Chou, M. C., Chua, G. A., & Teo, C.-P. (2010). On range and response: Dimensions of process flexibility. *European Journal of Operational Research, 207*(2), 711–724. doi:10.1016/j.ejor.2010.05.038.

Chou, M. C., Teo, C. P., & Zheng, H. (2008). Process flexibility: Design, evaluation, and applications. *Flexible Services and Manufacturing Journal, 20*(1–2), 59–94.

Çimen, M., Belbağ, S., & Soysal, M. (2016). Üretimde Esneklik ve Stok Yönetimi: Stok Optimizasyonu İçin Bir Karar Destek Modeli (Production Flexibility and Inventory Management: A Decision Support Model For Inventory Optimization). *İşletme Araştırmaları Dergisi, 8*(1), 360–379.

Çimen, M., & Kirkbride, C. (2017). Approximate dynamic programming algorithms for multidimensional flexible production-inventory problems. *International Journal of Production Research, 55*(7), 2034–2050. doi:10.1080/00207543.2016.1264643.

Fine, C. H., & Freund, R. M. (1990). Optimal investment in product-flexible manufacturing capacity. *Management Science, 36*(4), 449–466. doi:10.1287/mnsc.36.4.449.

Goyal, S., & Giri, B. C. (2001). Recent trends in modeling of deteriorating inventory. *European Journal of Operational Research, 134*(1), 1–16. doi:10.1016/S0377-2217(00)00248-4.

Hwang, H., & Hahn, K. H. (2000). An optimal procurement policy for items with an inventory level dependent demand rate and fixed lifetime. *European Journal of Operational Research, 127*(3), 537–545. doi:10.1016/S0377-2217(99)00337-9.

Iravani, S. M., Kolfal, B., & Van Oyen, M. P. (2014). Process flexibility and inventory flexibility via product substitution. *Flexible Services and Manufacturing Journal, 26*(3), 320–343.

Jaikumar, R. (1984). *Flexible Manufacturing Systems: A Managerial Perspective.* Boston, MA: Division of Research, Graduate School of Business Administration, Harvard University.

Janssen, L., Claus, T., & Sauer, J. (2016). Literature review of deteriorating inventory models by key topics from 2012 to 2015. *International Journal of Production Economics, 182*, 86–112. doi:10.1016/j.ijpe.2016.08.019.

Jordan, W. C., & Graves, S. C. (1995). Principles on the benefits of manufacturing process flexibility. *Management Science, 41*(4), 577–594. doi:10.1287/mnsc.41.4.577.

Mendoza, A., & Ventura, J. A. (2010). A serial inventory system with supplier selection and order quantity allocation. *European Journal of Operational Research, 207*(3), 1304–1315. doi:10.1016/j.ejor.2010.06.034.

Nahmias, S. (1982). Perishable inventory theory: A review. *Operations Research, 30*(4), 680–708. doi:10.1287/opre.30.4.680.

Nemhauser, G. L., & Wolsey, L. A. (1988). *Integer and Combinatorial Optimization. Interscience Series in Discrete Mathematics and Optimization.* Hoboken, NJ: John Wiley & Sons.

Olsson, F., & Tydesjö, P. (2010). Inventory problems with perishable items: Fixed lifetimes and back logging. *European Journal of Operational Research, 202*(1), 131–137. doi:10.1016/j.ejor.2009.05.010.

Sethi, A. K., & Sethi, S. P. (1990). Flexibility in manufacturing: A survey. *International Journal of Flexible Manufacturing Systems, 2*(4), 289–328. doi:10.1007/BF00186471.

Soysal, M., & Çimen, M. (2017). Süreç esnekliği tasarımlarının farklı talep yapıları altında envanter maliyeti performansları üzerine. *Finans Politik & Ekonomik Yorumlar, 54*(633), 79–91.

Soysal, M., Çimen, M., & Belbağ, S. (2018). Sabit Üretim Hazirlik Maliyetinin Süreç Esnekliğinde Stok Optimizasyonuna Etkisi. *Hacettepe Üniversitesi İktisadi ve İdari Bilimler Fakültesi Dergisi, 36*(1), 117–139.

Stadtler, H., & Kilger, C. (2002). *Supply Chain Management and Advanced Planning: Concepts, Models, Software and Case Studies.* Berlin, Germany: Springer.

18 Sustainability of Advance Machine Learning Algorithms in Manufacturing

Somvir Singh Nain

CONTENTS

18.1 INTRODUCTION

The advance machining process WEDM has been used enormously in the aeronautics and automobile industries. The process of material-removing in Wire-Cut Electric Discharge Machining (WEDM) relies upon the amount of thermal energy created between the wire electrodes and work material. The brass wire of diameter 0.05 mm to 0.3 mm is employed for the cutting process. The electrical conductivity of the work material is the prime requisite condition for the WEDM process.

Distinct approaches like simple regression [1–4], fuzzy-logic [5,6], artificial neural networking (ANN) [7–13], artificial fuzzy inference system (ANFIS) [14,15], Taguchi integrated with utility concept and GREY relational and PSO approaches [16–19] have been practiced by different researchers to evaluate the WEDM process. Recently, a qualitative work related to an influence of polarity on the EDM performance has been reported [20]. Recently, some work has been reported on the use of a support vector machine algorithm to evaluate the WEDM process [21–23].

The use of advanced algorithms like Gaussian process, random forest, and M5P tree has been used in the manufacturing and other fields also [24–27]. The literature study indicates that rare research work has been published on use of these advanced machine learning algorithms and to examine the sustainability of these algorithms in manufacturing. A high-quality work published on sustainability in supply chain management also inspired me to work on the sustainability of the advanced machine learning algorithm [28,29].

Therefore, the current study has been made to check the applicability and sustainability of an advanced machine algorithm like support vector machines using different kernels, Gaussian process using different kernels, M5P tree, and random forest in manufacturing, which were hardly being used before this study.

18.2 EXPERIMENTAL DETAILS

First, the experimental work has been planned based on the Taguchi approach using the L27 orthogonal array. Total 27 experiments have been performed on WEDM and repeated three times. Six process variables and two response variables have been taken to evaluate the WEDM process. Thereupon, the sustainability of different machine learning algorithms is checked to get the better prediction outcome for the WEDM process. The surface roughness is checked by using surface contact profilometer, and Wire Wear Ratio (WWR) is calculated as [23]:

$$WWR = \frac{WWB - WWA}{WWB} \qquad (18.1)$$

where WWB indicates the weight of the wire before machining, and WWA is the weight of the wire after machining.

18.3 MODELING OF THE PROCESS

The predicted result for the experimental outcome of the Surface Roughness (SR) and WWR are calculated using Support Vector Machine (SVM), Gaussian Process (GP), random forest, and M5P tree with prune and unpruned methods. The evaluation parameters, like correlation coefficient (R) and root mean square error (RMSE), are used to check the importance of each model for the training and testing set of SR and WWR. Out of 81 observations, data is separated as train set (66.67%) and test set (33.33%). Seven models are formulated for each response parameter and the performance of each model is compared for the training set of SR and WWR. The proficiency of each model for each response characteristic is checked on the test data set. The graphs for SR and WWR have been portrayed among the experimental and anticipated outcome for the train and test data sets. Thereupon, the validation and error graphs for SR and WWR have been plotted. The predicted outcome for all the models for the training and testing of SR and WWR is illustrated in Tables 18.1 and 18.2, respectively.

TABLE 18.1

Prediction Outcome of the Entire Models for the Training Set of SR and WWR

Sr. No.	Ton (µS)	Toff (µS)	IP (A)	WT (gm)	SV (V)	WF mmin⁻¹	Act. SR	SVM Poly	SVM RBF	GP Poly	GP RBF	SR by RF	SR by M5P Prune	SR by M5P Unprune	Act. WWR	SVM Poly	SVM RBF	GP Poly	GP RBF	RF	M5P Prun	M5P Unprun
1	0.4	09	130	1020	36	06	1.925	1.991	2.111	2.131	2.000	1.972	1.911	1.911	0.0802	0.081	0.082	0.093	0.090	0.082	0.070	0.070
2	0.4	09	160	1260	58	08	1.796	1.889	2.035	1.751	1.857	1.823	1.796	1.796	0.0839	0.076	0.084	0.077	0.090	0.082	0.071	0.071
3	0.4	13	130	1260	58	10	1.800	1.804	2.049	1.690	1.835	1.812	1.697	1.697	0.0309	0.056	0.073	0.058	0.087	0.033	0.059	0.059
4	0.4	13	160	1500	80	06	1.272	1.755	2.063	1.495	1.348	1.320	1.502	1.502	0.0607	0.08	0.089	0.076	0.091	0.061	0.062	0.062
5	0.4	22	130	1500	80	08	0.908	1.642	2.075	1.352	0.954	0.944	1.225	1.225	0.0603	0.061	0.078	0.053	0.087	0.060	0.062	0.062
6	0.4	22	160	1020	36	10	1.570	1.925	2.001	1.724	1.573	1.558	1.549	1.549	0.0733	0.073	0.076	0.068	0.089	0.073	0.072	0.072
7	0.8	09	130	1260	80	08	2.294	2.068	2.239	2.084	2.268	2.377	2.333	2.334	0.0927	0.078	0.091	0.084	0.092	0.094	0.097	0.097
8	0.8	09	160	1500	36	10	2.452	2.439	2.277	2.251	2.375	2.448	2.352	2.353	0.0948	0.095	0.096	0.100	0.094	0.095	0.102	0.102
9	0.8	13	130	1500	36	06	2.459	2.407	2.383	2.376	2.425	2.445	2.373	2.371	0.1143	0.104	0.100	0.114	0.095	0.115	0.101	0.101
10	0.8	13	160	1020	58	08	2.369	2.217	2.193	2.201	2.271	2.357	2.240	2.240	0.1154	0.094	0.094	0.098	0.094	0.114	0.100	0.100
11	0.8	22	130	1020	58	10	2.372	2.104	2.205	2.058	2.289	2.371	2.240	2.240	0.0748	0.075	0.082	0.076	0.09	0.076	0.089	0.089
12	0.8	22	160	1260	80	06	2.057	2.054	2.220	1.863	2.095	2.131	2.190	2.190	0.0988	0.099	0.099	0.093	0.094	0.100	0.096	0.096
13	1.2	09	130	1500	58	10	2.478	2.617	2.480	2.584	2.573	2.556	2.606	2.605	0.1281	0.097	0.103	0.107	0.096	0.126	0.121	0.121
14	1.2	09	160	1020	80	06	2.483	2.48	2.384	2.595	2.473	2.440	2.479	2.479	0.1158	0.116	0.112	0.124	0.098	0.114	0.113	0.113
15	1.2	13	130	1020	80	08	2.221	2.396	2.396	2.535	2.385	2.300	2.463	2.463	0.0963	0.096	0.101	0.105	0.095	0.096	0.107	0.107
16	1.2	13	160	1260	36	10	2.193	2.766	2.433	2.701	2.327	2.242	2.414	2.414	0.1097	0.113	0.107	0.121	0.098	0.110	0.116	0.116
17	1.2	22	130	1260	36	06	2.682	2.706	2.538	2.743	2.691	2.666	2.544	2.544	0.1219	0.123	0.109	0.132	0.098	0.122	0.116	0.116
18	1.2	22	160	1500	58	08	2.618	2.604	2.461	2.363	2.627	2.652	2.549	2.550	0.1217	0.118	0.111	0.116	0.098	0.122	0.123	0.123
19	0.4	09	130	1020	36	06	1.993	1.991	2.111	2.131	2.000	1.972	1.911	1.911	0.0811	0.081	0.082	0.093	0.09	0.082	0.070	0.070

(Continued)

TABLE 18.1 (Continued)
Prediction Outcome of the Entire Models for the Training Set of SR and WWR

Sr. No.	Ton (μS)	Toff (μS)	IP (A)	WT (gm)	SV (V)	WF mmin⁻¹	Act. SR	SVM Poly	SVMRBF	GP Poly	GP RBF	SR by RF	SR by M5P Prune	SR by M5P Unprune	Act. WWR	SVMPoly	SVMRBF	GPPoly	GPRBF	RF	M5P Prun	M5P Unprun
20	0.4	09	160	1260	58	08	1.769	1.889	2.035	1.751	1.857	1.823	1.796	1.796	0.0822	0.076	0.084	0.077	0.09	0.082	0.071	0.071
21	0.4	13	130	1260	58	10	1.820	1.804	2.049	1.690	1.835	1.812	1.697	1.697	0.0317	0.056	0.073	0.058	0.087	0.033	0.059	0.059
22	0.4	13	160	1500	80	06	1.327	1.755	2.063	1.495	1.348	1.320	1.502	1.502	0.0609	0.080	0.089	0.076	0.091	0.061	0.062	0.062
23	0.4	22	130	1500	80	08	0.905	1.642	2.075	1.352	0.954	0.944	1.225	1.225	0.0608	0.061	0.078	0.053	0.087	0.060	0.062	0.062
24	0.4	22	160	1020	36	10	1.487	1.925	2.001	1.724	1.573	1.558	1.549	1.549	0.0757	0.073	0.076	0.068	0.089	0.073	0.072	0.072
25	0.8	09	130	1260	80	08	2.362	2.068	2.239	2.084	2.268	2.377	2.333	2.334	0.0945	0.078	0.091	0.084	0.092	0.094	0.097	0.097
26	0.8	09	160	1500	36	10	2.440	2.439	2.277	2.251	2.375	2.448	2.352	2.353	0.0939	0.095	0.096	0.100	0.094	0.095	0.102	0.102
27	0.8	13	130	1500	36	06	2.406	2.407	2.383	2.376	2.425	2.445	2.373	2.371	0.1146	0.104	0.100	0.114	0.095	0.115	0.101	0.101
28	0.8	13	160	1020	58	08	2.357	2.217	2.193	2.201	2.271	2.357	2.240	2.240	0.1154	0.094	0.094	0.098	0.094	0.114	0.100	0.100
29	0.8	22	130	1020	58	10	2.291	2.104	2.205	2.058	2.289	2.371	2.240	2.240	0.0753	0.075	0.082	0.076	0.09	0.076	0.089	0.089
30	0.8	22	160	1260	80	06	2.222	2.054	2.220	1.863	2.095	2.131	2.190	2.190	0.0993	0.099	0.099	0.093	0.094	0.100	0.096	0.096
31	1.2	09	130	1500	58	10	2.546	2.617	2.480	2.584	2.573	2.556	2.606	2.605	0.1273	0.097	0.103	0.107	0.096	0.126	0.121	0.121
32	1.2	09	160	1020	80	06	2.410	2.48	2.384	2.595	2.473	2.440	2.479	2.479	0.1147	0.116	0.112	0.124	0.098	0.114	0.113	0.113
33	1.2	13	130	1020	80	08	2.355	2.396	2.396	2.535	2.385	2.300	2.463	2.463	0.0959	0.096	0.101	0.105	0.095	0.096	0.107	0.107
34	1.2	13	160	1260	36	10	2.299	2.766	2.433	2.701	2.327	2.242	2.414	2.414	0.1105	0.113	0.107	0.121	0.098	0.110	0.116	0.116
35	1.2	22	130	1260	36	06	2.706	2.706	2.538	2.743	2.691	2.666	2.544	2.544	0.1226	0.123	0.109	0.132	0.098	0.122	0.116	0.116
36	1.2	22	160	1500	58	08	2.691	2.604	2.461	2.363	2.627	2.652	2.549	2.550	0.1209	0.118	0.111	0.116	0.098	0.122	0.123	0.123

(Continued)

TABLE 18.1 (Continued)
Prediction Outcome of the Entire Models for the Training Set of SR and WWR

Sr. No.	Ton (µS)	Toff (µS)	IP (A)	WT (gm)	SV (V)	WF mmin⁻¹	Act. SR	SVM Poly	SVMRBF	GP Poly	GP RBF	SR by RF	SR by M5P Prune	SR by M5P Unprune	Act. WWR	SVMPoly	SVMRBF	GPPoly	GPRBF	RF	M5P Prun	M5P Unprun
37	0.4	09	130	1020	36	06	2.033	1.991	2.111	2.131	2.000	1.972	1.911	1.911	0.0814	0.081	0.082	0.093	0.09	0.082	0.070	0.070
38	0.4	09	160	1260	58	08	1.893	1.889	2.035	1.751	1.857	1.823	1.796	1.796	0.0823	0.076	0.084	0.077	0.09	0.082	0.071	0.071
39	0.4	13	130	1260	58	10	1.809	1.804	2.049	1.690	1.835	1.812	1.697	1.697	0.0319	0.056	0.073	0.058	0.087	0.033	0.059	0.059
40	0.4	13	160	1500	80	06	1.337	1.755	2.063	1.495	1.348	1.320	1.502	1.502	0.062	0.08	0.089	0.076	0.091	0.061	0.062	0.062
41	0.4	22	130	1500	80	08	0.903	1.642	2.075	1.352	0.954	0.944	1.225	1.225	0.0592	0.061	0.078	0.053	0.087	0.060	0.062	0.062
42	0.4	22	160	1020	36	10	1.565	1.925	2.001	1.724	1.573	1.558	1.549	1.549	0.073	0.073	0.076	0.068	0.089	0.073	0.072	0.072
43	0.8	09	130	1260	80	08	2.502	2.068	2.239	2.084	2.268	2.377	2.333	2.334	0.0942	0.078	0.091	0.084	0.092	0.094	0.097	0.097
44	0.8	09	160	1500	36	10	2.445	2.439	2.277	2.251	2.375	2.448	2.352	2.353	0.0945	0.095	0.096	0.100	0.094	0.095	0.102	0.102
45	0.8	13	130	1500	36	06	2.459	2.407	2.383	2.376	2.425	2.445	2.373	2.371	0.1146	0.104	0.100	0.114	0.095	0.115	0.101	0.101
46	0.8	13	160	1020	58	08	2.360	2.217	2.193	2.201	2.271	2.357	2.240	2.240	0.1151	0.094	0.094	0.098	0.094	0.114	0.100	0.100
47	0.8	22	130	1020	58	10	2.420	2.104	2.205	2.058	2.289	2.371	2.240	2.240	0.0749	0.075	0.082	0.076	0.090	0.076	0.089	0.089
48	0.8	22	160	1260	80	06	2.066	2.054	2.220	1.863	2.095	2.131	2.190	2.190	0.0995	0.099	0.099	0.093	0.094	0.100	0.096	0.096
49	1.2	09	130	1500	58	10	2.619	2.617	2.480	2.584	2.573	2.556	2.606	2.605	0.1271	0.097	0.103	0.107	0.096	0.126	0.121	0.121
50	1.2	09	160	1020	80	06	2.479	2.48	2.384	2.595	2.473	2.440	2.479	2.479	0.1145	0.116	0.112	0.124	0.098	0.114	0.113	0.113
51	1.2	13	130	1020	80	08	2.306	2.396	2.396	2.535	2.385	2.300	2.463	2.463	0.0958	0.096	0.101	0.105	0.095	0.096	0.107	0.107
52	1.2	13	160	1260	36	10	2.214	2.766	2.433	2.701	2.327	2.242	2.414	2.414	0.1095	0.113	0.107	0.121	0.098	0.110	0.116	0.116
53	1.2	22	130	1260	36	06	2.686	2.706	2.538	2.743	2.691	2.666	2.544	2.544	0.1227	0.123	0.109	0.132	0.098	0.122	0.116	0.116
54	1.2	22	160	1500	58	08	2.661	2.604	2.461	2.363	2.627	2.652	2.549	2.550	0.1228	0.118	0.111	0.116	0.098	0.122	0.123	0.123

TABLE 18.2
Prediction Outcome of the Entire Models for the Testing Set of SR and WWR

Ton (μS)	Toff (μS)	IP (A)	WT (gm)	SV (V)	WF mmin⁻¹	Actual SR	SVMPoly	SVMRBF	GPPoly	GPRBF	PrunSR	UnprunSR	RF	Act. WWR	SVMPoly	SVM RBF	GPPoly	GP RBF	M5P. UPF	M5P. UNT	RF
0.4	9	190	1500	80	10	1.531	1.787	1.967	1.37	1.681	1.625	1.621	1.316	0.0799	0.071	0.086	0.061	0.091	0.071	0.071	0.075
0.4	13	190	1020	36	08	2.074	2.038	1.993	1.868	1.543	1.812	1.812	1.905	0.0777	0.093	0.087	0.091	0.093	0.069	0.069	0.074
0.4	22	190	1260	58	06	1.589	1.876	2.019	1.529	1.523	1.446	1.446	1.600	0.0789	0.097	0.092	0.086	0.092	0.073	0.073	0.067
0.8	9	190	1020	58	06	2.476	2.302	2.182	2.262	2.294	2.126	2.137	2.368	0.0995	0.114	0.105	0.117	0.097	0.099	0.099	0.108
0.8	13	190	1260	80	10	2.379	2.115	2.121	1.821	2.178	2.191	2.191	2.282	0.1159	0.089	0.096	0.082	0.094	0.106	0.106	0.094
0.8	22	190	1500	36	08	2.075	2.425	2.256	2.029	2.245	2.295	2.288	2.473	0.1179	0.116	0.104	0.109	0.096	0.103	0.103	0.110
1.2	9	190	1260	36	08	2.327	2.851	2.416	2.762	2.535	2.392	2.392	2.338	0.1339	0.133	0.117	0.140	0.100	0.122	0.122	0.110
1.2	13	190	1500	58	06	2.515	2.717	2.445	2.506	2.982	2.539	2.538	2.505	0.1211	0.137	0.122	0.138	0.101	0.126	0.126	0.118
1.2	22	190	1020	80	10	2.673	2.414	2.273	2.188	2.599	2.298	2.298	2.349	0.0841	0.108	0.106	0.100	0.097	0.116	0.116	0.101
0.4	9	190	1500	80	10	1.605	1.787	1.967	1.37	1.681	1.625	1.621	1.316	0.0795	0.071	0.086	0.061	0.091	0.071	0.071	0.075
0.4	13	190	1020	36	08	1.993	2.038	1.993	1.868	1.543	1.812	1.812	1.905	0.0783	0.093	0.087	0.091	0.093	0.069	0.069	0.074
0.4	22	190	1260	58	06	1.487	1.876	2.019	1.529	1.523	1.446	1.446	1.600	0.0796	0.097	0.092	0.086	0.092	0.073	0.073	0.067
0.8	9	190	1020	58	06	2.527	2.302	2.182	2.262	2.294	2.126	2.137	2.368	0.1001	0.114	0.105	0.117	0.097	0.099	0.099	0.108

(Continued)

TABLE 18.2 (Continued)

Prediction Outcome of the Entire Models for the Testing Set of SR and WWR

Ton (µS)	Toff (µS)	IP (A)	WT (gm)	SV (V)	WF mmin⁻¹	Actual SR	SVMPoly	SVMRBF	GPPoly	GPRBF	PrunSR	UnprunSR	RF	Act. WWR	SVMPoly	SVM RBF	GPPoly	GP RBF	M5P. UPF	M5P. UNT	RF
0.8	13	190	1260	80	10	2.395	2.115	2.121	1.821	2.178	2.191	2.191	2.282	0.1161	0.089	0.096	0.082	0.094	0.106	0.106	0.094
0.8	22	190	1500	36	08	2.193	2.425	2.256	2.029	2.245	2.295	2.288	2.473	0.1176	0.116	0.104	0.109	0.096	0.103	0.103	0.110
1.2	9	190	1260	36	08	2.244	2.851	2.416	2.762	2.535	2.392	2.392	2.338	0.1347	0.133	0.117	0.140	0.100	0.122	0.122	0.110
1.2	13	190	1500	58	06	2.478	2.717	2.445	2.506	2.982	2.539	2.538	2.505	0.1203	0.137	0.122	0.138	0.101	0.126	0.126	0.118
1.2	22	190	1020	80	10	2.69	2.414	2.273	2.188	2.599	2.298	2.298	2.349	0.0842	0.108	0.106	0.100	0.097	0.116	0.116	0.101
0.4	9	190	1500	80	10	1.549	1.787	1.967	1.37	1.681	1.625	1.621	1.316	0.0800	0.071	0.086	0.061	0.091	0.071	0.071	0.075
0.4	13	190	1020	36	08	2.224	2.038	1.993	1.868	1.543	1.812	1.812	1.905	0.0786	0.093	0.087	0.091	0.093	0.069	0.069	0.074
0.4	22	190	1260	58	06	1.513	1.876	2.019	1.529	1.523	1.446	1.446	1.600	0.0788	0.097	0.092	0.086	0.092	0.073	0.073	0.067
0.8	9	190	1020	58	06	2.453	2.302	2.182	2.262	2.294	2.126	2.137	2.368	0.0989	0.114	0.105	0.117	0.097	0.099	0.099	0.108
0.8	13	190	1260	80	10	2.426	2.115	2.121	1.821	2.178	2.191	2.191	2.282	0.1154	0.089	0.096	0.082	0.094	0.106	0.106	0.094
0.8	22	190	1500	36	08	2.077	2.425	2.256	2.029	2.245	2.295	2.288	2.473	0.117	0.116	0.104	0.109	0.096	0.103	0.103	0.110
1.2	9	190	1260	36	08	2.313	2.851	2.416	2.762	2.535	2.392	2.392	2.338	0.1334	0.133	0.117	0.140	0.100	0.122	0.122	0.110
1.2	13	190	1500	58	08	2.526	2.717	2.445	2.506	2.982	2.539	2.538	2.505	0.1204	0.137	0.122	0.138	0.101	0.126	0.126	0.118
1.2	22	190	1020	80	10	2.605	2.414	2.273	2.188	2.599	2.298	2.298	2.349	0.0837	0.108	0.106	0.100	0.097	0.116	0.116	0.101

18.3.1 SUPPORT VECTOR MACHINE (SVM)

SVM is a classification and regression algorithms rely upon the optimal classification of classes. It prefers the linear classifier among the infinite number of classifiers to minimize the generalization error or to establish the upper bound on error. It entails the structure risk minimization. The hyperplane that has the highest margin amid two classes has been selected. Support Vector Regression (SVR) relies upon the insensitive loss function to permit the notion of margin. The prime intention of SVR is to come across a function that is smooth and has the largest variation from the target vector for all the training data. The general regression function can be written as:

$$f(x) = \sum_{i=1}^{p} \left(\alpha_i' - \alpha_i\right) K\langle x_i, x_j \rangle + b' \qquad (18.2)$$

where b represents the offset of discriminating plane with respect to the origin. Distinct kernels can be used for creating the SVM models. We preferred the polynomial and radial basis kernels for this proposed work. They are given as:

$$RBF = K\left(x, x'\right) = e^{\gamma|x-x'|^2} \qquad (18.3)$$

$$Polynomial\ kernel = K\left(x, x'\right) = \left(x, x'\right) = \left(\left(x, x'\right) + 1\right)^{d^*} \qquad (18.4)$$

For complete details of the SVM process, reviewers are suggested to read the articles [21,22].

18.3.2 GAUSSIAN PROCESS (GP)

Gaussian process is the assortment of a number of random variables—any finite number of which have a combined Gaussian distribution, which is given by the mean function $p(x)$ and covariance function $q(x, x')$ [30]. Literally, the mean function is considered as zero function; accordingly, the Gaussian process is given as $f(x) \sim GP\ [0, q(x, x')]$. GP is the process of creating a model to get the prediction result for the response variables dependent upon the number of process variables and their different values. The GP model can be shown as:

$$y_p = \emptyset(X_i)^T w + \in_n \qquad (18.5)$$

where y_p represents the response weight tardiness for the pth sequence, X_i is the column vector for the ith sequence of the inputs set, w represents the vector weight of the variables, and $\in_n$ represents the Gaussian noise distributed as $\in_n \sim Normal\ (0, \sigma_n^2)$ where σ_n^2 is the noise variance for n observations. For complete details of the GP regression and function, see "Gaussian process models for robust regression, classification, and reinforcement learning" [31]. The kernels used in present work are polynomial and radial basis kernels, which are shown in Equations (18.3) and (18.4).

18.3.3   RANDOM FOREST (RF)

The random forest algorithm is used to create a model that consists of a group of many trees. Each tree represents a particular classification and vote of the classification. The forest selects the classification, which has the highest voting in the forest. The tree is mature if N is the number of cases in the training set—N cases at random with replacements from original data may be the input data set to mature the tree. The user defined variables (m) are selected randomly out of total input variables (K) for the best split and number of m variables should be less than K. The m variables are given some constant value. The Tree is developed without pruning up to the largest extent. RF can work efficiently and accurately with large complex data having many thousands of input variables. For details of the RF approach, read the article [27].

18.3.4   M5P

M5P algorithm is used to develop a decision tree by engaging the linear regression function approach at nodes with an aim of constructing a model that proposes a correlation among the target values of the training cases and the values of input attributes. The splitting approach is applied at each node to gain the maximum information to minimize the variation in the intra subset class value down to each branch. The splitting process will be converged when there are diminutive variations among the class values of the instances or only few instances left or when the tree is pruned back. The developed tree shows very good structure and prediction accuracy due to showing more potential linearity at the leaf node.

The prune and unprune model structure with distinct numbers of rules has been developed using the M5P algorithm. The M5P tree with prune and unprune structure is developed with a different number of leaves. For example, Figures 18.1 and 18.2 show that prune trees for SR and WWR are matured with 6 and 8 leaves or rules, respectively.

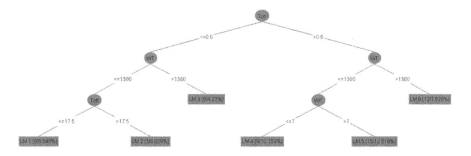

FIGURE 18.1   M5P prune tree model of SR.

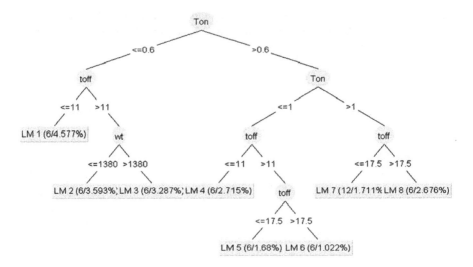

FIGURE 18.2   M5P prune tree model of WWR.

Similarly, Figures 18.3 and 18.4 show that unprune trees for SR and WWR are matured with 17 and 13 leaves or rules, respectively. To better understand the M5P tree, see "Learning with continuous classes" [32]. Tables 18.3 and 18.4 show the number of rules considered to mature the prune and unprune decision tree for SR and WWR.

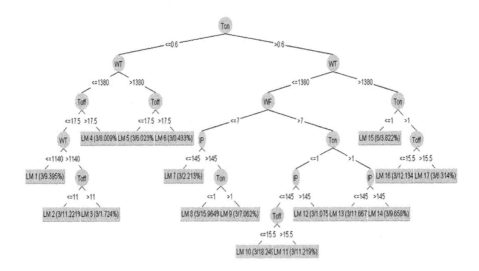

FIGURE 18.3   M5P unprune tree model of SR.

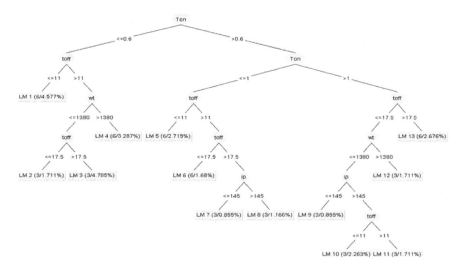

FIGURE 18.4   M5P unprune tree model of WWR.

TABLE 18.3

Splitting Rules Considered in M5P Prune Tree Model for SR and WWR

Sr. No.	M5P prune tree rules for SR
LM_1	SR=0.5219 * Ton – 0.0248 * Toff – 0.0005 * WT + 2.4125
LM_2	SR=0.5219 * Ton – 0.0264 * Toff – 0.0004 * WT + 2.3616
LM_3	SR=0.5219 * Ton – 0.0308 * Toff – 0.0005 * WT + 2.3841
LM_4	SR=0.6298 * Ton – 0.0045 * Toff – 0.0034 * IP + 0.0001 * WT – 0.0093 * WF + 2.2711
LM_5	SR=0.4203 * Ton – 0.0045 * Toff – 0.0018 * IP + 0.0001 * WT – 0.0075 * WF + 2.2239
LM_6	SR=0.5615 * Ton – 0.0021 * Toff – 0.001 * IP + 0.0001 * WT + 1.889

Sr. No.	M5P prune tree rules for WWR
LM_1	WWR=0.0286 * Ton +0.0001IP+0.0572
LM_2	WWR=0.0286 * Ton +0.0007Toff+0.0001IP+0.0357
LM_3	WWR=0.0286 * Ton +0.0004Toff+0.0001IP+0.042
LM_4	WWR=0.0319 * Ton – 0.0003Toff+0.0001IP+0.0516
LM_5	WWR=0.0319 * Ton – 0.0006Toff+0.0001IP+0.0515
LM_6	WWR=0.0319 * Ton – 0.0006Toff+0.0002IP+0.0444
LM_7	WWR=0.0319 * Ton – 0.0006Toff+0.0001IP+0.0381
LM_8	WWR=0.0319 * Ton – 0.0001IP+0.0415

TABLE 18.4
Splitting Rules Considered in M5P Unpruned Tree Modeling for SR and WWR

Sr. No.	M5P unpruned tree rules for SR
LM$_1$	SR=0.5219 * Ton − 0.0248 * Toff − 0.0005 * WT + 2.4016
LM$_2$	SR=0.5219 * Ton − 0.0248 * Toff − 0.0005 * WT + 2.3894
LM$_3$	SR=0.5219 * Ton − 0.0248 * Toff − 0.0005 * WT + 2.3894
LM$_4$	SR=0.5219 * Ton − 0.0264 * Toff − 0.0004 * WT + 2.3616
LM$_5$	SR=0.5219 * Ton − 0.0296 * Toff − 0.0005 * WT + 2.3689
LM$_6$	SR=0.5219 * Ton − 0.0296 * Toff − 0.0005 * WT + 2.3583
LM$_7$	SR=0.6065 * Ton − 0.0045 * Toff − 0.0032 * IP + 0.0001 * WT − 0.0093 * WF + 2.2714
LM$_8$	SR=0.6231 * Ton − 0.0045 * Toff − 0.003 * IP + 0.0001 * WT − 0.0093 * WF + 2.2183
LM$_9$	$SR = 0.6231$ * Ton − 0.0045 * Toff − 0.003 * IP + 0.0001 * WT − 0.0093 * WF + 2.221
LM$_{10}$	SR=0.4416 * Ton − 0.0045 * Toff − 0.0018 * IP + 0.0001 * WT − 0.0075 * WF + 2.207
LM$_{11}$	SR=0.4416 * Ton − 0.0045 * Toff − 0.0018 * IP + 0.0001 * WT − 0.0075 * WF + 2.2069
LM$_{12}$	SR=0.4416 * Ton − 0.0045 * Toff − 0.0018 * IP + 0.0001 * WT − 0.0075 * WF + 2.2067
LM$_{13}$	SR= 0.4365 * Ton − 0.0045 * Toff − 0.0018 * IP + 0.0001 * WT − 0.0075 * WF + 2.2047
LM$_{14}$	SR= 0.4365 * Ton − 0.0045 * Toff − 0.0018 * IP + 0.0001 * WT − 0.0075 * WF + 2.2041
LM$_{15}$	SR= 0.5336 * Ton − 0.0028 * Toff − 0.001 * IP + 0.0001 * WT + 1.9188
LM$_{16}$	SR= 0.5336 * Ton − 0.0021 * Toff − 0.001 * IP + 0.0001 * WT + 1.9225
LM$_{17}$	SR= 0.5336 * Ton − 0.0021 * Toff − 0.001 * IP + 0.0001 * WT + 1.9241

Sr. No.	M5P unpruned tree rules for WWR
LM$_1$	WWR=0.0286 * Ton + 0.0001 * IP - 0.000 * WT + 0.0572
LM$_2$	WWR=0.0286 * Ton + 0.0007 * Toff + 0.0001 * IP+0.0364
LM$_3$	WWR=0.0286* Ton + 0.0007 * Toff + 0.0001 * IP + 0.0369
LM$_4$	WWR= 0.0286 * Ton + 0.0004 * Toff + 0.0001 * IP - 0 * wt + 0.042
LM$_5$	WWR=0.0319 * Ton − 0.0003 * Toff + 0.0001 * IP + 0 * WT + 0.0516
LM$_6$	WWR=0.0319 * Ton − 0.0006 * Toff + 0.0001 * IP + 0 * WT + 0.0515
LM$_7$	WWR=0.0319 * Ton − 0.0006 * Toff + 0.0002 * IP + 0 * WT + 0.0453
LM$_8$	WWR=0.0319 * Ton - 0.0006 * toff + 0.0002 * ip + 0 * wt + 0.0455
LM$_9$	WWR=0.0319 * Ton − 0.0004 * Toff + 0.0001 * IP + 0 * WT + 0.0367
LM$_{10}$	WWR=0.0319 * Ton − 0.0004 *+ 0 * WT + 0.0375
LM$_{11}$	WWR=0.0319 * Ton − 0.0004 * Toff + 0.0001 * IP + 0 * WT + 0.0375
LM$_{12}$	WWR=0.0319 * Ton − 0.0005 * Toff + 0.0001 * IP + 0 * WT + 0.0394
LM$_{13}$	WWR=0.0319 * Ton + 0.0001 * IP + 0 * WT 0.0415

18.4   RESULT AND DISCUSSION

Figure 18.5a and b show the scattering plot for the predicted outcome of SR obtained by employing distinct modeling approaches like SVM using poly kernel and RBF kernel, GP with poly kernel and RBF kernel, M5P tree with prune or unprune

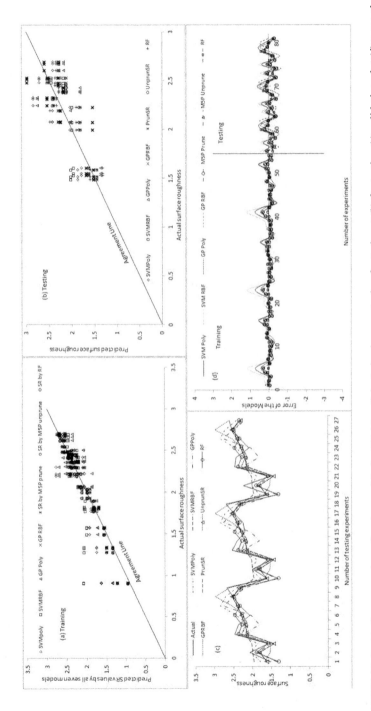

FIGURE 18.5 Performance plots of entire models for SR (a) anticipated values scattering for train set (b) for testing (c) validation plot (d) error plots of entire models.

methods, and RFfor the training and testing set. It has been observed that all the models play a significant role to get the better prediction result for the SR. But overall, the random forest model is more sustainable in comparison to the other models, as confirmed by Figure 18.5a and b.

The validation plot for SR also assures that the RF model is more sustainable against the other models, as shown in Figure 18.5c. Thereupon, the error plots have been plotted also prove the maximum sustainability of the RF model over the others, as shown in Figure 18.5d. The evaluation parameter performances of the models also reveal that the RF model is more sustainable among the others, as given in Table 18.5. The RF model has the maximum correlation coefficient and minimum RMSE value in comparison to the other models, as shown in Table 18.5.

Similarly, the sustainability of each model is checked for the evaluation of the WWR of the WEDM. The scattering plot for the anticipated outcome of each model for the train and test set of WWR is represented in Figure 18.6. Figure 18.6a and b show that the anticipated outcome obtained for the training and testing set are lying closely to the agreement line in comparison to the anticipated outcome derived by other models. The anticipated outcome path of each model is portrayed collectively, assuring that the RF model line travels the path

TABLE 18.5

Performance Evaluation of Entire Models for SR and WWR

SR Data Set

Model	Train set		Test Set	
	R	RMSE	R	RMSE
SVM Poly	0.8337	0.2809	0.6964	0.2959
SVM RBF	0.7956	0.3755	0.6811	0.2963
GP Poly	0.8622	0.2405	0.7472	0.2946
GP RBF	0.9891	0.072	0.8021	0.2849
M5P.UPF	0.9633	0.1376	0.8545	0.2250
M5P.UPT	0.9633	0.0612	0.8592	0.2223
RF	0.9955	0.0460	0.8751	0.2036

WWR Data Set

Model	Train set		Test Set	
	R	RMSE	R	RMSE
SVM Poly	0.8836	0.0125	0.7133	0.0178
SVM RBF	0.874	0.0158	0.77	0.0155
GP Poly	0.8881	0.0117	0.7251	0.0173
GP RBF	0.874	0.0158	0.7723	0.0157
M5P.UPF	0.9125	0.0105	0.7913	0.0136
M5P.UPT	0.9125	0.0105	0.7896	0.0137
RF	0.9995	0.001	0.8034	0.0134

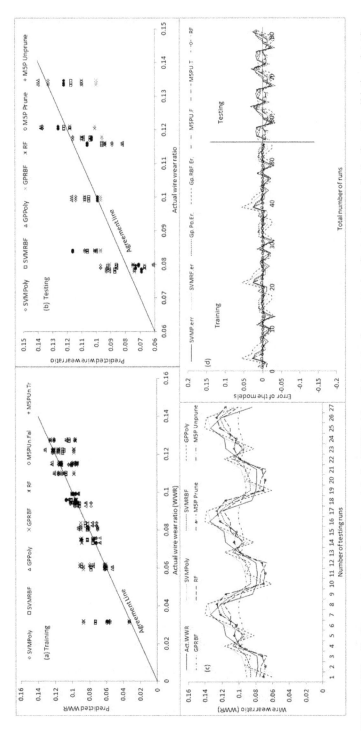

FIGURE 18.6 Performance plots of entire models for WWR (a) anticipated values scattering for train set (b) for testing (c) validation plot (d) error plots of entire models.

closely with the experimental outcome path of WWR, as shown in Figure 18.6c. The error plots for the train and test set of each model also confirm that the RF model is more sustainable in comparison to the other models, as shown in Figure 18.6d. The evaluation parameters performance of each model also confirm that the RF model is more sustainable in comparison to the other models, as shown in Table 18.5.

18.5 UNIQUE CONTRIBUTION

The work related to check the sustainability of advanced machine learning algorithms in manufacturing has not been reported before this research was made. The use of advanced machine learning algorithms in manufacturing is a unique work. This chapter divulges the procurement made to check the sustainability of seven sophisticated approaches in manufacturing.

18.6 CONCLUSION

All the models present the sustainable prediction result for SR and WWR of the WEDM of aeronautics super alloy. All the models are sustainable and can easily be employed in the manufacturing field. Out of the seven models formulated for SR and WWR, the RF model proves its sustainability over the others. The decreasing order of sustainability of the models for SR is shown as: RF > M5P Unprune > M5P Prune > GP RBF > GP Poly > SVM Poly > SVM RBF. The decreasing order of sustainability of the models for WWR is shown as: RF > M5P Prune > M5P Unprune > GP RBF > SVM RBF > GP Poly > SVM Poly.

REFERENCES

1. Huang, J. T., Liao, Y. S., & Hsue, W. J. (1999). Determination of finish-cutting operation number and machining-parameters setting in wire electrical discharge machining. *Journal of Materials Processing Technology, 87*(1–3), 69–81.
2. Özdemir, N., & Özek, C. (2006). An investigation on machinability of nodular cast iron by WEDM. *The International Journal of Advanced Manufacturing Technology, 28*(9–10), 869.
3. Ikram, A., Mufti, N. A., Saleem, M. Q., & Khan, A. R. (2013). Parametric optimization for surface roughness, kerf and MRR in wire electrical discharge machining (WEDM) using Taguchi design of experiment. *Journal of Mechanical Science and Technology, 27*(7), 2133–2141.
4. Nain, S. S., Garg, D., & Kumar, S. (2017). Prediction of the performance characteristics of WEDM on Udimet-L605 using different modelling techniques. *Materials Today: Proceedings, 4*(2), 546–556.
5. Abhishek, K., Datta, S., Biswal, B. B., & Mahapatra, S. S. (2017). Machining performance optimization for electro-discharge machining of Inconel 601, 625, 718 and 825: An integrated optimization route combining satisfaction function, fuzzy inference system and Taguchi approach. *Journal of the Brazilian Society of Mechanical Sciences and Engineering, 39*(9), 3499–3527.
6. Nain, S. S., Sihag, P., & Luthra, S. (2018). Performance evaluation of fuzzy-logic and BP-ANN methods for WEDM of aeronautics super alloy. *MethodsX, 5*, 890–908.

7. Spedding, T. A., & Wang, Z. Q. (1997). Parametric optimization and surface characterization of wire electrical discharge machining process. *Precision Engineering, 20*(1), 5–15.

8. Su, C. T., Wong, J. T., & Tsou, S. C. (2005). A process parameters determination model by integrating artificial neural network and ant colony optimization. *Journal of the Chinese Institute of Industrial Engineers, 22*(4), 346–354.

9. Mandal, D., Pal, S. K., & Saha, P. (2007). Modeling of electrical discharge machining process using back propagation neural network and multi-objective optimization using non-dominating sorting genetic algorithm-II. *Journal of Materials Processing Technology, 186*(1–3), 154–162.

10. Guven, O., Esme, U., Kaya, I. E., Kazancoglu, Y., Kulekci, M. K., & Boga, C. (2010). Comparative modeling of wire electrical discharge machining (Wedm) process using Back propagation (BPN) and general regression neural networks (GRNN). *Materials and Technology, 44*(3), 147–152.

11. SAGBAS, A., Kahraman, F., & Esme, U. (2012). Optimization of wire electrical discharge machining process using taguchi method and back propagation neural network. *Journal of Engineering and Architecture Faculty of EskişehirOsmangazi University, 25*(1).

12. Dhuria, G. K., Singh, R., & Batish, A. (2017). Application of a hybrid Taguchi-entropy weight-based GRA method to optimize and neural network approach to predict the machining responses in ultrasonic machining of Ti–6Al–4V. *Journal of the Brazilian Society of Mechanical Sciences and Engineering, 39*(7), 2619–2634.

13. Tamang, S. K., & Chandrasekaran, M. (2017). Integrated optimization methodology for intelligent machining of inconel 825 and its shop-floor application. *Journal of the Brazilian Society of Mechanical Sciences and Engineering, 39*(3), 865–877.

14. Çaydaş, U., Hasçalık, A., & Ekici, S. (2009). An adaptive neuro-fuzzy inference system (ANFIS) model for wire-EDM. *Expert Systems with Applications, 36*(3), 6135–6139.

15. Aldas, K., Ozkul, I., & Akkurt, A. (2014). Modelling surface roughness in WEDM process using ANFIS method. *Journal of the Balkan Tribological Association, 20*(4), 548–558.

16. Karsh, P. K., & Singh, H. (2018). Multi-characteristic optimization in wire electrical discharge machining of inconel-625 by using Taguchi-Grey Relational Analysis (GRA) Approach: Optimization of an existing component/product for better quality at a lower cost. In *Design and Optimization of Mechanical Engineering Products* (pp. 281–303). IGI Global.

17. Kandpal, B. C., Kumar, J., & Singh, H. (2017). Optimization and characterization of EDM of AA 6061/10% Al2O3 AMMC using Taguchi's approach and utility concept. *Production & Manufacturing Research, 5*(1), 351–370.

18. Kumar, D., Payal, H. S., & Beri, N. (2017). Taguchi-grey established optimisation for M2-tool steel with conventional/PM electrodes on EDM with and without powder mixing dielectric. *Pertanika Journal of Science and Technology, 25*(4), 1331–1342.

19. Nain, S. S., Garg, D., & Kumar, S. (2018). Investigation for obtaining the optimal solution for improving the performance of WEDM of super alloy Udimet-L605 using particle swarm optimization. *Engineering Science and Technology, an International Journal, 21*(2), 261–273.

20. Schulze, H. P. (2017). Importance of polarity change in the electrical discharge machining. In *AIP Conference Proceedings* (Vol. 1896, No. 1, p. 050001). AIP Publishing.

21. Zhang, L., Jia, Z., Wang, F., & Liu, W. (2010). A hybrid model using supporting vector machine and multi-objective genetic algorithm for processing parameters optimization in micro-EDM. *The International Journal of Advanced Manufacturing Technology, 51* (5–8), 575–586.

22. Nain, S. S., Garg, D., & Kumar, S. (2017). Modeling and optimization of process variables of wire-cut electric discharge machining of super alloy Udimet-L605. *Engineering Science and Technology, an International Journal, 20*(1), 247–264.

23. Nain, S. S., Garg, D., & Kumar, S. (2018). Evaluation and analysis of cutting speed, wire wear ratio, and dimensional deviation of wire electric discharge machining of super alloy Udimet-L605 using support vector machine and grey relational analysis. *Advances in Manufacturing, 6*(2), 225–246.

24. Pal, M., & Deswal, S. (2010). Modelling pile capacity using gaussian process regression. *Computers and Geotechnics, 37*(7–8), 942–947.

25. Sihag, P., Singh, B., Sepah Vand, A., & Mehdipour, V. (2018). Modeling the infiltration process with soft computing techniques. *ISH Journal of Hydraulic Engineering*, 1–15.

26. Sihag, P., Jain, P., & Kumar, M. (2018). Modelling of impact of water quality on recharging rate of storm water filter system using various kernel function based regression. *Modeling Earth Systems and Environment, 4*(1), 61–68.

27. Nain, S. S., Garg, D., & Kumar, S. (2018). Performance evaluation of the WEDM process of aeronautics super alloy. *Materials and Manufacturing Processes*, 1–16.

28. Luthra, S., Garg, D., & Haleem, A. (2015). An analysis of interactions among critical success factors to implement green supply chain management towards sustainability: An Indian perspective. *Resources Policy, 46*, 37–50.

29. Luthra, S., Mangla, S. K., Chan, F. T., & Venkatesh, V. G. (2018). Evaluating the drivers to information and communication technology for effective sustainability initiatives in supply chains. *International Journal of Information Technology & Decision Making, 17*(01), 311–338.

30. Rasmussen, C. E. (2006). *CKI Williams Gaussian Processes for Machine Learning MIT Press*. Cambridge, MA.

31. Kuss, M. (2006). Gaussian process models for robust regression, classification, and reinforcement learning (Doctoral dissertation, TechnischeUniversität).

32. Quinlan, J. R. (1992, November). Learning with continuous classes. In *5th Australian Joint Conference on Artificial Intelligence* (Vol. 92, pp. 343–348).

Index

Note: Page numbers in italic and bold refer to figures and tables, respectively.